CULTIVATION OF MICROALGAE

Edited by:

Bruno Augusto Amato Borges

Delve
PUBLISHING
www.delvepublishing.com

Cultivation of Microalgae

Bruno Augusto Amato Borges

Delve Publishing
2010 Winston Park Drive,
2nd Floor
Oakville, ON L6H 5R7
Canada
www.delvepublishing.com
Tel: 001-289-291-7705
001-905-616-2116
Fax: 001-289-291-7601
Email: orders@arclereducation.com

ISBN: 978-1-77361-040-5 (Hardcover)

DECLARATION

Some content or chapters in this book are open access copyright free published research work, which is published under Creative Commons License and are indicated with the citation. We are thankful to the publishers and authors of the content and chapters as without them this book wouldn't have been possible.

ABOUT THE EDITOR

Bruno Augusto Amato Borges

Bruno Augusto is Aquaculture Engineer from Federal University of Santa Catarina. His focus is on writing projects and development of new technologies in aquaculture. He is currently project service provider, consultant and develops research in the field of Biofloc Technology (BFT) for freshwater fish farming.

TABLE OF CONTENT

LIST OF CONTRIBUTORS

Bruno Augusto Amato Borges
Aquaculture Engineer Florianopolis, Brazil

Jan-Willem F. Zijffers
Bioprocess Engineering, Wageningen University, Wageningen, The Netherlands

Klaske J. Schippers
Bioprocess Engineering, Wageningen University, Wageningen, The Netherlands

Ke Zheng
Bioprocess Engineering, Wageningen University, Wageningen, The Netherlands

Marcel Janssen
Bioprocess Engineering, Wageningen University, Wageningen, The Netherlands

Johannes Tramper
Bioprocess Engineering, Wageningen University, Wageningen, The Netherlands

René H. Wijffels
Bioprocess Engineering, Wageningen University, Wageningen, The Netherlands

Junfeng Wang
Key Laboratory of Biofuels, Qingdao Institute of Bioenergy and Bioprocess Technology, Chinese Academy of Sciences, Qingdao, People's Republic of China

Jinli Liu
Low-Carbon Energy Conversion Science and Engineering Center,

Shanghai Advanced Research Institute, Chinese Academy of Sciences, Shanghai, People's Republic of China

Tianzhong Liu
Key Laboratory of Biofuels, Qingdao Institute of Bioenergy and Bioprocess Technology, Chinese Academy of Sciences, Qingdao, People's Republic of China

Wasana Arkronrat
Klongwan Fisheries Research Station, Faculty of Fisheries, Kasetsart University, Mueang, Prachuap Khiri Khan, 77000 Thailand.

Prapapron Deemark
Aquaculture Development and Certification Centre, Department of Fisheries, Chatuchak, Bangkok, 10900 Thailand.

Vutthichai Oniam
Klongwan Fisheries Research Station, Faculty of Fisheries, Kasetsart University, Mueang, Prachuap Khiri Khan, 77000 Thailand.

Hamed Safafar
Technical University of Denmark, National Food Institute (DTU Food), Søltofts Plads, Building 221, Kongens Lyngby 2800, Denmark

Michael Z. Hass
Kalundborg Municipality, Department Development, Torvet 3A, Kalundborg 4400, Denmark

Per Møller
Kalundborg Municipality, Department Development, Torvet 3A, Kalundborg 4400, Denmark

Susan L. Holdt
Technical University of Denmark, National Food Institute (DTU Food), Søltofts Plads, Building 221, Kongens Lyngby 2800, Denmark

Charlotte Jacobsen
Technical University of Denmark, National Food Institute (DTU Food), Søltofts Plads, Building 221, Kongens Lyngby 2800, Denmark

Md. Mahfuzur R. Shah
School of Environment and Energy, Peking University, Shenzhen Graduate School, Shenzhen, China

Yuanmei Liang
School of Environment and Energy, Peking University, Shenzhen Graduate School, Shenzhen, China

Jay J. Cheng
School of Environment and Energy, Peking University, Shenzhen Graduate School, Shenzhen, China
Department of Biological and Agricultural Engineering, North Carolina State University, Raleigh, NC, USA

Maurycy Daroch
School of Environment and Energy, Peking University, Shenzhen Graduate School, Shenzhen, China

Victor Marco Emmanuel N. Ferriols
Institute of Aquaculture, College of Fisheries and Ocean Sciences, University of the Philippines Visayas, Miagao, Iloilo, Philippines

Riza O. Aguilar
Institute of Aquaculture, College of Fisheries and Ocean Sciences, University of the Philippines Visayas, Miagao, Iloilo, Philippines

Theoneste Ndikubwimana
1. Department of Chemical and Biochemical Engineering, College of Chemistry and Chemical Engineering, Xiamen University, Xiamen, China

Xianhai Zeng
College of Energy, Xiamen University, Xiamen, China
The Key Laboratory for Synthetic Biotechnology of Xiamen City, Xiamen University, Xiamen, China

Theophile Murwanashyaka
Department of Chemical and Biochemical Engineering, College of Chemistry and Chemical Engineering, Xiamen University, Xiamen, China

Emmanuel Manirafasha
Department of Chemical and Biochemical Engineering, College of Chemistry and Chemical Engineering, Xiamen University, Xiamen, China

Ning He
Department of Chemical and Biochemical Engineering, College of Chemistry and Chemical Engineering, Xiamen University, Xiamen, China
The Key Laboratory for Synthetic Biotechnology of Xiamen City, Xiamen University, Xiamen, China

Wenyao Shao
Department of Chemical and Biochemical Engineering, College of Chemistry and Chemical Engineering, Xiamen University, Xiamen, China

Yinghua Lu
Department of Chemical and Biochemical Engineering, College of Chemistry and Chemical Engineering, Xiamen University, Xiamen, China
The Key Laboratory for Synthetic Biotechnology of Xiamen City, Xiamen University, Xiamen, China

Kevin J. Flynn
Swansea University, Swansea, UK

Philip Kenny
Swansea University, Swansea, UK

Aditee Mitra
Swansea University, Swansea, UK

Krystal K Lum,
Department of Animal Science, Cornell University, Ithaca, NY 14850, USA

Jonggun Kim
Department of Animal Science, Cornell University, Ithaca, NY 14850, USA

Xin Gen Lei
Department of Animal Science, Cornell University, Ithaca, NY 14850, USA

Edith Martinez-Guerra
Department of Civil and Environmental Engineering, Mississippi State University, Mississippi State, MS 39762, USA

Veera Gnaneswar Gude
Department of Civil and Environmental Engineering, Mississippi State University, Mississippi State, MS 39762, USA

Matteo Marsullo
Department of Industrial Engineering, University of Padova, Padova, Italy

Alberto Mian
Industrial Process and Energy System Engineering Group (IPESE), École Polytechnique Fédérale de Lausanne, Lausanne, Switzerland

Adriano Viana Ensinas
Industrial Process and Energy System Engineering Group (IPESE), École Polytechnique Fédérale de Lausanne, Lausanne, Switzerland
Universidade Federal do ABC, Santo Andre, Brazil

Giovanni Manente
Department of Industrial Engineering, University of Padova, Padova, Italy

Andrea Lazzaretto
Department of Industrial Engineering, University of Padova, Padova, Italy

François Marechal
Industrial Process and Energy System Engineering Group (IPESE), École Polytechnique Fédérale de Lausanne, Lausanne, Switzerland

Niels Hempel
Faculty of Natural Sciences, Lausitz University of Applied Sciences, Senftenberg, Germany

Ingolf Petrick
Faculty of Natural Sciences, Lausitz University of Applied Sciences, Senftenberg, Germany

Frank Behrendt
Department of Energy Engineering, Berlin Institute of Technology, Berlin, Germany

Sukarni
Doctoral Program of Mechanical Engineering, Faculty of Engineering, University of Brawijaya, Malang, Indonesia
Department of Mechanical Engineering, Faculty of Engineering, State University of Malang, Malang, Indonesia

Sudjito
Department of Mechanical Engineering, Faculty of Engineering, University of Brawijaya, Malang, Indonesia

Nurkholis Hamidi
Department of Mechanical Engineering, Faculty of Engineering, University of Brawijaya, Malang, Indonesia

Uun Yanuhar
Biotechnology Laboratory, Department of Water Resources Management, Faculty of Fisheries and Marine Sciences, University of Brawijaya, Malang, Indonesia

I. N. G. Wardana
Department of Mechanical Engineering, Faculty of Engineering, University of Brawijaya, Malang, Indonesia

Xiaobin Wen
Key Laboratory of Pant Germplasm Enhancement and Specialty Agriculture, Wuhan Botanical Garden, Chinese Academy of Sciences, Wuhan, China

Kui Du
Key Laboratory of Pant Germplasm Enhancement and Specialty Agriculture, Wuhan Botanical Garden, Chinese Academy of Sciences, Wuhan, China
University of Chinese Academy of Sciences, Beijing, China

Zhongjie Wang
Key Laboratory of Pant Germplasm Enhancement and Specialty Agriculture, Wuhan Botanical Garden, Chinese Academy of Sciences, Wuhan, China

Xinan Peng
Key Laboratory of Pant Germplasm Enhancement and Specialty Agriculture, Wuhan Botanical Garden, Chinese Academy of Sciences, Wuhan, China
University of Chinese Academy of Sciences, Beijing, China

Liming Luo
Key Laboratory of Pant Germplasm Enhancement and Specialty Agriculture, Wuhan Botanical Garden, Chinese Academy of Sciences, Wuhan, China
Department of Pathology and Immunology, Baylor College of Medicine, Houston, USA

Huanping Tao
Key Laboratory of Pant Germplasm Enhancement and Specialty Agriculture, Wuhan Botanical Garden, Chinese Academy of Sciences, Wuhan, China

Yan Xu
Key Laboratory of Pant Germplasm Enhancement and Specialty Agriculture, Wuhan Botanical Garden, Chinese Academy of Sciences, Wuhan, China
University of Chinese Academy of Sciences, Beijing, China

Dan Zhang
Key Laboratory of Pant Germplasm Enhancement and Specialty Agriculture, Wuhan Botanical Garden, Chinese Academy of Sciences, Wuhan, China

Yahong Geng
Key Laboratory of Pant Germplasm Enhancement and Specialty Agriculture, Wuhan Botanical Garden, Chinese Academy of Sciences, Wuhan, China

Yeguang Li
Key Laboratory of Pant Germplasm Enhancement and Specialty Agriculture, Wuhan Botanical Garden, Chinese Academy of Sciences, Wuhan, China

Junping Lv
School of Life Science, Shanxi University, Taiyuan 030006, China

Jia Feng
School of Life Science, Shanxi University, Taiyuan 030006, China

Qi Liu
School of Life Science, Shanxi University, Taiyuan 030006, China

Shulian Xie
School of Life Science, Shanxi University, Taiyuan 030006, China

M.H. Sayadi
Department of Environmental Science, Faculty of Natural Resources and Environment, University of Birjand, Birjand, Iran

N. Ahmadpour
Department of Environmental Science, Faculty of Natural Resources and Environment, University of Birjand, Birjand, Iran

M. Fallahi Capoorchali
Research Center, National Inland Water Aquaculture Institute, Anzali Port, Iran

M.R. Rezaei
Department of Environmental Science, Faculty of Natural Resources and Environment, University of Birjand, Birjand, Iran

Viswanath Kiron
Faculty of Biosciences and Aquaculture, Nord University, Bodø, Norway

Mette Sørensen
Faculty of Biosciences and Aquaculture, Nord University, Bodø, Norway

Mark Huntley
Department of Biological and Environmental Engineering, Cornell University, Ithaca, NY, USA
Marine Laboratory, Nicholas School of the Environment, Duke University, Beaufort, NC, USA

Ghana K. Vasanth
Faculty of Biosciences and Aquaculture, Nord University, Bodø, Norway

Yangyang Gong
Faculty of Biosciences and Aquaculture, Nord University, Bodø, Norway
Key Laboratory of East China Sea and Oceanic Fishery Resources Exploitation, East China Sea Fisheries Research Institute, Shanghai, China

Dalia Dahle
Faculty of Biosciences and Aquaculture, Nord University, Bodø, Norway

Anjana M. Palihawadana
Faculty of Biosciences and Aquaculture, Nord University, Bodø, Norway

Pallab K. Sarker
Environmental Studies Program, Dartmouth College, Hanover, NH 03755, United States of America

Anne R. Kapuscinski
Environmental Studies Program, Dartmouth College, Hanover, NH 03755, United States of America

Alison J. Lanois
Environmental Studies Program, Dartmouth College, Hanover, NH 03755, United States of America

Erin D. Livesey
Environmental Studies Program, Dartmouth College, Hanover, NH 03755, United States of America

Katie P. Bernhard
Environmental Studies Program, Dartmouth College, Hanover, NH 03755, United States of America

Mariah L. Coley
Environmental Studies Program, Dartmouth College, Hanover, NH 03755, United States of America

PREFACE

Algae are recognized as one of the oldest life-forms (Falkowski & Raven, 1997). The use of microalgae dates back around 2000 years to the Chinese, who used Nostoc to survive during famine (Spolaore et al., 2006), and to the Aztecs who collected and cultivated Spirulina (Henrikson, 2011). For the past 50 years, extensive research has been performed on microalgae and how they can be used in a wide variety of processes or to manufacture many practical and economic important products.

This group of individuals is present in several ecosystems, representing a big variety of species living in a wide range of environmental conditions. Microalgae can be autotrophic or heterotrophic; the autotrophic require only inorganic compounds such as CO2, salts and a light energy source for growth; the heterotrophic are nonphotosynthetic, therefore require an external source of organic compounds as well as nutrients as an energy source (Brennan & Owende, 2009).

The cultivation of microalgae is an activity that offers high productivity in dry biomass, compared the production of seaweeds. One important advantage of the cultivation of microalgae is that it can be performed in various locations, due to the use of closed systems of cultivation. In addition, can generate crops throughout the year and has high photosynthetic efficiency and bioremediation potential.

There are several groups of individuals who are part of the large group of microalgae; so many differences can be identified with respect to chemical and biological composition of each. Actually, the main genres worldwide cultured are *Skeletonema*, *Thalassiosira*, *Nannochloropsis*, *Phaeodactylum*, *Chaetoceros*, *Isochrysis*, *Tetraselmis*, *Chlamydomonas*, *Dunaliella* and *Spirulina*.

One of the great advantages present in the cultivation of microalgae is the positive appeal to your benefits with regard to the environment. This production plays in a variety of ways to promote sustainability. Microalgae biomass has been proven as a sustainable feedstock for biofuels, feed and numerous value added products that involves nutraceuticals and therapeutic industry (Guldhe, 2016).

Microalgae are a highly renewable resource. It can be grown and harvested all year round, in several environments. Production is low impact – microalgae cultivation needs no chemicals or pesticides, in addition to require no deforestation.

Knowing the many uses and importance of these organisms to the different sectors of the industry, and your environmental importance, it is essential to maintain the targeted efforts in pursuit of the development of new technologies and applications, as well as improvements in cropping systems and processes used currently.

Editor

SECTION I

HISTORY, DEVELOPMENT AND FUTURE PERSPECTIVES OF MICROALGAE CULTIVATION

CHAPTER

1

CULTIVATION OF MICROALGAE

Bruno Augusto Amato Borges

Aquaculture Engineer, Florianopolis, Brazil

CURRENT SITUATION OF CULTIVATION

Algae are recognized as one of the oldest life-forms (Falkowski & Raven, 1997). The use of microalgae dates back around 2000 years to the Chinese, who used Nostoc to survive during famine (Spolaore et al., 2006), and to the Aztecs who collected and cultivated Spirulina (Henrikson, 2011). For the past 50 years, extensive research has been performed on microalgae and how they can be used in a wide variety of processes or to manufacture many practical and economic important products. The first large-scale culture of microalgae started in the early 1960s in Japan, with the Chlorella culture (Spolaore et al., 2006). Only in the middle of the last century microalgal biotechnology began to develop, but it already has numerous commercial applications (Hasan & Chakrabarti, 2009).

This group of individuals is present in several ecosystems, representing a big variety of species living in a wide range of environmental conditions, also has a wide variety, including prokaryotes and eukaryotes and is present in about 14 phyla. According to Richmond (2004), it is estimated that more than 50,000 species exist, but only around 30,000 have been studied and analyzed. Microalgae can be autotrophic or heterotrophic; the autotrophic require only inorganic compounds such as CO2, salts and a light energy source for growth; the heterotrophic are nonphotosynthetic, therefore require an external source of organic compounds as well as nutrients as an energy source (Brennan & Owende, 2009).

In aquatic environments, microalgae are the base of the food chain and consequently, in aquaculture production chain. According to FAO (2016), the world aquaculture production has grown a lot over the last few years, in 2006 total production was nearly 45 million tons, and in the year 2014 already reaches 73 million tons. Since the production of most aquatic organisms depends on the microalgae in any stage of life, it is essential that the production of microalgae to grow along with the production of fish, crustaceans and mollusks, to meet the demand required. It is extremely important for later success in the cultivation of any aquatic organisms that depend on this food source during any stage of life. Therefore, it is considered a fundamental sector within any enterprise or laboratory that depends on this feature to production. On this way, cultivation of microalgae are essential in the commercial production of several aquatic species as a food source for all growth stages of bivalve molluscs, larval stages of some crustacean species, and very early growth stages of some fish species. Algae are furthermore used to feed zooplankton (rotifers, copepods, artemia, etc) which serve in turn as food for larval and early-juvenile stages of crustaceans and fish. The genus Nannochloropsis are widely used in the live food chain required for the cultivation of marine fish larvae, and are among the most commonly used microalgal species in many mariculture systems. One of their main uses is the cultivation of rotifers of the genus Brachionus that also are used as live feed for fish larvae. (Freire et al., 2016). According to Benemann (1992), the use of microalgae in aquaculture has various potential advantages over the production of microalgae for human foods or terrestrial animal feeds: among them the high conversion efficiencies, no need for harvesting, drying and storage, as the animals or food chains could use the algae as produced. The same research (Benemman, 1992) related that the main limitations to improve use of microalgae feeds are both technical and economic: the problem is how to mass culture desirable species or the cost of production must be reduced.

Besides, for "green water technique" culture method, microalgae are used directly in the larval tanks, acting in stabilizing the water quality, nutrition of the larvae, and microbial control (Lavens & Sorgeloos, 1996). This method is based on the fertilization of the tanks through the prior settlement with microalgae, which serve as food for the larvae that will be grown on site. In addition, if the system is well controlled and wielded, assists in maintaining water quality.

The cultivation of microalgae is an activity that offers high productivity in dry biomass, compared the production of seaweeds. One important advantage of the cultivation of microalgae is that it can be performed in various locations, due to the use of closed systems of cultivation. In addition, can generate crops throughout the year and has high photosynthetic efficiency and bioremediation potential.

So, you can see the current importance of culture and the immense potential that microalgae have in the various sectors of industry, with numerous possibilities of use. This group of individuals has an ecological, economic and social role that can hardly be replaced. In this way, it is very important to continue developing research in this subject.

MAIN CULTURED GENRES

There are several groups of individuals who are part of the large group of microalgae; so many differences can be identified with respect to chemical and biological composition of each. That way it is important to know the basics of each species for your utility and what will be done to enhance the benefits of your use. So, below are some of the main genres produced currently and its applications.

Actually, the main genres worldwide cultured are *Skeletonema*, *Thalassiosira*, *Nannochloropsis*, *Phaeodactylum*, *Chaetoceros*, *Isochrysis*, *Tetraselmis*, *Chlamydomonas*, *Dunaliella* and *Spirulina*.

Skeletonema sp.

Skeletonema usually dominates the diatom abundance in coastal waters (Balkis, 2003). It is increasingly used as live feed in the larviculture of penaeids, polychaetes, lobsters, marine ornamental prawns, barnacles, mussels, oysters, scallops and pearl spats (Kitto and Regunathan, 1997). *Skeletonema* cells form an excellent feed for copepods (Paffenhöfer, 1970). Kumar & Prabu (2014) reports that the production and management of *Skeletonema* cultures can be produced in a number of ways without modifying the quality of the algal cells produced.

Thalassiosira sp.

This genus is widely applied in the shrimp and shellfish larviculture, considered by several hatcheries to be the single best alga for larval shrimp, is widely also used for feeding copepods and *Artemia* (Rice, 2014). The advantages in the use of this genre of microalgae in marine shrimp larvicuture are, according to Derner (2006), increased speed of metamorphosis and growth and higher survival of larvae, this is probably related to your biochemical composition.

Nannochloropsis sp.

Nannochloropsis is a genus of unicellular and nonmobile marine microalgae. There are several advantages on this genus culture, among them a high grown rate, considerable lipid content, suitable fatty acid composition, good environmental adaptability and contamination resistance (Ma et al., 2016). In addition, it is an important oleaginous model microalga due to the great photosynthetic efficiency, high lipid productivity and relatively knowledge technology for outdoor cultivation on large-scale (Vieler et al., 2012). Also, Ma et al. (2016) reports that research and development priority should be given to the better understanding of molecular mechanism for lipid metabolism and the development of a more stable and robust genetic toolkit for *Nannochloropsis* to enable this organism to be a lipid source.

Phaeodactylum sp.

Phaeodactylum is an interesting microalgae for biodiesel production, because it grows well in culture and produces a significant quantity of lipids (Rodolfi et al. 2009). This genus is one of the few microalgae whose entire genome has been fully sequenced (Bowler et al. 2008), facilitating genetic modification way for the purpose of increasing lipid productivity.

Chaetoceros sp.

Chaetoceros is the largest genus of marine plank-tonic diatoms. Your species have a worldwide distribution, and they often dominate marine ecosystems (Hasle et al., 1996). It is widely used in aquaculture industries, as it is comprised of nutritional value proper for most marine feeders, mainly for penaeid shrimp larvae and bivalves.

Isochrysis sp.

Isochrysis is a small marine flagellate widely used in aquaculture as PUFA-rich microalgae (Grima et al., 1994). It is commercially produced as feed for the early larval stages of mollusks, fish, and crustaceans. Also, *Isochysis* produce antibacterial substances, which increase the toxicity of free fatty acids such as eicosapentanoic acid (EPA) to several pathogens, without the use of chemicals that might harm the environment and the cultured organisms (Molina-Cárdenas et al., 2011).

Tetraselmis sp.

The genus *Tetraselmis* is in the class Prasinophyceae, which are free-living, flagellated green microalgae. *Tetraselmis* has adapted to several environmental conditions; occurs in marine, brackish and freshwater habitats, and commonly blooms in summer when UV irradiance is highest (Kristoffersen et al., 2016).

Currently, *Tetraselmis* has been cited as a source in the production of bioethanol, due to your high biomass and production of starch, even in situations of deprivation of light (Yao et al., 2012).

Chlamydomonas sp.

Chlamydomonas is a genus of green algae which are part single celled flagellates beings used as models for research in molecular biology. *Chlamydomonas* has many commercial applications, including the production of biomass for food and animal health, environmental monitoring, bioremediation and production of hydrogen (Walker et al., 2005).

remove suspended solids. Also according to Christenson & Sims (2011), the first stage of flocculation is the aggregation of suspended solids into larger particles; the second stage involves the coalescing of aggregates into large flocs that settle out of suspension.

Thus, we can see how hard it is making decisions about what if when there is interest in starting microalgae cultivation commercially. Several factors must be considered for the elaboration of the project. Market, kind of cultivation and area used are just a few of these. I believe today, that the overall trend is to think more about a sustainable activity responsibly, in search of producing safe and healthy food for generations.

MICROALGAE AND SUSTAINABLE UTILIZATION

One of the great advantages present in the cultivation of microalgae is the positive appeal to your benefits with regard to the environment. This production plays in a variety of ways to promote sustainability.

The term sustainability is gaining enough space over the last decade and has been inserted into some new ventures in a variety of ways; claims for the use of ecological engineering tools towards the development of a sustainable aquaculture have been increasing (Chopin et al., 2008).

Microalgae biomass has been proven as a sustainable feedstock for biofuels, feed and numerous value added products that involves nutraceuticals and therapeutic industry (Guldhe, 2016). Also, have higher stability, which enables working in more concentrated and toxic environments. Microalgae Chlorella actively utilizes mineral elements, sugar, and amino acids and as compared with active slime enables higher purification rate: up to 96-98 % for organic and 80 % for mineral components. (Avagyan, 2017).

Microalgae can serve as the basis for producing various types of fuels such as biodiesel, methane, hydrogen, ethanol, etc. Microalgae have been identified as promising raw material for the supply of oil destined for the production of biodiesel as an alternative to conventional oilseed crops (Hossain et al., 2008). This production, whatever the scale, can serve for various proposals and brings several benefits to the environment. Can be acts in removing atmospheric carbon dioxide through photosynthesis (Fernández et al., 2012; Cheng et al., 2006; Wang et al., 2008). Also, it can be used in domestic or industrial wastewater treatment, and as fertilizer for the agricultural industry (Wang et al., 2009; Abdel-Raouf et al., 2012; Hammouda et al., 1995). Currently, due to the high variety of receivers that can be obtained from this feature, this sector has been inserted in many areas of biotechnology, including the production of cosmetics, pharmaceuticals, nutrition and food addictives.

Biodiesel production can be divided into three phases: extraction of the lipid of microalgal biomass, removal of excess solvent, and biodiesel production by transetherification catalyzed (homogeneous or heterogeneous catalyst). It is can also be made from the etherification.

In the current conditions, the biodiesel from microalgae is still not a favorable alternative to commercial production. But there are possibilities to be become more feasible with the improvement in the reduction of environmental impacts and improving technologies for extraction. Still don't know any industrial plant for the production of biodiesel from microalgae, but several searches are underway in the world with promising results. The main aspects to be improved to make the commercial production of biofuel are: the choice of species with high content of lipids that can achieve large-scale production, installation of energy-efficient extraction methods, and the recovery of materials, energy, and nutrients contained in the by-products, part of the concept of integrated biorefineries.

During the last decades, aquaculture has been growing rapidly when compared to other productive activities, leading to an increasing production of nutrient-rich waste and wastewater that need to be treated. So, efforts have been undertaken to remove excess of nutrients from aquaculture wastewater to keep satisfactory effluent quality that does not result in eutrophication of local ecosystems and/or enables water recycling (Váradi et al., 2009). Recirculating aquaculture systems (RAS) offer advantages related to reduce water consumption, and improve opportunities for waste management and nutrient recycling compared to conventional flow through aquaculture systems (Martins et al., 2010). According to Udom et al. (2013) the key factor that determines the economic viability of microalgae systems for wastewater treatment is the effective cost separation of microalgae from the treated water. In addition, some research show that microalgal-based treatment of aquaculture effluents is advantageous because it removes CO2, oxygenates water and produces more than one order of magnitude of high value biomass per unit of nitrogen removed than bacteria (Milhazes-Cunha & Otero, 2016).

The use of microalgae in the treatment of effluents, and later use for other purposes in the cultivation or by industry, constitutes an integrated system that has been growing and gaining strength currently in worldwide. This system, called IMTA (Integrated Multi-Trophic Aquaculture), has been widely used; your basic principle consists in diversifying cultivated species, so that each meets your ecological role in the environment, reducing losses and waste, with the benefit of diversifying production. According to Troell et al. (2005), the main purpose of IMTA systems is increasing long-term sustainability and profitability of aquaculture, as the wastes of one crop (fed animals) are converted into particulate matter – macro and microalgae – that can be used to feed low trophic level animals that can in turn be sold on the market. The concept of IMTA has been applied throughout human history. In China, microalgae have been used for millennia in integrated systems, alongside with macroalgae and cyanobacteria. In such systems, rice is cultured with fish and algae (Neori et al., 2004), potentially increasing production by 8 to 47% (Kangmin, 1988). According to Bosma & Verdegem (2011), the most effective solution is to intensify ecological and technological research on aquaculture for sustainable development, through optimizing aquaculture structure and practicing IMTA system to reduce organic and inorganic pollution.

Microalgae are a highly renewable resource. It can be grown and harvested all year round, in several environments. Production is low impact – microalgae cultivation needs no chemicals or pesticides, in addition to require no deforestation. Microalgae cultivation can still be added in different production systems as bioremediation and in treatment of effluents from industry, in order to contribute to the maintenance of the quality of the environment.

Knowing the many uses and importance of these organisms to the different sectors of the industry, and your environmental importance, it is essential to maintain the targeted efforts in pursuit of the development of new technologies and applications, as well as improvements in cropping systems and processes used currently.

REFERENCES

1. Abdel-Raouf, N., Al-Homaidan, A.A., Ibraheem, I.B.M. (2012). Microalgae and wastewater treatment. Saudi journal of biological sciences, v.19, issue 3, 257-275.
2. Acién Fernández, F.G., González-López, C.V., Fernández Sevilla, J.M. et al. (2012). Conversion of CO2 into biomass by microalgae: how realistic a contribution may it be to significant CO2 removal? Appl. Microbiol.Biotechnol., 96: 577. doi:10.1007/s00253-012-4362-z.
3. Avagyan, A. (2017). Microalgae production development: global prospects and profitable technology wastewater purification by using microalgae. http://www.waterworld.com/articles/2008/09/microalgae-production-development-global-prospects-and-profitable-technology-wastewater-purification-by-using-microalgae.html. (Accessed 25.03.2017)
4. Balkis, N. (2003). Seasonal variations in the phytoplankton and nutrient dynamics in the neritic water of Büyükçekmece Bay, Sea of Marmara. J. Plankton Res., 25: 703-717.
5. Benemann, J.R. (1992). Microalgae aquaculture feeds. Journal of applied phycology, 4: 233-245.
6. Bosma, R.H., Verdegem, M.C.J. (2011). Sustainable aquaculture in ponds: principles, practices and limits. Livest Sci 139(1–2):58–68.
7. Bowler, C., Allen, A.E., Badger, J.H. et al. (2008). The Phaeodactylum genome reveals the dynamic nature and multi-lineage evolutionary history of diatom genomes. Nature 456:239–244.
8. Brennan, L., Owende, P. (2010). Biofuels from microalgae – A reviews of technology productions, processing, and extractions of biofuels and co-products. Renewable and Sustainable Energy Reviews, 14, 557–577.
9. Cheng, L., Zhang, L., Chen, H., Gao, C. (2006). Carbon dioxide removal from air by microalgae cultured in a membrane-photobioreactor. Separation and purification technology, 50, 3, pp 324-329.
10. Chrinstenson, L., Sims, R. (2011). Production and harvesting of microalgae for wastewater treatment, biofuels, and bioproducts.

Biotechnology Advances, 29 (6): 686-702.

11. Chisti, Y. (2007). Biodiesel from microalgae. Biotechnology advances, 25, 294-306.
12. Chopin, T., Robinson, S.M.C., Troell, M., Neori, A., Buschmann, A.H., Fang, J., (2008). Multitrophic integration for sustainable marine aquaculture. Ecological Engineering. Vol. 3. Encyclopedia of Ecology, vol. 5. Elsevier, Oxford, pp. 2463–2475.
13. Colla, L. M., Reinehr, C. O., Reichert, C., Costa, J. A. V. (2007). Production of biomass and nutraceutical compounds by Spirulina platensis under different temperature and nitrogen regimes. Bioresource Technology, v. 98, n. 7, p. 1489-1493.
14. Derner, R. B. (2006). Efeito de fontes de carbono no crescimento e na composição bioquímica das microalgas Chaetoceros muelleri e Thalassiosira fluviatilis, com ênfase no teor de ácidos graxos poliinsaturados. 2006. Tese/Doutorado, Departamento de Ciências e Tecnologia de Alimentos, Universidade Federal de Santa Catarina, Florianópolis. 158 p.
15. Falkowski PG, Raven JA. Aquatic photosynthesis. London: Blackwater Science; 1997, 375.
16. FAO. 2016. FAO yearbook. Fishery and Aquaculture Statistics. 2014. Rome, Italy..
17. Freire, I. et al. (2016). Nannochloropsis limnetica: A freshwater microalga for marine aquaculture. Aquaculture, 459, 124-130.
18. Grima, E.M., Pérez, J.A.S., Camacho, F.G., J. L., Sanchez, J.L.G., Fernández, F.G.A., Alonso, D.L. (1994). J. Biotechnol., 1994, 37(2), 159–166.
19. Guldhe, A., Ansari, F.A., Singh, P., Bux, F. (2016). Heterotrophic cultivation of microalgae using aquaculture wastewater: A biorefinery concept for biomass production and nutrient remediation. Ecological Engineering, 99: 47-53.
20. Hammouda, O., Gaber, A., Abdelraouf, N. (1995). Microalgae and wastewater treatment. Ecotoxicology and Environmental Safety, 31, issue 3, 205-210.
21. Hasle, G.R., Syvertsen, E.E., Steidinger, K.A., Tangen, K. Identifying marine diatoms and dinoflagellates. Edited by Tomas,

C.R. Academic Press, Inc.

22. Hasan, M.R., Chakrabarti, R. (2009). Use of algae and aquatic
23. macrophytes as feed in small-scale aquaculture: a review.
24. FAO Fisheries and Aquaculture Technical Paper. No. 531. Rome, FAO. 2009. 123p.
25. Henrikson, R. (2011). Algae in historical legends. Adaped from "Spirulina world food: How this micro algae can transform your health and our planet. 6p.
26. Hossain, A. B. M. S., Salleh, A., Boyce, A. N., Chowdhury, P., Naqiuddin, M. (2008). Biodiesel fuel production from algae as renewable energy. American Journal of Biochemistry and Biotechnology, 4:250-254.
27. Kangmin, L. (1988). Rice-fish culture in China: a review, Aquaculture 71 (3), 173–186.
28. Kitto, M.R., Regunathan, C. (1997). Skeletonema kills luminosis in shrimp hatcheries. Fish Farmer, 11(6): 15 16.
29. Knuckey, R.M, Brown, M.R., Robert, R., Frampton, D.M.F. (2006). Production of microalgal concentrates by flocculation and their assessment as aquaculture feeds. Aquacultural Engineering, 35, 300-313.
30. Kristoffersen, A.S., Hamre, B., Frette, Ø. et al. (2016). Chlorophyll a fluorescence lifetime reveals reversible UV-induced photosynthetic activity in the green algae Tetraselmis Eur. Biophys. J., 45: 259. doi:10.1007/s00249-015-1092-z.
31. Kumar, C.S., Prabu, V.A. (2014). Culture of phytoplankton Skeletonema costatum, Cleve, 1873. Int. J. Curr. Microbiol. App. Sci., 3 (11), 129-136.
32. Lavens, P. & Sorgeloos, P. (eds.) (1996). Manual on the production and use of live food for aquaculture FAO Fisheries Technical Paper. No. 361. Rome, FAO. 1996. 295p.
33. Ma, X., Chen, T., Liu, B.Y.J., Chen, F., (2016). Lipid production from Nannochloropsis. Marine drugs, 14, 61. DOI: 10.3390/md14040061.
34. Martins, C.I., Eding, E.H., Verdegem, M.C.J., Heinsbroek, L.T.N., Schneider, O., Blancheton, J.P., Roque d'Orbcastel, E., Verreth,

J.A.J., (2010). New developments in recirculating aquaculture systems in Europe: a perspective on environmental sustainability. Aquacult. Eng. 43 (3), 83–93.

35. Milhazes-Cunha, H., Otero, A. (2016). Valorization of aquaculture effluents with microalgae: The Integrated Multi-Trophic Aquaculture concept, Algal Res., http://dx.doi.org/10.1016/j.algal.2016.12.011

36. Molina-Cárdenas, C.A., Sánchez-Saavedra, M.D.P., Lizárraga-Partida, M.L. (2014) J. Appl. Phycol., 2014, 26(6), 2347–2355.

37. Murthy, K.N.C., Vanitha, A., Rajesha, J., Swamy, M.M., Sowmya, P.R., Ravishankar, G.A. (2005). In vivo antioxidant activity of carotenoids from Dunaliella salina – a green microalga. Life Sciences, v. 76, Issue 12, 1381-1390.

38. Neori, A., Chopin, T., Troell, M., Buschmann, A.H., Kraemer, G.P., Halling, C. et al. (2004). Integrated aquaculture: rationale, evolution and state of the art emphasizing seaweed biofiltration in modern mariculture, Aquaculture 231 (1–4).361–391.

39. Paffenhöfer, G.A.(2002). An assessment of the effects of diatoms on planktonic copepods. Mar. Ecol. Prog. Ser., 227: 305 310.

40. Iba, W., Rice, M.A., Wikfors, G.H. (2014). Microalgae in western pacific white shrimp hatcheries: A review on roles and culture environments. Asian fisheries science, 27: 212-233.

41. Richmond A. (2004). Handbook of microalgal culture: biotechnology and applied phycology. Blackwell Science Ltd.

42. Rodolfi L. et al (2009). Microalgae for oil: strain selection, induction of lipid synthesis and outdoor mass cultivation in a low-cost photobioreactor. Biotech Bioeng 102:100–112.

43. Shoji, J., Ajithkumar, P.B. (2015). Microalgal culture and maintenance in marine hatcheries. 178-185.

44. Spolaore P, Joannis-Cassan C, Duran E, Isambert A. Commercial applications of microalgae. Journal of Bioscience and Bioengineering 2006;101(2): 87–96.

45. Troell, M., Neori, A., Chopin, T., Buschmann, A.H. (2005). Biological wastewater treatment in aquaculture more than just bacteria, WORLD Aquac. 36 (1), 27–29.

46. Udom, I., Zaribaf, B.H., Halfhide, T., Gillie, B., Dalrymple, O.,

Zhang, Q., Ergas, S.J., (2013). Harvesting microalgae grown on wastewater. Bioresour. Technol. 139, 101–106.

47. Váradi, L., Bardócz, T., Oberdieck, A. (2009). SustainAqua – Integrated approach for a sustainable and healthy freshwater aquaculture" (2009). SustainAqua handbook – A handbook for sustainable aquaculture.
48. Vieler, A., Wu, G., Tsai, C.H., Bullard, B., Cornish, A.J., Harvey, C., Reca, I.B., Thornburg, C., Achawanantakun, R., Buehl, C.J. et al. (2012). Genome, functional gene annotation, and nuclear transformation of the heterokont oleaginous alga Nannochloropsis oceanica CCMP1779. PLoS Genet., 8.
49. Vonshak, A. (2002). Spirulina platensis (Arthrospira) physiology, cell-biology and biotechnolo ugy. CRC Press, 252 p.
50. Walker, T.L., Purton, S., Becker, D.K., Collet, C. (2005). Microalgae as bioreactors. Plant Cell Rep 2n4:629–641.
51. Wang, B., Li, Y., Wu, N., Lan, C.Q. (2008). CO2 bio-mitigation using microalgae. Applied microbiology and biotechnology, v.79, issue 5, 707-718.
52. Wang, L., Min, M., Li, Y. et al. (2010). Cultivation of Green Algae Chlorella sp. in Different Wastewaters from Municipal Wastewater Treatment Plant. Appl Biochem Biotechnol, 162: 1174. doi:10.1007/s12010-009-8866-7.
53. Yao, C., Ai, J., Cao, X., Xue, S., Zhang, W. (2012). Enhancing starch production of a marine green microalga Tetraselmis subcordiformis through nutrient limitation. Bioresour Technol., 118:438–44.

SECTION II

MANAGEMENT, TECHNIQUES AND ASPECTS OF PRODUCTION

CHAPTER

2

MAXIMUM PHOTOSYNTHETIC YIELD OF GREEN MICROALGAE IN PHOTOBIOREACTORS

Jan-Willem F. Zijffers, Klaske J. Schippers, Ke Zheng, Marcel Janssen, Johannes Tramper and René H. Wijffels

Bioprocess Engineering, Wageningen University, Wageningen, The Netherlands

ABSTRACT

The biomass yield on light energy of Dunaliella tertiolecta and Chlorella sorokiniana was investigated in a 1.25- and 2.15-cm light path panel photobioreactor at constant ingoing photon flux density (930 μmol photons $m^{-2} s^{-1}$). At the optimal combination of biomass density and dilution rate, equal biomass yields on light energy were observed for both light paths for both microalgae. The observed biomass yield on light energy appeared to be based on a constant intrinsic biomass yield and a constant maintenance energy requirement per gram biomass. Using the model of Pirt (New Phytol 102:3–37, 1986), a biomass yield on light energy of 0.78 and 0.75 g mol photons^{-1} and a maintenance requirement of 0.0133 and 0.0068 mol photons $g^{-1} h^{-1}$ were found for D. tertiolecta and C. sorokiniana, respectively. The observed yield decreases steeply at low light supply rates, and according to this model, this is related to the increase of the amount of useable light energy diverted to biomass maintenance. With this study, we demonstrated that the observed biomass yield on light in short light path bioreactors at high biomass densities decreases because maintenance requirements are relatively high at these conditions. All our experimental data for the two strains tested could be described by the physiological models of Pirt (New Phytol 102:3–37, 1986). Consequently, for the design of a photobioreactor, we should maintain a relatively high specific light supply rate. A process with high biomass densities and high yields at high light intensities can only be obtained in short light path photobioreactors.

Keywords: Microalgae Light path Photosynthetic yield Photobioreactor design

INTRODUCTION

Microalgae receive a lot of attention presently because of their potential to be used as a feedstock for production of biodiesel. Worldwide research programs are initiated to develop technology for production of biodiesel from microalgae, and many new companies have been developed and most probably will be develop in the future.

Microalgae have potentially an areal productivity superior to traditional agricultural crops (Chisti 2007). Realistic estimates for areal productivity are in the order of magnitude of 40–80 t of dry matter per year depending on the technology used and the location of production (Wijffels 2008). For production of algal biomass for biodiesel purposes, it is essential that the production capacity (hectares of culture) increases and the cost of production decreases dramatically.

High volumetric productivities and high biomass yields on sunlight are needed to decrease production system volumes and to lower production costs. This can be fulfilled by cultivating the microalgae in photobioreactors with a high surface to volume ratio, i.e., a short light path, such as panel reactors. Panel photobioreactors have been demonstrated to be promising photobioreactors for the controlled cultivation of microalgae (Degen et al. 2001; Hu et al. 1998c; Hu et al. 1998b; Tredici and Zittelli 1998). Furthermore, such panels can be integrated in photobioreactors in which advanced optical engineering is used to dilute light (Gordon 2002; Zijffers et al., 2008).

Record productivities have been obtained during cultivations of the cyanobacterium Arthrospira platensis in short light path, 1.3 and 2.8 cm, panel reactors that were turbulently mixed (Hu et al. 1996, 1998a, b; Hu and Richmond 1996). After recalculation of the presented data, the photosynthetic efficiency showed a maximum value of around 1.5 g of biomass (dry matter) produced per mole of photons (g mol photons^{-1}) at incident light intensities up to 2,000 μmol photons m^{-2} s^{-1}. This high efficiency was attributed to short exposure times to (over-)saturating light intensities at the reactor surface due to turbulent mixing. The biomass yield of A. platensis reported by Hu et al. is close to the theoretical attainable maximum biomass yield on light energy. This can be calculated using the stoichiometric reaction equation for biomass formation on carbon dioxide, water, and nitrate or urea (Appendix).

Although the case for A. platensis seems to be solid, it is unclear if the outcome of this work can be extrapolated to the cultivation of eukaryotic microalgae in short light path photobioreactors. Only a limited amount of work has been done with Monodus, Chlorococcum, Chlorella, and Nannochloropsis (Degen et al. 2001; Hu et al. 1996, 1998c; Richmond and Cheng-Wu 2001; Cuaresma et al. 2009). The photosynthetic efficiency, or biomass yield on light energy, was not explicitly presented in these studies. Based on the data given, we estimated the biomass yields for microalgae to be in the range of 0.3 to 1.0 g mol photons^{-1}, significantly lower than obtained with the cyanobacterium A. platensis.

The objective of this study was to determine the effect of biomass density and light path on the biomass yield of green algae at saturating light intensities. The maximum yield could be determined because we are able to analyze the yield as a function of biomass concentrations. For this, the algae were cultivated under pseudo steady-state conditions in D-stat cultivations: continuous cultured algae were allowed to acclimate to the light regime at a wide range of dilution rates and consequently a wide range of biomass densities (Barbosa et al. 2005; Hoekema et al. 2006). The maximal microalgal biomass yield on light energy has been determined for two different model green microalgae: the marine microalga Dunaliella tertiolecta and the freshwater microalga Chlorella sorokiniana in reactors with two light paths (1.25 and 2.15 cm).

MATERIALS AND METHODS

Microalgae and Reactor Start-up

D. tertiolecta CCAP 19/6B was maintained as a suspended culture in Erlenmeyer flasks containing artificial seawater medium. C. sorokiniana CCAP 211/8k was maintained as a suspended culture in Erlenmeyer flasks containing M-8a medium. Prior to the D-stat experiments, the algae were cultivated batch wise in the panel photobioreactor, until a sufficient biomass density was reached to withstand the maximum incident light intensity used in the experiments. After increasing the light intensity to the maximum value, chemostat cultivation was started at 70% to 80% of the maximum growth rate. Table 1 shows the operational conditions of the different D-stat cultivations.

Table 1: Operational conditions of the different cultivations.

Alga	Incident light intensity (**μmol** photons m^{-2} s^{-1})	Illuminated surface area (m^2)	Light path (cm)	Aeration rate (L min^{-1})	Superficial gas velocity (m s^{-1})
D. tertio-lecta	933	0.103	1.25	1.0	6.41×10^{-3}
	990	0.103	2.15	2.0	7.75×10^{-3}
C. soroki-niana	871	0.103	1.25	1.0	6.41×10^{-3}
	940	0.103	2.15	2.0	7.75×10^{-3}

Medium

D. tertiolecta was cultivated in an enriched artificial seawater medium. The ratio between the salts, nutrient, and trace element concentrations were chosen according to the biomass composition determined by Ho et al. (2003). The D. tertiolecta medium was designed to support 100 g of dry weight biomass. C. sorokiniana was cultivated on a concentrated M-8a medium, which was based on M-8 medium developed by Mandalam and Palsson (1998). The M-8a medium was estimated to support at least 50 g of C. sorokiniana dry weight biomass. Sodium bicarbonate was added to both media after the pH was set. The medium used in the continuous cultivation experiments was kept under carbon dioxide atmosphere to prevent precipitation of salts. The following medium composition was used (M): D. tertiolecta: 0.42 NaCl, 5.6×10^{-3} $MgCl_2$, 3.6×10^{-4} $CaCl_2$,

5.3×10^{-3} Na_2SO_4, 0.10 KNO_3, 2.0×10^{-2} $NaHCO_3$, 6.3×10^{-3} NaH_2PO_4, 3.1×10^{-4} Na_2 EDTA, 1.1×10^{-4} $FeCl_3$, 6.1×10^{-6} $CuSO_4$, 1.4×10^{-5} $ZnSO_4$, 9.5×10^{-8} $CoCl_2$, 1.5×10^{-5} $MnCl_2$, and 1.0×10^{-7} Na_2MoO_4; C. sorokiniana: $4.8 \times 10^{-2}$$CO(NH_2)_2$, 1.6×10^{-2} KH_2PO_4, 4.4×10^{-3} Na_2HPO_4, 4.9×10^{-3} $MgSO_4$, 2.7×10^{-4} $CaCl_2$, 9.5×10^{-4} EDTA ferric sodium salt, 3.0×10^{-4} Na_2 EDTA, 3.0×10^{-6} H_3BO_3, 2.0×10^{-4} $MnCl_2$, 3.3×10^{-5} $ZnSO_4$, 2.2×10^{-5} $CuSO_4$, and 5.0×10^{-3} $NaHCO_3$.

Reactor Set-up and Control

Figure 1 shows a front and side view of the panel reactor. The reactor was built up by transparent polycarbonate sheets held together in a frame similar to the system used by Barbosa et al. (2005). The panel reactor had a width of 20 cm, a height of 60 cm, and a light path as shown in Fig. 1 and Table 1.

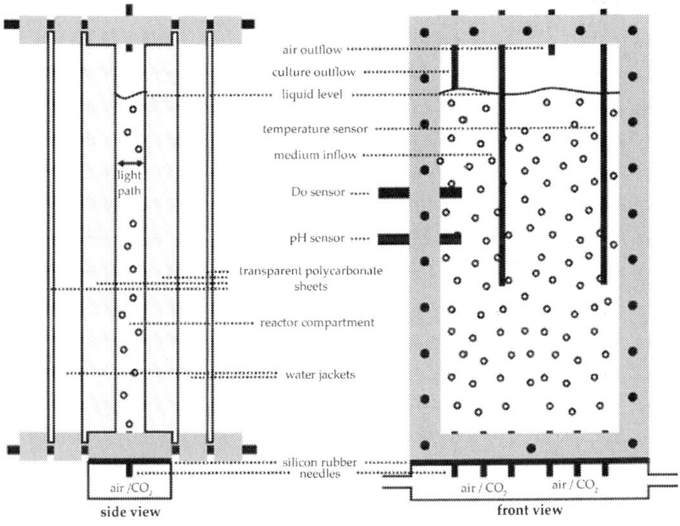

Figure 1: Front and side view of the flat panel photobioreactor. The light path is determined by the thickness of the transparent polycarbonate sheets placed in the frame and was 1.25 or 2.15 cm

The temperature inside the reactor was measured using a pt-100 and kept constant by an ADI 1030 Bio-controller (Applikon, Schiedam, the Netherlands) controlling the temperature of the cryostat connected to water jackets on both the front and back surface of the reactor. Temperature was maintained at 30°C and 37°C for D. tertiolecta and

C sorokiniana, respectively. The reactor content was mixed by sparging of air through needles pierced through a piece of silicon rubber at the bottom of the reactor. At least 24 needles were used to prevent cell death due to high bubble formation speeds (Barbosa et al. 2004). The airflow through the needles was controlled by mass flow controllers (Brooks Instrument, Hatfield, MA, USA). Before entering the reactor, the air was passed through a 20-l carboy with water to humidify the air to prevent evaporation of water from reactor. The pH was measured using pH gel sensors (Applisens, Schiedam, the Netherlands) connected to the ADI 1030 Bio-controller. The pH was maintained at 7.8 and 6.7 for D. tertiolecta and C sorokiniana, respectively, by addition of carbon dioxide to the air through mass flow controllers, controlled by the ADI 1030 Bio-controller.

Illumination

The reactors were illuminated on one side using ten compact fluorescent tubes (Lynx LE 55 W, 535 mm, Sylvania, Danvers, MA, USA). The PAR photon flux density on the reactor surface was measured several times throughout the experiment using a Li-190sa quantum sensor (LI-COR, USA). The light absorption and reflection in passing the first water jacket was determined in an empty reactor and was corrected for in determining the photon flux density into the photobioreactor. The spectral composition (Fig. 2) of the light was determined behind the water jacket (IRRAD 2000 fiber-optic spectroradiometer, TOP sensor systems, Eerbeek, the Netherlands).

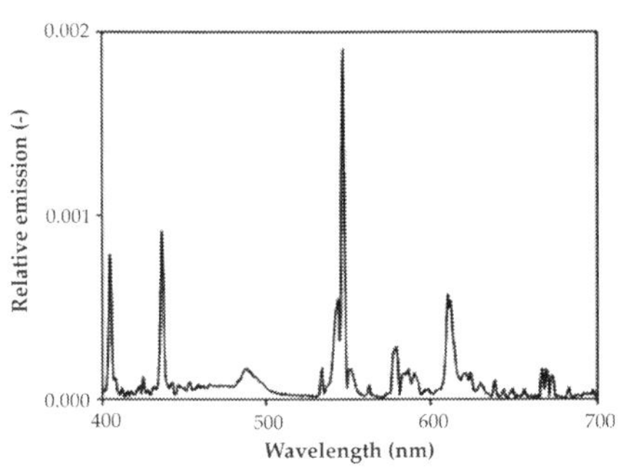

Figure 2: Spectral composition of the fluorescent tubes after passing the first water jacket (Lynx LE 55 W, 535 mm, Sylvania, Danvers, MA, USA). The emission spectrum was normalized on a quantum basis

Continuous Cultivation, the A/D-stat Strategy

The yield of biomass on light energy is best investigated in chemostat cultivations where the organisms are acclimated to the conditions in the bioreactor with no biomass accumulation. In this case, the biomass yield can be directly related to a specific light availability. A disadvantage of chemostat cultures is the number of conditions that can be determined in time. Achieving steady-state conditions can take a considerable amount of time when large changes in dilution rates are made. By slowly changing the dilution rate in a photobioreactor using the A-stat method as described by Paalme et al. (1995), the microalgal productivity can be determined for a wide range of dilution rates in less time compared to performing a number of chemostat cultivations. The continuous rate of change of the dilution rate has to be chosen such that the microalgae are able to acclimate to the continuously changing conditions in the photobioreactor. In this way, the system can be considered to be in a so-called pseudo steady state, and results represent a real steady-state situation when corrected for biomass accumulation.

The flow of fresh medium into the photobioreactor was programmed and controlled through BioXpert software (Applikon, Schiedam, the Netherlands), which determined the rotation speed of a peristaltic pump (Watson Marlow 205u, Cheltenham, UK). The actual dilution rate was determined from the decrease in weight of the medium vessel. The volume inside the reactor was kept constant by pumping excess algal suspension out through an overflow (Fig. 1).

D. tertiolecta was cultivated using a linear decrease in dilution rate using a similar rate of change as Barbosa et al. (2005). The dilution rate during the D. tertiolecta cultivations was calculated according to Eq. 1.

$$D = D_0 - d \times t \quad (\mathrm{h}^{-1}) \qquad (1)$$

Applying a constant deceleration rate, however, causes a rapid relative decrease in dilution rate at low dilution rates and larger deviations of the growth rate from the applied dilution rate. Because the relative change in dilution rate increases, the algae need to adapt more rapidly in order to maintain a situation representing steady state. C. sorokiniana was therefore cultivated using a proportional decrease in dilution rate according to Eqs. 2 and 3. Eq. 2 will never reach zero; therefore, an extra linear part was added to reach zero (Eq. 3) in the final part of the experiment. The linear part starts at t_{linear}, a predefined time at which a smooth transition to a linear decrease in dilution was made.

If $t < t_{linear}$

$$D = D_0 \cdot e^{-d_1 \cdot t} \quad (\mathrm{h}^{-1}) \tag{2}$$

Else

$$D = D_0 \cdot e^{-d_1 \cdot t_{\mathrm{linear}}} - d_2 \cdot (t - t_{\mathrm{linear}}) \quad (\mathrm{h}^{-1}) \tag{3}$$

Figure 3 shows the dilution rate development during the D-stat cultivations for the two algae. The proportional D-stat was designed to have the same duration as the linear D-stat cultivations. The dilution rate shown was determined based on the actual fresh medium supply into the reactor, measured by the decrease in weight of the medium vessel. The points shown correspond to sampling times. Using these samples, the following parameters were analyzed: dry weight biomass concentration and spectrally averaged light absorption cross section.

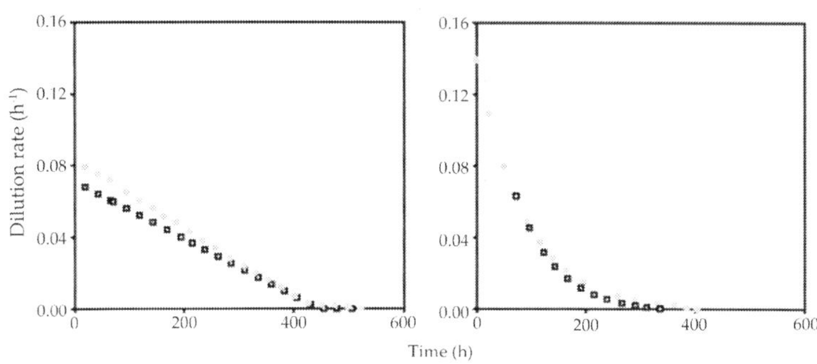

Figure 3: Dilution rate profiles during the Dunaliella tertiolecta (left) and Chlorella sorokiniana(right) D-stat cultivations. Squares: 1.25 cm light path; circles: 2.15 cm light path

Dry Weight

The biomass concentration was determined by a triplicate measurement of the dry weight of biomass in a reactor sample according to Zhu and Lee (1997). Glass microfiber filters (Whatman GF/F, Kent, UK) were used to filter the microalgae. The filter weight was determined on a 0.01-

mg resolution balance (Sartorius ME 235P, Goettingen, Germany). Prior to filtration, the D. tertiolecta sample was diluted ten times using a 0.5-M ammonium formate solution to dissolve any possible salt depositions in the sample. Demineralized water was used to dilute the freshwater C. sorokiniana sample. An additional volume of 50 ml of 0.5 M ammonium formate solution (D. tertiolecta) or demineralized water (C. sorokiniana) was filtered to further remove any small salt particles attached to the filter cake.

Absorption Cross Section

The spectrally averaged light absorption of the microalgae, a measure of the pigment content of the microalgae, was determined by measuring the wavelength-dependent light absorption of the microalgae. The absorption was measured between 400 and 750 nm using an integrating sphere (Labsphere RSA-BE-65, North Sutton, NH, USA) in a Beckman DU-640 spectrophotometer (Beckmann Coulter, Fullerton, CA, USA). The results showed residual absorption of light above 720 nm. This was attributed to backward scattering and was subtracted from the absorbance measured at all wavelengths. The wavelength-dependent specific absorption coefficient combined with the spectral composition of the radiant flux entering the reactor (Fig. 2) was used to calculate the spectrally averaged light absorption cross section according to Dubinsky et al. (1986). This spectrally averaged light absorption cross section is a measure for the specific light absorption, weighed for the light source used.

Specific Growth Rate

The specific growth rate of the organism in both the linear and the proportional D-stat cultivation was calculated from the dilution rate corrected for biomass accumulation as shown in Eq. 4 (Barbosa et al. 2005; Hoekema et al. 2006).

$$\mu_t = \frac{\left[\left(\frac{\left(C_{x,t}-C_{x,t+1}\right)}{\left(t_t-t_{t+1}\right)\times C_{x,t}}+D_t\right)+\left(\frac{\left(C_{x,t}-C_{x,t-1}\right)}{\left(t_t-t_{t-1}\right)\times C_{x,t}}+D_t\right)\right]}{2} \quad \left(\mathrm{h}^{-1}\right) \qquad (4)$$

Observed Biomass Yield on Light Energy

The observed biomass yield on light energy was calculated by Eq. 5. The yield was based on the biomass production rate and light supply rate. The yield was not corrected for unused light passing through the flat panel photobioreactor at very low biomass densities.

$$Y_{x,E(\text{obs})} = \frac{C_x \times \mu \times V}{\text{PFD}_{\text{in}} \times A \times 3.600 \times 10^{-6}} \quad (\text{g mol photons}^{-1}) \tag{5}$$

Specific Light Supply Rate

The cultivation results were compared based on the specific light supply rate, calculated by Eq. 6. The photon flux entering the panel photobioreactor was divided by the amount of biomass present in the reactor.

$$r_{E,x} = \frac{\text{PFD}_{\text{in}} \times A}{C_x \times V} \quad (\mu\text{mol photons g}^{-1}\ \text{s}^{-1}) \tag{6}$$

RESULTS AND DISCUSSION

Biomass Concentration

The biomass concentration of D. tertiolecta and C. sorokiniana in flat panel bioreactors with light paths of 1.25 and 2.15 cm as a function of the dilution rate were determined (Fig. 4). In continuous culture, the maximum biomass concentration was about 18 g/l and decreased as a function of dilution rate as expected. Higher biomass concentrations were obtained in the reactors with shorter light paths. The biomass concentrations obtained were about equal for both algae, even though the two microalgae differ in maximum growth rate and require different cultivation conditions. The two different microalgae were cultivated under controlled and optimal conditions. Concentrations of nutrients and trace elements were available in excess, pH and temperature were kept optimal, and produced oxygen was removed by aeration. Hence, the microalgae were cultivated under light-limited conditions, and the productivity obtained at a dilution rate is a direct measure of the efficiency of use of the constant available light intensity.

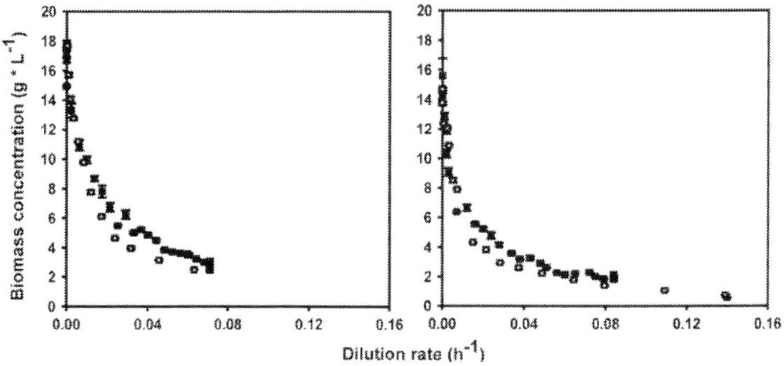

Figure 4: Biomass concentration of Dunaliella tertiolecta and Chlorella sorokiniana during the D-stat cultivations. Left: 1.25 cm light path. Right: 2.15 cm light path. Closed squares: D. tertiolecta biomass; open squares: C. sorokiniana biomass. Error bars represent standard errors from the mean

Observed Biomass Yield on Light Energy

The biomass yield on light was calculated with Eq. 4 and plotted as a function of the light supply rate (the amount of light available per amount of biomass) as calculated with Eq. 5. In Figs. 5 and 6, the observed biomass yield on light energy as a function of the light supply rate at the different light paths are shown for D. tertiolecta and C. sorokiniana cultivations, respectively. In Fig. 5, we also added the data for D. tertiolecta in a photobioreactor with a light path of 3 cm from Barbosa et al. (2005) At higher photon flux densities, a relatively constant yield between 0.6 and 0.8 g/mol photons is obtained for all cases. At lower photon flux densities, i.e., at higher biomass concentrations, the yield drops considerably. At a given light supply rate, the highest biomass concentration is present in the 1.25-cm light path reactor and the lowest in the 3-cm light path reactor, leading to different light penetration depths in the reactors but the observed yields as a function of light supply rate in the different photobioreactors are independent of the light path used. The different biomass concentrations in the different biooreactors result in different light gradients and different exposure patterns of the algae to the light, ranging from an exposure to a steep gradient in the 1.25-cm reactor to an exposure to a more gradual gradient in the 3-cm reactor. Apparently, this gradient and the corresponding light exposure pattern does not influence the biomass yield in these reactors since the biomass yield observed at a given specific light supply rate was equal for all three reactors.

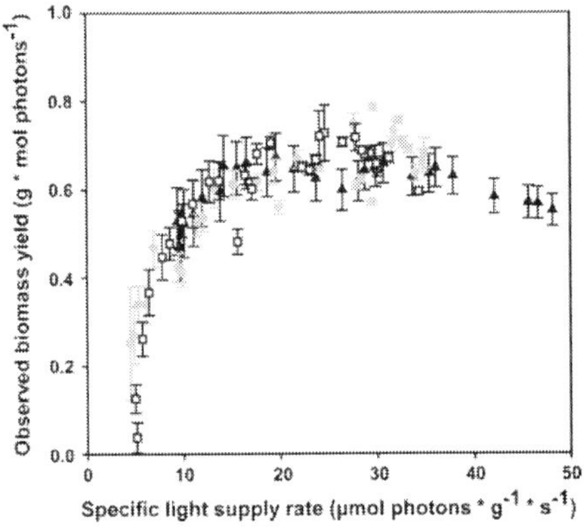

Figure 5: Observed biomass yield on light energy during the Dunaliella tertiolecta cultivations. Squares: 1.25 cm light path; circles: 2.15 cm light path; triangles: 3 cm light path (Barbosa et al. 2005). Error bars represent standard errors from the mean

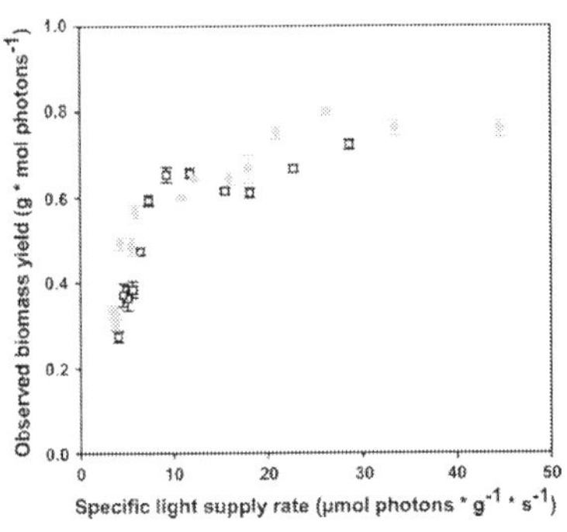

Figure 6: Observed biomass yield on light energy during the Chlorella sorokiniana cultivations. Squares: 1.25 cm light path; circles: 2.15 cm light path. Error bars represent standard errors from the mean

A biomass yield of 0.65 ± 0.10 g mol photons^{-1} was obtained for D. tertiolecta in all panel reactors ranging from a light supply rate of 35 down to 13 µmol photons g^{-1} s^{-1}. The yields obtained during the C. sorokiniana cultivations were similar to the values obtained during the D. tertiolecta cultivations (Fig. 6). Yields between 0.6 and 0.8 g mol photons^{-1} were obtained between light supply rates of 45 down to 10 µmol photons g^{-1} s^{-1}.

Absorption Cross Section

To determine the absorption characteristics at the different light supply rates, the spectrally averaged light absorption cross sections were determined as shown in Figs. 7 and 8. The light absorption cross section of D. tertiolecta (Fig. 7) increased with this decreasing light supply rate, indicating a photo acclimation response generally observed for microalgae subjected to decreasing light availability. The cross section increases to a maximum value until it decreases slightly at the end of the D-stat cultivation.

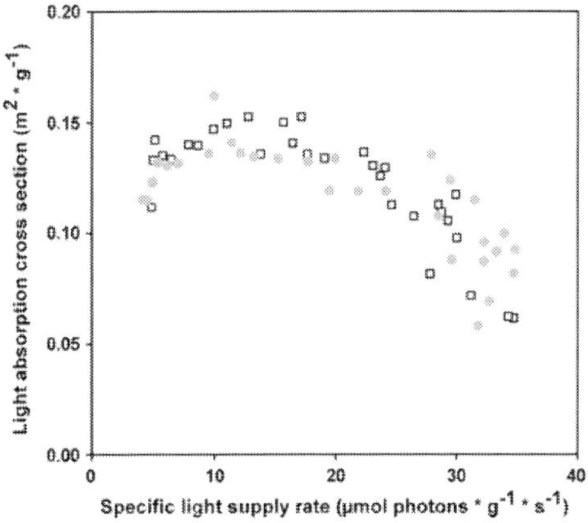

Figure 7: Spectrally averaged light absorption cross section of Dunaliella tertiolecta during the cultivations. Squares: 1.25 cm light path; circles: 2.15 cm light path

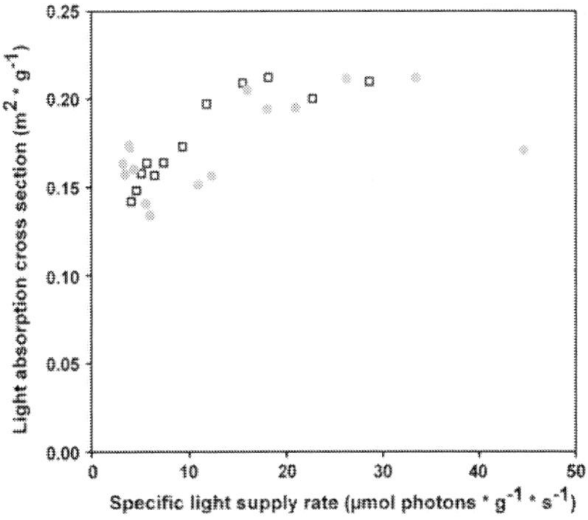

Figure 8: Spectrally averaged light absorption cross section of Chlorella so-

rokiniana during the cultivations. Squares: 1.25 cm light path; circles: 2.15 cm light path

Figure 8 shows the light absorption cross section for C. sorokiniana during the two D-stat cultivations. We did not observe an increase in absorption cross section while decreasing the light supply rate. Unpublished results in a similar set-up show a smaller absorption cross section at higher light supply rates than the ones tested here. Possibly, photoacclimation of C. sorokiniana already occurs at higher specific light supply rates and is therefore not clearly visible in these experiments. The decrease in absorption cross section at very low specific light supply rates, on the other hand, was more pronounced during the C. sorokiniana cultivation. It started at a light supply rate of 20 as compared to 10 μmol photons g^{-1} s^{-1} for D. tertiolecta.

Figures 7 and 8 show the microalgae adapted to the changing light supply rate by changing the absorption cross section. The decreasing light supply rate caused an increase in the light absorption cross section until the maximum size was reached. Eventually, the light supply rate was insufficient to maintain the maximum absorption cross section causing the cross section to decrease. Both the response in absorption cross section and the small difference in growth rate compared to the dilution rate (data not shown) showed that the microalgae were able to acclimate to the changing light regime. The results obtained during the D-stat were therefore considered to be representative for a steady-state situation.

Biomass Yield on Light Energy

The biomass yield on light energy of the acclimated microalgae is determined by the ability of the algae to use the supplied light energy for biomass formation. The light energy distribution in the reactor is characterized by a high PFD on the reactor surface, which decreases rapidly due to light absorption by the microalgae, creating a photic zone and a dark zone in the reactor. The photic zone is characterized by a strong light gradient with a PFD of about 930 μmol $photons^{-1}$ m^{-2} s^{-1} at the surface that quickly decreases to a light intensity below the compensation point at the interface with the dark zone. The compensation point is the light intensity at which the photosynthesis and respiration compensate each other, i.e., net oxygen production is zero.

Assuming a steady liquid flow, a decrease of the size of the photic zone will lead to a shorter exposure time to (over-)saturating light levels. Also, if panel width is reduced, the absolute residence time in the photic zone can be reduced, since at constant PFD, a higher biomass concentration, and thus a smaller photic zone, is obtained. Based on existing knowledge, such a reduction in exposure time to (over-)saturating light could lead to higher photosynthetic efficiencies, since less light-induced damage will build up, and possibly also, non-photochemical quenching (heat dissipation) of light energy will be less pronounced (Horton and Ruban 2005; Muller et al. 2001). Increased exposure times to (over-)saturating PFDs are therefore expected to lead to reduced photosynthetic efficiency and vice versa.

The strong decrease of biomass yield on light energy observed at low specific light supply rates, i.e., low specific growth rates and high biomass densities, in the panel reactors, however, does not correlate to the change in light regime, i.e., the decrease in size of the photic zone. The yield decrease, on the other hand, can be explained assuming that a significant fraction of the available light energy is not used for biomass formation but it is used for basic physiological maintenance, which includes processes such as osmoregulation, cell motility, defense mechanisms, and proofreading and internal turnover of macromolecular compounds (van Bodegom 2007).

To demonstrate that maintenance requirements determined the reduction of observed yield at lower light supply rates, we fitted our data through the model of Pirt (1986), as shown by Eq. 7, for both C. sorokiniana and D. tertiolecta for the 1.25- and 2.15-cm light path panel reactors. The model assumes that the light energy available to a microalga ($r_{E,x}$; µmol photons g^{-1} s^{-1}) is either used for maintenance processes or for formation of new biomass. The model is based on a constant light energy requirement for maintenance ($m_{E,x}$; mol photons^{-1} g^{-1} h^{-1}). The model is also based on a constant energy requirement for the formation of biomass ($Y_{x,E,}$; g mol photons^{-1}), which is different from the observed biomass yield on light energy shown in Figs 5and 6.

$$r_{E,x} \times 0.0036 = \frac{\mu_t}{Y_{x,E}} + m_{E,x} \quad (\text{mol photons g}^{-1}\,\text{h}^{-1}) \tag{7}$$

During the D-stat experiment, the dilution rate through the short light path reactor decreases, and biomass accumulates. The increase in biomass concentration reduces the light energy available for the individual microalgae cells ($r_{E,x}$; μmol photons g^{-1} s^{-1}) and affects the specific growth rate (μ_t; h^{-1}). When the light supply rate is thus plotted as a function of the obtained growth rate, a straight line should be obtained according to Eq. 7. The slope of this line represents the inverse of the yield, and the offset represents the maintenance coefficient. Figure 9 shows that the D-stat results obtained for D. tertiolecta and C. sorokiniana can be well described by a constant yield and maintenance requirement. Since no clear influence of the light path on the observed biomass yield on light energy was observed, the data for the 1.25- and 2.15-cm reactors were taken together, and the following relations were found:

D.tertiolecta:C.sorokiniana:rE,x×0.0036=μt0.782+0.0133rE,x ×0.0036=μt0.751+0.0068R2=0.977R2=0.988(molphotonsg−1h−1) (molphotonsg−1h−1)D.tertiolecta:rE,x×0.0036=μt0.782+0.0133R2=0.977(molphotonsg−1h−1)C.sorokiniana:rE,x×0.0036=μt0.751+0.0068R2 =0.988(molphotonsg−1h−1)

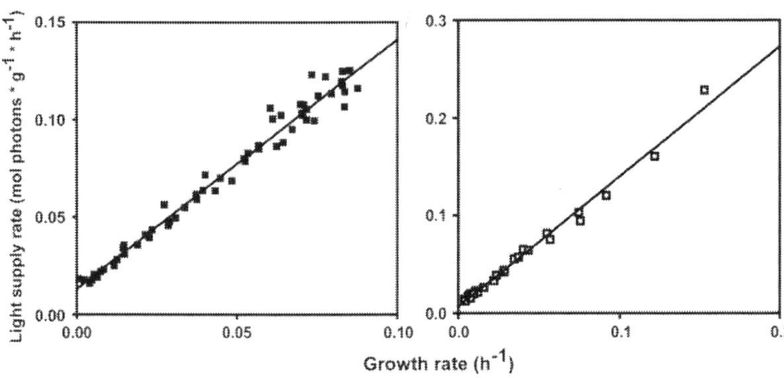

Figure 9: Specific light supply rate in the panel reactor as a function of the growth rate. Closed squares: Dunaliella tertiolecta; open squares: Chlorella sorokiniana

The constant value for the biomass yield on light energy clearly shows that the steep decrease in observed biomass yield at high biomass concentration (Figs. 5 and 6) is thus a direct result of a considerable cellular maintenance requirement. Photosynthetic activity is constant throughout the experiment, but an increasing fraction of the generated chemical energy is used for maintenance processes, decreasing net photosynthesis and biomass formation. The decrease in biomass yield commonly found at high biomass concentrations might thus not be caused by a decrease in efficiency of light use, i.e., a decrease in yield, but by an increase in maintenance energy requirement because of the increase in biomass concentration.

Cellular maintenance energy requirements of microalgae could have major implications for the development of systems for mass cultivation of microalgae. In order to optimize microalgae yield on light energy while obtaining high volumetric productivities, high growth rates need to be combined with high biomass densities. Research should thus focus on further reducing the reactor light path, such that high light supply rates can be obtained at high biomass densities. In this way, also at high biomass densities, high specific growth rates can be maintained reducing the losses associated to maintenance.

Based on the Pirt model, a constant biomass yield on light energy of 0.78 and 0.75 g mol photons^{-1} was found for D. tertiolecta and C. sorokiniana, respectively, in the current study. The yield differed from the theoretical maximum of 1.5 and 1.8 g mol photons^{-1} as calculated in the Appendix. The difference between the biomass yield and this theoretical maximum is most likely related to heat dissipation of absorbed light energy in the photosynthetic antenna complex by processes collectively called non-photochemical quenching (Muller et al. 2001). These processes are activated when the rate of light absorption exceeds the maximal capacity of the photosynthetic machinery and prevents photodamage. In case of D. tertiolecta, a calculated 48% of the available light energy was not used for biomass formation, and in case of C. sorokiniana, it increased to 58%. Figures 5 and 6 do not directly show a difference in observed biomass yield on light energy, but the microalgae do differ in the efficiency of light use. D. tertiolecta has a higher maintenance energy requirement, whereas C. sorokiniana has a slightly lower biomass yield on light energy. The higher maintenance energy requirement causes the yield of D. tertiolecta to decrease faster at high biomass densities.

In literature, yield values are generally not explicitly mentioned, requiring recalculation of presented biomass production and assumptions on reactor geometry and light intensities entering the reactors and making an exact comparison of the values difficult. Our studies presented here show observed yields at different biomass concentrations and different dilution rates. The maximum observed yields that we have found are comparable to maximum yields reported in literature or which were re-calculated from literature. Meiser et al. (2004) found a maximum yield of 0.5 g mol photons^{-1} for the diatom Phaeodactylum tricornutum at a light intensity of 1,000 μmol photons m^{-2} s^{-1}. Hu et al. (1998c) obtained a yield of 0.5 g mol photons^{-1} cultivating the microalga Chlorococcum littorale at a light intensity of 2,000 μmol photons m^{-2} s^{-1}. Richmond et al. (2003) obtained a maximum yield of 0.6 g mol photons^{-1}cultivating the microalga Nannochloropsis sp. at a light intensity of 2,000 μmol photons m^{-2} s^{-1}. Monodus subterraneus showed a maximum yield of 1.0 g mol photons^{-1} (Hu et al. 1996). This productivity was obtained at an increased level of turbulence compared to the cultivations performed in this study.

Increasing turbulence by increasing the superficial gas velocity in our cultures did not result in a clear improvement of culture productivity (data not shown). High superficial gas velocities (0.04 m s^{-1}) even lead to unstable cultures due to increased cell death and considerable foaming and, as such, biomass loss via the gas phase. High gassing rates are also not attractive for large-scale microalgae cultivation because of the large expenditures (energy and cost) associated with it.

With this study, we demonstrated that the observed biomass yield on light in short light path bioreactors at high biomass densities decreases because maintenance requirements are relatively high at these conditions. This is apparently in contradiction with the work of Richmond et al. (2003) but supported by the physiological models of Pirt (1986) and others hereafter. We demonstrated that all our experimental data for the two strains tested could be described by this model. Consequently, for the design of a photobioreactor, we should maintain a relatively high specific light supply rate. A process with high biomass densities and high yields at high light intensities can only be obtained in short light path photobioreactors

CONCLUSIONS

The microalgae biomass yield on light energy obtained during D-stat experiments in short light path photobioreactors appeared to be determined by the specific light supply rate, which depends on biomass density, light path, and the light intensity at the reactor surface. In reactors with different light paths (1–3 cm), equal biomass yields were observed at equal specific light supply rates. This response was observed for two different species of microalgae, the marine strain D. tertiolecta and the freshwater strain C. sorokiniana. The optimal biomass density, on the other hand, was highest in the short light path reactor. The only rationale to decrease light path then is related to the requirement of high biomass concentration for more efficient processing of the biomass in industrial applications.

With this study, we demonstrated that the observed biomass yield on light in short light path bioreactors at high biomass densities decreases because maintenance requirements are relatively high at these conditions. For the design of a photobioreactor with a maximum productivity, we should maintain a relatively high specific light supply rate.

NOTES

Acknowledgements

This work was financially supported by the EU/Energy Network project SOLAR-H (FP6 contract 516510).

Appendix

The theoretical biomass yield on light energy can be calculated based on the stoichiometric reaction equations for the formation of biomass on carbon dioxide, water, and the nitrogen source used in the cultivation.

For growth on nitrate:

$$CO_2(g) + 0.95 \cdot H_2O(l) + 0.12 \cdot NO_3^-(aq)$$
$$\rightarrow CH_{1.78}O_{0.36}N_{0.12}(s) + 1.415 \cdot O_2(g) + 0.12 \cdot OH^-(aq)$$

For growth on urea:

$$0.94 \cdot CO_2(g) + 0.77 \cdot H_2O(l) + 0.06 \cdot CH_4ON_2(aq)$$
$$\rightarrow CH_{1.78}O_{0.36}N_{0.12}(s) + 1.175 \cdot O_2(g)$$

The yield of the light reactions is assumed to be 0.1 mol of oxygen per mole of photons within the PAR spectrum. This value represents the maximal quantum yield as determined under low light by several independent researchers over the past decades. The theoretical maximal based on the Z-scheme of photosynthesis would be 0.125. The difference with 0.1 is probably related to intrinsic inefficiencies of photosynthesis.

Assuming D. tertiolecta and C sorokiniana have the same elemental composition of $C_{1.78}O_{0.36}N_{0.12}$ as found for Chlorella Spain sp. (Duboc et al. 1999), the molecular mass of a C-mol biomass is 21.25 g mol^{-1}. To form one C-mol of biomass, 14.15 or 11.75 mol of photons are needed to evolve the required amount of oxygen for growth on nitrate and urea, respectively, following the stoichiometric reaction equations. This leads to a theoretical biomass yield of 1.5 and 1.8 g mol photons^{-1} for growth on nitrate and urea, respectively.

REFERENCES

1. Barbosa MJ, Hadiyanto S, Wijffels RH (2004) Overcoming shear stress of microalgae cultures in sparged photobioreactors. Biotechnol Bioeng 85:78–85
2. Barbosa MJ, Zijffers JW, Nisworo A, Vaes W, Van Schoonhoven J, Wijffels RH (2005) Optimization of biomass, vitamins, and carotenoid yield on light energy in a flat-panel reactor using the A-stat technique. Biotechnol Bioeng 89:233–242
3. Chisti Y (2007) Biodiesel from microalgae. Biotechnol Adv 25:294–306
4. Cuaresma M, Janssen M, Vilchez C, Wijffels RH (2009) Productivity of Chlorella sorokinianain a short light-path (SLP) panel photobioreactor under high irradiance. Biotechnol Bioeng 104:352–359
5. Degen J, Uebele A, Retze A, Schmid-Staiger U, Trosch W (2001) A novel airlift photobioreactor with baffles for improved light utilization through the flashing light effect. J Biotechnol 92:89–94
6. Dubinsky Z, Falkowski PG, Wyman K (1986) Light harvesting and utilization by phytoplankton. Plant Cell Physiol 27:1335–1349
7. Duboc P, Marison I, Von Stockar U (1999) Handbook of thermal analysis and calorimetry. Elsevier, Amsterdam
8. Gordon JM (2002) Tailoring optical systems to optimized photobioreactors. Int J Hydrogen Energy 27:1175–1184

9. Ho TY, Quigg A, Finkel ZV, Milligan AJ, Wyman K, Falkowski PG, Morel FMM (2003) The elemental composition of some marine phytoplankton. J Phycol 39:1145–1159
10. Hoekema S, Douma RD, Janssen M, Tramper J, Wijffels RH (2006) Controlling light-use by Rhodobacter capsulatus continuous cultures in a flat-panel photobioreactor. Biotechnol Bioeng 95:613–626
11. Horton P, Ruban A (2005) Molecular design of the photosystem II light-harvesting antenna: photosynthesis and photoprotection. J Exp Bot 56:365–373
12. Hu Q, Richmond A (1996) Productivity and photosynthetic efficiency of Spirulina platensis as affected by light intensity, algal density and rate of mixing in a flat plate photobioreactor. J Appl Phycol 8:139–145
13. Hu Q, Guterman H, Richmond A (1996) A flat inclined modular photobioreactor for outdoor mass cultivation of photoautotrophs. Biotechnol Bioeng 51:51–60
14. Hu Q, Faiman D, Richmond A (1998a) Optimal tilt angles of enclosed reactors for growing photoautotrophic microorganisms outdoors. J Ferment Bioeng 85:230–236
15. Hu Q, Zarmi Y, Richmond A (1998b) Combined effects of light intensity, light-path and culture density on output rate of Spirulina platensis (Cyanobacteria). Eur J Phycol 33:165–171
16. Hu Q, Kurano N, Kawachi M, Iwasaki I, Miyachi S (1998c) Ultrahigh-cell-density culture of a marine green alga Chlorococcum littorale in a flat-plate photobioreactor. Appl Microbiol Biotechnol 49:655–662
17. Mandalam RK, Palsson BO (1998) Elemental balancing of biomass and medium composition enhances growth capacity in high-density Chlorella vulgaris cultures. Biotechnol Bioeng 59:605–611
18. Meiser A, Schmid-Staiger U, Trosch W (2004) Optimization of eicosapentaenoic acid production by Phaeodactylum tricornutum in the flat panel airlift (FPA) reactor. J Appl Phycol 16:215–225
19. Muller P, Li XP, Niyogi KK (2001) Non-photochemical quenching. A response to excess light energy. Plant Physiol 125:1558–1566
20. Paalme T, Kahru A, Elken R, Vanatalu K, Tiisma K, Vilu R (1995)

The computer-controlled continuous culture of Escherichia coli with smooth change of dilution rate (A-stat). J Microbiol Methods 24:145–153

21. Pirt SJ (1986) Tansley review No-4—the thermodynamic efficiency (quantum demand) and dynamics of photosynthetic growth. New Phytol 102:3–37
22. Richmond A, Cheng-Wu Z (2001) Optimization of a flat plate glass reactor for mass production of Nannochloropsis sp outdoors. J Biotechnol 85:259–269
23. Richmond A, Zhang CW, Zarmi Y (2003) Efficient use of strong light for high photosynthetic productivity: interrelationships between the optical path, the optimal population density and cell-growth inhibition. Biomol Eng 20:229–236
24. Tredici MR, Zittelli GC (1998) Efficiency of sunlight utilization: tubular versus flat photobioreactors. Biotechnol Bioeng 57:187–197
25. Van Bodegom P (2007) Microbial maintenance: a critical review on its quantification. Microb Ecol 53:513–523
26. Wijffels RH (2008) Potential of sponges and microalgae for marine biotechnology. Trends Biotechnol 26:26–31
27. Zhu CJ, Lee YK (1997) Determination of biomass dry weight of marine microalgae. J Appl Phycol 9:189–194
28. Zijffers JWF, Janssen M, Tramper J, Wijffels RH (2008) Design process of an area-efficient photobioreactor. Mar Biotechnol 10:404–415

CHAPTER

3

THE DIFFERENCE IN EFFECTIVE LIGHT PENETRATION MAY EXPLAIN THE SUPERIORITY IN PHOTOSYNTHETIC EFFICIENCY OF ATTACHED CULTIVATION OVER THE CONVENTIONAL OPEN POND FOR MICROALGAE

Junfeng Wang[1], Jinli Liu[2] and Tianzhong Liu[1]

[1]Key Laboratory of Biofuels, Qingdao Institute of Bioenergy and Bioprocess Technology, Chinese Academy of Sciences, Qingdao, People's Republic of China

[2]Low-Carbon Energy Conversion Science and Engineering Center, Shanghai Advanced Research Institute, Chinese Academy of Sciences, Shanghai, People's Republic of China

ABSTRACT

Background

The 'attached cultivation' technique for microalgae production, combining the immobilized biofilm technology with proper light dilution strategies, has shown improved biomass production and photosynthetic efficiency over conventional open-pond suspended cultures. However, how light is transferred and distributed inside the biofilm has not been clearly defined yet.

Results

In this research, the growth, photosynthetic oxygen evolution, and specific growth rate for microalgal cells in both open-pond and attached cultivation were studied to determine the effective light penetration at different phases of the cultivation. As a result, the light conditions inside the culture broth as well as the biofilm were revealed for the first time. Results showed that outdoor, in a conventional 20-cm deep open pond, all of the algal cells were fully illuminated in the first 3 days of cultivation. As the biomass concentration increased from day 4 to day 10, the light could only effectively penetrate 45.5% of the open-pond depth, and then effective light penetration gradually decreased to 31.1% at day 31, when the biomass density reached a maximum value of 0.45 g L^{-1} or 90 g m^{-2}. In the attached cultivation system, under nitrogen-replete condition, almost 100% of the immobilized algal cells inside the biofilm were effectively illuminated from day 0 through day 10 when the biomass density increased from 8.8 g m^{-2} to 107.6 g m^{-2}.

Conclusion

Higher light penetration efficiency might be the reason why, using attached cultivation, observed values for photosynthetic efficiency were higher than those recorded in conventional open-pond suspended cultures.

Keywords: Microalgae Photosynthetic efficiency Open pond Attached cultivation Effective light penetration

BACKGROUND

Microalgae are a group of photosynthetic microorganisms represented by more than 40,000 species, many of which can produce high-value bioactive compounds [1]. In recent years, due to the price increase of fossil fuel as well as the concerns on environment deterioration, the global R&D investment on microalgae biofuel vastly increased. The reasons for this are that microalgae have high photosynthetic efficiency and high oil content and are believed to be the most promising feedstock for environment-friendly renewable liquid biofuel thanks to the fact that they do not compete with food crops for arable land and water supply [2-7]. However, until now, the success of microalgae cultures only happened in some small-scale tests and no success in biofuel production or CO_2 mitigation has yet been achieved on a commercial scale anywhere in the world with microalgae cultivation [8,9]. The primary reason for this is that even when state-of-the-art techniques and designs are adopted on a large scale for open ponds or closed photobioreactors (PBRs), microalgal cell growth rate values currently cannot reach those obtained in highly controlled lab environments. The highest recorded biomass productivities for open ponds and PBRs are around 40 g dry mass m^{-2} day^{-1} and long-term averaged productivity is 10~20 g dry mass m^{-2} day^{-1} [3,10,11], which are far less than the theoretical maximum of 120~150 g m^{-2} day^{-1}, or photosynthetic efficiency of 12.4% (based on total solar radiation spectrum) and 28% (based on visible light spectrum, 400~700 nm) [12,13].

The biofilm cultivation of microalgae, in which the algal cells are generally immobilized and fixed onto/into supporting materials in high density and fed with nutrient solutions, is a different cultivation method than conventional aqueous suspended cultures. Open ponds and PBRs have existed for a long time now; however, they often exhibited relatively low biomass productivity [14-21]. Recently, our research group proposed an improved cultivation method based on a reactor where both the immobilized biofilm technology and the light dilution theory work together in a patented design called 'attached cultivation' [22]. The basic principles of this 'attached cultivation' technique as well as the related photobioreactors include the following: (i) a highly dense wet algal paste attached onto artificial supporting materials to form a thin layer of the algal biofilm and (ii) many layers of these biofilms arranged in an array fashion to dilute the high light so that the light intensity impinging the algal biofilm is much lower than the light that reaches the photobioreactor footprint area (Figure 1A). The biomass productivity potential of this 'attached cultivation' method was evaluated both under indoor and outdoor conditions. Results showed that with indoor light of 700 μmol m^{-2} s^{-1} and a light dilution rate of 10, the maximum biomass productivity for *Scenedesmus* was close to 120 g m^{-2} day^{-1}, with photosynthetic efficiency of *ca.* 18% (based on visible light). Under outdoor conditions, the maximum biomass productivity reached *ca.* 80 g m^{-2} day^{-1} corresponding to a photosynthetic efficiency of 17.3% (based on visible light) and 8.3% (based on total solar irradiation), which were both seven times higher than the data reported for a conventional open pond under the same environment conditions [22]. Similar biomass productivity and photosynthetic efficiency were also achieved with *Botryococcus braunii* which grows slowly with aqueous suspended cultivation but exhibits a high biomass productivity of *ca.* 50 g m^{-2} day^{-1} with attached cultivation, corresponding to a photosynthetic efficiency of *ca.* 15% (based on visible light) [23].

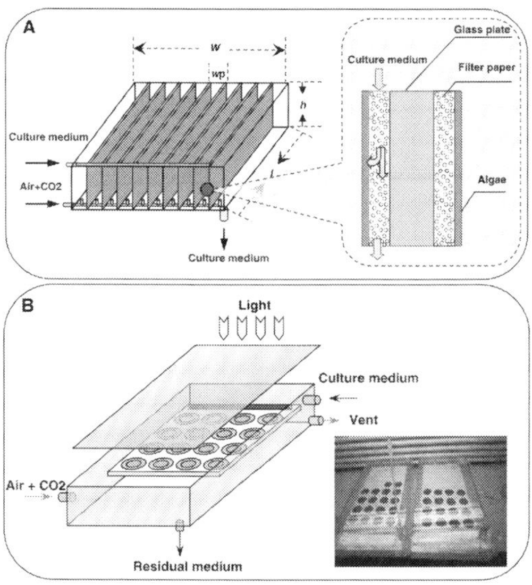

Figure 1: The schematic diagrams of the photobioreactors for attached cultivation. (A)The schematic diagram for the multiple-layer photobioreactor for attached cultivation (adapted from Liu *et al.* [22]). (B) The schematic diagram and the actual photograph for the single-layer photobioreactor for attached cultivation used in the research.

As shown in these results, the 'attached cultivation' technique greatly improves the biomass productivity as well as photosynthetic efficiency of microalgae. However, the reason(s) and mechanism(s) for the superiorities of this method over suspended cultivation have not been fully understood yet. Maybe the secret of its superior performances over traditional suspended cultures lies in the differences in delivery of light and nutrients and in CO_2 transfer between the two systems [24,25]. In this research, we especially focused on how the illumination worked, because light is always considered as the most important environmental factor that determines the behaviors of the microalgal cells [12,26]. A basic but important aspect to consider is how deep light penetrates during the microalgal cultivation. For open-pond suspended cultures, it is obvious that with the increases of biomass density, the light penetration depth decreases [12] and results in less and less percentage of algal cells being illuminated with light intensity which is higher than the compensation point (effectively illuminated) where the biomass accumulation rate due to photosynthesis equilibrates with the biomass loss rate due to respiration at a specific biomass density [11]. However, until to now, the dynamics of the 'light' *vs.* 'dark' for the open pond has not been quantitatively assessed. The information on illumination properties of the immobilized biofilm cultivation is scarce. How does the light transfer and penetrate inside the immobilized microalgal biofilm? Does light transfer and penetrate differently in the microalgal biofilm and in a suspended culture? How is the difference(s) related with the performances in photosynthetic efficiency? To answer these questions, the growth, photosynthetic oxygen evolution property, and specific growth rate for microalgal cells in open-pond and 'attached cultivation' systems were studied to determine the dynamics of effective light penetration (d_E, depth from the surface where the light intensity is equal to the light compensation point) during cultivation. The light condition inside the outdoor open pond as well as the biofilm was revealed, and the effect of light distribution on the photosynthetic efficiency was discussed.

RESULTS AND DISCUSSION

The light distribution inside the suspended cultivation system

The light intensity inside the open pond declined with the increase of broth depth as well as biomass concentration (Figure 2B). Similar results had been reported by Tredici [12]. According to the oxygen evolution properties, light intensity of 12.5 ± 3.9 μmol m^{-2} s^{-1} was considered as the light compensation point (LCP) for *Scenedesmus dimorphus*; under these circumstances, the oxygen consumption due to respiration already surpassed the oxygen evolution due to photosynthesis (Figure 2C). It should be noticed that the photosynthesis-light intensity (PI) curve of Figure 2C was measured on healthy vegetative algal cells that had been cultivated in the outdoor open pond for 5 days. According to our pilot experiment, from day 0 to day 30, there were non-significant differences among LCPs of algal cells from the outdoor open pond supplied with a non-nitrogen-deficient nutrient solution. The d_E for any given biomass density can be estimated from the results of light attenuation and LCP (Figure 2D). Accordingly, the d_E was only 0.57 cm at a biomass density of 5.0 g L^{-1}, which means that only 40.4% of cells were effectively illuminated for a glass column having a diameter of 5 cm.

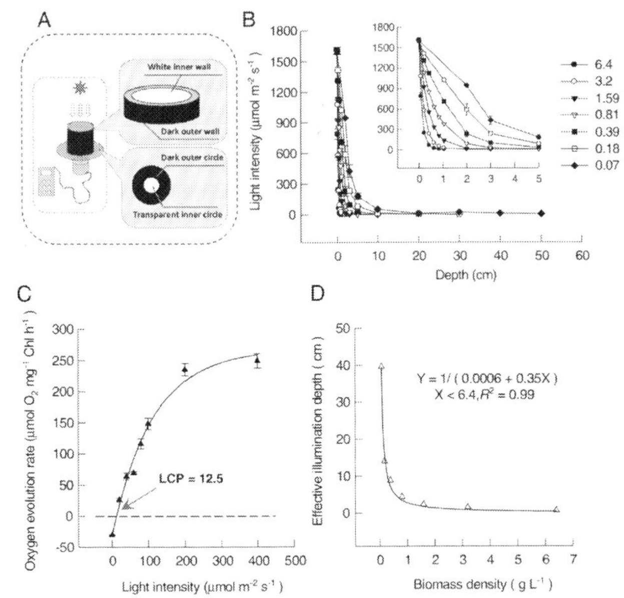

Figure 2: The light attenuation inside the culture broth of suspended cultivation of *Scenedesmus dimorphus*. (A) The schematic diagrams of light intensity measurements under different depths of the culture broth. (B) The dependence of light attenuation on culture density as well as depth. Seven different culture densities were tested, *viz.* 6.4 (black circle), 3.2 (white circle), 1.59 (black down-pointing triangle), 0.81 (white down-pointing triangle), 0.39 (black square), 0.18

(white square), and 0.07 (black diamond). The measurements were carried out outdoors with natural light. Data were mean±standard deviation of three measurements. (C) The relationship of oxygen evolution rate *versus* light intensity (black triangle). Data were mean±standard deviation of three measurements. The light compensation point (LCP) was indicated by arrows. (D) The effective illumination depth of aqueous suspended *S. dimorphus* culture broth at different biomass densities (white triangle).

In these conditions, the maximum biomass density would be 5.0 g L^{-1} after 10 days with continuous illumination of 100 μmol m^{-2} s^{-1} under indoor conditions. This d_E value might also be affected by light conditions, for example, the d_E might decrease in cloudy weather. To avoid this uncertainty, in this experiment, d_E measurement was performed under outdoor conditions with natural sunlight intensity of 1,500~1,600 μmol m^{-2} s^{-1} so that this equation could be directly applied to outdoor cultivation.

In this study, we proved that for a conventional 20-cm deep open pond, the maximum biomass density would reach 0.45 g L^{-1} in 10 days and remain stable in the following days (Figure 3A). This maximum biomass density of 0.45 g L^{-1} in the open pond, corresponding to 90 g m^{-2}, was similar with the results of Chisti [11] where the biomass density for open pond could not exceed 0.5 g L^{-1}. The pH during the cultivation slightly fluctuated in the range of 6.7~7.5, indicating that the carbon source was not limited during the experiment (Figure 3A). According to the equation in Figure 2D, thanks to the low biomass concentration, during the first 3 days of cultivation, light could easily penetrate the suspended culture delivering optimal light intensity to each algal cell. After 31 days, only 31.1% of the algal cells were effectively illuminated, with the effective d_E being only 6.2 cm. In this research, the fastest biomass accumulation for the open pond was recorded from day 3 to day 7, corresponding to an increase of the biomass concentration from 0.1 to 0.3 g L^{-1} or areal biomass density from 20 to 60 g m^{-2}. During the same time window, biomass productivity reached *ca.* 10 g m^{-2} day^{-1}, and the effective light penetration continuously decreased from 100% to 45.5% (Figure 3B). The biomass productivity did not decrease with the reduced effective illumination depth during these times but remained at 10 g m^{-2} day^{-1}. In microalgal suspended cultures, light distribution is not uniform at all. Light intensity decreases exponentially as we move farther away from the illuminated surface; for this reason, a thin layer on the water surface usually receives oversaturating light intensity while the bottom of the pond, as the algae cells reproduce and the biomass concentration increases, lies in total darkness. The algal cells traveling through the dark portion of the reactor consume biomass by dark respiration, if they spend a long enough time in this kind of environment. An appropriate depth of the water layer in an open pond is required to avoid biomass loss by cellular respiration. The optimal depth of any suspended culture of algae cells should be equal to the maximum depth at which a sufficient amount of solar light can penetrate. The aforementioned values for how 'light' and 'dark' regions expand and contract as the culture grows thicker are very likely to change depending on the different geographic locations considered other than Qingdao, China, where this experiment was conducted and the particular algal strain considered or the reactor design adopted. Identifying the upper limit of biomass density and the maximum depth at which algae cells are still effectively illuminated depending on the specific design of the reactor (different thicknesses of the water layer) has been an interesting task for our research team. Obtaining this information will definitely help to better understand how light travels through the water layer of an open pond, and it will also be useful knowledge for enhancing biomass productivity.

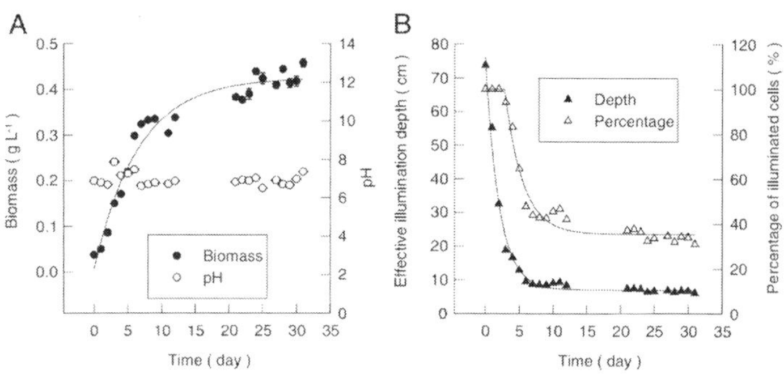

Figure 3: The biomass increase, pH, and effective illumination depth of conventional aqueous suspended open ponds outdoors. (A) The biomass density (black circle) and pH (white circle) changes of the open pond outdoors. Data for biomass were mean ± standard deviation of six measurements for two independent open ponds (three measurements for each). Data for pH were mean ± standard deviation of two measurements for two independent open ponds (one measurement for each). (B) The effective illumination depth (black triangle) and

percentage of effective illuminated algal cells (white triangle) at different days of cultivation in outdoor open ponds.

The light distribution inside the biofilm of the attached cultivation system

One of the methods to determine the d_E of the attached microalgal biofilm is to directly measure the light intensity from underneath the biofilm. If the light intensity was equal to or higher than the LCP, then the corresponding depth could be effectively illuminated. However, we must also consider that the photosynthetically active wavelengths (red light) are more prone to be absorbed by chlorophyll than other less desired wavelengths (green light). The method we just described to measure light intensity might overestimate the d_E if only the quantum flux intensity is measured because the undesired wavelengths travel farther away through layers of cells than the desired ones. To avoid the aforementioned problem in this research, we used a different method to estimate the d_E by monitoring the specific growth rate of a thin layer of microalgal cells, which was called 'marker' layer, inserted just beneath the re-constructed biomass layer. An optimum 'marker' layer should have two main characteristics, *viz.* 1) thin enough to avoid the formation of light gradient inside the layer and 2) sensitive enough to the changes in light intensity. Algal cells from this thin layer should experience the same conditions as the upper layer. However, in pilot experiments, we found that such thin marker layers failed to meet requirement #2. In other words, the biomass changes for this kind of thin layers were too slight to be detected by regular gravimetric method, especially under lower light conditions. As an alternative, the algal cells used to inoculate the attached cultivation were chosen as the 'marker' layer because they were photosynthetically active even at low biomass densities. However, the risk of over- or underestimating the d_E with this method was still present if the LCP of the inoculum algal cells was lower or higher than the LCPs of daughter cells that only experienced attached cultivation conditions. The LCP for the newly formed microalgal cells was firstly measured at different days. Typical light curves for oxygen evolution rate were obtained for all of these detached and re-suspended samples (Figure 4A). The LCPs of the daughter cells from day 0 to day 10 were in the range of $12.6 \pm 2.2 \sim 19.8 \pm 3.2$ μmol m^{-2} s^{-1}, without significant differences ($P > 0.05$) (Figure 4B). These results indicated that it is safe to determine the d_E of the attached algal biofilm with this double-layer method under nutrient-replete condition without worrying about over- or underestimation. We also tested the LCP changes for the attached algal biofilm under nitrogen-depleted conditions and found that it increased dramatically to *ca.* 60 μmol m^{-2} s^{-1} at day 2. Because of this, a 'marker' layer consisting of the cells used to inoculate the attached cultivation surface cannot be used to estimate the d_E of the nitrogen-starved attached cultivation.

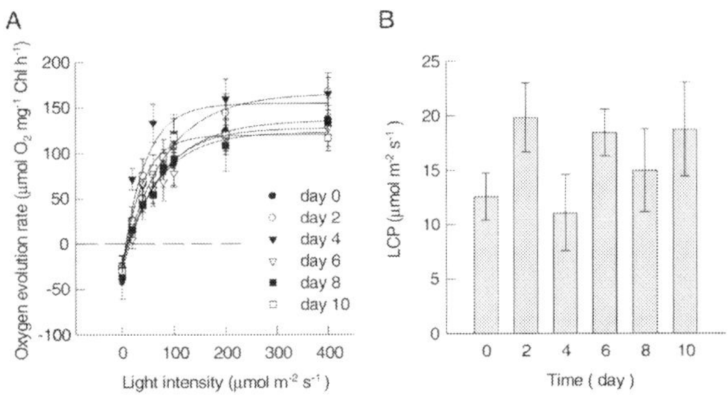

Figure 4: The oxygen evolution characters of *Scenedesmus dimorphus*under attached cultivation for different days of cultivation. (A) The oxygen evolution rate versus light intensity (light curve) of *S. dimorphus* under attached cultivation for 0 (black circle), 2 (white circle), 4 (black down-pointing triangle), 6 (white down-pointing triangle), 8 (black square), and 10 (white square) days of cultivation. Data were mean ± standard deviation of six measurements for three independent experiments (two measurements for each). **(B)** The light compensation point of *S. dimorphus* under attached cultivation for different days of cultivation. The ANOVA (SPSS, Chicago, IL, USA) result indicated that there were no differences for different days.

The results of the double-layer experiments for attached cultivation of *S. dimorphus* under nitrogen-replete condition are shown in Figure 5. For the biofilm layer that has been attached cultivated for several days, the specific growth rate of the 'marker' layer (μ) decreased as the biomass of the upper layer increased, mainly because the light traveling through the upper layer became weaker and weaker (Figure 5A). Meanwhile, the rate of this decrease for μgradually slowed down as the days went by, resulting in the increase of the thicknesses for the effectively illuminated upper layers (Figure 5A). According to the results in Figure 5A, the d_E for attached cultivation under nitrogen-replete condition increased from 33.9±6.5 μm for day 0 to 237.3±11.8 μm for day 10, corresponding respectively to biomass densities of 14.6±1.6 g $m^{-2}$and 111.5±4.1 g m^{-2}, whereas the actual biomass density increased from 8.8±1.3 g m^{-2} to 107.6±2.6 g m^{-2} in 10 days, corresponding to thicknesses of 21.7±5.9 μm and 229.1±8.6 μm (Figure 5B), respectively. The ratios of d_E *vs.* actual biomass were almost equal to 1 throughout the experiments except for day 0 and day 2 which were 1.66 and 0.68, respectively. The fact that d_E was similar to the actual biomass density indicated that nearly 100% of the algal cells inside the immobilized biofilm were effectively illuminated under attached cultivation.

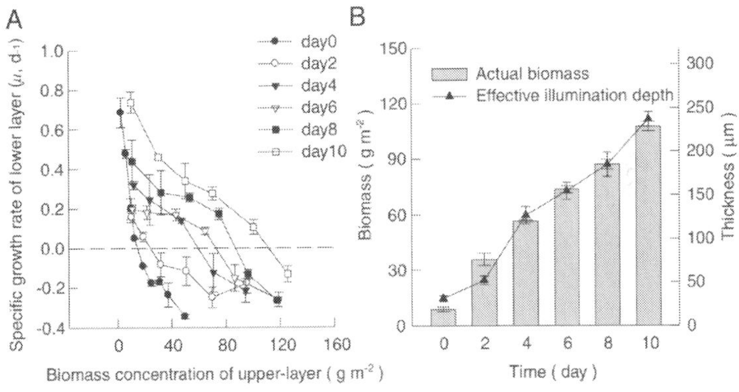

Figure 5: The estimation of effective illumination depth for the attached cultivation of*S.dimorphus*with twin-layer method. (A) The specific growth rate of lower layer *versus* the upper layer biomass density. Data were mean±standard deviation of nine measurements for three independent experiments (three measurements for each). The upper layers were consisted with biomass grown with attached cultivated technique for 0 (black circle), 2 (white circle), 4 (black down-pointing triangle), 6 (white down-pointing triangle), 8 (black square), and 10 (white square) days. **(B)** The actual biomass density for the attached cultivation from day 0 to day 10 (closed bar), and the effective illumination depth for the attached cultivation from day 0 to day 10 (black triangle).

Chlorophyll is the main light-harvesting molecule for photosynthetic organism, but some of the carotenoid molecules can also capture light and pass the excited energy to chlorophyll [27]. In this research, the content of chlorophyll as well as carotenoid was measured and values have been expressed as both a function of the dry biomass and the cultivated surface area. From day 0 to day 10, the dry mass-based chlorophyll content decreased from 3.3% to 0.4% and the carotenoid content decreased from 0.5% to 0.2%. The areal chlorophyll content increased from 0.29 to 0.70 g m^{-2} during the first 8 days of attached cultivation and then decreased to 0.45 g m^{-2} at day 10. The areal carotenoid content increased from 0.05 to 0.26 g m^{-2} from day 0 to day 8 and then remained constant. In general, during the attached cultivation, the increase of the d_E was accompanied by the decrease in biomass-based chlorophyll and carotenoid contents as well as the increase in areal chlorophyll and carotenoid contents (Figure 6). Figure 7suggests how the light transfer and distribution inside the immobilized microalgal biofilm are thought to be happening. In the earlier phase of attached cultivation, the pigment content in every single cell is relatively high, so that the light intensity decreased sharply when passing through each algal layer and can penetrate only a small depth of the algal biofilm. With the increase of the cultivation days, the algal layers as well as the pigment content along the light path increased but the pigment content in each single cell decreased, and only a moderate decrease in light intensity happens when passing through each algal layer. In the end, light can penetrate a bigger depth of the algal biofilm (Figure 7).

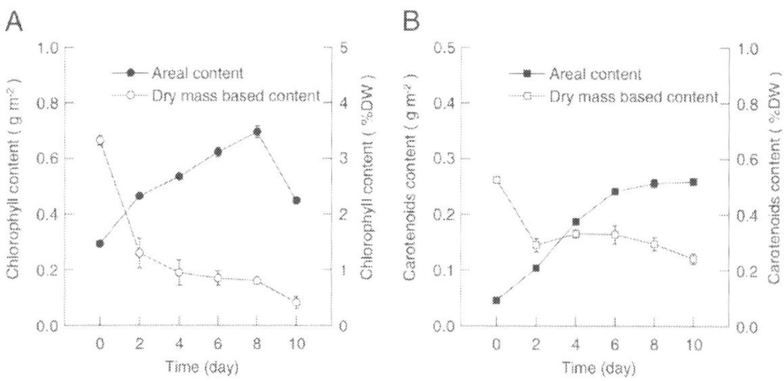

Figure 6: The chlorophyll and carotenoid contents for the attached cultivation of*S.dimorphus*. (A) The areal (black circle) and dry mass-based (white circle) chlorophyll contents for attached cultivation of *S. dimorphus* for different days of cultivation. Data were mean ± standard deviation of six measurements for three independent experiments (two measurements for each). (B) The areal (black square) and dry mass-based (white square) carotenoid contents for attached cultivation of *S. dimorphus* for different days. Data were mean ± stan-

dard deviation of six measurements for three independent experiments (two measurements for each).

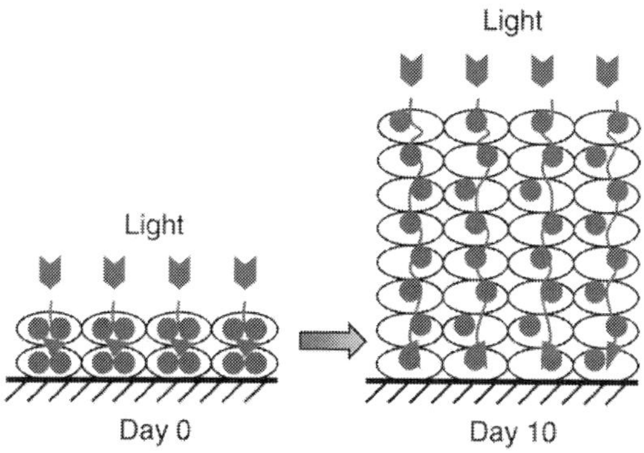

Figure 7: The schematic model for the chlorophyll and light distribution inside the attached cultivation biofilm of*S.dimorphus*. The small oval represents the algal cells inside the biofilm. The green dots indicate the chlorophyll content of algal cells. The suggested light penetration path inside the biofilm is indicated by red arrows.

Comparison of the illumination aspects of the suspended and attached cultivation

The obvious difference between the two cultivation systems was that full illumination only happened for a short period of time (day 0 to day 3) in the open-pond system, when biomass concentration was very dilute (<20 g m^{-2}), whereas in the attached cultivation system, the full illumination happened during a prolonged interval of 10 days as well as a wider range of biomass density (up to 100 g m^{-2} for a single layer). This high percentage of photosynthetic active biomass might explain the high biomass productivity and photosynthetic efficiency of attached cultivation under high light conditions. For example, in our previous research with a multiple-plate attached cultivation bioreactor with a light dilution rate of 10 (refer to Figure 1A), the land areal biomass density increased from 163.7 to 484.5 g m^{-2} in 5 days [22] under outdoor high light conditions (1,500~2,000 μmol m^{-2} s^{-1}). Since the averaged light intensity impinging the cultivated surface was only about 100~150 μmol m^{-2} s^{-1}, we can assume that nearly 100% of the biomass in this case was effectively illuminated. Similar results were also achieved under indoor conditions. The high photosynthetic efficiency of *ca.* 18% (based on visible light) was reached for *Scenedesmus* indoors with a light intensity of 700 μmol m^{-2} s^{-1} in the land areal biomass density range of 192.5 to 828.1 g m^{-2} [22]. Photosynthetic efficiency of *ca.* 15% (based on visible light) was also reached for *Botryococcus braunii* indoors with a light intensity of 500 μmol m^{-2} s^{-1} in the standing areal biomass density range of *ca.* 70 to *ca.* 580 g m^{-2} [23]. It is hard to imagine an open pond working with such high percentage of photosynthetic active cells and consequently such a condensed biomass concentration range. In general, for the attached cultivation, the high light intensity is diluted and absorbed by a cultivated area hosting an extremely dense biomass, where there is virtually no space left between adjacent cells, which is totally photosynthetically active, whereas in conventional open ponds, the high light intensity is only absorbed by a much smaller amount of algal cells so that the average light energy received by each photosynthetically active cell is higher in the open pond and a large portion of light is dissipated in a non-photosynthetic pathway before reaching other cells.

CONCLUSIONS

In a conventional 20-cm deep open pond, all of the algal cells were fully illuminated for the first 3 days of outdoor cultivation. With the increase of the biomass concentration, the effective illuminated portion of the water layer decreased steeply to 45.5% from day 4 to day 10 and then gradually decreased to 31.1% at day 31, when the biomass density reached 0.45 g L^{-1} or 90 g m^{-2}. In the attached cultivation system under nitrogen-replete condition, almost 100% of the immobilized algal cells inside the biofilm were effectively illuminated from day 0 through day 10, with biomass density increasing from 8.8 to 107.6 g m^{-2}. Attached cultivation facilitates super high biomass density (up to *ca.* 800 g m^{-2}) and super high efficient illumination (100%) when the reactor consists of multiple panels hosting the biofilm surface which might be the key to the high biomass productivity as well as high photosynthetic efficiency.

METHODS

Microalgae strain and medium

The freshwater microalgal species *S. dimorphus* was locally screened in

Qingdao, China. The algal cells were maintained in BG11 medium [28], each liter of which contains 1.5 g $NaNO_3$, 0.075 g $MgSO_4 \cdot 7H_2O$, 0.036 g $CaCl_2 \cdot 2H_2O$, 0.04 g $KH_2PO_4 \cdot H_2O$, 0.02 g Na_2CO_3, 6.0 × 10^{-3} g citric acid, 1.0 × 10^{-3} g Na_2EDTA, 6.0 × 10^{-3} g ferric ammonium citrate, 2.22 × 10^{-4} g $ZnSO_4 \cdot 7H_2O$, 6.9 × 10^{-5} g $CuSO_4 \cdot 5H_2O$, 1.81 × 10^{-3} g $MnCl_2 \cdot 4H_2O$, 3.9 × 10^{-4} g $Na_2MoO_4 \cdot 2H_2O$, 4.94 × 10^{-5} g $Co(NO_3)_2 \cdot 6H_2O$ and 2.86 × 10^{-3} g H_3BO_3.

Inoculum preparation

The algal inoculum was prepared in glass columns (diameter = 0.05 m, working volume = 0.7 L) under indoor conditions of continuous illumination (100 μmol m^{-2} s^{-1}; cold fluorescent lamp, white light; FSL T8 36W, Foshan Electrical Lighting Co., Ltd., Foshan, China) and continuous aeration (1.0 vvm; 98% air + 2% CO_2, *v*/*v*). The temperature of the culture broth was 25°C ± 1°C. The algal biomass at late exponential phase (5~7 days after inoculation, 2~3 g L^{-1}) was used to inoculate the experimental systems.

Experimental design

The d_E for the algal biofilm of the attached cultivation and conventional aqueous suspended open pond were determined under indoor and outdoor conditions, respectively. For the attached cultivation, the d_E was measured with a single-layer attached cultivation system (Figure 1B) following the flow chart of Figure 8. Firstly, the algal cells were attached cultivated for 10 days with BG-11 (step 1; refer to the 'Cultivation systems' section for details). Every other day, some of the algal disks were sampled and the attached cells on it were detached and washed three times with de-ionized water (step 2). A 'marker' layer was built by filtering some fresh inoculum onto new cellulose acetate/nitrate membranes at the biomass density of s(g m^{-2}). Then, some aliquots of the re-suspended algal culture from step 2 were gently filtered onto the 'marker' layers to form 'double-layer' attached algal disks. The corresponding 'single-layer' attached algal 'disks' were also prepared at the same time as a control using identical aliquots of re-suspended algal culture but without the 'marker' layer (step 3). Both of the newly prepared 'double-layer' and 'single-layer' algal disks were cultivated for 24 h under the same environmental conditions using the PBR depicted in Figure 1B. The biomass increase for the 'double-layer' algal disk was denoted by a (g m^{-2}) and the biomass increase for the corresponding 'single-layer' algal disk was a' (g m^{-2}), so that the net biomass gain for the 'marker' layer was calculated as $\delta = a - a'$ (g m^{-2}), and μ was calculated as:

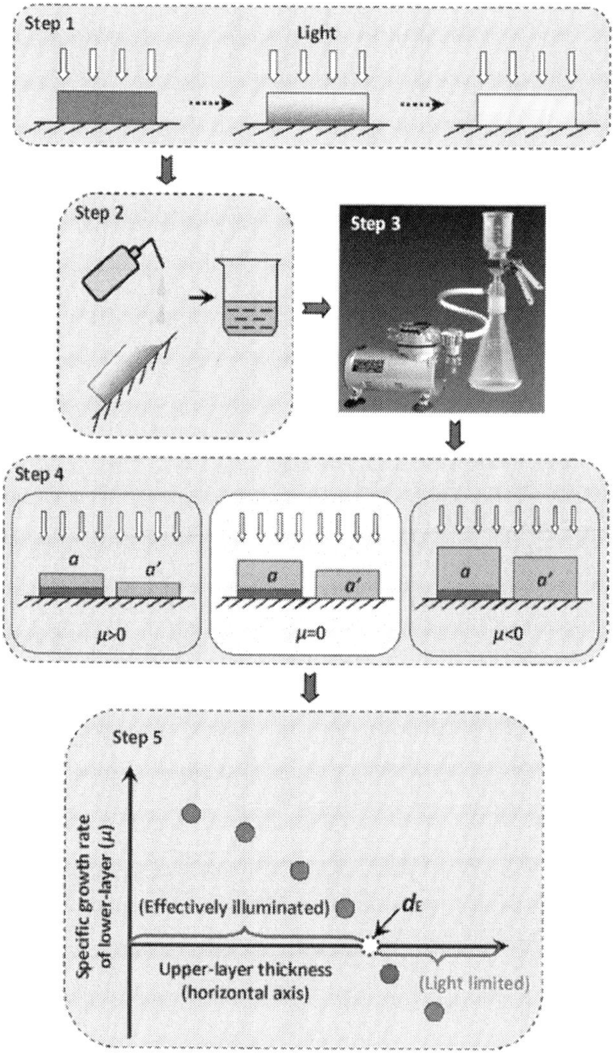

Figure 8: The schematic flow chart of the experiment design to determine the effective illumination depth for the attached cultivation.

$$\mu = \ln(s + \delta) - \ln(s) \quad (1)$$

If $\mu > 0$, it means that the 'marker' layer is receiving a light intensity higher than the LCP. In other words, the upper layer is not thick enough to fully attenuate the impinging light and the depth of the upper layer (d_A) was only a fraction of the d_E, $d_A < d_E$.

If $\mu < 0$, it means that the light intensity reaching the 'marker' layer is lower than the LCP and the upper layer is too thick to allow the 'marker' layer to accumulate biomass, $d_A > d_E$.

If $\mu = 0$, the light intensity at the 'marker' layer is equal to the LCP and the biomass increase of the biofilm due to photosynthesis is equal to the biomass loss due to cellular respiration. The thickness of the upper layer could be considered as the maximum distance the light can effectively penetrate, $d_A = d_E$ (Figure 8, step 4). In actual experiments, however, considering the difficulties in precisely manipulating the amount of the upper-layer biomass to make $d_A = d_E$, the d_E was generally approximately estimated as the intercept of the *x*-axis (Figure 8, step 5).

The percentage of the effectively illuminated cells (P_{light}, %) for attached cultivation was calculated as:

$$P_{light} = d_E \times 100/d_G \qquad (2)$$

where d_G represents the actual thickness of the attached algal 'disk' at the considered sampling day.

The d_E in the suspended cultivation system was determined with a conventional open-pond system and a set of homemade equipment with different light paths (Figure 2A; refer to the sections 'Equipments for determining the effective illumination depth for suspended cultures' and 'Cultivation systems' for detailed specifications and operations). Firstly, the relationships of biomass concentrations *vs.* penetrated light intensities were measured (Figure 2B). Secondly, according to the LCP determined by oxygen evolution rates at different light intensities (Figure 2C; refer to the 'Determination of LCP' section for details), the d_E were calculated for each tested biomass densities. An equation was proposed to calculate the d_E at any given biomass density in the range of 0.07~6.4 g L^{-1} (Figure 2D). Finally, the changes in biomass density of *S. dimorphus* in open ponds were studied outdoors (Figure 3A), and the d_E was estimated according to the proposed equation. The percentage of the effectively illuminated cells (P_{light}, %; Figure 3B) was calculated as:

$$P_{light} = d_E \times 100/d_P \qquad (3)$$

where the d_p represents the depth of the open pond, 0.2 m in this research.

Equipments for determining the effective illumination depth for suspended cultures

A series of white polyvinyl chloride (PVC) tubes (inner diameter = 0.03 m) with different lengths (2, 3, 5, 10, 20, 30, 40, and 50 cm; three tubes for each length) were prepared, and one of the open ends of these PVC tubes was covered with a piece of highly transparent polystyrene plate (thickness = 0.8 mm; >98% light transmittance). The outer wall of the PVC tubes and the outer circles of the polystyrene sheets were painted black to prevent the unwanted penetration of outside irradiance, whereas the inner wall and the inner circles (diameter = 1 cm) were not painted but thoroughly cleaned with water (Figure 2A). Algal suspensions with different biomass densities, that is, 0.07, 0.18, 0.39, 0.81, 1.59, 3.20 and 6.4 g L^{-1}, were prepared in beakers and then used to fill the PVC tubes, so that the heights of the algal broth, *viz.* light paths, were equal to the lengths of the PVC tubes. The PVC pipes were placed vertically in open air, under direct sunlight with an intensity of approximately 1,500 ~ 1,600 μmol m^{-2} s^{-1}. A photosynthetic quantum sensor (Li-250A, Li-Cor, Lincoln, NE, USA) was placed underneath each tube and tightly attached to the polystyrene sheets to measure the penetrated photon flux density.

When low water layer thicknesses were to be tested, that is, 0.084 ~ 1.05 cm, the algal suspension was poured into a Petri dish made of highly transparent polystyrene (thickness = 0.8 mm; >98% light transmittance) and then the penetrated light intensity was measured. The thickness of the water layer in the Petri dish was calculated according to the volume of algal suspension poured into and the diameter of the Petri dish used. The external surface of the Petri dish was also painted with black paint with the exception of a small circular window (diameter = 1 cm) where the quantum sensor was to be attached.

Cultivation systems

Two different cultivation systems were adopted in this research to evaluate the active proportions for photosynthesis. The first one was a mini open-pond system which measured 2 m in length, 1 m in width, and 0.2 m in culture depth. The culture broth was propelled by a paddle wheel at a flow speed of 0.3 m s^{-1}. A gas stream of pure CO_2 (>99%, *v*/*v*) was injected into the bottom of the pond through a ventilation stone at an aeration speed of 4 L min^{-1} (0.01 vvm) from 8 am to 4 pm to supply the carbon source for the algal cells and maintain a pH of 7.5 ± 0.5. The measured maximum light intensity at the water surface was *ca.* 2,000 μmol m^{-2} s^{-1}during the daytime. The temperature of the culture broth was 30°C~35°C from 12 pm to 2 pm in the afternoon on a sunny day and 10°C~25°C for the rest of the time. Two identical ponds were operated synchronously in outdoor conditions at Qingdao, China (35°35′ N, 119°30′ E) from 17 September to 17 October 2012, with the same inoculated biomass density of 0.04 g L^{-1}(8 g m^{-2}).

The second adopted cultivation system was the single-layer attached biofilm cultivation system (Figure 1B) which has been introduced in detail in our previous research as a 'type 1' photobioreactor [22]. In brief, algal cells were evenly filtered onto a cellulose acetate/nitrate membrane (pore size 0.45 μm) to form algal disks with 10 ± 0.5 cm^2 of footprint at an inoculated areal biomass density of 8 g m^{-2}. These algal disks were then put onto a layer of regular filter paper and finally onto a 0.2 m × 0.4 m glass plate (3 mm thick). The glass plate as well as the filter paper and algal disks on it were sealed in a 0.5 m × 0.3 m × 0.05 m chamber made out of 5-mm-thick transparent glass plates. During the cultivation, the algal disks were continuously illuminated with 100 ± 10 μmol m^{-2} s^{-1} of white light from cold fluorescent lamps. A continuous air stream that contained 1.5% (*v*/*v*) of CO_2 was injected into the chamber with a flow rate of 0.75 L min^{-1} (0.1 vvm). The BG11 medium was dripped onto the filter paper with a speed of 0.5 mL min^{-1} to keep the algal disks attached to the filter paper and provide nutrients for the growth of the algal cells. Only fresh medium was used, and any excess was immediately discarded. The temperature inside the glass chamber was kept at 25°C ± 2°C during the experiment.

Analysis

Growth analysis

In this research, the growth of the algal culture was investigated by measuring the changes in dry biomass concentration (DW). Some aliquots of the culture broth (open pond) or detached biomass from the 'algal disk' (attached cultivation) were first washed with 50 mL of de-ionized water and then filtered onto a pre-weighted 0.45-μm GF/C filter membrane (Whatman, Little Chalfont, England; DW_0). The membrane was oven dried at 105°C for 24 h and then weighted (DW_1).

The DW for open pond (g L^{-1}) was calculated as:

$$DW = (DW_1 - DW_0) \times 1,000/v \quad (4)$$

where v represents the sample volume expressed in milliliters.

The DW for attached cultivation was calculated as:

$$DW = (DW_1 - DW_0)/0.001 \quad (5)$$

where 0.001 represents the footprint area (m^2) of attached algal cells.

Determination of LCP

The LCP was determined according to the photosynthetic oxygen evolution rate *vs*. light intensity curve (PI curve); this was measured with a Chlorolab-2 liquid-phase Clark-type oxygen electrode (Hansatech, Norfolk, England). The newly prepared inoculum for the algal disks (refer to the 'Inoculum preparation' section) was collected by centrifugation at 5,000 × gfor 30 s (Allegra X-22R, Beckman Coulter, Brea, CA, USA). The obtained pellet was then re-suspended with BG-11 medium containing 100 mM $NaHCO_3$ and adjusted to a proper biomass concentration so that the chlorophyll content was as close as possible to 10 μg per milliliter. One milliliter of the sample was added to the reaction cuvette and then bubbled with nitrogen gas for 30 s to expel the dissolved oxygen gas. Light intensities of 400, 200, 100, 80, 60, 40, 20 and 0 μmol m^{-2} s^{-1} were applied to drive the photosynthetic oxygen evolution. The measurement was repeated three times for each light intensity. The LCP was obtained as the intercept on the horizontal axis of the PI curve [29].

Determination of the thickness of the attached algal disk

Prior to the experiments, the thickness of the algal disks at different cultivation days was measured by means of a surface profiler (Veeco Dektak150, Veeco Instruments Inc., Plainview, NY, USA) and the corresponding areal biomass density was also measured (refer to the 'Growth analysis' section). Accordingly, the correlation of the thickness (d, expressed in μm) of the algal film with the areal biomass density (DW, in g m^{-2}) was obtained as follows:

$$d = 2.1 \times DW + 3.2 \tag{6}$$

For later experiments, the thickness (μm) of the attached algal disks during cultivation was estimated by this equation. According to the pilot experiments, this equation could be applied in the biomass range of 0~120 g m^{-2} for *S. dimorphus* grown with the attached cultivation technique for up to 10 days under both nitrogen-replete or nitrogen-depleted conditions ($R^2 = 0.99$).

Determination of the chlorophyll and carotenoid contents

The chlorophyll and carotenoid contents were determined according to the methods described by Wellburn [30]. The algal biomass was harvested and washed with plenty of de-ionized water and then mixed with methanol under 60°C for 12 h in darkness until the algal cells whitened completely. The optical densities of the extraction were measured with a spectrophotometer at 666, 653, and 470 nm (Cary 50, Varian Inc., Palo Alto, CA, USA). The chlorophyll a (Chl_a), chlorophyll b (Chl_b), and carotenoid (Car) concentrations in the extraction were calculated as follows (mg L^{-1}):

$$Chl_a = 15.65 \times OD_{666} - 7.34 \times OD_{653} \tag{7}$$

$$Chl_b = 27.05 \times OD_{653} - 11.21 \times OD_{666} \tag{8}$$

$$Car = (1,000 \times OD_{470} - 2.86 \times Chl_a - 129.2 \times Chl_b)/221 \tag{9}$$

NOTES

Acknowledgements

This work was supported by the National Natural Science Foundation of China (41276144), the Director Innovation Foundation of Qingdao Institute of Bioenergy and Bioprocess Technology, CAS, and the Science and Technology Development Planning of Shandong Province (2013GHY11520).

Competing interests

The authors declare that they have no competing interests.

Authors' contributions

JW and TL proposed the idea and hypothesis. JW carried out the experiment design and outdoor open-pond experiment and drafted the manuscript. JL carried out the biofilm experiment. TL performed the statistical analysis and helped to draft and revise the manuscript. All authors read and approved the final manuscript for publication.

REFERENCES

1. Hu Q, Sommerfeld M, Jarvis E, Ghirardi M, Posewitz M, Seibert M, et al. Microalgal triacylglycerols as feedstocks for biofuel production: perspectives and advances. Plant J. 2008;54:621–39.
2. Chisti Y. Biodiesel from microalgae. Biotechnol Adv. 2007;25:294–306.
3. Mata TM, Martins AA, Caetano NS. Microalgae for biodiesel production and other applications: a review. Renew Sustain Energ Rev. 2010;14:217–32.
4. Sostaric M, Klinar D, Bricelj M, Golob J, Berovic M, Likozar B. Growth, lipid extraction and thermal degradation of the microalga Chlorella vulgaris. New Biotechnol. 2012;29:325–31.
5. Lam MK, Lee KT. Microalgae biofuels: a critical review of issues, problems and the way forward. Biotechnol Adv. 2012;30:673–90.
6. Larkum AWD, Ross IL, Kruse O, Hankamer B. Selection, breeding and engineering of microalgae for bioenergy and biofuel production.

Trends Biotechnol. 2012;30:198–205.

7. Weisz PB. Basic choices and constraints on long-term energy supplies. Phys Today. 2004;57:47–52.
8. Stephens E, Ross IL, Mussgnug JH, Wagner LD, Borowitzka MA, Posten C, et al. Future prospects of microalgal biofuel production systems. Trends Plant Sci. 2010;15:554–64.
9. Wijffels R, Barbosa M. An outlook on microalgal biofuels. Science. 2010;329:796–9.
10. Brennan L, Owende P. Biofuels from microalgae - a review of technologies for production, processing, and extractions of biofuels and co-products. Renew Sustain Energ Rev. 2010;14:557–77.
11. Chisti Y. Raceways-based production of algal crude oil. In: Posten C, Walter C, editors. Microalgal biotechnology: potential and production. Berlin: de Gruyter; 2012. p. 113–46.
12. Tredici MR. Photobiology of microalgae mass cultures: understanding the tools for the next green revolution. Biofuels. 2010;1:143–62.
13. Zhu XG, Long SP, Ort DR. What is the maximum efficiency with which photosynthesis can convert solar energy into biomass? Curr Opin Biotechnol. 2008;19:153–9.
14. Shi J, Podola B, Melkonian M. Removal of nitrogen and phosphorus from wastewater using microalgae immobilized on twin layers: an experimental study. J Appl Phycol. 2007;19:417–23.
15. Mulbry W, Kondrad S, Pizarro C, Kebede-Westhead E. Treatment of dairy manure effluent using freshwater algae: algal productivity and recovery of manure nutrients using pilot-scale algal turf scrubbers. Bioresour Technol. 2008;99:8137–42.
16. Boelee NC, Temmink H, Janssen M, Buisman CJN, Wijffels RH. Nitrogen and phosphorus removal from municipal wastewater effluent using microalgal biofilms. Water Res. 2011;45:5925–33.
17. Naumann T, Çebi Z, Podola B, Melkonian M. Growing microalgae as aquaculture feeds on twin-layers: a novel solid-state photobioreactor. J Appl Phycol. 2013;25:1413–20.
18. Zamalloa C, Boon N, Verstraete W. Decentralized two-stage sewage treatment by chemical–biological flocculation combined with

microalgae biofilm for nutrient immobilization in a roof installed parallel plate reactor. Bioresour Technol. 2013;130:152–60.

19. Bayless DJ, Kremer G, Vis M, Stuart B, Shi L. Photosynthetic CO2 mitigation using a novel membrane-based photobioreactor. J Environ Eng Manag. 2006;16:209–15.
20. Johnson MB, Wen Z. Development of an attached microalgal growth system for biofuel production. Biotechnol Prod Process Eng. 2010;85:525–34.
21. Ozkan A, Kinney K, Katz L, Berberoglu H. Reduction of water and energy requirement of algae cultivation using an algae biofilm photobioreactor. Bioresource Technol. 2012;114:542–8.
22. Liu T, Wang J, Hu Q, Cheng P, Ji B, Liu J, et al. Attached cultivation technology of microalgae for efficient biomass feedstock production. Bioresour Technol. 2013;127:216–22.
23. Cheng P, Ji B, Gao L, Zhang W, Wang J, Liu T. The growth, lipid and hydrocarbon production of Botryococcus braunii with attached cultivation. Bioresour Technol. 2013;138:95–100.
24. Ji C, Wang J, Zhang W, Liu J, Wang H, Gao L, et al. An applicable nitrogen supply strategy for attached cultivation of Aucutodesmus obliquus. J Appl Phycol. 2014;26:173–80.
25. Chisti Y. Constraints to commercialization of algal fuels. J Biotechnol. 2013;167:201–14.
26. Wang J, Han D, Sommerfeld M, Lu C, Hu Q. Effect of initial biomass density on growth and astaxanthin production of Haematococcus pluvialis in an outdoor photobioreactor. J Appl Phycol. 2013;25:253–60.
27. Green BR, Durnford DG. The chlorophyll-carotenoid proteins of oxygenic photosynthesis. Annu Rev Plant Physiol Plant Mol Biol. 1996;47:685–714.
28. Stanier RY, Kunisawa R, Mandel M, Cohenbaz G. Purification and properties of unicellular blue-green algae (order Chroococcales). Bacteriol Rev. 1971; 35:171–205.
29. Ye Z. A new model for relationship between light intensity and the rate of photosynthesis in Oryza sativa. Photosynthetica. 2007;45:637–40.

30. Wellburn AR. The spectral determination of chlorophylls a and b, as well as total carotenoids, using various solvents with spectrophotometers of different resolution. J Plant Physiol. 1994;144:307–13.

CHAPTER

4

GROWTH PERFORMANCE AND PROXIMATE COMPOSITION OF MIXED CULTURES OF MARINE MICROALGAE (NANNOCHLOROPSIS SP. & TETRASELMIS SP.) WITH MONOCULTURES

Wasana Arkronrat[1], Prapapron Deemark[2], and Vutthichai Oniam[1]

[1] Klongwan Fisheries Research Station, Faculty of Fisheries, Kasetsart University, Mueang, Prachuap Khiri Khan, 77000 Thailand.

[2] Aquaculture Development and Certification Centre, Department of Fisheries, Chatuchak, Bangkok, 10900 Thailand.

ABSTRACT

The growth performance and proximate composition of two marine microalgae, Nannochloropsis sp. and Tetraselmis sp., in monoculture and mixed culture of both species under laboratory conditions were compared. Microalgae were cultured in enriched seawater (28 ppt) in 8 L plastic bottles using Conway medium. Microalgae were grown in a temperature-controlled room (25°C) under continuous illumination at a light intensity of about 3,000 Lux for 240 hours (10 days). Results showed that maximum cell density and growth rate of both species ($45.06\pm2.67\times10^5$ cells ml^{-1} and 0.43 ± 0.02 day-1, reps.) in mixed culture had a significantly higher than Tetraselmis sp. ($5.09\pm0.99\times10^5$ cells ml^{-1} and 0.16 ± 0.03 day^{-1}, reps.), but not significantly different with Nannochloropsis sp. ($29.25\pm22.70\times10^5$ cells ml-1 and 0.37 ± 0.14 day^{-1}, reps.) in monoculture. For proximate composition, protein content of both species in mixed culture and Nannochloropsis sp. in monoculture was significantly higher than for Tetraselmis sp. in monoculture after 120 and 240 hours of cultivation. Carbohydrate and lipid contents of each treatment were found to be significantly different after 120 hours of cultivation but not significantly different after 240 hours of cultivation. Therefore, this study indicates that an alternative option for commercial microalgal production is to use a mixed culture to obtain higher production.

Keywords: microalgae, mixed culture, growth performance, proximate composition

INTRODUCTION

Commercial culture of microalgae has been done for more than 30 years with the main microalgal species cultured such as Chlorella sp. and Spirulina sp. for healthy food, Dunaliella salina for b-carotene and several species for aquaculture (Borowitzka, 1999; Araújo and Garcia, 2005). Marine microalgae such as the green algae Chlorella sp., Tetraselmis sp. and Nannochloropsis sp., the diatom Chaetoceros sp. and Thalassiosira sp., and the flagellate Isochrysis sp., were commonly used as important live food for crustacean, fish and bivalve larvae in Thailand, and therefore they have been produced commercially (Wongrat, 2000; Arkronrat and Oniam, 2012a; 2012b; 2012c). The success of commercial laboratory-scale production of microalgae depends on many factors, mainly those which control microalgae growth, such as temperature, nutrients, light, salinity and pH (Tzovenis et al., 1997; Zhu et al., 1997; Araujo and Garcia, 2005). Currently, mixed cultures of microalgae are another important factor for the increase of the production of microalgae (Huang et al., 2011; Ahmad et al., 2012; Brito et al., 2013; Arkronrat and Oniam,

2014; Sureshkumar et al., 2014). Nannochloropsis sp. and Tetraselmis sp. are widely used in aquaculture industries as they are comprised of nutritional values suitable for rearing of marine animal larvae. However, the growth characteristics and biochemical composition of microalgal species could also be altered when two or more microalgal species were cultured together in the same culture medium (Cai and Duan, 2008; Huang et al., 2011). Therefore, the aims of this study were to compare the cell density, growth rate and proximate composition of Nannochloropsis sp. and Tetraselmis sp. in monoculture and mixed culture of both species under laboratory conditions as an alternative for commercial microalgal production.

MATERIALS AND METHODS

Experimental design and set-up

The experiment on growth performance of microalgae was conducted at the Phytoplankton Laboratory of Klongwan Fisheries Research Station (KFRS), Prachuap Khiri Khan Province, Thailand. Stock cultures of the two species of commercial microalgae, the green algae Nannochloropsis sp. and Tetraselmis sp., were obtained from the Prachuap Khiri Khan Coastal Fisheries Research and Development Center, Department of Fisheries.

These microalgae were cultured under laboratory conditions, 1) monoculture of Nannochloropsis sp., 2) monoculture of Tetraselmis sp., and 3) mixed culture of both species. Microalgae were cultured in enriched seawater (28 ppt) in 8 L plastic bottles using Conway medium without silicate (AQUACOP, 1984), grown in a temperature-controlled room (25°C) under continuous illumination with cool white fluorescent lamps at a light intensity of about 3,000 Lux, for 240 hours (10 days). All cultures were started with equal inoculums (about 10^5 cell ml^{-1}) of respective species while the inoculum of mixed culture consisted of half the number of each species. The experiment was performed with four replicates and followed a completely randomized design

During the growth experiment, algal cell samples were collected every 12 hours for estimation of cell density and growth pattern. Cells were fixed with 5% formalin then counted using a haemacytometer under a compound microscope at 40× magnification. Growth rate (K) of the culture was calculated by the following equation (Phatarpekar et al., 2000):

$$K = \frac{InN_t - InN_o}{T}$$

where Nt is the cell count at time "t", No is the initial cell count at time "o" and T is the time (day).

For proximate composition analysis, the dry biomass was evaluated on the same dates, concentrating known volumes of each culture in precalibrated Whatman GF-C glass fiber filters. These were washed with 5-6 ml of a 3% ammonium formate solution to eliminate salt and dried to constant weight at 60°C (Sánchez-Saavedra and Voltolina, 2006). Protein, total carbohydrate and total lipid were determined according to the methods of Lowry et al. (1951), Dubois et al. (1956) and Bligh and Dyer (1959), respectively.

Statistical analysis

At the end of the experiments, data on the growth performance (cell density and growth rate) and proximate composition (protein, carbohydrate and lipid) were analyzed using analysis of variance and the difference between means was tested using Duncan's multiple range test at the 95% level of confidence using the SPSS program.

RESULTS

Cell density and growth rate

Initial cell densities of Nannochloropsis sp., Tetraselmis sp. in monoculture and both species in mixed culture were 1.12±0.09, 1.18±0.14 and 1.13±0.13×10^5 cells ml^{-1} , respectively. The cell density of both species in mixed culture increased rapidly to 15.87±1.76×10^5 cells ml^{-1}, at hour 60 of cultivation without any apparent lag phase of growth. Nannochloropsis sp. culture cell density increased to 17.13 ±26.24×10^5 cells ml^{-1} at hour 156 of cultivation (lag phase). Meanwhile Tetraselmis sp. culture cell density increased to 3.90±1.35×10^5 cells ml^{-1} at hour 132 of cultivation (lag phase). The cell density of both species in mixed culture and Nannochloropsis sp. in monoculture decreased considerably at hour

216 and 180 of cultivation, respectively, until the end of the experimental period. In contrast, the cell densities of Tetraselmis sp. in monoculture continued increasing until the end of the experimental period. In addition, growth patterns of Nannochloropsis sp. and Tetraselmis sp. improved appreciably in the mixed culture than momoculture (Figure 1).

Maximum cell density of both speciesin mixed culture had a significantly higher than maximum cell density of Tetraselmis sp. in monoculture (P0.05). Also, the growth rates of both species in mixed culture had a significantly higher than growth rates of Tetraselmis sp. in monoculture (P0.05) (Table 1).

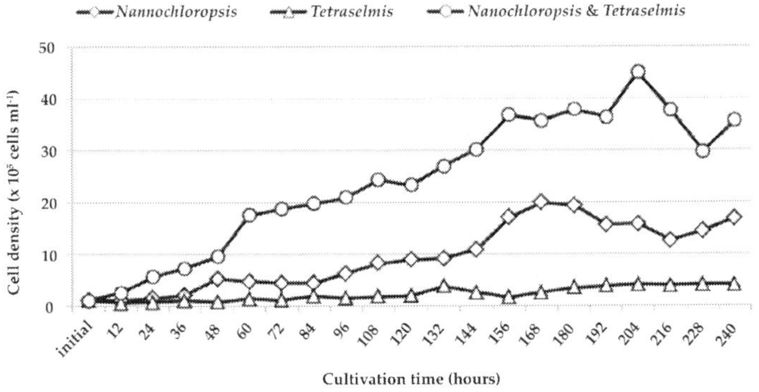

Figure 1: Growth of Nannochloropsis sp. and Tetraselmis sp. in monoculture and mixed culture of both species under laboratory conditions.

Table 1: Maximum cell density, growth rate and proximate composition of Nannochloropsis sp., Tetraselmis sp. in monoculture and mixed culture of both species under laboratory conditions (n = 4).

Parameter	*Nannochloropsis*	*Tetraselmis*	Mixed culture of both species
Maximum cell density($\times 10^5$ cells ml^{-1})	29.25 ± 22.70^{ab}	$5.09 + 0.99^{b}$	45.06 ± 2.67^{a}
Growth rate (day^{-1})	0.37 ± 0.14^{a}	0.16 ± 0.03^{b}	0.43 ± 0.02^{a}
120 hours of cultivation			
Protein (% dw)	35.8 ± 2.6^{a}	25.7 ± 1.3^{b}	33.7 ± 0.9^{a}
Carbohydrate (% dw)	19.9 ± 1.1^{a}	16.6 ± 1.2^{b}	17.5 ± 1.5^{b}
Lipid (% dw)	11.7 ± 0.5^{a}	9.4 ± 0.9^{b}	10.8 ± 1.0^{ab}
240 hours of cultivation			
Protein (% dw)	32.6 ± 3.2^{a}	21.7 ± 0.7^{b}	29.7 ± 3.0^{a}
Carbohydrate (% dw)	17.6 ± 1.7^{a}	14.5 ± 1.6^{a}	16.2 ± 1.2^{a}
Lipid (% dw)	11.1 ± 0.9^{a}	9.4 ± 0.8^{a}	10.0 ± 1.4^{a}

Note: Data in the same row with different superscripts are significantly different (P< 0.05), dw = dry weight.

Proximate composition

After 120 hours of cultivation, protein content of both species in mixed culture and Nannochloropsis sp. in monoculture were found to be significantly higher than protein content of Tetraselmis sp. in monoculture (P<0.05). For Nannochloropsis sp. in monoculture, carbohydrate content had a significantly higher than Tetraselmis sp. in monoculture and both species in mixed culture (P>0.05) but had a significantly higher than Tetraselmis sp. in monoculture (P<0.05) (Table 1).

After 240 hours of cultivation, protein content of both species in mixed culture and Nannochloropsis sp. in monoculture had a significantly higher than protein content of Tetraselmis sp. in monoculture (P0.05) (Table 1).

DISCUSSION

Nannochloropsis sp. and Tetraselmis sp. were widely used in aquaculture as a source of protein, lipid and carbohydrate (Khatoon et al., 2014). The growth performance and the growth patterns of Nannochloropsis sp. and Tetraselmis sp. in monoculture of this studied were different. The monoculture of Nannochloropsis sp. grew rapidly, while Tetraselmis sp. had a relatively slow and steady growth. This is because of the size difference of the microalgae had a affect growth rate (Nannochloropsis sp. – 2 μm diameter, Tetraselmis sp. – 8 μm diameter), smaller size species grow faster than larger ones because of greater surface or volume ratio of smaller sized cells which facilitates assimilation of nutrients at relatively faster rate (Phatarpekar et al., 2000). In the mixed culture, the growth performance and the growth patterns of both species were better than monoculture. The results suggest that mixed culture of microalgae can increased the production of microalgae as was similarly reported by Cai and Duan (2008), Huang et al. (2011), Ahmad et al. (2012), Arkronrat and Oniam (2014) and Sureshkumar et al. (2014). In order to support this result, Brito et al. (2013) reported that mixed culture of microalgae was not simply a mixture of individual monocultures; the growth characteristics, biochemical compositions, nutritional compositions and other growth factors of mixed culture can be substantially different from those of monocultures. For proximate composition in this study, protein content of Nannochloropsis sp. and Tetraselmis sp. in mixed culture had a significantly higher than protein content of Tetraselmis sp. in monoculture after 120 and 240 hours of cultivation. Carbohydrate and lipid contents of each treatment were found to be significantly different after 120 hours of cultivation, which exhibited the similar trends as other microalgal species. Phatarpekar et al. (2000) compared the growth performance, biochemical and nutritive values of Isochrysis galbana and Chaetoceros calcitrans in mixed culture with those in monoculture under laboratory condition and found that cellular concentrations of chlorophyll a, protein, carbohydrate, lipid and particulate organic carbon were significantly higher in mixed culture when compared with monocultures during all the growth phases. Huang et al. (2011) reported that chlorophyll a and protein content of mixed culture of Dunaliella salina and Phaeodactylum tricornutum were increased with the increasing age of the culture when compared with that obtained with monoculture of D. salina and P. tricornutum. However, the biochemical content of microalgae can vary with changes in the environ mental conditions. For example, Khatoon et al. (2014) reported that Nannochloropsis sp. and Tetraselmis sp. had signicantly higher cell density, lipid and carbohydrate contents under control condition at 30 ppt. High cell density, protein, lipid and carbohydrate content was obtained when cultured at pH 7.5 and 8.5 for both species. In addition, the microalgae growth was limited by several factors such as the type and concentration of nutrients, light intensity, photoperiod, genetic deficiency and CO2 all this modifies the biochemical composition of microalgae (Tzovenis et al., 1997; Zhu et al., 1997; Phatarpekar et al., 2000; Araujo and Garcia, 2005).

CONCLUSIONS

The results indicated that the growth performance and the growth patterns of Nannochloropsis sp. and Tetraselmis sp. improved appreciably in the mixed culture than monoculture. Additionally, protein, carbohydrate and lipid contents were significantly higher in mixed culture when compared with monocultures. Therefore, an alternative option for commercial microalgal production is to mixed culture to higher production. Moreover, there is a possibility to enhance growth performance and proximate composition by environmental factor optimization. This is an interesting point for future study

REFERENCES

1. Ahmad, F., Khan, A.U. and Yasar, A. 2012. Uptake of nutrients from municipal wastewater and biodiesel production by mixed algae culture. Pakistan Journal of Nutrition. 11, 550-554.
2. AQUACOP. 1984. Aquaculture en milieu tropical. IFREMER Service documentation Publication, Cedex, France.
3. Araújo, S.D.C. and Garcia, V.M.T. 2005. Growth and biochemical composition of the diatom Chaetoceros cf. wighamii brightwell under different temperature, salinity and carbon dioxide levels. I. Protein, carbohydrate and lipids. Aquaculture. 246, 405-412.
4. Arkronrat, W. and Oniam, V. 2012a. Effects of Light Color on Microalgae Production (Isochrysis sp., Thalassiosira spp., Chlorella sp. and Tetraselmis spp.) in Laboratory. pp.574-581. Proceeding of 50th Kasetsart University Annual Conference: Animals, Veterinary Medicine, Fisheries, Bangkok, Thailand, January 31 – February 2, 2012
5. Arkronrat, W. and Oniam, V. 2012b. Effects of a Diatom Thalassiosira spp. on survival rate and development of Blue Swimming Crab (Portunus pelagicus) larvae from Zoea I to IV stages. Khon Kaen Agriculture Journal. 40, 61-68.
6. Arkronrat, W. and Oniam, V. 2012c. Growth performance and production cost of commercial microalgae cultured under laboratory conditions with different aeration. Kasetsart University Fisheries Research Bulletin. 36, 31-38.
7. Arkronrat, W. and Oniam, V. 2014. Growth performance of mixed culture of microalgae (Chlorella & Tetraselmis) with monocultures under laboratory conditions. p. 200. Proceeding of 4th International Fisheries Symposium (IFS 2014), Surabaya, Indonesia, October 30-31, 2014.
8. Blight, E.G. and Dyer, W.J. 1959. A rapid method of total lipid extraction and purification. Canadian Journal of Biochemistry and Physiology. 37, 911-917.
9. Borowitzka, M.A. 1999. Commercial production of microalgae:

ponds, tanks, tubes and fermenters. Journal of Biotechnology. 70, 313-321.

10. Brito, D., Castro, A., Guevara, M., Gómez, E., Villarroel, A. and Aron, N.M. 2013. Biomass and pigments production of the mixed culture of microalgae (Hyaloraphidium contortum and Chlorella vulgaris) by cultivation in media based on commercial fertilizer. AUDJG-Food Technology. 37, 85-97.
11. Cai Z.P. and Duan, S.S. 2008. Growth in polyculture of algae, Dunaliella salina and Platymonas subcordifomis. Fisheries Science. 27, 330-333.
12. Dubois, M., Gilles, K.A., Hamilton, J.K., Robers, P.A. and Smith, F.1956. Colorimetric method for the determination of sugars and related substances. Analytical Chemistry. 18, 350-356.
13. Huang, W.W., Dong, B.Z., Cai, Z.P. and Duan, S.S. 2011. Growth effects on mixed culture of Dunaliella salina and Phaeodactylum tricornutum under different inoculation densities and nitrogen concentrations. African Journal of Biotechnology. 10, 13164-13174.
14. Khatoon, H., Rahman, N.A., Banerjee, S., Harun, N., Suleiman, S.S., Zakaria, N.H., Lananan, F., Abdul Hamid, S.H. and Endut, A. 2014. Effects of different salinities and pH on the growth and proximate composition of Nannochloropsis sp. and Tetraselmis sp. isolated from South China Sea cultured under control and natural condition. International Biodeterioration and Biodegradation. 95, 11-18.
15. Lowry, O.H., Rosebrough, N.J., Farr, A.L. and Randall, R.J. 1951. Protein measurement with the Folin phenol reagent. Journal of Biological Chemistry. 193, 265-275.
16. Phatarpekar, P.V., Sreepada, R.A., Pednekar, C. and Achuthankutty, C.T. 2000. A comparative study on growth performance and biochemical composition of mixed culture of Isochrysis galbana and Chaetoceros calcitrans with monocultures. Aquaculture. 181, 141- 155.
17. Sánchez-Saavedra, M.P. and Voltolina, D. 2006. The growth rate, biomass production and composition of Chaetoceros sp. grown with different light sources. Aquacultural Engineering. 35, 161-165.
18. Sureshkumar, S., Jasmin, B., Mujeeb Rahiman, K.M. and Hatha

Mohammed, A.A. 2014. Growth enhancement of micro algae, Chaetoceros calcitrans and Nannochloropsis oculata, using selected bacterial strains. International Journal of Current Microbiology and Applied Sciences. 3, 352-359.

19. Tzovenis, I., De Pauw, N. and Sorgeloos, P. 1997. Effect of different light regimes on the docosahexaenoic acid (DHA) content of Isochrysis aff. galbana (clone TISO). Aquaculture International. 5, 489-507.
20. Wongrat, L. 2000. Manual of Plankton Culture. Kasetsart University, Bangkok.
21. Zhu, C.J., Lee, Y.K. and Chao, T.M. 1997. Effects of temperature and growth phase on lipid and biochemical composition of Isochrysis galbana TK1. Journal of Applied Phycology. 9, 451-457.

CHAPTER

5

HIGH-EPA BIOMASS FROM NANNOCHLOROPSIS SALINA CULTIVATED IN A FLAT-PANEL PHOTO-BIOREACTOR ON A PROCESS WATER-ENRICHED GROWTH MEDIUM

Hamed Safafar [1], Michael Z. Hass [2], Per Møller [2], Susan L. Holdt[1] and Charlotte Jacobsen [1]

[1]Technical University of Denmark, National Food Institute (DTU Food), Søltofts Plads, Building 221, Kongens Lyngby 2800, Denmark

[2]Kalundborg Municipality, Department Development, Torvet 3A, Kalundborg 4400, Denmark

ABSTRACT

Nannochloropsis salina was grown on a mixture of standard growth media and pre-gasified industrial process water representing effluent from a local biogas plant. The study aimed to investigate the effects of enriched growth media and cultivation time on nutritional composition water, which were close to the ideal atomic ratio of 16 for *N. salina* [30]. Some previous studies reported that *Nannochloropsis* sp. can utilize ammonium as well as nitrate [14, 24, 25, 28, 29, 30, 31]. The growth interruption of *Nannochloropsis salina* on ICW as main nutrient source might be related to different factors including concentration of ammonia in growth media (Table 2). Ammonia toxicity is mainly attributed to NH_3 at pH > 9 and at pH < 8, toxicity is likely associated with the ammonium ion rather than ammonia. During growth of algae on a growth media containing ammonium, the pH decreases, as H^+ ions released to the medium. In contrast, growth on nitrate causes an increase in the pH due to the release of OH^- ions [31]. This phenomenon can result in less available carbon, as the main source of the carbon was CO_2, which was injected automatically when the pH raised over set value (e.g., pH = 8). Cultivation of *Nannochloropsis salina* at various pH values (8.3 ± 0.2, 7.3 ± 0.2 and 6.3 ± 0.2) with the mixed nutrient composition (20% ICW + 80% F/2) showed that at pH 6.3, the growth rate was significantly lower than at higher pH (Figure 1b). It has also been reported previously that optimum pH for *Nannochloropsis salina* ranged from 7.5 to 8.5 [32]. In this study pH 8.3 ± 0.2 was selected as the optimum pH for further experiments.

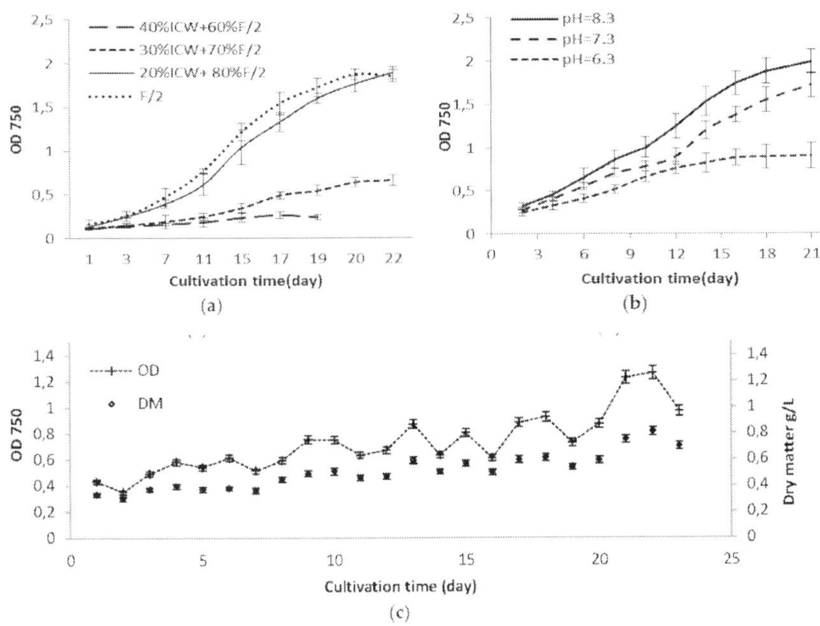

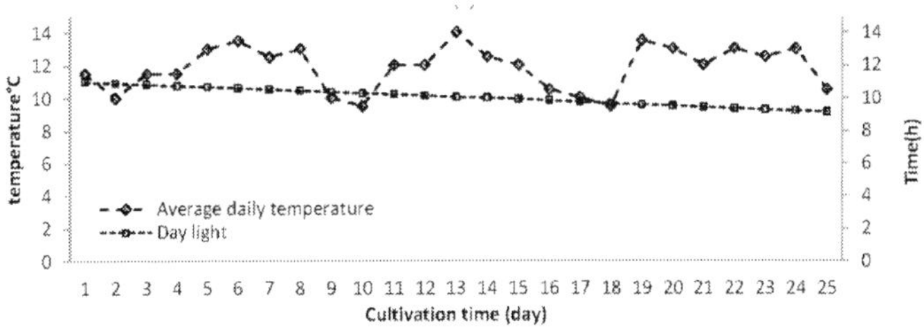

(d)

Figure 1: Growth curves for: (**a**) Effect of different levels of industrial process water on growth of *Nannochloropsis salina* at pH = 8.5; (**b**) *Nannochloropsis salina* cultivated on 20% ICW + 80% F/2 at 3 different pH; (**c**) large scale cultivated *Nannochloropsis salina*; and (**d**) variation in daily light hours and temperature during large scale cultivation. Large scale cultivation was done at pH = 8.3 ± 0.2 using 20% ICW + 80% F/2 as growth media at October–November 2015 in Kalundborg, Denmark. The standard errors are presented as bars (n = 2).

Table 1: Chemical composition of industrial wastewater.

Item	Unit	Amount
pH	-	8.1
Suspended solids	mg/L	20
Total N	mg/L	190
Ammonia + ammonium-N	mg/L	150
Nitrite + nitrate	mg/L	<0.1
Total P	mg/L	11
Sulphate	mg/L	3.6
Total Alkalinity	mmol/L	62.5
EDTA	mg/L	<0.5
Sodium(Na)	mg/L	1500
Copper (Cu)	µg/L	3.4
Iron (Fe)	mg/l	0.23

Table 2: Type and amounts of nitrogen in each growth media.

Growth Media *	NO_3^-	NH_4^+	Total N	Total P
F/2	75	-	75	5.0
20% ICW + 80% F/2	60	30	90	6.2
30% ICW + 70% F/2	52	45	97	6.8
40% ICW + 60% F/2	45	60	105	7.7

* All values are in mg/L.

Changes in Biomass Composition in Response to Growth Media

Protein, Lipids and Tocopherols

Lipid content increased during the cultivation from 7.9% ± 0.50% DW to 29.0% ± 1.50% DW and from 6.15% ± 1% DW to 26.0% ± 1.75% DW, for ICW + F/2 and F/2 cultivated microalgae, respectively (Figure 2a,b). Effect of growth media was evaluated as not significant while cultivation time significantly affected the oil accumulation in both experiments ($p < 0.05$). Amounts of lipids in biomass depend on several factors such as nutrient composition and environmental factors, so comparison of the results with other studies is not possible. Various amounts of lipids in *N. salina* were previously reported [11,12,14,24,33]. One study reported that under nitrate limitation, total fatty acids could exceed 70% of the biological dry mass in this specie [11], while in another study lipid accumulation in *N. salina* was the lowest (36.95% ± 0.91% DW) among 9 different *Nannochloropsis* species [12]. Several studies demonstrated that accumulation of lipids is enhanced in nitrogen-starved or deprived cultures of microalgae [11,14,32]. In our study, relatively high N/P ratio of the growth media may have impacted the lipid contents. However, our findings indicated that nitrogen starvation most likely had not occurred during cultivation for 21 days.

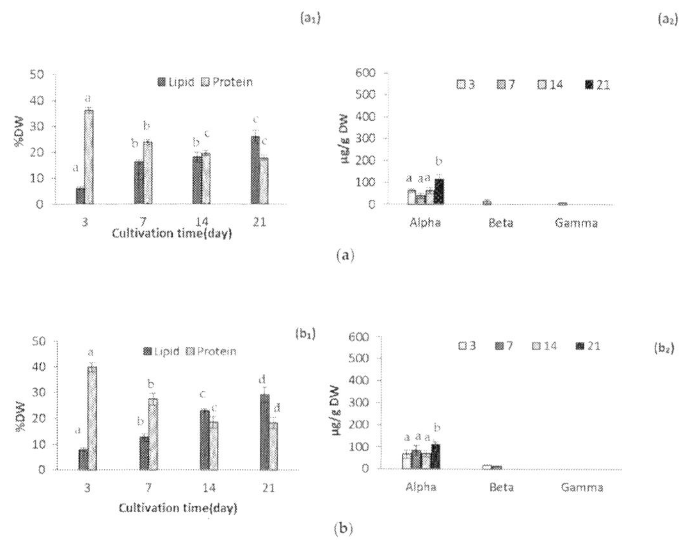

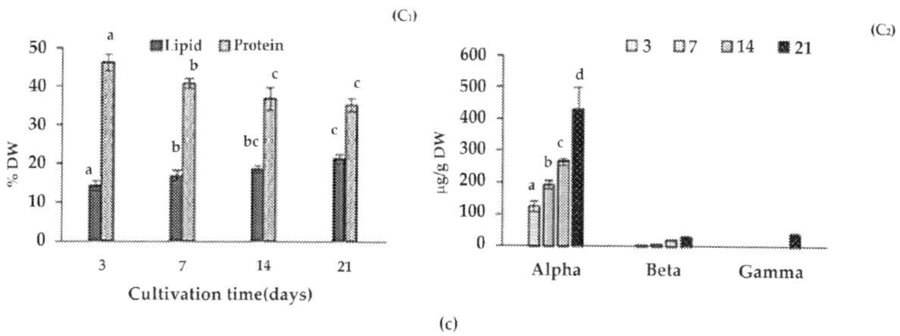

Figure 2: Lipid and protein content (a_1,b_1,c_1) and tocopherol composition (a_2,b_2,c_2) for (**a**) F/2 experiment; (**b**) ICW + F/2 experiment and (**c**) large scale experiment, respectively. The standard errors are presented as bars ($n = 2$), and different letters indicate significant differences ($p < 0.05$).

In general, microalgae have a limited ability to produce nitrogen storage materials when growing under nitrogen-sufficient conditions, with some exceptions in cyanobacteria. When microalgae are grown under nitrogen starvation, the most striking effect is the active and specific degradation of phycobilisomes [32]. Until available nitrogen in cell falls below a threshold value, photosynthesis still continues at lower rate. Under these circumstances, photosynthetically fixed carbon is then diverted from the protein synthesis into the pathways for carbohydrate and lipid synthesis [32], resulting in gradual decrease of protein during the cultivation. Our results confirm this phenomenon as protein content decreased during the cultivation from 39.7% DW to 18.1% DW and from 36.2% DW to 17.8% DW for ICW + F/2 and F/2 cultivated microalgae, respectively (Figure 2a_1,b_1). Both growth media and cultivation time had a statistically significant effect on protein content at the 95.0% confidence level ($p < 0.05$). Difference in protein contents was, however, not significant after day 14.

Tocopherol composition mostly included α-tocopherol as reported by several studies [34,35]. Total tocopherol content was not significantly different in ICW + F/2 and F/2 experiments during the cultivation (Figure 2a_2,b_2). Amounts of α-tocopherol ranged from 61.2 ± 8.4 µg/g to 113 ± 24 µg/g and 67.3 ± 8.5 µg/g to 109.2 ± 14 µg/g for ICW + F/2 and F/2 cultivated microalgae, respectively. The effects of cultivation time and growth media were not found statistically significant ($p = 0.1034$). Durmaz et al. [35] showed that decreasing nitrogen concentrations could lead to an increase in α-tocopherol accumulation in *N. oculata*. Concentration of α-tocopherol could be attributed to different factors such as; nitrogen source (e.g., ammonium or nitrate), nitrogen concentration, growth phase and light (L)/dark (D) photoperiod. Microalgae *Nannochloropsis* sp. produce significantly lower quantities of α-tocopherol at 12:12 h (L:D) photo period compared to 24:0 h (L:D) [35]. It could be a reason for lower amounts of α-tocopherol in our experiments compared to the values reported by Durmaz et al. [35].

Fatty Acid Composition

The major fatty acids in *Nannochloropsis* sp. are 14:0, 16:0, 16:1, *n*-7 and EPA, but C18 fatty acids and 20:4 (*n*-6) are present in lower quantities [11,12,13,15]. Table 3 shows the variation of the fatty acid compositions of *N. salina* grown on two growth media. During the 21 days of cultivation, EPA increased significantly from 4.95% ± 0.78% TFA to 32.0% ± 0.82% TFA and from 7.47% ± 0.23% TFA to 37.1% ± 0.77% TFA, for ICW + F/2 and F/2 cultivated microalgae, respectively. Both cultivation time and growth media had a statistically significant effect on EPA variations at the 95.0% confidence level. Amounts of C 20:4 (*n*-6) were higher in F/2 cultivated microalgae during the course of cultivation. Palmitic acid was the major saturated fatty acid in both experiments. Highest amounts of palmitic acid were detected in samples from the 3rd day of cultivation, which decreased from 48.8% ± 2.59% TFA to 21.3% ± 0.30% TFA and from 43.3% ± 0.98% TFA to 19.1% ± 0.40% TFA after 21 days, for ICW + F/2 and F/2 cultivated microalgae, respectively. Only cultivation time had a statistically significant effect on variation of C16:0 at the 95.0% confidence level. C16:1 (*n*-7) was detected as the main *n*-7 fatty acid in *N. salina* in this study, and it decreased during the cultivation for F/2 cultivated microalgae, but variations were not significant. Effects of cultivation time and growth media on the variations of C18:1 (*n*-7) were not significant at 95.0% confidence level. Total saturates varied significantly from 51.6% ± 3.25% TFA to 25.2% ± 1.63% TFA and from 48.3% ± 2.33% TFA to 25.9% ± 0.62% TFA for ICW + F/2 and F/2 cultivated microalgae, respectively. Variation in total saturated fatty acids was attributed to the cultivation time while both cultivation time and growth media had a statistically significant effect on total *n*-3 PUFA content at the 95.0% confidence level. EPA belongs to a group of fatty acids that are part of the phospholipids, which serve as structural components in the cell wall. Under nutritional limitations, such as nitrogen, cells are unable to resynthesize them or keep the concentration of these components constant [15]. However, with adequate nutrition, cells are capable of synthesizing high amounts of energy rich PUFAs, such as EPA [36], so enrichment of growth media with nitrogen (ammonia) successfully enhanced the EPA contents in *N. salina*.

Table 3: Fatty acid composition of *N. salina* grown on F/2 and ICW + F/2 growth media in laboratory scale. Different letters indicate significant differences ($p < 0.05$).

Cultivation Time (Day)	3		7		14		21	
Fatty Acid	F/2	ICW + F/2	F/2	ICW + F/2	F/2	ICW + F/2	F/2	ICW + F/2
14:0	1.45 ± 0.44	3.22 ± 0.50	3.52 ± 0.12	0.51 ± 0.05	1.40 ± 0.13	0.56 ± 0.30	2.05 ± 0.50	3.53 ± 0.13
15:0	0.04 ± 0.00	nd	nd	nd	0.54 ± 0.01	1.10 ± 0.40	0.30 ± 0.09	1.46 ± 0.03
16:0	48.8 ± 0.59 ^{a}	43.3 ± 0.98 ^{b}	32.9 ± 0.53 ^{c}	31.9 ± 0.77 ^{c}	23.7 ± 0.40 ^{d}	20.6 ± 0.64 ^{de}	21.3 ± 0.30 ^{de}	19.1 ± 0.40 ^{e}
16:1 (*n*-7)	30.3 ± 0.40 ^{a}	30.9 ± 0.10 ^{a}	24.0 ± 0.09 ^{b}	20.4 ± 0.02 ^{c}	25.1 ± 0.51 ^{b}	23.0 ± 0.02 ^{b}	31.9 ± 0.66 ^{a}	17.7 ± 0.51 ^{c}
16:2 (*n*-4)	0.18 ± 0.01	0.23 ± 0.02	1.24 ± 0.00	0.65 ± 0.00	0.33 ± 0.01	0.76 ± 0.07	0.64 ± 0.02	0.38 ± 0.01
16:3 (*n*-4)	0.11 ± 0.03	3.57 ± 0.09	0.61 ± 0.02	0.56 ± 0.03	0.75 ± 0.04	0.63 ± 0.08	0.24 ± 0.05	1.01 ± 0.04
16:4 (*n*-1)	0.36 ± 0.22	3.35 ± 0.08	0.45 ± 0.04	0.14 ± 0.19	0.44 ± 0.01	0.28 ± 0.10	0.43 ± 0.03	0.66 ± 0.06
18:0	2.35 ± 0.20	1.69 ± 0.19	5.32 ± 0.09	1.82 ± 0.06	2.20 ± 0.01	2.10 ± 0.06	0.51 ± 0.10	1.86 ± 0.07
18:1 (*n*-9)	7.03 ± 0.23	3.31 ± 0.06	9.43 ± 0.08	5.61 ± 0.06	9.06 ± 0.20	6.53 ± 0.06	4.22 ± 0.06	4.38 ± 0.20
18:1 (*n*-7)	0.75 ± 0.20	0.00 ± 0.06	0.45 ± 0.08	6.31 ± 0.60	7.52 ± 0.21	7.50 ± 0.45	0.57 ± 0.06	3.32 ± 0.21
18:2 (*n*-6)	1.13 ± 0.01	0.74 ± 0.06	1.45 ± 0.03	1.35 ± 0.06	2.23 ± 0.08	1.67 ± 0.07	1.63 ± 0.03	1.46 ± 0.09
18:3(*n*-3)	0.64 ± 0.10	0.50 ± 0.01	0.49 ± 0.04	0.46 ± 0.03	0.04 ± 0.02	nd	0.30 ± 0.01	nd
20:1(*n*-7)	0.10 ± 0.00	nd	1.20 ± 0.12	nd	0.10 ± 0.01	nd	0.10 ± 0.01	nd
20:4 (*n*-6)	1.21 ± 0.20	0.73 ± 0.06	1.92 ± 0.05	4.82 ± 0.08	3.82 ± 0.07	5.63 ± 0.08	2.33 ± 0.06	4.91 ± 0.07
20:3 (*n*-3)	0.22 ± 0.08	0.33 ± 0.08	1.57 ± 0.04	0.26 ± 0.02	nd	0.31 ± 0.02	0.17 ± 0.08	0.93 ± 0.02
20:4 (*n*-3)	0.10 ± 0.00	0.33 ± 0.02	0.43 ± 0.03	nd	0.31 ± 0.08	nd	0.10 ± 0.00	nd
20:5 (*n*-3)	4.95 ± 0.78 ^{a}	7.47 ± 0.23 ^{ab}	10.9 ± 0.11 ^{b}	22.2 ± 0.21 ^{c}	19.9 ± 0.42 ^{c}	26.2 ± 0.25 ^{d}	32.0 ± 0.82 ^{e}	37.1 ± 0.77 ^{f}
22:5 (*n*-3)	0.66 ± 0.05	nd	0.40 ± 0.01	0.26 ± 0.04	1.35 ± 0.05	0.14 ± 0.03	0.66 ± 0.03	0.27 ± 0.05
22:6 (*n*-3)	0.46 ± 0.09	0.15 ± 0.04	0.24 ± 0.02	1.04 ± 0.40	nd	0.69 ± 0.40	0.19 ± 0.02	0.16 ± 0.05
∑ SAFA	51.6 ± 3.25 ^{a}	48.3 ± 2.33 ^{a}	41.8 ± 0.63 ^{b}	34.2 ± 1.25 ^{b}	27.9 ± 0.51 ^{c}	24.4 ± 1.44 ^{c}	24.2 ± 1.63 ^{c}	25.9 ± 0.62 ^{c}
∑ *n*-3	6.92 ± 2.41 ^{a}	8.77 ± 0.45 ^{a}	13.6 ± 0.23 ^{b}	24.3 ± 0.70 ^{c}	21.66 ± 0.66 ^{c}	27.4 ± 0.74 ^{d}	33.37 ± 1.0 ^{e}	38.6 ± 1.00 ^{f}

Data are shown as % of fatty acid in total fatty acid composition. Total saturated fatty acids (Σ SAFA), n-3 poly unsaturated fatty acid (ř n-3); nd, not detected. Different letters in the same row represent significant difference ($p < 0.05$) for 16:0, 16:1 (n-7), 20:5 (n-3), total SAFA and n-3 fatty acids.

Hu et al. [13] showed an increase of the EPA concentration of *N. salina* in batch cultures supplied with both high and low nitrate concentrations, so effect of environmental factors such as light intensity and temperature should also be considered in development of EPA in this species. It is generally assumed that microalgae have common biosynthetic pathway for production of EPA, that is, desaturation of 18:2 (*n*-6) is conducted by either Δ6 or Δ15 (*n*-3) desaturase trails, resulting in either 18:3 (*n*-6) or 18:3 (*n*-3), which lead to 20:4 (*n*-6) or EPA, respectively [36]. In high-CO_2-grown cells, 18:2 (*n*-6) was of higher relative amount, and available for Δ6 desaturation, which gave rise to an increased production of 18:3 (*n*-6) and subsequently 20:5 (*n*-3) [37]. In the present study, 18:3 (*n*-6) was not detected, suggesting rapid turnover of 18:3 (*n*-6) and higher contents of EPA in the presence of CO_2 as described in previous studies [13,37].

Amino Acid Composition

The amino-acid compositions of *Nannochloropsis salina* grown on two growth media is shown in Figure 3a,b. Nearly all of the amino acids were present in higher concentrations in ICW + F/2 cultivated samples. Concentration of amino acids decreased in the course of cultivation for both experiments. Aspartate, glutamate and arginine were generally found in the highest concentrations. Amino acid profiles are also similar to previous reports on *Nannochloropsis* genus and even microalgae species from different classes [38]. The similarity in the amino acid compositions suggests that the structure of microalgal protein may also be similar. This could indicate that many of the proteins performing specific functions and common to all species (e.g., C-fixation enzymes, membrane proteins associated with light-harvesting pigments) are highly conserved in their amino acid composition throughout the different species [38]. Notable difference was significantly higher amounts of lysine in F/2 cultivated microalgae samples at early stages of cultivation, while amounts of proline and glutamine were higher in ICW + F/2 cultivated samples. Ammonia is generally thought to directly assimilate into the amino acid glutamine [32], which can explain higher concentrations of glutamine in ICW + F/2 cultivated samples (58 mg/g vs. 48 mg/g). In general total amino acid content was higher in ICW + F/2 cultivated samples (Figure 3d), which could be contributed to higher amounts of nitrogen in growth media.

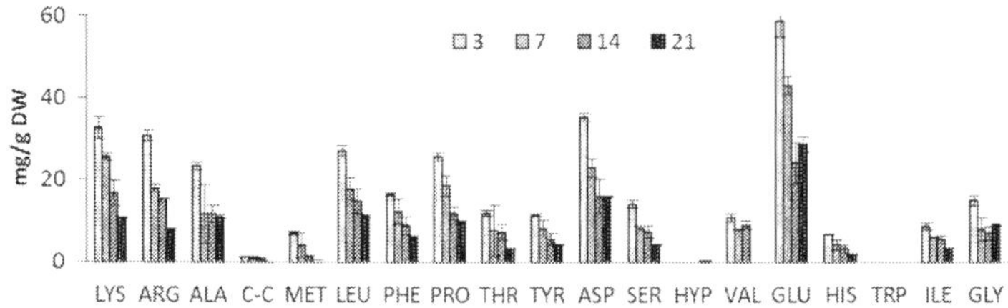

(a)

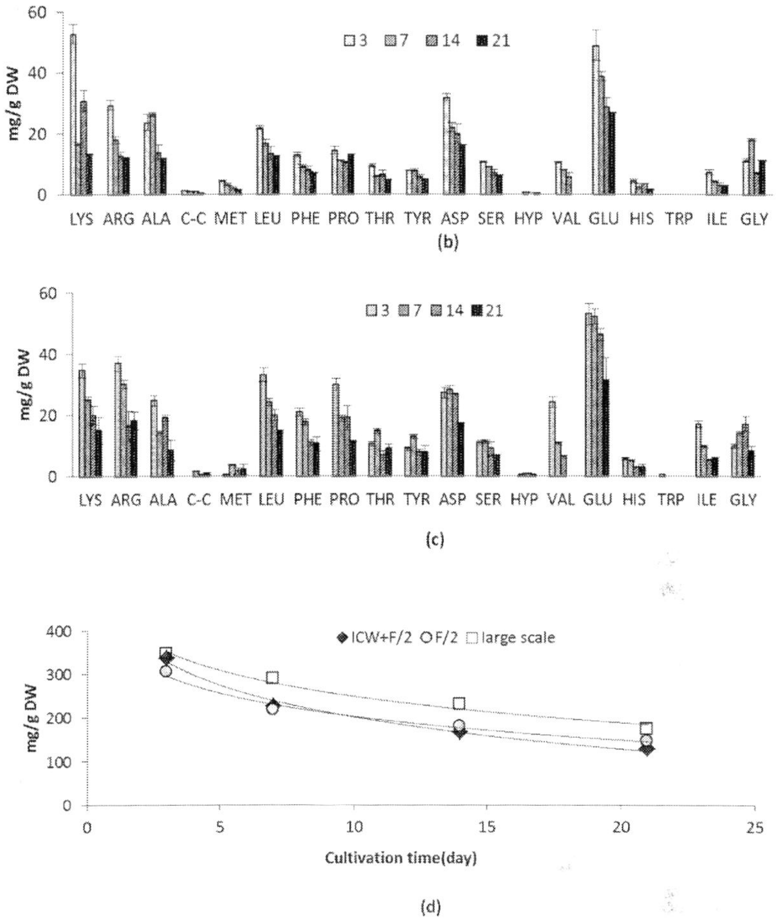

Figure 3: Amino acid composition for: (**a**) ICW + F/2 experiment; (**b**) F/2 experiment; (**c**) large scale experiment and (**d**) total amino acids variations during 21 days of cultivation for ICW + F/2, F/2 and large scale experiments. The standard errors are presented as bars ($n = 2$).

There was a statistically significant correlation at 95.0% confidence level ($p < 0.05$), between results of protein content and total amino acids in all experiments. The correlation coefficient equals 0.86, indicating a moderately strong relationship between the variables.

Pigments

Pigment composition in *Nannochloropsis* sp. include chlorophyll a (not b or c), and also violaxanthin and vaucheriaxanthin esters, which play a major role in light harvesting as main accessory pigments [34,39]. Other carotenoids include xanthophylls such as canthaxanthin, anteraxanthin, zeaxanthin and carotenes such as β-carotene were also reported in *Nannochloropsis* sp. [39].

Variations in pigment composition of ICW + F/2 and F/2 experiments are shown in Figure 4a,b. Chlorophyll a (Chl) increased during the course of cultivation from 867 ± 84.0 μg/g to 10,030 ± 239 μg/g and from 1048 ± 22.1 μg/g to 8427 ± 419 μg/g for ICW + F/2 and F/2 cultivated microalgae, respectively. Carotenoids include violaxanthin, vaucheriaxanthin esters, alloxanthin, antheraxanthin, cantaxanthin, zeaxanthin and beta carotene in both experiments. During 21 days of experiment, total carotenoids increased from 749 ± 81 μg/g to 11387 ± 874 μg/g and from 797 ± 55 μg/g to 7761 ± 419 μg/g for F/2 and ICW + F/2 cultivated microalgae, respectively. Both cultivation time and growth media significantly influenced the variations of violaxanthin as main carotenoid, while effect of growth media on variations of vaucheriaxanthin was not found significant. The presence of xanthophyll fatty acid esters (mainly vaucheriaxanthin) is characteristic for carotenoid composition in some microalgae such as *Nannochloropsis* spp. [39], and the free-to-esterified vaucheriaxanthin ratio varies in different *Nannochloropsis* spp. from ca. 1:1 to 1:4 [40].

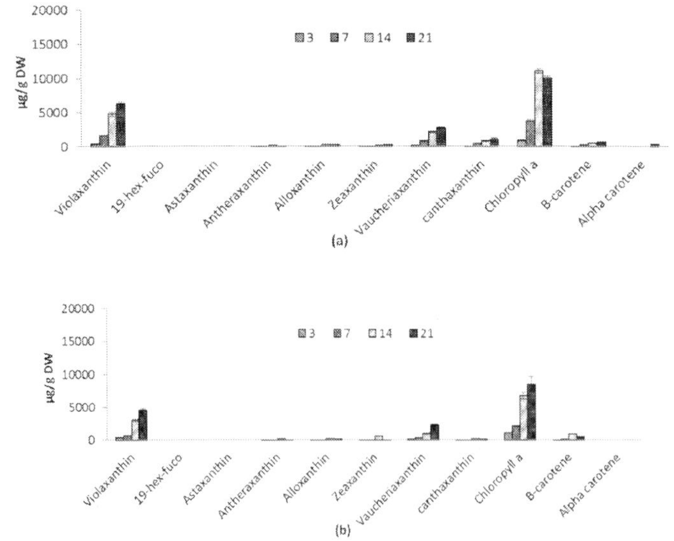

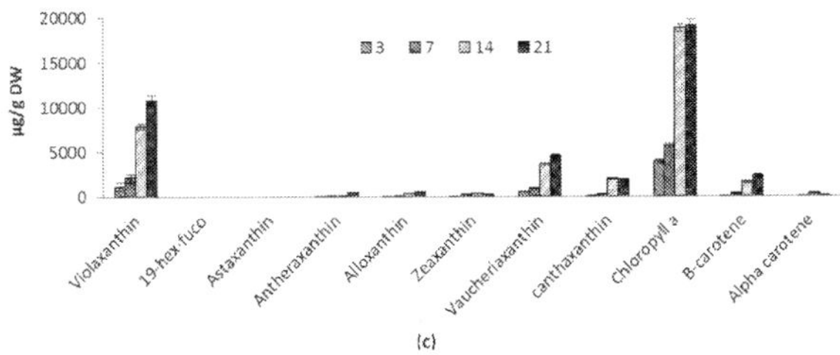

Figure 4: Pigment composition for: (**a**) F/2 experiment; (**b**) ICW + F/2 expeiment; and (**c**) large scale experiment. The standard errors are presented as bars ($n = 2$).

In this study the ratio of free to esterified vaucheriaxanthin was 1:3.9 for F/2 and 1:3.7 for ICW + F/2 experiments. Like non-esterified vaucheriaxanthin, esterified forms may be attributed to pigment-protein complexes of the thylakoid membranes, but they do not participate in light harvesting in *Nannochloropsis* [39], so vaucheriaxanthin will not follow the variations in light harvesting carotenoids. Car/Chl increased during the cultivation from 0.76 to 0.92 for ICW + F/2 and from 0.86 to 1.13 for F/2 experiments. It was reported previously that considerable rise in Car/Chl along with accumulation of FA takes place similarly and is induced by nitrogen starvation, while there were also reports about accumulation of carotenoids in stressed *Nannochloropsis* cells [39,41]. In this study, nitrogen concentration was higher in ICW + F/2 growth media and resulted in higher chlorophylls and lower carotenoids in the biomass, compared to F/2 growth media. So comparatively higher values of Car/Chl in F/2 cultivated samples can be attributed to the lower levels of nitrogen. It has been clearly demonstrated previously, that in the process of chlorophyll biosynthesis, nitrogen (in form of δ-aminolevulinic acid) is involved as a precursor in the synthesis of pyrol ring in chlorophyll structure [40]. In the presence of a carbon source, concentration of chlorophyll(s) can be directly attributed to the amount of nitrogen (inorganic form) in growth media.

Changes in Growth and Biomass Composition of Large Scale Cultivated Nannochloropsis salina

Growth Rate and Biomass Productivity

The growth curve for large scale cultivation of *Nannochloropsis salina* using nutrient composition as 20% ICW + 80% F/2 at pH 8.3 ± 0.2 is shown in Figure 1c. Growth started rapidly, but did not reach the stationary phase. Variation in optical density and biomass productivity can be attributed to temperature variations (Figure 1d), which resulted in sedimentation and inhomogeneity of the culture. The average solar radiation values from 2006 to 2010 in Denmark previously reported as 16 $KWh{\cdot}m^{-2}$ for October–November, which compared to June–July (75 $KWh{\cdot}m^{-2}$) is very low [42]. Daily light varied from 11 h to 9 h during the experiment. Both growth rate and biomass productivity can be affected by low temperature and low daily light hour. For example Sforza et al. [14] reported 30% less growth rate and 28% less biomass concentration for *Nannochloropsis salina* cultivated under light: dark cycle compared to continuous illumination. On the other hand, lower biomass concentration will provide a greater possibility for absorption of light energy in flat panel photo-bioreactor narrow tubes (32 mm). Olofsson et al. [18] reported a relatively constant biomass concentration during two years of cultivation of *Nannochloropsis oculata* in large scale outdoor photo bioreactor, with narrow light path (5 cm). In another study, an increase in the light path length of photo bioreactor from 5 to 10 cm significantly decreased the biomass productivity in *Nannochloropsis gaditana* [21]. Chini Zittelli et al. [17] reported that continuous illumination with artificial light resulted in 37% increase in volumetric productivity of *Nannochloropsis* sp. cultivated in flat panel photo bioreactor. Environmental and operational conditions highly affect the growth and lipid production in *Nannochloropsis* sp. [22] *Nannochloropsis salina* can grow in Denmark's winter time conditions and tolerate the temperature/light variations, but when high biomass productivity (and not chemical composition) is the main target, artificial light will be required. For efficient cultivation of microalgae in a photo bioreactor, optimizing light availability and light intensity is critically required [14,21,22].

Biomass Chemical Composition

Large scale cultivation of *Nannochloropsis salina* influenced the chemical composition of the biomass during the course of cultivation and before reaching the steady state conditions. Cultivation time significantly influenced variations in lipid and protein contents ($p < 0.05$).

Total lipid contents increased from 10.8% ± 1.2% DW to 21.1% ± 1.2% DW, but the rate of accumulation was lower comparing to the laboratory scale experiment with the same growth media, which can be explained with lower light and temperature. Olofsson et al. [18] also reported that variations in total lipid content of *Nannochloropsis oculata* can be explained with both light and temperature variations. Highest and lowest lipid productivity was reported in autumn and winter, respectively. Camacho-Rodríguez et al. [21] reported that an increase in the path length width of photo bioreactor from 5 to 10 cm significantly decreased both lipid contents in *Nannochloropsis gaditana* from 28% to 24% DW [21]. The geometry of flat-panel bioreactor is a key factor for lipid productivity along with environmental and operational parameters such as pH and dilution (harvest) rate [21,22].

In our study protein contents decreased from 46.0% ± 1.3% DW to 33.4% ± 1.7% DW (Figure 2c_1). In general variations in protein contents depends to the growth stage, environmental conditions such as light and temperature and composition of growth media. Protein content is also species-related, which can explain higher protein contents in our results compared to another study which reported the variations of protein contents in *Nannochloropsis occulata* [20]. San Pedro et al. [22] reported that protein content of biomass remained in the range of 49.3% and 30.9% DW for *Nannochloropsis gaditana*cultivated in outdoor pilot-scale raceway ponds. It has also been shown that light path lengths in a photo bioreactor can affect the protein contents in *Nannochloropsis gadiata,* as protein contents increased significantly when light path length increased from 5 to 10 cm [21]. In this study, average protein contents were reported as 45% DW with highest values in the microalgae cultivated at the summer. It can be concluded that several factors such as reactor design and specifications, dilution (harvest) rate, growth media and environmental factors such as temperature and light affect the protein accumulation. On the other hand, the protein accumulation in microalgae is species-specific [3]. In our study, rate of variation in protein content was lower compared to the results of laboratory scale cultivation (54.1% and 27.4%, respectively) as shown in Figure 1d. This could be explained by the fact that large scale cultivation was done during the autumn–winter season when environmental factors such as lower light and temperature and sufficient nitrogen in the growth media [20, 21, 22, 33] influence the growth and metabolism pathway.

Uptake and utilization of nitrogen are reported to be temperature sensitive [11], so nitrogen level available in the growth media may have to be optimized during the winter season to obtain optimal growth. It can be considered as benefit when protein is the target compound in the biomass.

Large scale cultivation enhanced the concentration of α-tocopherol as it increased to 431 ± 28 μg/g during 21 days of cultivation compared to 109.08 ± 11 in lab scale (Figure $2C_2$). This difference could be explained by higher availability of light (even in autumn-winter season) in flat panel photo-bioreactor because of short path length compared to the batch laboratory reactors. On the other hand our values were still low, compared to data reported by Durmaz et al. for *Nannochloropsis oculata*, which was reported to contain 2325.8 ± 39 μg/g α-tocopherol [35]. This lower concentration can be explained by seasonal low photoperiod, which was demonstrated as an important factor, so it could be suggested that in spring-summer season α-tocopherol content will be increased with the same growth media. Durmaz et al. [35] also reported that concentration of tocopherols increased significantly in stationary phase as a response to the nitrogen starvation.

Variations in fatty acid composition resembled the laboratory scale findings for C16:0 and total saturates, which decreased during the cultivation time from 24.9% ± 2.28% TFA to 17.4% ± 0.24% TFA and from 32.0% ± 3.10% TFA to 21.2% ± 0.74% TFA, respectively (Table 4). EPA and total *n*-3 fatty acids increased accordingly, from 23.9% ± 1.78% TFA to 44.2% ± 2.30% TFA and from 27.8% ± 2.30% TFA to 48.1% ± 2.43% TFA, respectively. Opposite of laboratory scale results, variation in C16:1 was not significant. Results of EPA reported in our study are among the highest ever reported for *N. salina* in both laboratory and large scale [11,12,13,15,18,19,20,25,32,33]. Several studies reported that low temperature and short day light, which is equal to less solar energy, could enhance the EPA accumulation in *Nannochloropsis* sp.

[15, 18, 19, 20, 21, 22]. The geometry of a photo-bioreactor can also affect the EPA productivity in *Nannochloropsis* sp. [21]. A significant decrease in EPA productivity was obtained when the light path length in the photo-bioreactor increased from 5 to 10 cm, while the highest EPA productivity was obtained in winter [21]. Low temperatures can decrease the cell membrane fluidity, and microalgae counteracted this environmental effect by increased synthesis of polyunsaturated fatty acids such as EPA [11]. It can be concluded that low temperature (12.0 ± 1.5 °C) during large scale cultivation enhanced the accumulation of EPA in *Nannochloropsis salina*, compared to laboratory scale results for the experiment with the same growth media (ICW + F/2). Laboratory scale experiments were carried out at 20 ± 2 °C.

Table 4: Fatty acid composition of *N. salina* cultivated in flat-panel photo-bioreactor. Different letters indicate significant differences ($p < 0.05$).

Cultivation Time (Day)	3	7	14	21
Fatty Acid				
14:0	5.28 ± 0.57	4.20 ± 0.50	3.58 ± 0.37	3.37 ± 0.37
15:0	0.67 ± 0.44	0.38 ± 0.04	0.26 ± 0.04	0.24 ± 0.04
16:0	24.9 ± 2.28 ^{a}	19.9 ± 0.59 ^{b}	18.5 ± 0.19 ^{b}	17.4 ± 0.24 ^{b}
16:1 (*n*-7)	26.5 ± 2.59 ^{a}	28.7 ± 0.30 ^{a}	28.1 ± 0.44 ^{a}	26.4 ± 0.19 ^{a}
16:2 (*n*-4)	0.18 ± 0.03	0.23 ± 0.02	0.10 ± 0.02	0.10 ± 0.00
16:3 (*n*-4)	0.20 ± 0.01	0.45 ± 0.02	0.22 ± 0.03	0.10 ± 0.02
16:4 (*n*-1)	0.44 ± 0.03	1.01 ± 0.02	0.50 ± 0.02	0.21 ± 0.02
18:0	1.91 ± 0.22	0.37 ± 0.08	0.26 ± 0.03	0.47 ± 0.02
18:1 (*n*-9)	6.06 ± 0.76	3.93 ± 0.19	1.87 ± 0.10	1.76 ± 0.21
18:1 (*n*-11)	2.62 ± 0.20	1.31 ± 0.06	0.94 ± 0.03	0.24 ± 0.00
18:2 (*n*-6)	2.39 ± 0.23	1.30 ± 0.06	0.16 ± 0.01	0.88 ± 0.03
18:3 (*n*-3)	0.10 ± 0.01	nd	nd	nd
20:1 (*n*-9)	0.10 ± 0.02	0.11 ± 0.01	nd	nd
20:1(*n*-7)	0.24 ± 0.02	nd	0.14 ± 0.01	nd
20:4 (*n*-6)	0.05 ± 0.02	nd	nd	0.13 ± 0.01
20:3 (*n*-3)	1.10 ± 0.03	6.45 ± 0.06	3.38 ± 0.05	2.75 ± 0.03
20:4 (*n*-3)	0.61 ± 0.01	0.10 ± 0.08	1.02 ± 0.34	0.91 ± 0.01
20:5 (*n*-3)	23.9 ± 1.78 ^{a}	30.1 ± 1.21 ^{b}	39.3 ± 1.10 ^{c}	44.2 ± 2.30 ^{d}
22:1 (*n*-11)	nd	0.16 ± 0.03	0.10 ± 0.03	0.12 ± 0.03
21:5 (*n*-3)	1.82 ± 0.39	0.81 ± 0.35	1.03 ± 0.01	0.09 ± 0.01
22:5 (*n*-3)	0.32 ± 0.05	0.23 ± 0.03	0.07 ± 0.01	0.06 ± 0.02
22:6 (*n*-3)	nd	0.12 ± 0.20	0.10 ± 0.04	0.10 ± 0.01
∑ SAFA	32.0 ± 3.10 ^{a}	24.5 ± 1.21 ^{b}	22.3 ± 0.72 ^{b}	21.2 ± 0.74 ^{b}
∑ *n*-3	27.8 ± 2.30 ^{a}	37.7 ± 2.00 ^{b}	44.9 ± 1.52 ^{bc}	48.1 ± 2.43 ^{c}

Data are shown as % of fatty acid in total fatty acid composition. Total saturated fatty acids (Σ SAFA), n-3 poly unsaturated fatty acid (ř n-3); nd, not detected. Different letters in the same row represent significant difference ($p < 0.05$) for 16:0, 16:1 (n-7), 20:5 (n-3), total SAFA and n-3 fatty acids.

Variations in amino acid composition of large scale cultivated *N. salina* are shown in Figure 3c. To our knowledge not the same study concerning the variations in amino acid composition in the *Nannochloropsis* sp. cultivated in large scale. Glutamate was rated as the main amino acid, while concentration of lysine, arginine, and leucine were nearly the same. It could be concluded that the variations in amino acids in large scale followed the variation in protein contents in laboratory scale experiments. The presented amino acid composition for large scale cultivated *Nannochloropsis salina* was an ideal composition for aquaculture feed formulas [38]. Total amino acid content also decreased during the cultivation with a gentler slope than ICW + F/2 laboratory scale experiment (Figure 3d).

Pigment composition of large scale cultivated *Nannochloropsis salina* resembled the results, which were discussed previously for laboratory scale cultivated species. Variations in pigment composition are shown in Figure 4c. Chlorophyll content increased during the course of cultivation from 3944 ± 200 μg/g to 19,020 ± 720 μg/g. During 21 days of experiment, total carotenoids increased from 2038 ± 48 μg/g to 20,797 ± 1048 μg/g. Amounts of beta-carotene increased significantly from 26.9 ± 48 μg/g to 2284 ± 234 μg/g. Other carotenoids including vaucheriaxanthin, canthaxanthin, and zeaxanthin also increased during the cultivation time. Free to esterified vaucheriaxanthin ratio was 1:4.2. Car/Chl increased during the cultivation from 0.5 to 1.1. This finding can be explained by environmental stresses such as low temperature and photoperiod, which may influence the light harvesting pigments, i.e., mainly violaxanthin in *Nannochloropsis* sp. [15,36].

MATERIALS AND METHODS

Chemicals and Reagents

Standards of fatty acids, amino acids and tocopherols were purchased from Sigma (St. Louis, MO, USA) and Fluka (Deisenhofen, Germany). Standards of pigments were purchased from DHI (Hørsholm, Denmark). HPLC grade acetonitrile, heptane, isopropanol, methanol and acetone were purchased from Sigma and Fluka. HPLC grade water was prepared at DTU Food using Milli-Q® Advantage A10 water deionizing system from Millipore Corporation (Billerica, MA, USA).

Growth Media

Industrial process water (ICW) was collected from effluent stream of an anaerobic methanogenic conversion reactor (Novozymes plant, Kalundborg, Denmark). Batches of industrial process water was filtered using an out-side-in dynamic cross flow microfiltration BioBooster system from Grundfos A/S (Bjerringbro, Denmark) equipped with 0.2 μm ceramic disc filters and stored at −20 °C prior to use. Chemical composition of the industrial process water is shown in Table 1. The basic culture medium was a conventional algae F/2 growth media powder based on Guillard & Ryther [24], which was purchased from Varicon Aqua solutions (Worcs, UK). Salt brine was prepared from refined salt (Scan salt A/S, Kolding, Denmark) and UV-sterilized and micro-filtrated tap water, which was added to the growth media in order to achieve a final salinity of 25 g/L.

Laboratory Scale Cultivation of Microalgae

Microalgae strain *Nannochloropsis salina* (Strain Number: 40.85) was obtained from culture collection of algae (SAG), University of Gottingen. The strain was cultivated in 1–5 L Schott bottles, while shaking gently in an orbital platform shaker (Heidolf instruments GmbH, Schwabach, Germany). During the cultivation, all reactors were continuously aerated with 2% carbon dioxide/air mixture under fluorescent lights with an intensity of 100 μmol photon $m^{-2} \cdot s^{-1}$ and 12:12 (day:night) photoperiod. Light intensity was measured using a Li-190 quantum sensor (LI-COR, Inc., Lincoln, NE, USA). Online monitoring and control of pH was performed by Milwaukee MC-122-pH controller (Milwaukee Electronics, Szeged, Hungary) equipped with solenoid valve to control the stream of CO_2. Control of temperature was done by an aluminum plate connected to cold water circulation system and the temperature was kept constant at 23 ± 2 °C. Preliminary experiments were performed to explore growth rates at different levels of substitution of F/2 stand growth media with ICW (20%, 30%, and 40%), and pH (6.3 ± 0.2, 7.3 ± 0.2, and 8.3 ± 0.2). Table 2 shows the type and amounts of nitrogen in each growth media. Cultivation was repeated for *N. salina* on 20% ICW + 80% F/2 and standard F/2 media in 10 L Schott bottles at pH 8.3 ± 0.2 and with the same conditions as mentioned before. During the cultivation and at each sampling point, 0.25 L of culture was taken and samples were centrifuged at 10,000 g. Then resulting biomass pellet was washed twice with deionized water. Resulting biomasses were freeze dried immediately to moisture content less than 1% DW. Measurement of moisture was done by an AD 4714A moisture analyzer (A&D Company, Tokyo, Japan). Samples stored at −20 °C prior to chemical analysis.

N. salina cultivated on 20% ICW + 80% F/2 was transferred to Kalundborg microalgae facility (Kalundborg, Denmark) for large scale experiment.

Large-Scale Cultivation of Microalgae

Large scale cultivation was done at the Kalundborg microalgae facility using flat panel photo-bioreactor system, type hanging garden from Ecoduna produktions GmbH (Bruck/Leitha, Austria). The photo-bioreactors were designed to track the sun thereby allowing the algae to grow most efficiently on natural sunlight. No artificial light was used during the large scale experiment. The system base unit includes 12 hollow chamber sheets (width 2100 mm; height 5600 mm and depth 32 mm) which are assembled on guide rails. The microalgae suspension was transferred into the sheets and was continuously circulated through the module. Each sheet was sealed off with a plastic cover on the top and with a deflecting profile on the bottom and supplied with nutrient stream from the bottom deflecting profile on the lateral input. The nutrient stream was pumped through the sheets by means of hydrostatic pressure and the gas lift effect, as a result, no additional energy is required. In these specific reactors, surface hit by sunlight is multiplied and the light is being distributed so the irradiation will never be too high. All of the sheets of a module were interconnected via a hose system thus ensuring the obstruction-free transfer of the nutrient solution. Details of the flat panel bioreactor are shown in Figure 5. Air stream was filtered through 0.2 μm membrane filter and mixed with pure CO_2 from Cylinder (purity 99.5% *v*/*v*). A mixture of air and 2% *v*/*v*CO_2 was then introduced from below to ensure gentle but effective mixing of nutrients and algae to control the pH which was adjusted on 8.5 ± 0.5. This Gas stream drives a gentle liquid flow (1 L·min^{-1} via an "air-lift effect" so the system does not require pump unit. Cultivation was done in batch mode and two separate units were used. At each sampling point, 2 L of culture was taken and samples were centrifuged at 10,000 g. Biomass samples were washed twice with distilled water and then immediately freeze dried and stored at −20 °C prior to the analysis. Cultivation was done during the period of October–November 2015. The culture was kept at steady state basis, by continuous biweekly harvest and feeding of new growth media after day 14.

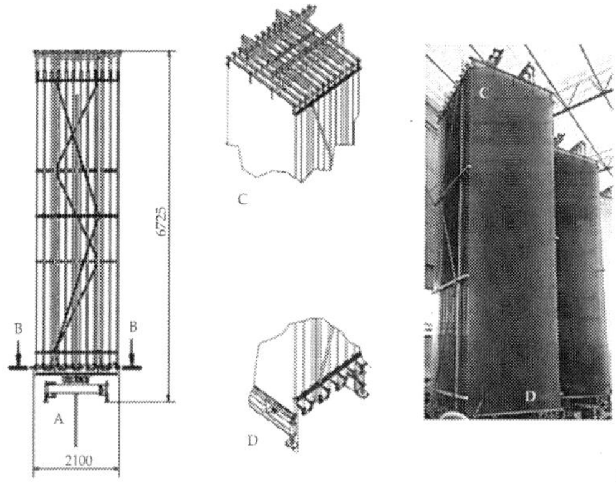

Figure 5: Schematic of flat panel photo-bioreactor which was used for the large scale cultivation experiment. (**A**) Rotating system; (**B**) ground level; (**C**) top part; (**D**) bottom part. Every unit includes 12 sheets. Sheet dimensions: width 2100 mm; height 5600 mm and depth 32 mm.

Analytical Methods

Growth Rate

Growth rate was monitored by daily detection of optical density at 750 nm.

Protein and Amino Acids

The protein content in the microalgae samples was estimated using a modified Micro biuret method [43] with some modification. One mL of 0.5 M NaOH aqueous solution was added to 10 mg of sample and extraction was carried out at 80 °C for 10 min. Samples were centrifuged at 5000 g and supernatants were moved to new tubes. Extraction was repeated 3 times and all extracts were mixed together before analysis. For protein content estimation, 200 μL of copper sulphate (0.21% $CuSO_4{\cdot}5H_2O$ in 30% NaOH aqueous solution) was mixed with 300 μL of the extract and 500 μL of water. Color formation was monitored at 310 nm. Bovine serum albumin (BSA) was used as the standard for calibration curves.

The amino acid composition was analyzed using EZ:faast™ amino acid analysis kit (Phenomenex Inc., Torrance, CA, USA). For protein hydrolysis around 30 mg of microalgae was weighted in microwave glass vials. The samples were first overlaid with nitrogen and then hydrolyzed in 6 M HCl in microwave oven (Anton Paar GmbH, Graz, Austria) for 60 min at 110 °C. Following individual cleanup-step for removing matrix interference, amino acids derivatization was done. The amino acid composition was determined by liquid chromatography with a Agilent 1100 series LC/MSD Trap mass spectrometry (Agilent technologies, Hørsholm, Denmark) using EZ:faast™ LC-MS column (250 × 3.0 mm, Phenomenex, Torrance, CA, USA).

Lipids, Fatty Acid and Tocopherols

Lipids were extracted in chloroform, methanol and water, as described by Bligh and Dyer using 200 mg of dried microalgae biomass [44].

Fatty acid profile was analyzed according to the method (FAME) based on the AOCS official method [45]. Around 1 g of Bligh and Dyer extract was weighted in methylation glass tube and was evaporated to dryness under a gentle stream of nitrogen. 100 μL of internal standard solution (2% *w/v* C23:0 in heptane), 200 μL of heptane with BHT (0.01% *w/v*), 100 μL of toluene and 1 mL of borontriflouride in methanol (BF_3-MeOH) was added. Samples were mixed and methylated in microwave oven (Microwave 3000 SOLV, Anton Paar, Ashland, VA, USA) for 10 min at 100 °C and power of 500 watts and then cooled down for 5 min. 1 mL of saturated salt water (NaCl) and 0.7 mL of heptane with BHT were added. After the separation of heptane the upper phase of the sample (around 0.7 mL) was transferred into vials. Samples were analyzed by gas chromatography system (HP-5890 A, Agilent Technologies, Santa Clara, CA, USA). Fatty acid methyl esters were separated and detected by the GC column Agilent DB wax 127-7012 (10 μm × 100 μm × 0.1 μm), from Agilent technologies (Santa Clara, CA, USA).

Analysis of tocopherols and toco-trienols was done by using LC-FLD. 3 g of Bligh and Dyer extract were weighted and evaporated to dryness under a gentle stream of nitrogen. Dry sample was mixed with 1 mL of heptane and then transferred to HPLC vials. Analysis was done based on the AOCS official method [45] using Agilent 1100 Liquid Chromatograph (Agilent Technologies, Santa Clara, CA, USA), equipped with a fluorescence detector, with the excitation wavelength set at 290 nm and emission wavelength at 330. The separation was carried out by a Spherisorb column 150 mm × 46 mm × 3 μm particle size (Waters Corporation, Milford, MA, USA), using mixture of isopropanol and heptane (0.5:99.5) as mobile phase.

Pigments

Extraction and analysis of the pigments was done using the method described by Safafar et al. [34]. Samples were extracted by methanol containing BHT (butylated hydroxyl toluene) in sonication bath (Branson ultrasonics, Danbury, CA, USA) at a temperature lower than 5 °C for 15 min. Samples were analyzed by HPLC using Agilent 1100 Liquid Chromatograph equipped with a DAD. The separation was carried out in a Zorbax Eclipse C8 column 150 mm × 46 mm × 3.5 μm from Phenomenex. The mobile phase was a mixture of solvent A (70% methanol + 30% of 0.028 M tertiary butyl ammonium acetate in water) and solvent B (methanol) at a flow rate of 1.1 $mL{\cdot}min^{-1}$. Total acquisition time was 40 min. Identification of peaks and calibration was done by individual standard for each pigment. Detection for carotenoids, chlorophylls and internal standard (BHT) was done at 440 nm, 660 nm, and 280 nm, respectively.

Statistical Analysis

All experiments were repeated two times independently, and data were recorded as the mean. Results were evaluated using ANOVA to test the effect of time and growth media for lab scale and time for large scale experiments. Multiple comparison procedure based on Fisher's least significant difference (LSD) was used to discriminate among the means at the 95.0% confidence level. All statistical analyses were done by STATGRAPHICS software, version Centurion XVI (Stat point Technologies Inc., Warrenton, VA, USA).

CONCLUSIONS

Cultivation of *N. salina* in a mixture of industrial process water and

F/2 standard growth media enhanced the EPA content compared to F/2 growth media. Large-scale cultivation of *Nannochloropsis salina* in a flat panel photo-bioreactor confirmed the laboratory-scale findings. The algae growth rate at winter condition of Denmark was slow, but results revealed that large-scale cultivation of *Nannochloropsis salina* in these conditions could improve the nutritional properties such as EPA, protein, α-tocopherol and carotenoids compared to culturing in lab scale. *Nannochloropsis salina* is a good candidate for large-scale cultivation in the autumn–winter climate in Denmark. Resulting biomass from *Nannochloropsis salina* could be a good source of EPA, amino acid, tocopherols and carotenoids for the aquaculture feed industry. The biomass could also be a rich source of these bioactive compounds for the food industry, but this may require further extraction and fractionation of the biomass.

ACKNOWLEDGMENTS

The authors wish to acknowledge Inge Holmberg, Anita Ljubic and Patrick Uldall Nørregaard for their technical assistance. This study was funded by the Green Development and Demonstration program (GUDP) under the Danish Food 371 Ministry and E4Water under EU's 7th framework programme. We would like to thank Kalundborg WWTP (Kalundborg Utilities) and Novozymes A/S for providing access to industrial wastewater.

REFERENCES

1. Mostafa, S. Microalgal biotechnology: Prospects and applications. In Agricultural and Biological Sciences; Dhal, N.K., Sahu, S.C., Eds.; InTech Open publication: Rijeka, Croatia, 2012.
2. Benemann, J.R. CO2 mitigation with microalgae systems. Energy Convers. Manag. 1997, 38, 475–479.
3. Se-Kwon, K. Chapter 1: Marine microalgae biotechnology present trends and future advances. In Handbook of Marine Microalgae; Elsevier Publications: Amsterdam, The Netherlands, 2015; p. 27.
4. Becker, E.W. Microalgae for aquaculture: Nutritional Aspects. In Handbook of Microalgal Culture: Applied Phycology Biotechnology; Wiley-Blackwell: Hoboken, NJ, USA, 2013; pp. 671–691.

5. Mourente, G.; Lubian, L.M.; Odriozola, J.M. Total fatty acid composition as a taxonomic index of some marine microalgae used as food in marine aquaculture. Hydrobiologia 1990, 203, 147–154.
6. Roy, S.S.; Pal, R. Microalgae in Aquaculture: A Review with Special References to Nutritional Value and Fish Dietetics. Proc. Zool. Soc. 2015, 68, 1–8.
7. Muller-Feuga, A. Microalgae for aquaculture: The current global situation and future trends. In Handbook of Microalgal Culture: Applied Phycology Biotechnology; Wiley-Blackwell: Hoboken, NJ, USA, 2013; pp. 613–627.
8. Muller-Feuga, A. The role of microalgae in aquaculture: Situation and trends. J. Appl. Phycol. 2000, 12, 527–534.
9. Fawley, K.P.; Fawley, M.W. Observations on the Diversity and Ecology of Freshwater Nannochloropsis(Eustigmatophyceae), with Descriptions of New Taxa. Protist 2007, 158, 325–336.
10. Andersen, R.A.; Brett, R.W.; Potter, D.; Sexton, J.P. Phylogeny of the Eustigmatophyceae Based upon 18S rDNA, with Emphasis on Nannochloropsis. Protist 1998, 149, 61–74.
11. Hoffmann, M.; Marxen, K.; Schulz, R.; Vanselow, K.H. TFA and EPA productivities of Nannochloropsis salinainfluenced by temperature and nitrate stimuli in turbidostatic controlled experiments. Mar. Drugs 2010, 8, 2526–2545.
12. Ma, Y.; Wang, Z.; Yu, C.; Yin, Y.; Zhou, G. Evaluation of the potential of 9 Nannochloropsis strains for biodiesel production. Bioresour. Technol. 2014, 167, 503–509.
13. Hu, H.; Gao, K. Response of growth and fatty acid compositions of Nannochloropsis sp. to environmental factors under elevated CO2 concentration. Biotechnol. Lett. 2006, 28, 987–992.
14. Sforza, E.; Bertucco, A.; Morosinotto, T.; Giacometti, G.M. Photo-bioreactors for microalgal growth and oil production with Nannochloropsis salina: From lab-scale experiments to large-scale design. Chem. Eng. Res. Des. 2012, 90, 1151–1158.
15. Tibaldi, E.; Zittelli, G.C.; Parisi, G.; Bruno, M.; Giorgi, G.; Tulli, F.; Venturini, S.; Tredici, M.R.; Poli, B.M. Growth performance and quality traits of European sea bass (D. labrax) fed diets including increasing levels of freeze-dried Isochrysis sp. (T-ISO) biomass as a

source of protein and n-3 long chain PUFA in partial substitution of fish derivatives. Aquaculture 2015, 440, 60–68.

16. Zitteli, G.C.; Lavista, F.; Bastianini, A.; Rodolfi, L.; Vincenzini, M.; Tredici, M.R. Production of eicosapentaenoic acid by Nannochloropsis sp. cultures in outdoor tubular photo-bioreactors. J. Biotechnol. 1999, 70, 299–312.
17. Zitteli, G.C.; Pastorelli, R.; Tredici, M.R. A Modular Flat Panel Photo-bioreactor (MFPP) for indoor mass cultivation of Nannochloropsis sp. under artificial illumination. J. Appl. Phycol. 2000, 12, 521–526.
18. Olofsson, M.; Lamela, T.; Nilsson, E.; Bergé, J.P.; del Pino, V.; Uronen, P.; Legrand, C. Seasonal variation of lipids and fatty acids of the microalgae Nannochloropsis oculata grown in outdoor large-scale photo-bioreactors. Energies 2012, 5, 1577–1592.
19. Zou, N.; Richmond, A. Effect of light-path length in outdoor fiat plate reactors on output rate of cell mass and of EPA in Nannochloropsis sp. Prog. Ind. Microbiol. 1999, 35, 351–356.
20. Olofsson, M.; Lamela, T.; Nilsson, E.; Bergé, J.P.; del Pino, V.; Uronen, P.; Legrand, C. Combined effects of nitrogen concentration and seasonal changes on the production of lipids in Nannochloropsis oculata. Mar. Drugs 2014, 12, 1891–1910.
21. Camacho-Rodríguez, J.; González-Céspedes, A.M.; Cerón-García, M.C.; Fernández-Sevilla, J.M.; Acién-Fernández, F.G.; Molina-Grima, E. A quantitative study of eicosapentaenoic acid (EPA) production by Nannochloropsis gaditanafor aquaculture as a function of dilution rate, temperature and average irradiance. Appl. Microbiol. Biotechnol. 2014, 98, 2429–2440.
22. San Pedro, A.; González-López, C.V.; Acién, F.G.; Molina-Grima, E. Outdoor pilot production of Nannochloropsis gaditana: Influence of culture parameters and lipid production rates in raceway ponds. Algal Res. 2015, 8, 205–213.
23. De Sousa, L.L.; da Hora, D.S.; Sales, E.A.; Perelo, L.W. Cultivation of Nannochloropsis sp. in brackish groundwater supplemented with municipal wastewater as a nutrient source. Braz. Arch. Biol. Technol. 2014, 57, 171–177.
24. Cai, T.; Park, S.Y.; Racharaks, R.; Li, Y. Cultivation of Nannochloropsis salina using anaerobic digestion effluent as a

nutrient source for biofuel production. Appl. Energy 2013, 108, 486–492.

25. Dong, B.; Ho, N.; Ogden, K.L.; Arnold, R.G. Cultivation of Nannochloropsis salina in municipal wastewater or digester centrate. Ecotoxicol. Environ. Saf. 2014, 103, 45–53
26. Emdadi, D.; Berland, B. Variation in lipid class composition during batch growth of Nannochloropsis salina and Pavlova lutheri. Mar. Chem. 1989, 26, 215–225.
27. Guillard, R.R.L.; Ryther, J.H. Studies of marine planktonic diatoms. Can. J. Microbiol. 1962, 8, 229–239.
28. Hii, Y.S.; Soo, C.L.; Chuah, T.S.; Mohd-Azmi, A.; Abol-Munafi, A. Interactive effect of ammonia and nitrate on the nitrogen uptake by Nannochloropsis sp. J. Sustain. Sci. Manag. 2011, 6, 60–68.
29. Hii, Y.S.; Soo, C.L.; Chuah, T.S.; Mohd-Azmi, A.; Abol-Munafi, A. The effects of temperature on the growth of and ammonia uptake by marine microalgae. Bot. Stud. 2012, 53, 125–133.
30. Hecky, R.E.; Kilham, P. Nutrient limitation of phytoplankton in freshwater and marine environments: A review of recent evidence on the effects of enrichment. Limnol. Oceanogr. 1988, 33, 796–822.
31. Collos, Y.; Harrison, P.J. Acclimation and toxicity of high ammonium concentrations to unicellular algae. Mar. Pollut. Bull. 2014, 80, 8–23.
32. Hu, Q. Environmental effects on cell composition. In Handbook of Microalgal Culture: Applied Phycology Biotechnology; Wiley-Blackwell: Hoboken, NJ, USA, 2013; pp. 114–122.
33. Chen, Y.; Tang, X.; Kapoore, R.V.; Xu, C.; Vaidyanathan, S. Influence of nutrient status on the accumulation of biomass and lipid in Nannochloropsis salina and Dunaliella salina. Energy Convers. Manag. 2015, 106, 62–72.
34. Safafar, H.; van Wagenen, J.; Møller, P.; Jacobsen, C. Carotenoids, phenolic compounds and tocopherols contribute to the antioxidative properties of some microalgae species grown on industrial wastewater. Mar. Drugs 2015, 13, 7339–7356.
35. Durmaz, Y. Vitamin E (α-tocopherol) production by the marine microalgae Nannochloropsis oculata(Eustigmatophyceae) in

nitrogen limitation. Aquaculture 2007, 272, 717–722.

36. Solovchenko, A.; Khozin-Goldberg, I.; Recht, L.; Boussiba, S. Stress-Induced Changes in Optical Properties, Pigment and Fatty Acid Content of Nannochloropsis sp.: Implications for Non-destructive Assay of Total Fatty Acids. Mar. Biotechnol. 2011, 13, 527–535.
37. Hoshida, H.; Ohira, T.; Minematsu, A.; Akada, R.; Nishizawa, Y. Accumulation of eicosapentaenoic acid in Nannochloropsis sp. in response to elevated CO2 concentrations. J. Appl. Phycol. 2005, 17, 29–34.
38. Brown, M.R. The amino-acid and sugar composition of 16 species of microalgae used in mariculture. J. Exp. Mar. Biol. Ecol. 1991, 145, 79–99.
39. Owens, T.G.; Gallager, J.C.; Alberte, R.S. Photosynthetic light harvesting function of viloaxanthin in Nannochloropsisspp. (Eustigmatophyceae). J. Phycol. 1987, 23, 79–85.
40. Mulders, K.J.M.; Lamers, P.P.; Martens, D.E.; Wijffels, R.H. Phototrophic Pigment Production with Microalgae: Biological Constraints and Opportunities. J. Phycol. 2014, 50, 229–242.
41. Rebolloso-Fuentes, M.M.; Navarro-Pérez, A.; García-Camacho, F.; Ramos-Miras, J.J.; Guil-Guerrero, J.L. Biomass nutrient profiles of the microalga Nannochloropsis. J. Agric. Food Chem. 2001, 49, 2966–2972. [PubMed]
42. Dragsted, J.; Furbo, S. Solar Radiation and Thermal Performance of Solar Collectors for Denmark. Available online: http://orbit.dtu.dk/files/51295325/Solar_radiation_and_thermal_performance_of_solar_collectors_for_Denmark.pdf(accessed on 28 July 2016).
43. Chen, Y.; Vaidyanathan, S. Simultaneous assay of pigments, carbohydrates, proteins and lipids in microalgae. Anal. Chim. Acta 2013, 776, 31–40.
44. Bligh, E.G.; Dyer, W.J. A rapid method of total lipid extraction and purification. Can. J. Biochem. Physiol. 1959, 37, 911–917.
45. Firestone, D. Official Methods and Recommended Practices of the AOCS, 5th ed.; American Oil Chemist's Society: Urbana, IL, USA, 2009.

CHAPTER

6

ASTAXANTHIN-PRODUCING GREEN MICROALGA HAEMATOCOCCUS PLUVIALIS: FROM SINGLE CELL TO HIGH VALUE COMMERCIAL PRODUCTS

Md. Mahfuzur R. Shah[1], Yuanmei Liang1Jay J. Cheng[1,2] and Maurycy Daroch[1]

[1]School of Environment and Energy, Peking University, Shenzhen Graduate School, Shenzhen, China

[2]Department of Biological and Agricultural Engineering, North Carolina State University, Raleigh, NC, USA

Many species of microalgae have been used as source of nutrient rich food, feed, and health promoting compounds. Among the commercially important microalgae, Haematococcus pluvialis is the richest source of natural astaxanthin which is considered as "super anti-oxidant." Natural astaxanthin produced by H. pluvialis has significantly greater antioxidant capacity than the synthetic one. Astaxanthin has important applications in the nutraceuticals, cosmetics, food, and aquaculture industries. It is now evident that, astaxanthin can significantly reduce free radicals and oxidative stress and help human body maintain a healthy state. With extraordinary potency and increase in demand, astaxanthin is one of the high-value microalgal products of the future.This comprehensive review summarizes the most important aspects of the biology, biochemical composition, biosynthesis, and astaxanthin accumulation in the cells of H. pluvialis and its wide range of applications for humans and animals. In this paper, important and recent developments ranging from cultivation, harvest and postharvest bio-processing technologies to metabolic control and genetic engineering are reviewed in detail, focusing on biomass and astaxanthin production from this biotechnologically important microalga. Simultaneously, critical bottlenecks and major challenges in commercial scale production; current and prospective global market of H. pluvialis derived astaxanthin are also presented in a critical manner. A new biorefinery concept for H. pluvialis has been also suggested to guide toward economically sustainable approach for microalgae cultivation and processing. This report could serve as a useful guide to present current status of knowledge in the field and highlight key areas for future development of H. pluvialis astaxanthin technology and its large scale commercial implementation.

INTRODUCTION

"Green microalgae" comprise more than 7000 species growing in a variety of habitats. Haematococcus pluvialis (Chlorophyceae, Volvocales) is unicellular fresh water microalga distributed in many habitats worldwide. It is considered as the best natural source of astaxanthin and the main producing organism of this commercial product (Lorenz, 1999; Ranga Rao et al., 2010). Astaxanthin (3,3′-dihydroxy-ß-carotene-4,4′-dione) is a bright red secondary carotenoid from the same family as lycopene, lutein, and β-carotene, synthesized de novo by some microalgae, plants, yeast, bacteria, and present in many of our favored seafood including salmon, trout, red sea bream, shrimp, lobster, and fish eggs (Higuera-Ciapara et al., 2006; Ranga Rao et al., 2014). Astaxanthin contains two chiral centers and can exist in three different stereoizomers, (3S, 3′S); (3R, 3′S), and (3R, 3′R). The ratio of 1:2:1 of these isomers is obtained during chemical synthesis of this compound. H. pluvialis biosynthesizes predominantly the 3S, 3′S stereoisomer, the most valuable one (Yang et al., 2013; Al-Bulishi et al., 2015). Astaxanthin synthesis in H. pluvialis is directly correlated in space and time with deposition of cellular reserves in lipid droplets under conditions of cellular stress (Solovchenko, 2015). From biochemical perspective astaxanthin is synthesized through carotenoid pathway from glyceraldehyde-3-phosphate and pyruvate. Both of these compounds are products of photosynthesis and/or glycolysis depending on cultivation conditions. These two key metabolic intermediates then enter non-mevalonate (MEP) pathway to generate IPP—key intermediate for the synthesis of all carotenoids including astaxanthin. Astaxanthin has a wide range of applications in the food, feed, cosmetic, aquaculture, nutraceutical, and pharmaceutical industries because of its free radical scavenging capacity. In terms of antioxidant activity astaxanthin is 65 times more powerful than vitamin C, 54 times stronger than β-carotene, 10 times more potent than β-carotene, canthaxantin, zeaxanthin, and lutein; and 100 times more effective than α-tocopherol (Miki, 1991; Borowitzka, 2013; Koller et al., 2014; Pérez-López et al., 2014; Cyanotech, 2015). Currently, over 95% of the astaxanthin available in the market is produced synthetically; while H. pluvialis derived natural astaxanthin corresponds to < 1% of the commercialized quantity (Koller et al., 2014). Synthetic astaxanthin, synthesized from asta-C15 -triarylphosphonium salt and the C10 -dialdehyde in a Wittig reaction (Krause et al., 1997), has 20 times lower antioxidant capacity than its natural counterpart and to date has not been approved for human consumption (Lorenz and Cysewski, 2000; Koller et al., 2014). Moreover, there are concerns about the safety of using synthetic astaxanthin for direct human consumption due to both different stereochemistry and potential carryover of synthesis intermediates. These concerns make natural astaxanthin from H. pluvialis a preferred choice for high-end markets (Li et al., 2011). In addition, H. pluvialis has been already approved as a color additive in salmon feeds and as a dietary-supplement ingredient for human consumption in the USA, Japan, and several European countries (Yuan et al., 2011). Nevertheless, there is no EFSA (European Food Safety Authority) approval for the therapeutic application so far. Supercritical CO2 extracts from H. pluvialis have been granted "novel food" Status by the UK Food Standards Agency (FSA), whilst US FDA (Food and Drug Administration) granted astaxanthin from H. pluvialis "GRAS" status (Generally Recognized As Safe) (Grewe and Griehl, 2012; Capelli and Cysewski, 2013). The increasing demand for natural astaxanthin and its high price raises interest in efficient systems to produce astaxanthin from H. pluvialis. Various cultivation and astaxanthin production methods in photoautotrophic, heterotrophic, and mixotrophic growth conditions, both indoors; in open raceway ponds or closed photobioreactors using batch or fed-batch approach have been reported (Kang et al., 2005, 2010; Kaewpintong et al., 2007; Ranjbar et al., 2008; García-Malea et al., 2009; Issarapayup et al., 2009; Zhang et al., 2009; Li et al., 2011; Han et al., 2013; Wang et al., 2013a,b). The most recent advances in H. pluvialis cultivation for astaxanthin production include a two-stage mixotrophic culture system (Park et al., 2014) and an "attached cultivation" system using the immobilized biofilm (Zhang et al., 2014). Along with the cultivation process, the induction of carotenoid synthesis in H. pluvialis has a direct correlation with the astaxanthin content of cells and total astaxanthin productivity. The accumulation of astaxanthin is affected by both environmental factors such as light (Saha et al., 2013; Park et al., 2014); temperature (Yoo et al., 2012); pH (Hata et al., 2001); salt concentration (Kobayashi et al., 1993); and nutritional stresses (Boussiba et al., 1999; Chekanov et al., 2014), as well as various plant hormones and their derivatives (Yu et al., 2015). Attempts were made to genetically enhance the capacity of H. pluvialisto produce astaxanthin using both classical mutagenesis (Hu et al., 2008), and more recently genetic engineering (Sharon-Gojman et al., 2015). Once biomass is successfully grown and achieved high cell density, efficient harvesting, cell disruption, dehydration, and recovery/extraction of astaxanthin from H. pluvialis biomass are important factors. Harvesting have been so far achieved through a combination of passive settling and subsequent centrifugation (Han et al., 2013; Pérez-López et al., 2014), or flotation and centrifugation (Panis, 2015). Mechanical processes such as expeller pressing and bead milling are commonly used cell disruption methods to enhance recovery of astaxanthin from H. pluvialis (Razon and Tan, 2011). For the dehydration of H. pluvialis biomass, spray drying (Li et al., 2011; Panis, 2015), freeze drying, lyophilization, and cryodesiccation (Milledge, 2013) methods have been utilized. There are several methods of astaxanthin extraction utilizing solvents, acids, edible oils, enzymes, pressurized liquids (Jaime et al., 2010; Zou et al., 2013; Dong et al., 2014), supercritical carbon dioxide (SC-CO2) (Wang et al., 2012; Reyes et al., 2014), and liquefied dimethyl ether (DME) (Boonnoun et al., 2014).

Although many studies have explored various methods of extraction and downstream processing of astaxanthin from H. pluvialis biomass, more comprehensive investigation is required to take an advantage of the biological potential of this microalga and its highly valuable product. Since astaxanthin has a great potential in the global market (280 mt, $447 million in 2014 for both synthetic and natural astaxanthin) and high market value ($2500–7000/kg); (Koller et al., 2014; Pérez-López et al., 2014; Industry Experts, 2015), in depth investigation of H. pluvialis biology, physiology, efficient culture techniques, downstream bioprocessing, and product formation are highly desired for further development of this sector. Even though currently several commercial companies (Cyanotech Corporation, Mera Pharmaceuticals Inc, Algatechnologies, Fuji Chemical Industry Co. Ltd etc.) are involved in large scale production of H. pluvialis and astaxanthin, the production capacity is far beyond the global demand of natural astaxanthin.

This review summarizes both classical knowledge and most recent advances in the cell biology, physiological, and biochemical characteristics, responses to environmental stresses, and their effect on astaxanthin accumulation, genetic engineering, growth conditions, and different cultivation techniques, harvesting, and post harvest downstream bioprocessing of H. pluvialis. The biorefinery concept, global market potential, challenges, and future direction for development of H. pluvialis and astaxanthin production in commercial scale also are discussed.

BIOLOGY OF H. PLUVIALIS

Taxonomy and Occurrence

The freshwater unicellular biflagellate green microalgae H. pluvialis Flotow belongs to the class Chlorophyceae, order Volvocales and family Haematococcaseae (Bold and Wynne, 1985; Eom et al., 2006). It is also known as Haematococcus lacustris or Sphaerella lacustris. Haematococcus was first described by J. Von Flotow in 1844 and later in 1899 Tracy Elliot Hazen extensively presented its biology and life cycle (Hazen, 1899; Leonardi et al., 2011). H. pluvialis is common in small transient freshwater bodies and widely distributed in many habitats worldwide. It occurs primarily in temporary water bodies like ephemeral rain pools, artificial pools, natural and man-made ponds, and birdbaths

(Czygan, 1970; Burchardt et al., 2006). This microalga can be usually found in temperate regions around the world and has been isolated from Europe, Africa, North America, and Himachal Pradeslv India (Pringsheim, 1966; Suseela and Toppo, 2006). It has been also found across diverse environmental and climate conditions: in brackish water on the rocks on the seashore (Chekanov et al., 2014); freshwater basin in the rock filled with melted snow on Blomstrandhalvøya Island (Norway) (Klochkova et al., 2013); dried fountain near Rozhen, Blagoevgrad in Bulgaria Gacheva et al., 2015, freshwater fishpond in Bihor, Romania (Dragos et al., 2010); rooftop surface of a building of KIOST in Seoul Korea (Kim et al., 2015). It is well suited for survival under conditions of expeditious and extreme in light, temperature, and salt concentration that would be deleterious to many other microalgae, due to its ability to encyst (become enclosed by thick membrane) in a rapid manner (Proctor, 1957).

Cellular Morphology and Life Cycle

Cellular structure of H. pluvialis is similar to most of other members of volvocalean unicellular green algae. The life cycle of H. pluvialis consists of four types of distinguishable cellular morphologies: macrozooids (zoospores), microzooids, palmella, and hematocysts (aplanospores) (Hazen, 1899; Elliot, 1934). Macrozooids (zoospores), microzooids, and palmella stages are usually called "green vegetative phase" (Figures 1A,B). Hematocysts (aplanospores) are referred as "red nonmotile astaxanthin accumulated encysted phase" of the H. pluvialis life cycle (Figures 1C,D). Macrozooids (zoospores) are spherical, ellipsoidal, or pear-shaped cells with two flagella of equal length emerging from anterior end, and a cup-shaped chloroplast with numerous, scattered pyrenoids (Figure 1A). The macrozooid cells are between 8 and 20 μm long with a distinct gelatinous extracellular matrix of variable thickness. Numerous contractile vacuoles are irregularly distributed near the protoplast surface of the cell (Hagen et al., 2002). These flagellated fast-growing vegetative cells predominate under favorable culture conditions in the early vegetative growth stage (Figure 1A) Macrozooids may divide into 2–32 daughter cells by mitosis (Wayama et al., 2013) (Figures 2A,B). Under unfavorable environmental or culture conditions, macrozooids start losing flagella, and expand their cell size. They form an amorphous multilayered structure in the inner regions of the extracellular matrix or the primary cell wall as they develop into non-motile "palmella" and become resting vegetative cells (Hagen et al., 2002) (Figure 1B). With the continued environmental stress (i.e., nutrient deprivation, high light irradiance, high salinity) and cessation of cell division, "palmella" transform into the asexual "aplanospores" (Figures 1C,D). At this stage, cells contain two distinct structures, a thick and rigid trilaminar sheath, and secondary cell wall of acetolysis-resistant material. Such cells become resistant to prevailing extreme environmental conditions (Santos and Mesquita, 1984; Boussiba and Vonshak, 1991). Mature aplanospores; accumulate large amounts of secondary carotenoids, particularly astaxanthin, in lipid droplets deposited in the cytoplasm, which results in a characteristic bright red color of these cells (Hagen et al., 2002). Some H. pluvialis strains are reported to be capable of accumulating astaxanthin without forming aplanospores (Brinda et al., 2004). Once environmental or culture conditions return to optimal, red aplanospores germinate to form flagellated zoospores to initiate a new vegetative growth cycle (Figure 1A). In some cases, gametogenesis may occur in aplanospores (Figures 3A–D). Such process requires an exposure to extreme adverse conditions (freezing, desiccation, or nutrient starvation) followed by return to favorable culture conditions. During gametogenesis, aplanospore cells can produce up to 64 gametes which are known as microzooids. The microzooids are smaller in size (< 10 μm) compared to the zoospores (20–50 μm), and exhibit high-speed motility after their release from gametocysts. Sexual reproduction is rarely observed in H. pluvialis, and has been largely replaced by vegetative reproduction (Triki et al., 1997).

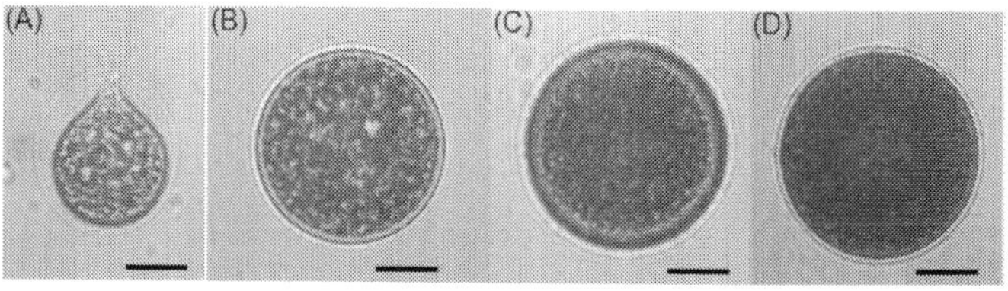

Figure 1: Light microscopic images of H. pluvialis cells in life cycle. (A) Green vegetative motile cell; (B) Green vegetative palmella cell; (C) Astaxanthin accumulating palmella cell in transition to aplanospore; (D) Astaxanthin accumulated aplanospore cell. Scale bar: 10 μm.

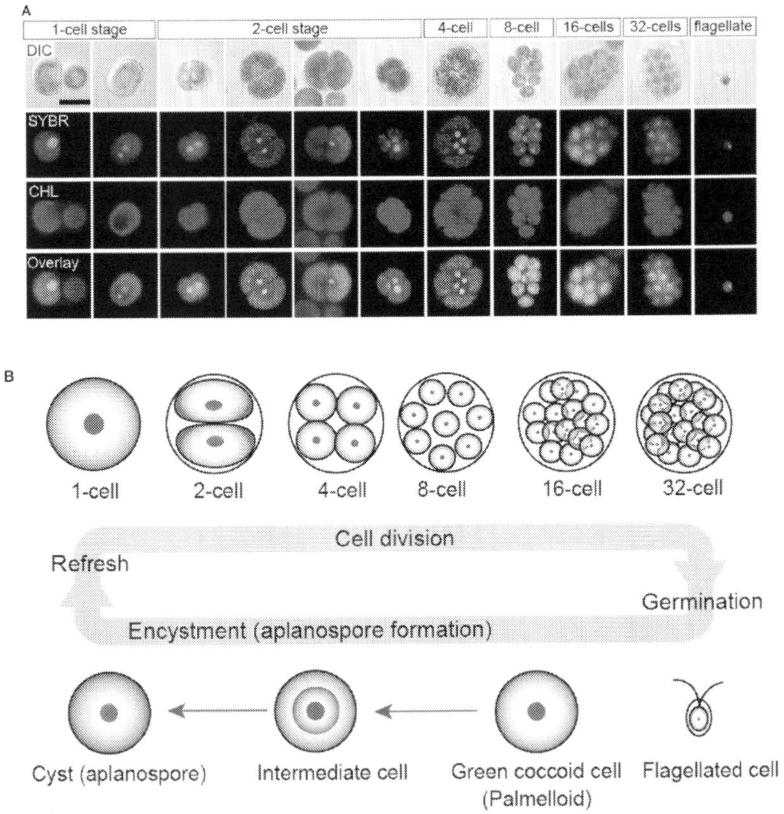

Figure 2: Life cycle of H. pluvialis. (A) Fluorescence microscopy images, showing the 1- to 32-cell stages, and the flagellated stage. DIC, differential interference contrast image; SYBR, SYBR Green I-stained cells (green); CHL, chlorophyll autofluorescence (red); and Overlay, overlaid images of SYBR and CHL. (B) Illustration of life cycle of H. pluvialis. Refresh, when old cultures are transplanted into fresh medium, coccoid cells undergo cell division to form flagellated cells within the mother cell wall. Germination, Flagellated cells settle and become coccoid cells. Continuous and/or strong light accelerate the accumulation of astaxanthin during encystment (red arrows). Figure reproduced from Wayama et al. (2013) distributed under the terms of the Creative Commons Attribution License.

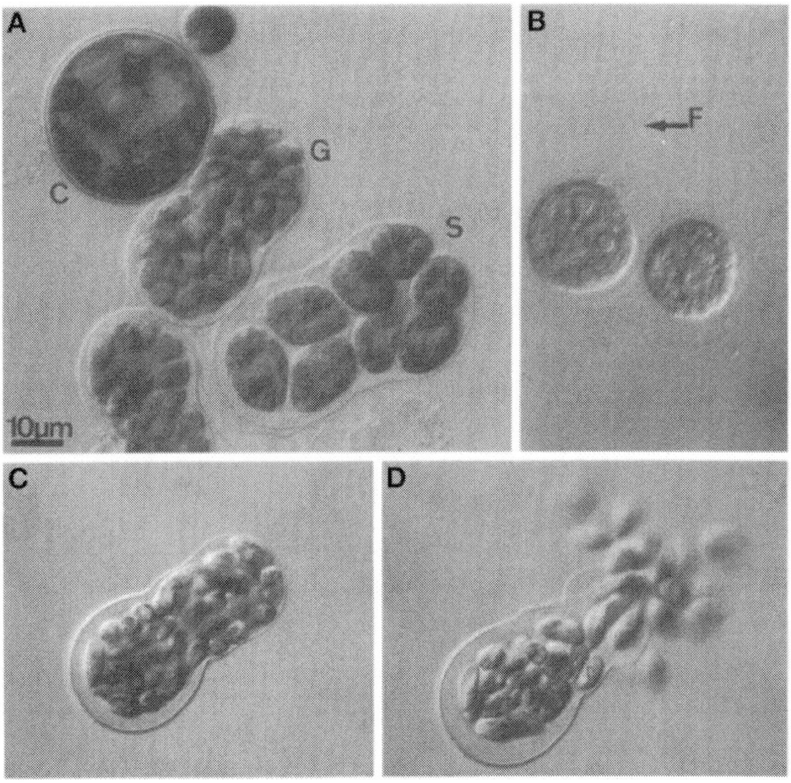

Figure 3: Gametogenesis in H. pluvialis. (A) C, cyst; G, gametocyst; S, a sporocyst; (B)vegetative zoospore, F, flagella (indicated by arrow); (C): gametocyst before releasing gametes; (D): release of gametes from gametocyst. Reproduced with permission from Triki et al., 1997, Phycologia, Allen Press Publishing Services. Copyright (1997) Allen Press Publishing Services.

Ultrastructural Changes of H. pluvialis during the Life Cycle

In H. pluvialis cells, large amount of ultrastructural changes occurs during their life cycle which is frequently associated with responses to stress conditions in the environment. In the green vegetative palmella cells, a thick cell wall surrounds the cell, and two layers of extracellular matrix present near the cell wall (Figure 4A). Nucleus is located in the center of the cell, and highly developed chloroplasts are located at the periphery (Figure 4A). Few astaxanthin granules are present which located around the nucleus (Figure 4A). Conspicuous pyrenoids with electron-dense matrix can be found in the stroma (Figures 4A,B) (Wayama et al., 2013). During the starting of encystment process, Haematococcus turned into greenish-orange cells (with some astaxanthin accumulation) which can be referred as intermediate stage cells. In this stage, conspicuous pyrenoids with electron-dense matrix located in the stroma, remain surrounded by thick starch capsules and many starch grains are located around the pyrenoids (Figure 5). Round oil droplets with various sizes, containing astaxanthin located around the nucleus (Figure 5). At this stage, thylakoids become partial degraded (Wayama et al., 2013). With the gradual accumulation of astaxanthin chloroplast reduce in volume but photosynthetic activity remains. In the aplanosopre or cyst stage, astaxanthin accumulates and cells form cysts. Oil droplets and astaxanthin accumulation patterns may differ among cyst cells. For example, electron-dense astaxanthin granules in oil droplets (Figures 6A,B) occurred in some cells. In other cells, relatively large oil droplets occurred throughout the cell (Figures 6C,D). Chloroplasts are degenerated and localized in the interspace between oil droplets (Wayama et al., 2013).

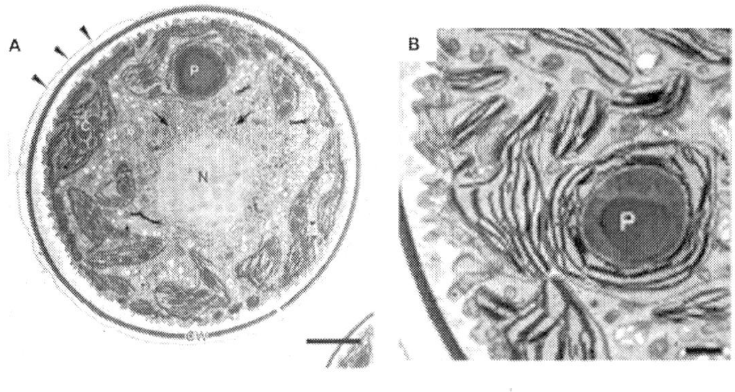

Figure 4: Transmission electron micrographs of green vegetative cells of H. pluvialis. (A) General ultrastructure. The cell wall is surrounded by extracellular matrix (arrowheads). Arrows indicate astaxanthin granules. (B) Chloroplast and pyrenoid. C, chloroplast; CW, cell wall; N, nucleus; P, pyrenoid. Scale bars in (A,B): 5 and 1 μm, respectively. Figure reproduced from Wayama et al. (2013) distributed under the terms of the Creative Commons Attribution License.

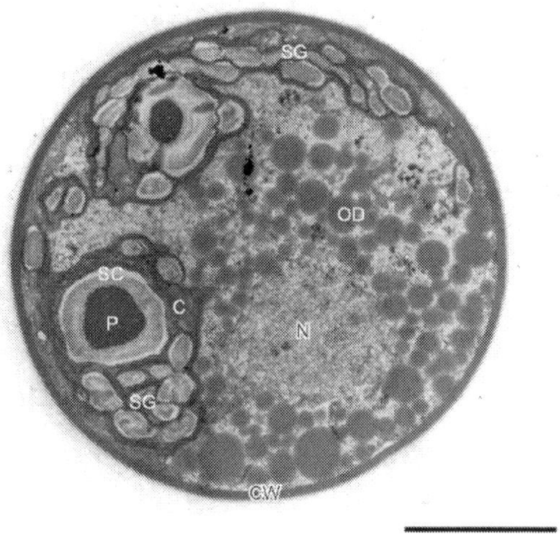

Figure 5. Transmission electron micrograph of intermediate H. pluvialis cell (general ultrastructure). C, chloroplast; CW, cell wall; N, nucleus; OD; oil droplet; P, pyrenoid; SC, starch capsule; SG, starch grain. Scale bar: 5 μm. Figure reproduced from Wayama et al. (2013) distributed under the terms of the Creative Commons Attribution License.

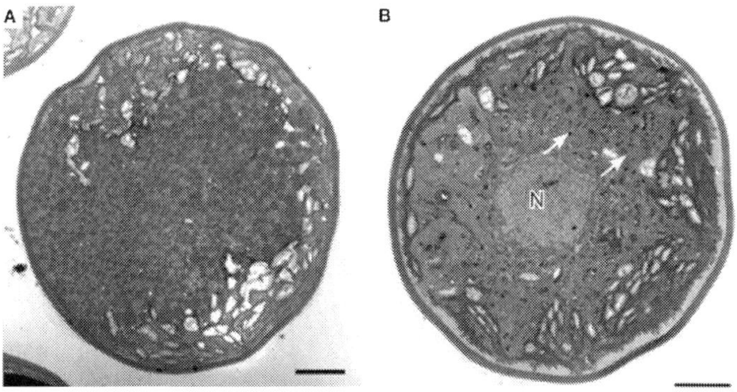

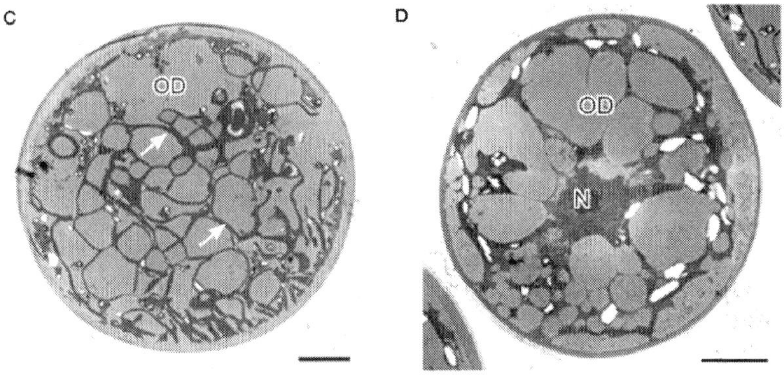

Figure 6: Transmission electron micrographs of H. pluvialis cyst cells. (A) General ultrastructure of cyst cells, showing small granules that contain astaxanthin. (B) General ultrastructure of a cyst cell, showing astaxanthin accumulation in oil droplets. (C) General ultrastructure of a cyst cell, showing large oil droplets. Chloroplasts localize in the interspace between oil droplets (arrows). (D) Some oil droplets are fused. C, chloroplast; N, nucleus; OD, oil droplet. Scale bars in (A–D): 5 μm. Figure reproduced from Wayama et al. (2013) distributed under the terms of the Creative Commons Attribution License.

BIOCHEMICAL COMPOSITION OF H. PLUVIALIS

Because of the unique life cycle of H. pluvialis, cellular composition of this microalga varies tremendously between its "green" and "red" stages of cultivation (Table 1).

Table 1: Typical composition of H. pluvialis biomass in green and red cultivation stages.

Composition content (% of DW)	Green stage	Red stage
Proteins	29–45	17–25
Lipids (% of total)	20–25	32–37
Neutral lipids	59	51.9–53.5
Phospholipids	23.7	20.6–21.1
Glycolipids	11.5	25.7–26.5
Carbohydrates	15–17	36–40
Carotenoids (% of total)	0.5	2–5
Neoxanthin	8.3	n.d
Violaxanthin	12.5	n.d
β -carotene	16.7	1.0
Lutein	56.3	0.5
Zeaxanthin	6.3	n.d
Astaxanthin (including esters)	n.d	81.2
Adonixanthin	n.d	0.4
Adonirubin	n.d	0.6
Canthexanthin	n.d	5.1
Echinenone	n.d	0.2
Chlorophylls	1.5–2	0

Adapted from Grewe and Griehl (2012). n.d., no data.

Protein

In green stage, during favorable growth conditions most H. pluvialis strains are rich in protein (29–45) (Table 1), lower protein content (23.6%) have been however observed in a Bulgarian strain Haematococcus cf. pluvialis Rozhen-12 during green stage cultivation (Gacheva et al., 2015). It was estimated that proteins contribute to 21 (Kim et al., 2015) and 23% (Lorenz, 1999) of cellular content during red stage cultivation of H. pluvialis. Amino acid composition of proteins in the red stage indicated that proteins were mainly composed of aspartic acid, glutamic acid, alanine, and leucine with total amino acid content of 10.02/100 mg, 46.0% of which belonged to essential amino acids (Lorenz, 1999; Kim et al., 2015).

Carbohydrates

In green stage, carbohydrate content approximates to 15–17%, about a half of the red stage (Table 1). In the red stage, under conditions of stress (e.g., nutrient starvation, light stress, high acidity, temperature variations etc.), H. pluvialis accumulates higher content of carbohydrates (starch), for example 38 (Lorenz, 1999), 60 (Recht et al., 2012), and 74% (Boussiba and Vonshak, 1991). Under prolonged stress conditions starch is consumed in the cell.

Lipid

In green stage, total lipid content varies from 20 to 25%, with approximately 10% lipids composed predominantly of short (C16, C18) polyunsaturated fatty acids deposited in the chloroplasts. Neutral lipids are predominant lipid class in both green and red cells (Table 1). In red stage, prolonged stress conditions direct larger flux toward the synthesis of neutral lipids—triacylglycerols (TAG). Red stage cells can accumulate up to 40% of their cell weight as cytoplasmic lipid droplets (LD), and considerable amount of secondary metabolites including up to 4% of the ketocarotenoid astaxanthin (Boussiba et al., 1992, 1999; Saha et al., 2013). The phospholipid content does not change compared to the green stage, while the glycolipid fraction nearly doubles in red cells when compared with green vegetative cells (Damiani et al., 2010). The total fatty acid profile of H. pluvialis is relatively complex. Palmitic (16:0), linoleic (18:2), and linolenic (18:3) acids are predominant components of the profile with highly polyunsaturated species also present in considerable amounts (Table 2). Based on the comparative studies on fatty acids profile of two different H. pluvialis, it was revealed that both strains varied in composition, especially of palmitic (16:0), oleic (18:1), linoleic (18:2), and linolenic (18:3) acids. This variation might be associated with several factors such as culture environment, stress conditions, culture parameters, variation of strain origin, nutrient etc. Higher lipid content of H. pluvialis grown under nutrient starvation and the suitable profile of its fatty acids indicate a possibility of biodiesel production from this microalga (Damiani et al., 2010; Saha et al., 2013). The massive astaxanthin accumulation in H. pluvialis is a cellular response to stress conditions and is accompanied by the enhanced biosynthesis of triacylglycerols (TAG) (Zhekisheva et al., 2002, 2005; Cerón et al., 2007), and the reduction in photosynthetic activity of PSII, loss of cytochrome f, and subsequent reduction in electron transport, and increased respiration rate (Boussiba, 2000). During transition from green vegetative cells to red aplanospores after exposure to stress conditions astaxanthin start to accumulate as fatty acid mono- or diesters in cytoplasmic lipid droplets (LD) (Aflalo et al., 2007). As cells undergo transition to red stage, both chlorophyll and protein contents drop.

Table 2: Comparison of fatty acid composition (%) of two different H. pluvialis strains.

Fatty acids	*Haematococcus* sp. KORDI03 (Kim et al., 2015)	*H. pluvialis* (Lorenz, 1999)
C12:0 lauric	N/A	0.1
C14:0 myristic	0.1	0.5
C15:0 pentadecanoic acid	0.1	N/A
C16:0 palmitic	13.7	29.0
C16:1 palmitoleic	0.5	0.6
C16:2	0.4	N/A
C16:3	3.5	N/A
C16:4	3.3	N/A
C17:0 margaric	N/A	0.2
C17:1 margaroleic	N/A	1.3
C18:0 stearic	0.7	2.1
C18:1 oleic	4.9	25.9
C18:2 linoleic	24.9	20.8
C18:3 linolenic	39.7	12.8
C18:4 octadecatetraenoic	5.8	1.4
C20:0 arachidic	N/A	0.6
C20:1 gadoleic	0.5	0.3
C20:2 eicosadenoic	N/A	1.2
C20:3 eicosatrienoic gamma	N/A	0.5
C20:4 arachidonic	0.9	1.4
C20:5 eicosapentaenoic	0.6	0.6
C22:0 behenic	N/A	0.4
C24:0 lignoceric	0.3	0.2
C24:1 nervonic acid	0.1	0.1
∑ SFAs	15.0	33.2
∑ MUFAs	6.0	28.1
∑ PUFAs	79.1	38.7
Total	100.0	100.0

Carotenoid

The carotenoid fraction of green vegetative cells consists of mostly lutein (75–80%), β-carotene (10–20%) and others, including chlorophyll a and b, primary carotenoids, violaxanthin, neoxanthin, lactucaxanthin, and zeaxanthin (Renstrøm et al., 1981; Harker et al., 1996a). In the red stage, the total carotenoid content is markedly enhanced, and the characteristic primary carotenoid pattern of vegetative stage is replaced by secondary carotenoids, mainly astaxanthin (80–99% of total carotenoids) (Harker et al., 1996a; Dragos et al., 2010). The ratio of carotenoids to chlorophylls is about 0.2 in the green stage and increases in the red stage by an order of magnitude and reaches about 2–9. The majority of astaxanthin is not deposited in its free form but it exists within the cell as fatty acid esters of astaxanthin, usually mono- or diesters of palmitic (16:0), oleic (18:1), or linoleic (18:2) acids. This type of modification is required for the deposition of this highly polar molecule within non-polar matrix of lipid droplets. Approximately 70% monoesters, 25% diesters, and only 5% of the free ketocarotenoid is present in the mature "red" cells of H. pluvialis(Zhekisheva et al., 2002; Solovchenko, 2015). Under certain conditions of stress H. pluvialis has been shown to accumulate up to 3–5% DW of astaxanthin (Han et al., 2013; Chekanov et al., 2014).

H. PLUVIALIS-DERIVED ASTAXANTHIN

H. pluvialis as a Major Source of Astaxanthin

H. pluvialis can accumulate up to 5% DW of astaxanthin and is considered as the best natural source of this high-value carotenoid pigment (Wayama et al., 2013). Dietary supplements containing Haematococcus astaxanthin has proved to be safe to humans and widely used for over 15 years as a nutraceutical supplement with no adverse side-effects of its supplementation (Capelli and Cysewski, 2013; Yang et al., 2013). Natural astaxanthin from H. pluvialis or krill oil is available in the market as a dietary supplement in dosages from 3.8 to 7.6 mg per day due to potential health benefits (Yang et al., 2013). As societies nowadays are looking toward "green" solutions, natural astaxanthin form H. pluvialis seems to be more favorable than its synthetic counterpart due to structure, function, application, and security (Choubert and Heinrich, 1993; Capelli and Cysewski, 2013; Pérez-López et al., 2014).

Biosynthesis of Astaxanthin in H. pluvialis

Biosynthesis of astaxanthin in H. pluvialis is a complex process that is highly up-regulated in conditions of stress and which coincides with the accumulation of triacylglycerols (TAGs). Both compounds are deposited in the cytosolic lipid bodies during the "red" stage of H. pluvialis cultivation. Astaxanthin belongs to carotenoids, a C_{40} tetraterpenes, synthesized from isoprene units. Isopentenyl pyrophosphate (IPP) is a key intermediate of carotenoid synthesis. In principle, IPP can originate from two dissimilar pathways: mevalonate pathway (MVA) located in cytosol and non-mevalonate (MEP) located in the chloroplast (Lichtenthaler et al., 1997; Lichtenthaler, 1999; Eisenreich et al., 2001). Alternative name for MEP is DOXP, due to the formation of 1-deoxy-D-xylulose-5-phosphate in the first stage of the pathway. Comparative transcriptomic analysis of astaxanthin biosynthesis in H pluvialis have shown that the key intermediate-IPP is most likely synthesized using the DOXP pathway. H. pluvialis lacks three key enzymes of the mevalonate pathway (MVA) catalyzing the formation of isopentenyl pyrophosphate (IPP) from acetoacetyl-CoA (Gwak et al., 2014). There has been numerous evidence of the full set of enzymes required for the conversion of photosynthesis-derived products i.e., pyruvate and glyceraldehyde-3-phosphate into isopentenyl pyrophosphate through DOXP pathway inside H. pluvialis chloroplasts (Gwak et al., 2014). It makes it the most likely source of IPP in H. pluvialis cells. The process of astaxanthin biosynthesis is presented on Figure 7. IPP derived from DOXP pathway is an initial building block of astaxanthin synthesis. In the subsequent step the IPP undergoes isomerization to dimethylallyl diphosphate (DMAPP). It has been long assumed that this conversion was catalyzed exclusively by isopentenyl pyrophosphate isomerase (IPI) (Sun et al., 1998; Lichtenthaler, 1999). However, recent transcriptomic studies suggest that neither of the two ipi genes of H pluvialis (ipi1 and ipi2IPI2) are up-regulated during cellular accumulation of astaxanthin (Gwak et al., 2014). Suggestions have been made that another enzyme of similar activity, 4-hydroxy-3-methylbut-2-enyl diphosphate reductase (HDR) was more likely to be responsible for catalyzing interconversion between IPP and DMAPP (Hoeffler et al., 2002; Rohdich et al., 2002; Gwak et al., 2014). Further studies are required to assess the contribution of these three enzymes to this key biosynthetic step of astaxanthin formation. Elongation of the isoprenoid chain is initiated with a molecule of DMAPP and a subsequent linear additions of three molecules of IPP catalyzed by an enzyme geranylgeranyl pyrophosphate synthase (GGPS) (Britton, 1993; Cunningham and Gantt, 1998). The final step of this process is the formation of a C_{20} compound, geranylgeranyl pyrophosphate (GGPP), a shared precursor with other isoprenoids. The first committed step of carotenoid synthesis is catalyzed by phytoene synthase (PSY) and results in a head-to-tail condensation of two GGPP molecules to form a C_{40} compound—phytoene that serves as a precursor for astaxanthin and other carotenoids (Cunningham and Gantt, 1998). The expression of the phytoene synthase gene (psy) was up-regulated in Haematococcus cells stressed with high light and undergoing transformation from "green" to "red" stage (Steinbrenner and Linden, 2001; Vidhyavathi et al., 2008; Gwak et al., 2014). The subsequent formation of highly unsaturated compound—lycopene proceeds through four desaturation steps catalyzed by two phytoene desaturases (PDS) and a ζ-carotene desaturase (ZDS) with two plastid terminal oxidase (PTOX 1, PTOX 2) acting as co-factors for electron transfer between C_{40} carotenoid intermediates, plastoquinone and final electron acceptor—oxygen (Li et al., 2010; Nawrocki et al., 2015). Of the two, PTOX 1 was found to be co-regulated with astaxanthin synthesis in H. pluvialis (Wang et al., 2009; Nawrocki et al., 2015). Desaturation reactions increase the number of conjugated carbon-carbon double bonds that form the chromophore in carotenoids and convert a colorless molecule of ζ-carotene to a pink colored lycopene (Cunningham and Gantt, 1998). Both termini of lycopene undergo cyclization catalyzed by lycopene cyclases (LCY-e and LCY-b). Cyclization is a branching point of the carotenoid biosynthesis in most organisms, yielding α-carotene (precursor of lutein) and β-carotene (precursor of other carotenoids including astaxanthin). In H. pluvialis vast majority of the carbon flux is directed into the latter (Gwak et al., 2014), and high level of LCY-b transcripts have been observed under stress conditions (Lorenz and Cysewski, 2000; Gwak et al., 2014). Final two oxygenation steps catalyzed by β-carotene ketolase (BKT) and β-carotene hydroxylase (CrtR-b) are rate limiting steps of astaxanthin synthesis (Linden, 1999; Steinbrenner and Linden, 2001; Vidhyavathi et al., 2008). Although in principle the reactions catalyzed by these two enzymes can proceed in any order, higher substrate specificity of BKT toward β-carotene than zeaxanthin favors initial addition of keto group before enantio-selective hydroxylation of canthaxanthin to astaxanthin is catalyzed by CrtR-b (Lotan and Hirschberg, 1995). Enantioselectivity of astaxanthin synthesis is of primary importance for the nutraceuticals market and the major advantage of H. pluvialisastaxanthin over its synthetic counterpart. Since astaxanthin has two identical chiral centers at the positions of 3 and 3′ it can exist in four different configurations which yield three different isomers: (3R, 3′S); (3R, 3′R); (3S, 3′S) depending on the spatial orientation of the hydroxyl (OH) groups in chiral carbon. During chemical synthesis these isomers are present in the ratio of 2:1:1, respectively, yielding only 25% of the naturally occurring (3S, 3′S) isoform. H. pluvialis synthesizes the (3S, 3′S) stereoisomer of astaxanthin and is therefore a much sought-off product in the nutraceutical market.

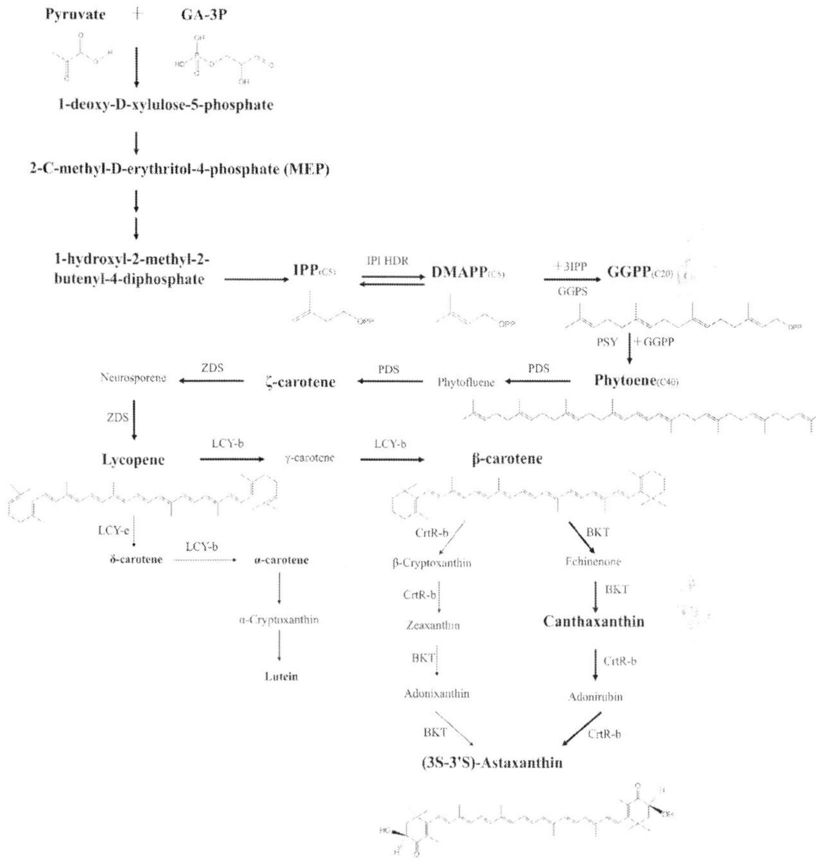

Figure 7: Pathway of (3S-3′S)-astaxanthin biosynthesis in H. pluvialis. Major carbon flux during the red stage of H. pluvialis cultivation is indicated with thick arrows, minor products are indicated with thin arrows. Major intermediates of biosynthesis are indicated in large fonts, minor intermediates in small fonts. Enzyme abbreviations are as follows: IPI, Isopentenyl pyrophosphate isomerase; HDR, 4-hydroxy-3-methylbut-2-enyl diphosphate reductase;

GGPS, geranylgeranyl pyrophosphate synthase; PSY, phytoenesysthase; PDS, phytoenedesaturase; ZDS, ζ-carotene desaturase; LCY-b, lycopene β-cyclase; LCY-e, lycopene ε-cyclase; BKT, β-carotene ketolase; CrtR-b, β-carotene 3,3′-hydroxylase; Intermediates: Phytofluene, Neurosporene, γ-carotene, β-Cryptoxanthin, Adonixanthin, Echinenone, Adonirubin.

Effect of Small Molecules on Astaxanthin Synthesis

Astaxanthin is a secondary metabolite, a carotenoid synthesized by H. pluvialis in a response to stress conditions such as high light, salinity, or carbon to nitrogen ratio (Gao et al., 2012a). Regulation of these pathways can be affected by numerous small molecules like plant hormones or their analogs. An array of such molecules has been explored to modulate astaxanthin accumulation by H. pluvialis. Plant hormones that are usually associated with stress response mechanisms e.g., abscisic acid (ABA), jasmonic acid (JA), methyl jasmonate (MJ) or growth regulators like gibberellic acid (GA3), salicylic acid (SA) or brassinosteroid—2, 4-epibrassinolide (EBR) were found particularly promising in increasing astaxanthin accumulation in H. pluvialis (Kobayashi et al., 1997b; Gao et al., 2012a,b, 2013a,b; Yu et al., 2015). It was found that these compounds affect numerous genes involved in astaxanthin biosynthesis and result in their even six to 10 fold up-regulation. All of these compounds were tested in various concentrations and the highest improvement of astaxanthin accumulation was achieved with salicylic acid. At relatively low concentrations of the hormone 50 mg L^{-1} and low light 25 μmol photons m^{-2} s^{-1} the content of astaxanthin raised seven fold from 0.391 mg L^{-1} to 2.74 mg L^{-1}. Higher levels of these hormones had however deleterious effect on both growth and astaxanthin accumulation (Gao et al., 2012a). Correlation between mRNA transcript levels of five key enzymes of astaxanthin synthesis pathway (ipi, psy, pds, crtO, and crtR-b encoding respectively isopentenyl-diphosphate δ-isomerase (IPI), phytoene synthase (PSY), phytoene desaturase (PDS), β-carotene oxygenase (CrtO), and β-carotene hydroxylase (CrtR-b) with alga growth and astaxanthin production suggested a complex, multiple regulatory mechanisms at transcriptional, translational, and post-translational levels controlling entire process of carotenoid synthesis in H. pluvialis (Li et al., 2010). Small molecules can have multiple effects on the regulation of each of these genes and more detailed investigation of the molecular responses to their application is required for both understanding gene regulation in H. pluvialis and enhancing its capacity as commercial astaxanthin producer.

Genetic Improvement of H. pluvialis for Astaxanthin Production

Green eukaryotic microalgae are among the organisms that are notoriously difficult to genetically engineer. In principle, genetic engineering of microalgae has been reported for over 30 strains of various species, including Haematococcus (Rosenberg et al., 2008; Radakovits et al., 2010; Forján et al., 2015). In majority of these reports however, only a transient transgene expression has been achieved and a desired, stable, hereditary, and efficient genetic modification existed only for model species such as Chlamydomonas reinhardtii and Volvox carteri. Due to these constrains genetic improvement of H. pluvialis strains have been long limited to classical mutagenesis. Combination of mutagenic treatment and various screening methods resulted in the development of numerous interesting mutants of H. pluvialis including those of higher astaxanthin accumulation capacity (Tjahjono et al., 1994a; Chumpolkulwong et al., 1997; Tripathi et al., 2001; Chen et al., 2003; Hu et al., 2008; Hong et al., 2012; Gomez et al., 2013). Various mutagenic treatments have been tested, and can be broadly divided into UV and chemical mutagenesis. Chemical mutagenesis has been generally found to be more suitable for H. pluvialis due to alga's intrinsic capacity to limit the damage from light. Typical chemicals used for mutagenesis include ethyl methanesulphonate (EMS) and N-methyl-N-nitro-N-nitrosoguanidine (MNNG). In all studies their concentrations were adjusted to target 85–95% cell mortality. Successful mutants are usually screened using a combination of factors that promote identification of mutants capable of high astaxanthin accumulation. Typically, herbicides that affect carotenoid synthesis pathway such as norflurazon, fluridone, nicotine, compactin, or diphenylamine are used (Tjahjono et al., 1994a; Chumpolkulwong et al., 1997; Tripathi et al., 2001; Chen et al., 2003; Gomez et al., 2013). Screening relies on identifying colonies that are capable of surviving and/or growing well in the presence of inhibitory concentrations of these compounds and high illumination. Surviving strains should in principle exhibit better capacity to synthesize carotenoids and divert larger, or more efficient carbon flux toward these compounds. A number of successful mutants have been isolated this way and typical improvement of astaxanthin accumulation ranges from several percent (Gomez et al., 2013) to two or three fold improvement (Hu et al., 2008). In former case the mutated strain was tested in commercial scale cultivation system (120,000 L) and retained the improved capacity for astaxanthin production. An alternative approach to strain improvement relying on selection of photosensitive mutants was recently attempted (Hong et al., 2012). Photosensitive mutant with an ability to grow under hetero or mixotrophic conditions should be in principle advantageous over wild type strains due to faster growth rates and more efficient stress trigger. Screening for successful mutants was performed in a three stage process. Since photosensitivity is connected with impaired photosynthesis, these impaired mutants were selected in the first screening. Secondary screening tested for the ability of heterotrophic growth of these photosensitive isolates. Tertiary screening involved mixotrophic conditions with moderate illumination to obtain mixotrophic photosensitive strain that accumulated 4.7% (w/w) of astaxanthin under much shorter induction time (Hong et al., 2012). The mutated strain was stable for at least 1.5 year and is an interesting example of using classical mutagenesis for improvement of H. pluvialis. Mutagenic strain improvement can be expanded by breeding or creating hybrid strains from previous genetic improvement efforts. Technique of protoplast fusion has been successfully applied to H. pluvialis (Tjahjono et al., 1994a). Two mutagenized strains, norflurazon-resistant and nicotine-resistant have been fused to create a hybrid containing genetic material of two initial strains and showed 30% higher astaxanthin accumulation than the initial wild type strain, when neither of the fused strains showed such characteristics (Tjahjono et al., 1994a). Till very recently H. pluvialis was one of the organisms in which genetic engineering of its nuclear genome was considered difficult due to lack of suitable shuttle vectors and satisfactory transformation efficiencies (Sharon-Gojman et al., 2015). A number of unsuccessful approaches have been tried that included various transformation methods (particle bombardment, electroporation, Agrobacterium), vectors, promoters, and strains (Sharon-Gojman et al., 2015). To address these limitations and open a new array of possibilities in H. pluvialis and astaxanthin biology and technology new developments were required. In recent years these developments emerged and stable transformations of H. pluvialischloroplast (Gutiérrez et al., 2012) and nuclear genomes (Sharon-Gojman et al., 2015) were achieved. Most recent nuclear transformation vectors are capable to transform one or two transgenes into the nuclear genome either 5′ or 3′ of the endogenous dominant selection marker, in the absence of any additional antibiotic resistance genes. The selection marker used in this system is a phytoene desaturase (pds) variant that confers resistance to a herbicide norflurazon due to a single point mutation (L504A). Successful transformation of H. pluvialis was obtained with particle bombardment and numerous constructs based on pds selection marker were delivered and incorporated to the genome showing stability of integration for over 16 months of subculturing (Sharon-Gojman et al., 2015). Genetic engineering of chloroplast genome of H. pluvialis have been also achieved relatively recently (Gutiérrez et al., 2012). So far these studies are limited to expressing exogenous antibiotic resistance gene (aadAcassette) between Internal Transcribed Spacer region and 16S gene of H. pluvialis chloroplast genome, but this technique may in the near future have significant impact on protein production in H. pluvialis(Gutiérrez et al., 2012) as higher protein yields are generally obtained during chloroplast expression of transgenes in other microalgal strains (Li et al., 2015). These new developments in genetic engineering of H. pluvialis can open a new chapter for the development of this organism as both industrial astaxanthin producer and an interesting model for carotenoid synthesis and accumulation studies.

APPLICATIONS OF H. PLUVIALIS ASTAXANTHIN

Astaxanthin in Human Health and as Nutraceutical

Astaxanthin possesses various human health benefits and nutraceutical applications and plenty of published information available with evidences, mainly from in vitro and animal models (Guerin et al., 2003; Chew et al., 2004; Higuera-Ciapara et al., 2006; Palozza et al., 2009; Yuan et al., 2011). The effect of Haematococcus derived astaxanthin on various physiological systems in animal and human subject is presented in Table 3. Astaxanthin is considered as “super anti-oxidant” which possesses one of the strongest known antioxidant effects. Its unique structure allows it to span biological membranes and act as an antioxidant by reducing and stabilizing free radicals (Hussein et al., 2006; Liu and Osawa, 2007; Ranga Rao et al., 2010). It is very good at protecting membrane phospholipids and other lipids against peroxidation (Naguib, 2000). There are several studies which showed high antioxidant activity of astaxanthin from H. pluvialis in rats supplemented with diet (Kamath et al., 2008; Ranga Rao et al., 2010, 2013; Augusti et al., 2012). Astaxanthin can terminate the induction of inflammation in biological systems. It can help to fight symptoms of ulcer disease from Helicobacter pylori (Liu and Lee, 2003); protect against gastric lesions (ulcers), improve gastrointestinal health (Nishikawa et al., 2005; Kamath et al., 2008); and treat gastrointestinal discomfort (Andersen et al., 2007; Kupcinskas et al., 2008). Astaxanthin offers protection against free radical damage to preserve immune-system defenses. The immunomodulating capacity of astaxanthin has been found to be superior to that of β-carotene and canthaxanthin (Chew and Park, 2004). Astaxanthin has shown significant effect on immune function in a number of in vitro and in vivo assays using both animal models (Chew et al., 2004) and humans (Park et al., 2010). Astaxanthin is a potential therapeutic agent against atherosclerotic cardiovascular disease (Fassett and Combes, 2011). Astaxanthin supplementation can be beneficial for people with enhanced risk for heart attacks. It is carried by VLDL, LDL, and HDL (high-density lipoprotein) in human blood and protects LDL-cholesterol against oxidation (Iwamoto et al., 2000); has a role in the reduction of blood plasma level (Karppi et al., 2007); and increases basal arterial blood flow (Miyawaki et al., 2008). Oxidative stress is a causative or at least ancillary factor in the pathogenesis of major neurodegenerative diseases (Alzheimer's, Huntington's, Parkinson's, and amyotrophic lateral sclerosis-ALS). Diets high in antioxidants offer the potential to lower the associated risks (Ferrante et al., 1997). Natural astaxanthin can cross the blood-brain barrier in mammals and can extend its antioxidant benefits beyond that barrier. Therefore, astaxanthin can help to alleviate the effects of Alzheimer's disease and other neurological diseases. Astaxanthin can improve respiratory and sympathetic nervous system activities (Nagata et al., 2006), inhibit the growth of fibrosarcoma, breast, and prostate cancer cells and embryonic fibroblasts (Palozza et al., 2009); cell death, cell proliferation and mammary tumors (Nakao et al., 2010). Astaxanthin supplementation can help to protect against UV-induced photooxidation; as an oral sun-protectant; can prevent skin thickening and reduce collagen reduction against UV induced skin damage (Ranga Rao et al., 2013) and can improve skin condition across its layers i.e., corneocyte, epidermis, basal, and dermis by combining oral supplementation and topical treatment (Seki et al., 2001; Yamashita, 2002; Tominaga et al., 2012). Results have shown that semen quality, pregnancy rate and sperm velocity in human subject can be improved (Elgarem et al., 2002; Comhaire et al., 2005) whereas idiopathic infertility can be decreased by astaxanthin (Andrisani et al., 2015).

Table 3: Effect of H. pluvialis-derived astaxanthin on various physiological systems in human and animal subjects.

Physiological System	Subject	Effect followed	Main outcome	References
Anti-oxidation	Rabbits	Thioredoxin reductase; Paraoxonase activity	Enhanced; No effect	Augusti et al., 2012
	Rats	Hepatoprotective and antioxidant activity	Improved	Ranga Rao et al., 2015
	Rats	Antioxidant enzymes, catalase, superoxide dismutase, peroxidase, and lipid peroxidation in plasma and liver	Increased	Ranga Rao et al., 2010, 2013
	Men (bilateral cataract)	Antioxidative effects through changes in superoxide scavenging activity, and hydroperoxides production in aqueous humor	Enhanced; Suppressed	Hashimoto et al., 2013
Eye function	18 healthy men	Deep vision	Improved	Sawaki et al., 2002
	10 healthy men	Eye function	Improved	Iwasaki and Tawara, 2006
	40 asthenopia patients	Eye accommodation power	Improved	Kenji et al., 2005
	49 healthy men	Uncorrected far visual acuity	Improved	Nakamura et al., 2004
	87 men (visual display terminal workers)	Eye accommodation amplitude (the adjustment in the lens of the eye that allows it to focus);	Improved;	Nagaki et al., 2002
		Eye soreness, dryness, tiredness, and blurred vision	Reduced	Nagaki et al., 2006
Skin	Healthy female or male	Skin wrinkle, corneocyte layer, epidermis, and dermis	Improved	Tominaga et al., 2012
	46 healthy women	Skin elasticity and moisture	Improved	Seki et al., 2001; Yamashita, 2002
Immune response	14 healthy women	Oxidative stress and inflammation markers; Immune response	Reduced; Improved	Park et al., 2010
Inflammation	Rats	Gastrointestinal health	Improved	Nishikawa et al., 2005
Gastric ulcer	*H. pylori*-infected mice	Bacterial load Gastric inflamation	Reduced	Liu and Lee, 2003
	Rats	Gastric ulcer markers	Reduced	Kamath et al., 2008
	44 patients with functional dyspepsia	Inflammatory markers; gastrointestinal discomfort	No effect; No effect	Andersen et al., 2007; Kupcinskas et al., 2008
Cardiovascular system	20 adult men	Blood flow time	Improved	Miyawaki et al., 2008
	Men	Blood plasma levels	Reduced	Karppi et al., 2007

Muscle endurance	16 non-trained men	Lactic acid accumulation after run	Reduced	Sawaki et al., 2002
	19 non-trained men	Respiratory and sympathetic nervous system activities	Improved	Nagata et al., 2006
	20 non-trained men	Strength/explosiveness test; strength/endurance test	No effect; Improved	Lignell, 2001
	20 resistance-trained men	Markers of skeletal muscle injury	No effect	Bloomer et al., 2005
Cancer	Rats	Growth of colon cancer cells	Inhibited	Palozza et al., 2009
Central nervous system	Healthy mice	Memory	Improved	Zhang et al., 2007
	10 healthy men (50–69 years)	Response time and accuracy of several tasks	Improved	Satoh et al., 2009
	Middle aged/elderly men and women	Cog Health battery scores (Nuropsychological memory test)	Improved	Katagiri et al., 2012
Male fertility	20 sub-fertile men	Semen quality, pregnancy rate	Improved	Elgarem et al., 2002
	30 sub-fertile men	Sperm velocity; oxidation markers; pregnancy rate	Improved; Reduced; Improved	Comhaire et al., 2005
	24 healthy men	Idiopathic infertility	Decreased	Andrisani et al., 2015
Metabolic Syndrome (MS)	Obese rats	Body weight; adipose tissue weight; MS markers	Reduced; Reduced; Improved	Ikeuchi et al., 2007

Astaxanthin in Aquaculture and Poultry Industry

During last 20 years, synthetic astaxanthin has been widely used for pigmentation of fish. Haematococcusastaxanthin has great potential in aquaculture industry, due to increasing consumer demands for natural products and ability of Haematococcus astaxanthin to provide necessary supplementation for adequate growth and reproduction of commercially valuable fishes (Salmonid, Red sea bream), rainbow trouts, and shrimps. Microalgae- derived astaxanthin has been demonstrated as safe and effective compound for flesh pigmentation of these fish (Torrissen and Naevdal, 1984; Tolasa et al., 2005). Utilization of H. pluvialismeal for pigmenting has resulted in significant astaxanthin deposition in flesh and skin, flesh coloration enhancement, enhanced antioxidant system, fish egg quality, better growth and survival of fry of Salmonid, sea bream, and rainbow trout (Arai et al., 1987; Sommer et al., 1991; Choubert and Heinrich, 1993; Sheikhzadeh et al., 2012a,b), ornamental fishes (Ako and Tamaru, 1999), and shrimp (Arai et al., 1987; Parisenti et al., 2011). A recent study indicated that diets supplemented with H. pluvialis can improve large yellow croaker fish growth more than diets supplemented with synthetic astaxanthin (Li et al., 2014). H. pluvialis-derived natural astaxanthin has shown to be efficient in pigmentation of egg yolks, egg production (Elwinger et al., 1997) in hen and breast muscle tissue improvement and higher feed efficiency in broiler chicken (Inborr and Lignell, 1997; Inbbor, 1998). It has also been proved to improve health and fertility of chicken and to decrease their mortality (Lignell and Inborr, 1999, 2000).

CULTIVATION AND PROCESSING OF H. PLUVIALIS FOR ASTAXANTHIN PRODUCTION

Culture Conditions and Requirements for Cell Growth and Astaxanthin Formation

Optimization of the various culture parameters, such as growth medium composition, light, pH, temperature etc. is necessary to achieve high biomass and astaxanthin production. Most of these parameters have different optima for biomass accumulation and astaxanthin production. For carotenogenesis induction, the stronger exposure to stress conditions, the higher astaxanthin accumulation. The origins of this stress can be diverse and successful astaxanthin accumulation has been induced with both, high levels of one stressor, or from a combination of multiple stress factors. In some cases, if cells are exposed to strong stress, cells growth completely ceases and cells begin to die in a relatively short time (Su et al., 2014). Various types of growth media are used for cultivation of H. pluvialis. The most commonly used media are BG- 11 (Rippka et al., 1979), BBM (Bischoff and Bold, 1963), OHM (Fábregas et al., 2000); KM1-basal medium with organic carbon sources in the form of sodium acetate (Kobayashi et al., 1993), and their modifications. An ideal composition of the medium to achieve high growth rate and biomass accumulation is different from ideal composition for high accumulation of astaxanthin. Sodium nitrate was found to be the most optimal inorganic nitrogen source (Sarada et al., 2002a), alternatively an organic source such as urea can be used. When culture is subjected to nutrient deficiency, it leads to accumulation of astaxanthin within the cells (Saha et al., 2013). Nitrogen limitation leads to approximately twice the rate of astaxanthin production than the limitation of phosphorus. It can be due to higher cellular damage resulting from a lack of nitrogen, which manifests in significant degradation of chlorophyll, compared to the phosphorus starvation (Boussiba et al., 1999). Micronutrients such as selenium and chromium result in an increased biomass and astaxanthin production (Tripathi et al., 1999; Fábregas et al., 2000; Domínguez-Bocanegra et al., 2004). Formation of astaxanthin can also be induced by adding NaCl (0.25–0.5% w/v) to the media. Also, when NaCl is added together with 2.2 mM sodium acetate, astaxanthin accumulation can be increased (Sarada et al., 2002b). Addition of 0.45 mM Fe^{2+} in the form of ferrous sulfate may significantly increase the biosynthesis of carotenoids in cysts due to formation of hydroxyl radicals. (Kobayashi et al., 1997b). This effect may be enhanced by combining Fe^{2+} treatment with an addition of sodium acetate and high temperature exposure (Kobayashi et al., 1993; Tjahjono et al., 1994b). According to most studies, the suitable temperature for the growth and astaxanthin accumulation of H. pluvialis is between 20 and 28°C (Fan et al., 1994; Hata et al., 2001; Lababpour et al., 2005; Kang et al., 2010; Yoo et al., 2012; Wan et al., 2014a). However, temperature above 30°C induces a transition from green vegetative stage to red stage and formation of red cysts can be observed within 2 days. This transition is combined with a significant slowdown in growth, while astaxanthin accumulation is 2–3 times higher than at 20°C. The increased temperature is likely to affect the synthesis of astaxanthin through stimulation of oxygen radicals formation and their higher reactivity (Tjahjono et al., 1994b). It is preferred that the temperature change takes place gradually, allowing better acclimation to the new conditions (Hata et al., 2001). pH can also significantly affect the cell growth and synthesis of chlorophyll and carotenoids in H. pluvialis. In terms of biomass and astaxanthin production optimal pH is within the range of 7.00–7.85 (Hata et al., 2001; Sarada et al., 2002a). The typical irradiation for H. pluvialis cultivation ranges between 40 and 50 µmol photons $m^{-2}s^{-1}$ (Hata et al., 2001; Chekanov et al., 2014; Park et al., 2014). Optimal irradiation to achieve a high growth rates tend to be higher, namely 70 (Zhang et al., 2014), 80 (Saha et al., 2013), 90 (Fan et al., 1994), or even up to 177 µmol photons $m^{-2}s^{-1}$ (Domínguez-Bocanegra et al., 2004). These different optimal values may be caused by other cultivation parameters such as media composition, temperature, or the strain of H. pluvialis. During vegetative stage cultivation of H. pluvialis, the regular cycles of alternating light and dark 12:12 or 16: 8 h are often used (Saha et al., 2013; Park et al., 2014), but the higher density cultures are achieved with continuous illumination (Domínguez-Bocanegra et al., 2004). The best practice to date appears to be white or blue LED lighting (Saha et al., 2013) or the mixture of both at the ratio of 3:1 at 7000 lx (~95 µmol photons $m^{-2}s^{-1}$). These conditions promote morphologic changes from green vegetative cells to red cyst cells (Sun et al., 2015). Carotenogenesis is induced in cells upon exposure to higher light intensity than the corresponding light saturation point (LSP). However, specific optimum value of LSP differ between studies. The lowest intensities that have been reported utilized irradiation of around 100–150 μmol photons $m^{-2}s^{-1}$ (Zhang et al., 2014) followed by 240 (Saha et al., 2013), 345 (Domínguez-Bocanegra et al., 2004), and 480 μmol photons $m^{-2}s^{-1}$ (Chekanov et al., 2014). Lower optimal irradiation was found to be influenced by other stress conditions, such as deficiency of nutrients (Saha et al., 2013; Zhang et al., 2014) or elevated temperature (Tjahjono et al., 1994b). It proves that for effective induction of carotenogenesis excessive irradiation may not be necessary if other stressors are present. With the reduced requirements of light for cultivation in photobioreactors, the costs of cultivation can be minimized which is essential for astaxanthin production in an industrial scale. Regarding the type of illumination, the highest carotenoid content was obtained by using a continuous PAR lighting (Saha et al., 2013). An interesting alternative to an immediate change in the radiation intensity to induce the transition from the vegetative phase to carotenoid production is gradually increasing the level of lighting. Gradual increase of light intensity can result in gradual transformation of cells to cysts and can also contribute to better accumulation of astaxanthin, because the cells are capable to cope with increasing higher levels of stress (Park et al., 2014).

Culture Systems

Astaxanthin-producing H. pluvialis is capable of growing in photoautotrophic, heterotrophic, or mixotrophic growth conditions in indoors, open raceway ponds or closed photobioreactors in batch, fed batch, or continuous modes.

PHOTOAUTOTROPHIC CULTURE

Photoautotrophic culture of H. pluvialis is mainly carried out in open raceway ponds or closed photobioreactors. Typical photobioreactors used for its cultivation include tubular, bubble column and airlift photobioreactors. As the culture conditions for maximum growth and maximum astaxanthin content are mutually exclusive, a two-step cultivation strategy is commonly adopted for the commercial production. The first step, green vegetative growth phase ("green stage") is to promote algal growth under favorable culture conditions (e.g., low light and nitrogen-replete) (Boussiba, 2000; Aflalo et al., 2007; Del Rio et al., 2007). When high cell density is reached, the culture enters into the second step, reddish inductive production phase ("red stage"), where algal cells are subjected to stress conditions such as high light intensity and nitrogen depletion, excess acetate addition, pH or salt stress, phosphate deficiency, or the addition of specific cell division inhibitors. These stress factors (either one or combination of more) induce the astaxanthin production in H. pluvialis (Fábregas et al., 2001; Torzillo et al., 2003; Orosa et al., 2005; He et al., 2007; Hu et al., 2008; Li et al., 2010; Choi et al., 2011). Therefore, carotenoid induction method has a direct correlation with both the astaxanthin content and total astaxanthin productivity. The optimal environmental and nutritional conditions for each stage are quite different (Del Rio et al., 2007). The reported biomass productivities in green stage and red stage ranged from 0.01 to 0.5 g L^{-1} d^{-1} and 0.01 to 4.8 g L^{-1} d^{-1}, respectively. Astaxanthin productivity and astaxanthin content ranged from 0.44 to 21 mg L^{-1} d^{-1} and 0.8 to 4.8% of DW, respectively (Table 4). Astaxanthin can be also produced efficiently by H. pluvialis using a simpler "one-step strategy." This strategy involves the administration of nitrate starvation and specific average irradiance in the culture medium, resulting in simultaneous algal cell growth and astaxanthin accumulation (Del Río et al., 2005; Del Rio et al., 2007; Del Río et al., 2008; García-Malea et al., 2009). At the laboratory scale and under continuous illumination, mean astaxanthin productivity of 20.8 mg L^{-1} d^{-1} has been reported for the one-step method (Del Río et al., 2008). The technical feasibility of this approach at a pilot scale was demonstrated in an outdoor tubular photobioreactor, which resulted in biomass and astaxanthin productivities of 0.7 g L^{-1} d^{-1} and 8 mg $L^{-1}d^{-1}$, respectively (García-Malea et al., 2009). One-stage cultivation seems attractive, since it is less complicated than the two-stage process and the production of astaxanthin takes place in a continuous mode. It has however two serious drawbacks. First, the actual astaxanthin production is significantly lower compared to the two-stage approach. Second, this cultivation is unsuitable for outdoor setting, since it requires incessant illumination during night as well what makes the process too expensive (Aflalo et al., 2007). An "attached cultivation" approach was successfully applied in the induction of H. pluvialis for astaxanthin production. In this method green cells are cultured in the conventional water column and then deposited on the membrane to increase light stress surface area in the second phase of cultivation. Under the optimized conditions, biomass, and astaxanthin productivities in the attached cultivation system were 2.8

(3.7 g m^{-2} d^{-1}) and 2.4-fold (65.8 mg m^{-2} d^{-1}) higher than those of the suspended bioreactor, respectively (Wan et al., 2014b). Other studies that used similar approach have reported higher astaxanthin productivities of 124 mg m^{-2} d^{-1} (Yin et al., 2015) and 164.5 mg m^{-2} d^{-1} (Zhang et al., 2014). Attached cultivation approach is superior to suspended induction in several aspects such as, lower water consumption and smaller risk of contamination. This indicates that attached induction approach can provide a promising way to boost economic benefits and considerably reduce production cost of astaxanthin from H. pluvialis (Zhang et al., 2014; Wan et al., 2014b). Recently, Park et al. (2014)established a two-stage "perfusion culture" system for H. pluvialis combining it with stepwise increase of light irradiance.

Table 4. Summary of various methods of H. pluvialis biomass cultivation and corresponding astaxanthin productivities.

Airlift column (30 L)	Indoor	Batch	Modified Bold's Basal medium	0.03	0.01	2.7	0.44^{b}	Harker et al., 1996b
Tubular /open pond (25,000 L)	Outdoor	--	Modified Bold's Basal Medium	0.036-0.052	N/A	2.8-3.0	N/A	Olaizola, 2000
Tubular (50 L)	Indoor	Semi continuous	BG-11 medium	N/A	0.05	3.6	$7.2^{a,d}$	Torzillo et al., 2003
Bubbling column (1.8)	Indoor	Batch	Basal inorganic culture medium	N/A	0.6	0.6	5.6^{a}	Del Río et al., 2005
Airlift Tubular (55 L)		--	Inorganic medium free of acetate	N/A	0.41	1.1	4.4^{a}	López et al., 2006
Bubbling column (0.5 L)	Indoor	Batch	BG-11 medium	0.5	0.21	4	11.5^{b}	Aflalo et al., 2007
Tubular (200 L)	Outdoor	Batch	BG-11 medium	0.37	0.21	3.8	$10.1^{b,c}$	Aflalo et al., 2007
Bubbling column (1.8 L)	Indoor	Batch	Basal inorganic culture medium	N/A	1.9	1.1	21^{a}	Del Río et al., 2008
Bubbling column (1 L)	Indoor	Batch	Standard inorganic medium	0.36	0.14	3.6	12^{b}	Ranjbar et al., 2008
Tubular (1.8 L), outdoor,	Outdoor	Continuous	Standard inorganic medium	N/A	0.7	1	8^{a}	García-Malea et al., 2009
Open pond	Indoor	Batch	BG-11 medium	N/A	0.15	2.79	4.3^{a}	Zhang et al., 2009
Flat type (1 L)	Indoor	Fed batch	NIES-C medium	0.33	0.44	4.8	14^{b}	Kang et al., 2010
Airlift column	Indoor	Batch	*Haematococcus* medium (OHM)	N/A	0.14	N/A	3.9^{a}	Choi et al., 2011
Bubbling column (6 L)	Indoor	Batch	NIES-C medium	N/A	0.047	N/A	1.4^{a}	Yoo et al., 2012
Bubbling column (0.6 L)	Outdoor	Batch	BG-11 medium	N/A	0.58	2.7	$17.1^{b,d}$	Wang et al., 2013a
Bubbling column (0.6 L)	Outdoor	Batch	BG-11 medium	N/A	0.30	3.8	$16.0^{b,d}$	Wang et al., 2013b

Approach is based on repeated replacement of the growth medium. Cells are grown in a photobioreactor and are periodically retained in the cell settling chamber whilst growth medium is being replaced in the photobioreactor. Cells are later recycled to the bioreactor and can efficiently utilize fresh growth medium which is free of inhibitory metabolic by-products. Perfusion culture can provide high biomass productivities of 0.18 g L^{-1} d^{-1}. Under stepwise increased light irradiance (150–450 μE/m^2/s), cellular density of 12.3 g L^{-1} of have been obtained which is 3.09 and 1.67 times higher than batch and fed-batch processes, respectively whilst the productivity of astaxanthin reached 602 mg L^{-1} (Park et al., 2014).

HETEROTROPHIC AND MIXOTROPHIC CULTURE

High light irradiance is often employed for enhancing astaxanthin formation in H. pluvialis cultures. However, light absorption and scattering caused by mutual shading of cells in large-scale cultures severely affects the productivity and quality of algal biomass and products. The high cost of illumination is another problem hindering the commercialization of Haematococcus products. To overcome this drawback, heterotrophic culture approach may be considered. Under heterotrophic growth conditions light is not needed as organic substrates serve as carbon and energy sources for growth and synthesis of secondary metabolites. Also, since lipid accumulation and astaxanthin biosynthesis are connected in space and time the effect of carbon source on lipid accumulation can have significant effect on overall productivity. It has been shown in Haematococcus and other microalgae lipid content and lipid profiles of microalgae are dependent on the cultivation conditions with various stress factors such as starvation or salt stress are efficient triggers of lipid accumulation, and can result in the alteration of fatty acid profiles due to cellular adjustment to particular stressor (Damiani et al., 2010; Lei et al., 2012; Saha et al., 2013; Chen et al., 2015). Various types of organic carbon sources have been used for heterotrophic cultivation and induction using acetate has been found effective for Haematococcus encystment and initiation of astaxanthin production (Kobayashi et al., 1991; Kakizono et al., 1992; Orosa et al., 2000; Hata et al., 2001; Kang et al., 2005). However, unlike many microalgae in which oversupply of easily accessible carbon in combination with nitrogen limitation yields diversion of the carbon flux toward lipid accumulation (Miao and Wu, 2006; Jia et al., 2014), H. pluvialis grows at a relatively low rate (0.22 d^{-1}) and accumulates negligible amount of astaxanthin, too low to be considered for commercial scale production if single step cultivation is considered (Kobayashi et al., 1992, 1997a; Moya et al., 1997). Additionally, heterotrophic cultivation of Haematococcus increases the risk of bacterial or fungal contamination (Hata et al., 2001; Olguín et al., 2012). H. pluvialis can be also produced indoors mixotrophically employing an organic acid (e.g., acetate) or carbohydrates as an additional carbon and energy source (Kobayashi et al., 1993). Studies have shown that both growth and astaxanthin production can be enhanced under mixotrophic culture conditions. A final cell density of 0.9–2.65 g L^{-1} and a maximum astaxanthin content of 1–2% DW were obtained from mixotrophic cultures of H. pluvialis (Chen et al., 1997; Zhang et al., 1999; Wang et al., 2003). A sequential, heterophotric-photoautotrophic culture mode was also explored. Heterotrophic culture was used in the green stage to produce algal biomass, while astaxanthin production was induced in photoautotrophic culture conditions. The induction of astaxanthin accumulation was performed under nitrogen deprivation conditions and whilst using bicarbonate or CO_2 as carbon sources. As a result, a very high cellular astaxanthin content of 7% (DW) was achieved, 3.4-fold higher than heterotrophic induction whilst astaxanthin productivity of 6.25 mg L^{-1} d^{-1} was obtained (Kang et al., 2005). Results indicate that photoautotrophic induction of astaxanthin production in H. pluvialis is more effective than heterotrophic one. Based upon the information obtained thus far, heterotrophic and mixotrophic culture modes are less cost-effective than the photoautotrophic one for Haematococcus mass culture.

Microbial Contamination and Possible Control Measures

Since interest in commercial microalgae cultivation is increasing, microbial contaminants that hamper production by resulting in reduced biomass yield and quality received great attention recently. Mass culture of H. pluvialis is reported to be contaminated by fungal parasites and zooplanktonic predators (e.g., amoebas, ciliates, and rotifers), as well as other microalgae and cyanobacteria (Han et al., 2013). A parasitic chytrid/blastoclad fungus Paraphysoderma sedebokerenses is found to be responsible for reduced astaxanthin productivity and frequent culture collapses in commercial Haematococcuscultivation facilities (Hoffman et al., 2008; Strittmatter et al., 2015). Detection of contaminants is prerequisite for preventing and controlling of microbial contamination in mass microalgal culture. Methods of detection usually include microscopy and staining, flow cytometry, molecular based detection and monitoring. In order to cope with microalgae culture contamination, the techniques that are generally used include abiotic stresses such as NO−3NO3- limitation, pH stress, temperature stress, light stress, toxic substances, and shear forces. There are some other techniques for parasite removal including salvage harvest, chemical agents (abscisic acid, copper sulfate), physical methods, biological methods (selective breeding and biological agents) (Carney and Lane, 2014). Recently, several patent applications relating to control of fungus P. sedebokerenses have been developed in the USA and China to protect production losses in commercial Haematococcus culture facilities across the world (McBride et al., 2013; Zhang et al., 2013; Carney and Sorensen, 2015).

Harvesting

Harvesting remains one of the most challenging issues and a limiting factor for commercial algal biomass production. Harvesting of H. pluvialis refers to the selection of appropriate techniques to recover the "red" biomass, after the accumulation of astaxanthin in the cells and also that can facilitate cost-efficient astaxanthin extraction in the extraction phase. For the large scale harvesting of H. pluvialis centrifugation is the most common method and combined with other processes. Usually haematocysts are separated from the water through passive settling and subsequently concentrated with centrifugation (Lorenz and Cysewski, 2000; Olaizola, 2000; Li et al., 2011; Han et al., 2013; Pérez-López et al., 2014). Through the combination of these processes total suspended solid of 13.5% in the algal cake is achieved (Li et al., 2011). Flotation and disk-stack centrifugation have been also reported as another alternative for H. pluvialisharvest. Both showed more than 95% biomass recovery efficiency (Panis, 2015).

Cell Disruption

Different techniques have been developed in order to disrupt the algal cell and recover the intracellular metabolites. The most appropriate cell disruption methods to enhance recovery of astaxanthin from H. pluvialis at a commercial scale involve mechanical processes and more specifically expeller pressing and bead milling (Lorenz and Cysewski, 2000; Olaizola, 2003; Mercer and Armenta, 2011; Razon and Tan, 2011). During pressing (pulverization) microalgae cells are squeezed under high pressure in order to rupture the thick sporopollenin wall. Main advantage of expeller pressing is simple operation and minimization of contamination from external sources. Algal oil recovery efficiency of 75% can be achieved in a single step. Bead milling utilizes vessels filled with tiny glass, ceramic or steel beads that are agitated at high speeds. The dried biomass is fed in these vessels, where continuous exposure of biomass to the grinding media (beads) leads to cell-wall rupture, and subsequent release of intracellular compounds. This method is most effective when biomass concentration in the algal cake after harvesting is between 100 and 200 g/l (Greenwell et al., 2010). Both methods are reliable and widely applied for the H. pluvialiscells disruption at a commercial scale.

Dehydration

In commercial scale astaxanthin production, dehydration (drying) ensures the quality of the pigment and leads to the formulation of the final product (Mata et al., 2010; Li et al., 2011). After algal cell walls have been disrupted, biomass must be processed rapidly within few hours to avoid spoilage. Thus, dehydration is a process applied prior to recovery of the desired metabolite, in order to extend the shelf-life of the algal biomass (Mata et al., 2010). The most known dehydration techniques that have been employed on microalgae are solar drying, spray drying, and freeze drying (Molina Grima et al., 2003; Brennan and Owende, 2010; Milledge, 2013). Spray drying has been considered as the most appropriate method to dry high-value microalgal products including H. pluvialis astaxanthin (Leach et al., 1998; Brennan and Owende, 2010; Li et al., 2011; Han et al., 2013; Milledge, 2013; Panis, 2015). The recovery efficiency of dry biomass (in powder) using this method exceeds 95% and in some occasions may approach 100% (Leach et al., 1998). After spray drying, the moisture content in "red" biomass is lowered to about 5% (Pérez-López et al., 2014). The main drawbacks of spray drying include high operational costs and the risk of microalgae pigments deterioration (Molina Grima et al., 2003). Freeze drying (lyophilization or cryodesiccation), involves the freezing of algal cake, the technique causes less damage than spray drying, but it is even more expensive, especially on a commercial scale (Milledge, 2013).

Recovery of Astaxanthin

Once the cell wall is disrupted and the biomass is fully dried, the recovery of the desired product is possible. Astaxanthin is a lipophilic compound and can be dissolved in solvents and oils. There is an abundance of astaxanthin extraction methods from H. pluvialis utilizing solvents, acids, edible oils, supercritical carbon dioxide (SC-CO_2) as well as microwave-assisted and enzyme-assisted approaches. Among the recovery methods used solvent extraction and supercritical carbon dioxide (SC-CO_2) extraction are considered as the most efficient, compatible, and widely used methods for astaxanthin extraction from H. pluvialis. The summary of various extraction methods of astaxanthin from H. pluvialiswith recent updates is presented in Table 5 Supercritical carbon dioxide (SC- CO_2) extraction has been widely used for industrial applications due to its many processing advantages. Due to low critical temperature of carbon dioxide, the SC-CO_2 system can be operated at moderate temperatures, preventing the degradation of valuable substances (Machmudah et al., 2006). Several studies have reported experiments on supercritical CO_2 extraction for the recovery of astaxanthin from H. pluvialis. Considering astaxanthin quality as the most important criterion, supercritical CO_2 extraction is the most favorable option. Supercritical CO_2 provides shorter extraction time and limits the use of toxic organic solvents. By contrast to most solvents, CO_2 is relatively cheap, chemically inert, non-toxic, and stable (Guedes et al., 2011). Supercritical fluid extraction has also been tested with Haematococcus, aiming at improving the extraction efficiency. For instance, supercritical carbon dioxide (SC- CO_2) coupled with ethanol or vegetable oil as a co-solvent can further increase the extraction efficiency of astaxanthin (80–90%) (Nobre et al., 2006; Krichnavaruk et al., 2008). There is an array of alternative approaches that can assist astaxanthin extraction from H. pluvialis such as solvents, acids, edible oils, enzymes, or pressurized liquids (Sarada et al., 2006; Kang and Sim, 2008; In, 2009; Jaime et al., 2010; Zou et al., 2013; Dong et al., 2014) Pressurized liquid extraction has several advantages over traditional solvent extraction. PLE requires shorter time, can be automated, uses less solvent, and retains the sample in an oxygen-free and light-free environment in contrast to traditional organic solvent extraction (Jaime et al., 2010). Recently, a simple method for the direct extraction of lipids from high moisture H. pluvialis microalgae was successfully achieved using liquefied dimethyl ether (Boonnoun et al., 2014).

Table 5. Summary of astaxanthin extraction methods from H. pluvialis.

Astaxanthin Extraction/Purification method	Astaxanthin yield/Extraction efficiency	References
SC-CO_2 at 20 MPa, 55°C and13% (w/w) ethanol for 120 min of extraction time	83% recovery	Reyes et al., 2014
CO_2 expanded ethanol (50% %w/w ethanol), 7 MPa, 45°C, 120 min of extraction time.	124.2% recovery	Reyes et al., 2014
SC- CO_2 at at 20 MPa, 60°C, 2 ml of ethanol for 1 h of extraction time	2.45 mg/g DW	Fujii, 2012
SC- CO_2, co-solvent 0.154–1% (v/v) ethanol, 7–34 MPa, 30–80°C, 0–100 min	74% recovery	Pan et al., 2012
SC- CO_2, co-solvent 1.25–8.75% (v/v) ethanol, 30–50 MPa, 35–75°C, 210 min	87.4% recovery	Wang et al., 2012
SC- CO_2, co-solvent 0–12% (v/v) vegetable oils, 30–60 MPa 50–80°C 300 min	51% recovery	Krichnavaruk et al., 2008
SC- CO_2, co-solvent 30–50 MPa 40–80° C 60–240 min	84% recovery	Thana et al., 2008
SC- CO_2, co-solvent,1.67–7.5% (v/v) ethanol 20–55 MPa 40–80°C, 240 min	80% recovery	Machmudah et al., 2006
SC- CO_2, co-solvent 0, 10% (v/v) ethanol 20–30 MPa, 40–60°C	90% recovery	Nobre et al., 2006
SC- CO_2, co-solvent 0, 9.4% (w/w) ethanol, 30 MPa at 60°C	97% recovery	Valderrama et al., 2003
Cell germination (12 h). Ionic liquid (1-ethyl-3- methylimidazolium ethylsulfate) (24 h).	32.5 pg/cell	Praveenkumar et al., 2015
Direct extraction using liquefid dimethyl ether (DME) at 0.59 MPa and 25°C, without drying, cell disruption, or heating.	1 (mg/g cell)	Boonnoun et al., 2014
HCl:acetone (5:5), 70°C, 20 min	19.8 mg/g cell	Dong et al., 2014
Ultrasound in solvent (EtOH and EA),16 min, 41°C, 40 kHz, 200 W. EtOH: ethyl acetate (20:1)	28 (mg/g)	Zou et al., 2013
Grinding three repetitions, pressurized hexane (10.3 Mpa)	35 (mg/g cell)	Jaime et al., 2010
Treating with enzymes (Viscozyme, Alcalase) at 50°C, 2 h	2649 ± 359 μg/g cell	In, 2009
Dodecane mixing 48 h, saponification with methanolic NaOH (0.02 M), sedimentation in darkness at 4°C, 12 h.	85% efficiency	Kang and Sim, 2008
Acid digestion, 2 N HCl, 70°C. Acetone extraction for 1 h	87% efficiency	Sarada et al., 2006
NaOH 30 min, Acetone (16 h)	7 (mg/g cell)	Mendes-Pinto et al., 2001
40% (v/v) acetone for 2 min at 80°C, followed by lyophilization or treatment with specific lytic enzymes	70% recovery	Kobayashi et al., 1997c

SC, supercritical; EA, ethyl acetate.

BIOREFINERY APPROACH FOR H. PLUVIALIS

Microalgae have been often proposed as third generation feedstock for biofuel production that does not compete for freshwater or land resources (Daroch et al., 2013a). However, despite significant advances in recent years it becomes apparent that cultivation of microalgae for the sole purpose of biofuel production is unlikely to be possible unless a major low-energy breakthrough technologies in algae cultivation, dewatering, and harvesting are developed (Li et al., 2015). In the meantime, microalgae are extremely important producers of many high-value nutraceutical compounds such as polyunsaturated fatty acids or astaxanthin that can justify high cost of microalgae cultivation and processing technologies. Integration of simultaneous production of numerous compounds within one system maximizing the benefits and limiting the costs is called biorefining (Li et al., 2015). Taking into consideration these findings H. pluvialis emerges as a very useful organism for the development of a dedicated microalgal biorefinery. It fits numerous requirements of for the development of first microalgal biorefineries especially the "high value product first" principle (Li et al., 2015). First, H. pluvialis is the best-known producer of astaxanthin-high value product worth in excess of several thousand US $ per kilogram. This product itself can easily justify expensive cultivation systems required for this organism. Second, H. pluvialis grown under nutrient starvation conditions induces both carotenogenesis (astaxanthin formation) and deposition of storage materials (triglycerides). It has been shown that these two responses are closely related and coincide in both space and time and triglycerides are essential for deposition of astaxanthin inside lipid bodies to confer its protective function (Solovchenko, 2015). In traditional approaches of microalgae to biofuels starvation-induced lipid accumulation is considered as significant challenge for commercialization of these systems as the overall lipid productivity of culture can drop significantly due to impaired growth rates under starvation conditions (Daroch et al., 2013b). In case of high value product like astaxanthin this drop becomes much less of the burden as the high value of the main product will compensate for this delay in final product formation. Due to the coexistence of astaxanthin and triglycerides in space and time it is possible to simultaneously obtain high value product (astaxanthin) and a biofuel feedstock (triglycerides) from a single algal feedstock. Since fatty acid content in the astaxanthin-containing 'red' cells can be as high as 30–60% of algae dry weight (Solovchenko, 2015) making H. pluvialis a very good candidate for biorefining strain. The fatty acid profiles of the alga have been evaluated by several studies and are summarized in Table 2, indicating that fatty acid profiles of the algae are suitable for biodiesel production (Damiani et al., 2010). Third, H. pluvialis have been found to be a mixotrophic alga what is highly advantageous for development of microalgae biorefineries. H. pluvialis is capable of utilizing carbon dioxide, carbonates, and carbohydrates as carbon sources, this opens a possibilities of lowering cultivation costs and/or speeding up the cultivation of the strain through using various waste streams like flue gasses or waste streams containing carbon and nutrient compounds (Wu et al., 2013). Auto-, hetero-, and mixo-trophic cultivation modes require energy and nutrients, both of which can be to an extent recycled from anaerobic digestion process. Carbon sources vary depending on cultivation mode. Photoautotrophic cultivation requires CO_2 that can be recycled from energy production at anaerobic digestion stage. Heterotrophic cultivation requires reduced carbon source—such as carbohydrates or acetate which need to be supplied from alternative source. These compounds can also originate from waste streams. For example, food industry is rich in carbohydrate-rich waste streams that can be used in heterotrophic cultivation of H. pluvialis (Wang, 2014). Mixotrophic cultivation can take advantage of both sources of carbon. After simultaneous extraction of both high value product-astaxanthin and biofuel product-triglycerides algal cake composed of residual biomass can be utilized as a supplementary feedstock for biogas production using anaerobic digestion that would further assist in extraction of residual energy from this integrated bioprocess. These three features of H. pluvialis make it a suitable strain for the development of algal biorefineries producing high value product (astaxanthin) and biofuel molecule (biodiesel and/or biogas). The proposed biorefinery scheme is presented on Figure 8 and employs a classical two stage cultivation of H. pluvialis in green and red stage.

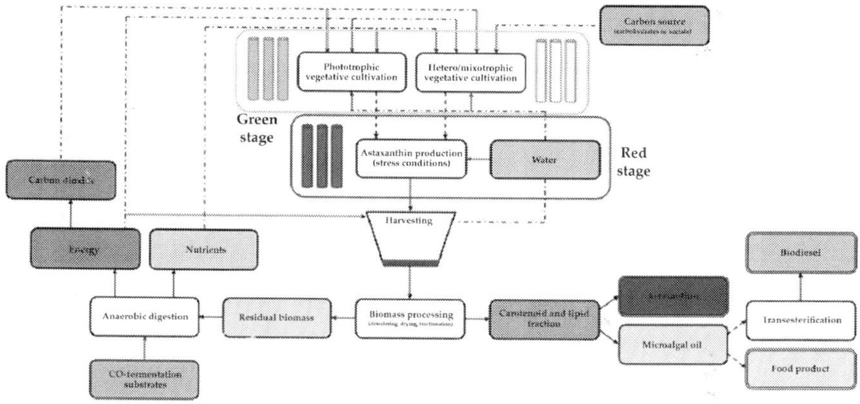

Figure 8: Scheme of two stage cultivation H. pluvialis biorefinery producing high value compound astaxanthin and either edible oil or biofuel compound-biodiesel. Green stage of cultivation can be performed using either photoauto-trophic cultivation (deep green section) or hetero/mixotrophic cultivation (pale green section) systems. Red stage cultivation (red section) takes place after green stage of H. pluvialis cultivation and is aimed to maximize astaxanthin content Recycling of waste is performed through anaerobic digestion process. Following annotations are used: solid arrows—subsequent steps; dashed arrows—optional steps; double lines—final products; double arrows—inputs; dotted lines—opportunities for recycling resources.

CURRENT GLOBAL MARKET AND MARKET PLAYERS OF H. PLUVIALIS ASTAXANTHIN

Synthetic astaxanthin dominates current commercial market, of the total value exceeding $200 million, corresponding to 130 metric tons of product per year (Li et al., 2011). According to recent reports, microalgae-derived astaxanthin corresponds to less than 1% of the commercialized quantity due to much lower price of synthetic astaxanthin and technological problems associated with large-scale algae cultivation (Koller et al., 2014; Pérez-López et al., 2014). In recent years, there has been a growing trend toward using natural ingredients in food, nutraceutical, and cosmetic markets, resulting from increasing concerns for consumer safety and regulatory issues over the introduction of synthetic chemicals into the human food chain. The demand for natural astaxanthin derived from H. pluvialis in the global market has been "sky-rocketing" in recent years owing to increasing consumer awareness of its health benefits. Global market for both synthetic and natural source astaxanthin in aquaculture feed, nutraceuticals, cosmetics, and food and beverages is estimated at 280 metric tons valued at $447 million in 2014. It is further projected to reach 670 metric tons valued at $1.1 billion by 2020 (Industry Experts, 2015; Panis, 2015). Synthetic astaxanthin, astaxanthin rich Phaffia yeast, and Paracoccus bacteria are predominantly used in the aquaculture sector, while the astaxanthin derived from H. pluvialis is the main source for human applications such as dietary supplements, cosmetics, and food and beverages. Nowadays, the estimated market value of astaxanthin depending on products' purity varies from $2500–7000/kg to about $15,000/kg pigment from H. pluvialis in some cases (Borowitzka, 2013; Koller et al., 2014; Pérez-López et al., 2014; Industry Experts, 2015), while the production cost is estimated at about $1000 per kg of astaxanthin from H. pluvialis (Li et al., 2011). Natural astaxanthin is three to four times more valuable than the synthetic alternative in nutraceutical and pharmaceutical markets (Han et al., 2013). Since there is growing market demand for natural astaxanthin for specific commercial applications (e.g., the nutraceuticals market) to replace the synthetic astaxanthin, mass cultivation of H. pluvialis in industrial scale has great potential and attractive business opportunity. However, current market demand for natural astaxanthin is not met. It is expected that in the foreseeable future after the optimization of the production technology, the production costs of the natural astaxanthin from H. pluvialis should be more competitive to these of the synthetic alternative (Pérez-López et al., 2014). Since the mid 1990›s, several leading companies are successfully producing H. pluvialis at commercial scale and marketing natural astaxanthin from H. pluvialis worldwide. The list of the leading companies with their products is presented in Table 6. The size of the nutraceutical astaxanthin market is growing day by day and this market is very attractive to Haematococcus astaxanthin producers since the price of these products is significantly higher than those of feed applications. Haematococcus producers need to invest their attention for increasing astaxanthin production capacity to meet the global demand. It is worthy to mention that several manufacturers already have doubled the cultivation capacities in recent years. Apart from an increase in existing players' capacities, new producers, such as BGG in China, have entered the market with significant production capacities.

Table 6. Leading commercial companies and their H. pluvialis-derived astaxanthin and related products in the world market.

Company Name	Country	Brand Name	Product particulars
Cyanotech Corporation (www.cyanotech.com)	USA	BioAstin®	Astaxanthin extract packaged in soft gel, beadlets: dietary supplement
		Naturose™	Algae meal; pigmentation source for ornamental fish and animals
Mera Pharmaceutigals Inc. (www.merapharma.com)	USA	AstaFactor®	Astaxanthin packaged as soft gel; dietary supplement
Stazen Inc. (www.stazen.com)	USA	Stazen®	Dietary supplement containing algae crushed and dried algae meal
Valensa international (www.valensa.com)	USA	Zanthin®	Astaxanthin extract, soft gel, beadlets
Algatechnologies Ltd. (www.algatech.com)	Israel	AstaPure™	Dry algal biomass, astaxanthin beadlets, and oleoresin
Fuji Chemical Industry Co. Ltd. (AstaReal Co Tld) (www.fujichemical.co.jp; www.astareal.ocm)	Japan, Sweden, USA	AstaREAL®	Astaxanthin oleoresin products, water dispersible, and soluble powders
BioReal (Sweden) AB (subsidiary of Fuji Chemical) (www.bioreal.se)	Sweden	AstaXine® AstaCaroxe®	Dietary supplement containing algae crushed, and dried algae meal
		AstaEquus®	Astaxanthin extract feed supplement for horses
		Novaasta®	Astaxanthin extract feed supplement for animals

Britannia Health Products Ltd. (www.britanniahealth.co.uk)	UK	Britaxan®	Astaxanthin complex with other carotenoids packaged as capsule- dietary supplement
Supreme Biotechnologies NZ Ltd. (www.supremebiotech.com)	New Zealand	AstaSupreme®	Algal biomass, Oleoresins, beadlets, and soft gels of Astaxanthin
Atacama Bio Natural (www.atacamabionatural.com)	Chile	Supreme Asta Oil™ Supreme Asta powder™	Oleoresin for food, nutraceutic, and cosmetic products, and powder for animal feed supplement
Jingzhou Natural Astaxanthin Inc. (www.asta.cn)	China	NaturAsta™	Dry algal biomass and astaxanthin soft gel
Kunming Biogenic Co. Ltd. (www.bgenic.com)	China	AstaBio®	Algal biomass, oleoresins, beadlets, and soft gels of Astaxanthin
Beijing Ginko Group (BGG) Biological Technology Co. Ltd. (www.gingkogroup.com.cn)	China	AstaZine®	Astaxanthin oil, powder, and beadlets
Wefirst Biotechnology Co. Ltd. (www.astawefirst.com)	China	AstaFirst™	Dried algal powder, Astaxanthin oleoresin, and soft gel
Algaetech International SDN BHD (www.algaetech.com.my)	Malaysia	Astaxanthin Premia-*Ex*	Algae biomass, oleoresin, and soft gel
Parry Nutraceuticals Ltd. (EID Parry) (www.eidparry.com/)	India	Zanthin®	Astaxanthin oleoresins, beadlets, and soft gel

Major Challenges for the Improvement of H. pluvialis Biomass and Astaxanthin Production

There are many challenges and problems for the development of large scale production of biomass and astaxanthin from H. pluvialis. Due to these obstacles the productivity can be hampered and in some cases a failure of the production system can make the production process economically unsustainable. Following issues are considered as most important challenges for the development of H. pluvialisastaxanthin production process:

- Lack of effective solution to prevent or treat microbial contaminations of mass cultures in a commercial scale.
- Slow cell growth rate, sensitivity of the cells to hydrodynamic stress, and changes in cell morphology under various environmental conditions.
- Inadequate and cost ineffective cultivation, drying, and astaxanthin extraction technologies at the commercial scale.
- Unavailability of genetically improved/engineered strains of H. pluvialis and genetic transformation tools for engineering astaxanthin biosynthesis pathways in this organism for improved astaxanthin production.
- Lack of sufficient number of skilled workers in production farms and insufficient collaboration between universities and commercial enterprises.
- Lack of adequate scientific research on the economic performance and viability of commercial scale astaxanthin production process.

CONCLUSION AND PERSPECTIVES

This review provides an insight about the latest scientific and technological advancements in various aspects of astaxanthin-producing microalga H. pluvialis such as cell biology, reproduction, biosynthesis pathway, stress mechanism, biomass production, and downstream processing. It also contemplates a broader image including potential benefits, global market opportunities and integration of astaxanthin production into biorefining. In recent years there is an increased interest for natural astaxanthin from green microalga H. pluvialis. Wide ranges of scientific improvements have been achieved during the last decade in terms of productivity and bioprocessing in order to obtain a refined astaxanthin product. Yet its commercial production, especially for low-end markets is too expensive for mass adoption of natural astaxanthin over its synthetic counterpart. H. pluvialis has been shown to be cultured in photoautotrophic, heterotrophic or mixotrophic growth conditions in various culture systems. Research have been conducted on the optimization of the various culture parameters, such as growth medium composition, light, pH, temperature etc. to achieve high biomass and astaxanthin production. Most of these parameters have been optimized and found different for biomass accumulation and astaxanthin production. Little can be done to address this limitatiation as it is funadamentally connected with the life cycle of this microalgae. We believe there exist three key areas where further improvements are required and interesting novel approaches have been recently developed: cultivation efficiency and cost; good cultivation practice and predator control; and astaxanthin isolation and purification.

First, due to complex life-cycle of H. pluvialis it is important to maximase cell densities of alga at "green stage" of cultivation to maximize astaxanthin yield from the "red stage." We think that a number of recent developments can make significant impact in maximizing cell densities. Especially attached cultivation approach and a two-stage "perfusion culture" system can be considered most promising due to the capability of producing several fold higher biomass and astaxanthin productivity and some other benefits such as lower water consumption and smaller risk of contamination. These improvements may boost economic benefits and reduce production cost of astaxanthin from H. pluvialis (Park et al., 2014; Wan et al., 2014b; Zhang et al., 2014). Alternatively, utilization of supplementary carbon source and adoption a two-stage sequential heterotrophic-photoautotrophic approach could improve biomass and astaxanthin production. Especially utilization of waste carbon and nutrient sources in biorefinery setup could help to decrease cultivation costs. Unfortunately, these researches are still in laboratory stage and need to be tested in large-scale commercial production for further validation.

Second, control of contaminants, parasites, and predators remains to be primary concern for Haematococcus growers and major issue in culture stability and astaxanthin productivity. Since there is very little that can be done once contamination takes place it is important to limit the possibility of such disruption and identify it as soon as possible, and avoid spreading to other parts of culture. Traditional detection methods such as microscopy and staining can be used to visualize algal parasites, however this technique may be too labor intensive to perform on a routine basis for most commercial operations. For routine detection, more automated systems such as flow cytometry would be ideal. Alternatively, molecular-based techniques that are considered as the most informative and sensitive for the detection and identification of parasites. Following techniques are worth further exploring DNA sequencing (Sanger, shotgun, or next generation) and then monitoring for these specifically using qPCR or phylochip technology. Decreasing costs of next generation DNA sequencing can make DNA sequencing for culture diagnostic purposes more accessible in the near future.

Third, combination of low cell densities and robust trilayer cell walls of astaxanthin-containing aplanospores make isolation of astaxanthin difficlut and expensive. Currentlly harvesting by centrifugation, cell wall disruption by expeller pressing and bead milling are the most common described methods for commercial scale astaxanthin production from H. pluvialis. After cell walls disruption, biomass is usually processed by spray drying or freeze drying. A number of astaxanthin extraction methods such as (solvents, acids, edible oils, supercritical carbon dioxide, microwave-assisted, and enzyme-assisted approaches have been reported for H. pluvialis and supercritical carbon dioxide (SC- CO2) extraction has been widely used for industrial applications. Two recently developed methods allow efficient extraction of astaxanthin-containing lipids from wet biomass at yields comparable to conventional drying-solvent extraction method. Efficient extraction of astaxanthin from wet H. pluvialisbiomass was achieved with liquefied dimethyl ether (Boonnoun et al., 2014) and also cell germination process in conjunction with ionic liquids treatment (Praveenkumar et al., 2015).

Despite significant advances in research and development of H. pluvialis astaxanthin production is still in laboratory stage and often faces difficulties to become implemented in large-scale commercial production. There is a number of other areas of improvement that will contribute to the expansion of Haematococcusproduction capacity, lowering the production cost, and increasing market penetration at low end applications. These include: next-generation culture systems along with advanced management practices; better understanding of astaxanthin biosynthesis, metabolic pathways and their regulation, genetic engineering, and omics-scale understanding of astaxanthin accumulation; development of genetic manipulation toolbox; exploration of integration of H. pluvialis cultivation with other processes. Yet, we firmly believe that three key areas of focus should be: cultivation efficiency and cost; good cultivation practice and predator control; and astaxanthin isolation and purification. Further developments in these fields can have a profound effect on the commercial deployment of H. pluvialis astaxanthin products and can act as a catalyst for the development of an entire microalgae industry in the near future.

Author Contributions

MS, collected data, participated in preparation of draft manuscript, participated in assembly and editing of the final manuscript; YML, collected data, participated in preparation of draft manuscript; JJC, participated in assembly and editing of the final manuscript; MD, collected data, participated in preparation of draft manuscript, participated in assembly and editing of the final manuscript.

Funding

Authors would like to acknowledge the support of National Natural Science Foundation of China for Young International Scientists Grant no. 31450110424 and 31550110497, Shenzhen Municipal Government for Special Innovation Fund for Shenzhen Overseas High-level Personnel KQCX20140521150255300 and Shenzhen Knowledge and Innovation Basic Research Grant JCYJ20150626110855791, State

Ocean Administration Grant 201305022, and National Thousand People Plan Grant of Jay Jiayang Cheng.

Conflict of Interest Statement

The authors declare that the research was conducted in the absence of any commercial or financial relationships that could be construed as a potential conflict of interest.

REFERENCES

1. Aflalo, C., Meshulam, Y., Zarka, A., and Boussiba, S. (2007). On the relative efficiency of two- vs. one-stage production of astaxanthin by the green alga Haematococcus pluvialis. Biotechnol. Bioeng. 98, 300–305. doi: 10.1002/bit.21391
2. Ako, H., and Tamaru, C. S. (1999). Are feeds for food fish practical for aquarium fish? Int. Aqua Feeds 2, 30–36.
3. Al-Bulishi, M. S. M., Changhu, X., and Tang, Q. J. (2015). Health aspects of astaxanthin: a review. Canad. J. Clin. Nutr. 3, 71–78. Doi: 10.14206/canad.j.clin.nutr.2015.02.08
4. Andersen, L. P., Holck, S., Kupcinskas, L., Kiudelis, G., Jonaitis, L., Janciauskas, D., et al. (2007). Gastric inflmmatory markers and interleukins in patients with functional dyspepsia treated with astaxanthin. FEMS Immunol. Med. Microbiol. 50, 244–248. doi: 10.1111/j.1574-695X.2007.00257.x
5. Andrisani, A., Donà, G., Tibaldi, E., Brunati, A. M., Sabbadin, C., Armanini, D., et al. (2015). Astaxanthin improves human sperm capacitation by inducing lyn displacement and activation. Mar. Drugs 13, 5533–5551. doi: 10.3390/md13095533
6. Arai, S., Mori, T., Miki, W., Yamaguchi, K., Konosu, S., Satake, M., et al. (1987). Pigmentation of juvenile coho salmon with carotenoid oil extracted from Antarctic krill. Aquaculture 66, 255–264. doi: 10.1016/0044-8486(87)90111-6
7. Augusti, P. R., Quatrin, A., Somacal, S., Conterato, G. M., Sobieskim, R., Ruviaro, A. R., et al. (2012). Astaxanthin prevents changes in the activities of thioredoxin reductase and paraoxonase in hypercholesterolemic rabbits. J. Clin. Biochem. Nutr. 51, 42–49. doi: 10.3164/jcbn.11-74

8. Bischoff, H. W., and Bold, H. C. (1963). "Phycological studies IV," in Some Soil Algae from Enchanted Rock and Related Algal Species.(Austin, TX: University of Texas Publications), 95.

9. Bloomer, R. J., Fry, A., Schilling, B., Chiu, L., Hori, N., and Weiss, L. (2005). Astaxanthin supplementation does not attenuate muscle injury following eccentric exercise in resistance-trained men. Int. J. Sport Nutr. Exerc. Metab. 15, 401–412.

10. Bold, H. C., and Wynne, M. J. (1985). Introduction to the Algae. Texas: Prentice-Hall Inc.

11. Boonnoun, P., Kurita, Y., Kamo, Y., Machmudah, S., Okita, Y., Ohashi, E., et al. (2014). Wet extraction of lipids and astaxanthin from Haematococcus pluvialis by liquefied dimethyl ether. J. Nutr. Food Sci. 4:305. doi: 10.4172/2155-9600.1000305

12. Borowitzka, M. A. (2013). High-value products from microalgae—their development and commercialisation. J. Appl. Phycol. 25, 743–756. doi: 10.1007/s10811-013-9983-9

13. Boussiba, S. (2000). Carotenogenesis in the green alga Haematococcus pluvialis: cellular physiology and stress response. Physiol. Plantarum 108, 111–117. doi: 10.1034/j.1399-3054.2000.108002111.x

14. Boussiba, S., and Vonshak, A. (1991). Astaxanthin accumulation in the green alga Haematococcus pluvialis. Plant Cell Physiol. 32, 1077–1082.

15. Boussiba, S., Bing, W., Yuan, J. -P., Zarka, A., and Chen, F. (1999). Changes in pigments profie in the green alga Haeamtococcus pluvialis exposed to environmental stresses. Biotechnol. Lett. 21, 601–604. doi: 10.1023/A:1005507514694

16. Boussiba, S., Fan, L., and Vonshak, A. (1992). "Enhancement and determination of astaxanthin accumulation in green alga Haematococcus pluvialis," in Carotenoids Part A: Chemistry, Separation, Quantitation, and Antioxidation, Enzymology (Academic Press), 386–391.

17. Brennan, L., and Owende, P. (2010). Biofuels from microalgae-A review of technologies for production, processing, and extractions of biofuels and co-products. Renew. Sust. Energ. Rev. 14, 557–577. doi: 10.1016/j.rser.2009.10.009

18. Brinda, B. R., Sarada, R., Kamath, B. S., and Ravishankar, G. A. (2004). Accumulation of astaxanthin in flgellated cells of Haematococcus pluvialis - cultural and regulatory aspects. Curr. Sci. 87, 1290–1294.

19. Britton, G. (1993). "Biosynthesis of carotenoids," in Carotenoids in Photosynthesis SE - 4, eds A. Young and G. Britton (London: Springer), 96–126. doi: 10.1007/978-94-011-2124-8_4

20. Burchardt, L., Balcerkiewicz, S., Kokocinski, M., Samardakiewicz, S., and Adamski, Z. (2006). Occurrence of Haematococcus pluvialisFlotow emend. Wille in a small artificial pool on the university campus of the collegium biologicum in poznan (Poland). Biodiv. Res. Conserv. 1, 163–166.

21. Capelli, G. C., and Cysewski, G. (2013). The Worlds' Best Kept Health Secret Natural Astaxanthin. Kailua-Kona, HI: Cyanotech Corporation.

22. Carney, L. T., and Lane, T. W. (2014). Parasites in algae mass culture. Front. Microbiol. 5:278. doi: 10.3389/fmicb.2014.00278

23. Carney, L. T., and Sorensen, K. (2015). Methods for Treating a Culture of Haematococcus pluvialis for Contamination using Hydrogen Peroxide. US Patent 9113607 B1.

24. Cerón, M. C., García-Malea, M. C., Rivas, J., Acien, F. G., Fernandez, J. M., Del Río, E., et al. (2007). Antioxidant activity of Haematococcus pluvialis cells grown in continuous culture as a function of their carotenoid and fatty acid content. Appl. Microbiol. Biotechnol. 74, 1112–1119. doi: 10.1007/s00253-006-0743-5

25. Chekanov, K., Lobakova, E., Selyakh, I., Semenova, L., Sidorov, R., and Solovchenko, A. (2014). Accumulation of astaxanthin by a new Haematococcus pluvialis strain BM1 from the White Sea coastal rocks (Russia). Mar. Drugs. 12, 4504–4520. doi: 10.3390/md12084504

26. Chen, F., Chen, H., and Gong, X. (1997). Mixotrophic and heterotrophic growth of Haematococcus lacustris and rheological behaviour of the cell suspensions. Bioresour. Technol. 62, 19–24. doi: 10.1016/S0960-8524(97)00115-6

27. Chen, G., Wang, B., Han, D., Sommerfeld, M., Lu, Y., Chen, F., et al. (2015). Molecular mechanisms of the coordination between

astaxanthin and fatty acid biosynthesis in Haematococcus pluvialis (Chlorophyceae). Plant J. 81, 95–107. doi: 10.1111/tpj.12713

28. Chen, Y., Li, D., Lu, W., Xing, J., Hui, B., and Han, Y. (2003). Screening and characterization of astaxanthin-hyperproducing mutants of Haematococcus pluvialis. Biotechnol. Lett. 25, 527–529. doi: 10.1023/A:1022877703008
29. Chew, B. P., Park, J. S., Hayek, M. G., Massimino, S., and Reinhart, G. A. (2004). Dietary astaxanthin stimulates cell-mediated and humoral immune response in cats. FASEB J. 18, 533.
30. Chew, B. P., and Park, J. S. (2004). Carotenoid action on the immune reponse. J. Nutr. 134, 257–261.
31. Choi, Y. E., Yun, Y. S., Park, J. M., and Yang, J. W. (2011). Determination of the time transferring cells for astaxanthin production considering two-stage process of Haematococcus pluvialis cultivation. Bioresour. Technol. 102, 11249–11253. doi: 10.1016/j.biortech.2011.09.092
32. Choubert, G., and Heinrich, O. (1993). Carotenoid pigments of the green alga Haematococcus pluvialis: assay on rainbow trout, Oncorhynchus mykiss, pigmentation in comparison with synthetic astaxanthin and canthaxanthin. Aquaculture 112, 217–226. doi: 10.1016/0044-8486(93)90447-7
33. Chumpolkulwong, N., Kakizono, T., Handa, T., and Nishio, N. (1997). Isolation and characterization of compactin resistant mutants of an astaxanthin synthesizing green alga Haematococcus pluvialis. Biotechnol. Lett. 19, 299–302. doi: 10.1023/A:1018330329357
34. Comhaire, F. H., El Garem, Y., Mahmoud, A., Eertmans, F., and Schoonjans, F. (2005). Combined conventional/ antioxidant "Astaxanthin" treatment for male infertility: a double blind randomized trial. Asian J. Androl. 7, 257–262. doi: 10.1111/j.1745-7262.2005.00047.x
35. Cunningham, F. X., and Gantt, E. (1998). Genes and enzymes of carotenoid biosynthesis in plants. Annu. Rev. Plant Physiol. Plant Mol. Biol. 49, 557–583. doi: 10.1146/annurev.arplant.49.1.557
36. Cyanotech (2015). BioAstin - Natural Astaxanthin. Available online at: http://www.cyanotech.com/bioastin.html (Retrieved 21 October, 2015).

37. Czygan, F. C. (1970). Blood-rain and blood-snow: nitrogendefiient cells of Haematococcus pluvialis and Chlamydomonas nivalis. Arch. Mikrobiol. 74, 69–76. doi: 10.1007/BF00408689
38. Damiani, M. C., Popovich, C. A., Constenla, D., and Leonardi, P. I. (2010). Lipid analysis in Haematococcus pluvialis to assess its potential use as a biodiesel feedstock. Bioresour. Technol. 101, 3801–3807. doi: 10.1016/j.biortech.2009.12.136
39. Daroch, M., Geng, S., and Wang, G. (2013a). Recent advances in liquid biofuel production from algal feedstocks. Appl. Energy 102, 1371–1381. doi: 10.1016/j.apenergy.2012.07.031
40. Daroch, M., Shao, C., Liu, Y., Geng, S., and Cheng, J. J. (2013b). Induction of lipids and resultant FAME profiles of microalgae from coastal waters of pearl river delta. Bioresour. Technol. 146, 192–199. doi: 10.1016/j.biortech.2013.07.048
41. Del Río, E., Acién, F. G., García-Malea, M. C., Rivas, J., Del Rio, E., Acién, F. G., et al. (2005). Efficient one-step production of astaxanthin by the microalga Haematococcus pluvialis in continuous culture. Biotechnol. Bioeng. 91, 808–815. doi: 10.1002/bit.20547
42. Del Rio, E., Acien, F. G., Garcia-Malea, M. C., Rivas, J., Molina-Grima, E., and Guerrero, M. G. (2007). Efficiency assessment of the one-step production of astaxanthin by the microalga Haematococcus pluvialis. Biotechnol. Bioeng. 100, 397–402. doi: 10.1002/bit.21770
43. Del Río, E., Acién, F. G., García-Malea, M. C., Rivas, J., Molina-Grima, E., and Guerrero, M. G. (2008). Efficiency assessment of the one-step production of astaxanthin by the microalga Haematococcus pluvialis. Biotechnol. Bioeng. 100, 397–402. doi: 10.1002/bit.21770
44. Domínguez-Bocanegra, A. R., Guerrero Legarreta, I., Martinez Jeronimo, F., and Tomasini Campocosio, A. (2004). Influence of environmental and nutritional factors in the production of astaxanthin from Haematococcus pluvialis. Bioresour. Technol. 92, 209–214. doi: 10.1016/j.biortech.2003.04.001
45. Dong, S., Huang, Y., Zhang, R., Wang, S., and Liu, Y. (2014). Four different methods comparison for extraction of astaxanthin from green alga Haematococcus pluvialis. Sci. World J. 7, 694–705. doi: 10.1155/2014/694305
46. Dragos, N., Bercea, V., Bica, A., Druga, B., Nicoara, A., and

Coman, C. (2010). Astaxanthin production from a new strain of Haematococcus pluvialis grown in batch culture. Ann. Roman. Soc. Cell Biol. 15, 353–361.

47. Eisenreich, W., Rohdich, F., and Bacher, A. (2001). Deoxyxylulose phosphate pathway to terpenoids. Trends Plant Sci. 6, 78–84. doi: 10.1016/S1360-1385(00)01812-4
48. Elgarem, Y., Lignell, A., and Comhaire, F. H. (2002). "Supplementation with Astaxanthin (Astacarox) improves semen quality in infertile men," in Proceedings of the 13th International Carotenoid Symposium (Honolulu, HI), 180–197.
49. Elliot, A. M. (1934). Morphology and life history of Haematococcus pluvialis. Arch. Protistenk. 82, 250–272.
50. Elwinger, K., Lignell, A., and Wilhelmson, M. (1997). "Astaxanthin rich algal meal (Haematococcus pluvialis) as carotenoid source in feed for laying hens," in Proceedings of the VII European Symposium on the Quality of Eggs and Egg Products (Poznan), 52–59.
51. Eom, H., Lee, C. G., and Jin, E. (2006). Gene expression profile analysis in astaxanthin-induced Haematococcus pluvialis using a cDNA microarray. Planta 223, 1231–1242. doi: 10.1007/s00425-005-0171-2
52. Fábregas, J., Domínguez, A., Regueiro, M., Maseda, A., and Otero, A. (2000). Optimization of culture medium for the continuous cultivation of the microalga Haematococcus pluvialis. Appl. Microbiol. Biotechnol. 53, 530–535. doi: 10.1007/s002530051652
53. Fábregas, J., Otero, A., Maseda, A., and Domínguez, A. (2001). Two-stage cultures for the production of astaxanthin from Haematococcus pluvialis. J. Biotechnol. 89, 65–71. doi: 10.1016/S0168-1656(01)00289-9
54. Fan,L.,Vonshak,A.,andBoussiba,S.(1994).Effectoftemperatureand irradiance on growth of Haematococcus pluvialis(Chlorophyceae). J. Phycol. 30, 829–833. doi: 10.1111/j.0022-3646.1994.00829.x
55. Fassett, R. G., and Combes, J. S. (2011). Astaxanthin: a potential therapeutic agent in cardiovascular disease. Mar. Drugs 9, 447–465. doi: 10.3390/md9030447
56. Ferrante, R. J., Browne, S. E., Shinobu, L. A., Bowling, A. C., Baik, M. J., MacGarvey, U., et al. (1997). Evidence of increased

oxidative damage in both sporadic and familial amyotrophic lateral sclerosis. J. Neurochem. 69, 2064–2074. doi: 10.1046/j.1471-4159.1997.69052064.x

57. Forján, E., Navarro, F., Cuaresma, M., Vaquero, I., Ruíz-Domínguez, M. C., Gojkovic, Z., et al. (2015). Microalgae: fast-growth sustainable green factories. Crit. Rev. Environ. Sci. Technol. 45, 1705–1755. doi: 10.1080/10643389.2014.966426
58. Fujii, K. (2012). Process integration of supercritical carbon dioxide extraction and acid treatment for astaxanthin extraction from a vegetative microalga. Food Bioprod. Process. 90, 762–766. doi: 10.1016/j.fbp.2012.01.006
59. Gacheva, G., Dimitrova, P., and Pilarski, P. (2015). New strain Haematococcus cf. pluvialis Rozhen-12 - growth, biochemical characteristics and future perspectives. Genet. Plant Physiol. 5, 29–38.
60. Gao, Z., Meng, C., Gao, H., Li, Y., Zhang, X., Xu, D., et al. (2013b). Carotenoid genes transcriptional regulation for astaxanthin accumulation in fresh water unicellular alga Haematococcus pluvialis by gibberellin A3 (GA3). Ind. J. Biochem. Biophys. 50, 548–553.
61. Gao, Z., Meng, C., Gao, H., Zhang, X., Xu, D., Su, Y., et al. (2013a). Analysis of mRNA expression profi les of carotenogenesis and astaxanthin production of Haematococcus pluvialis under exogenous. Biol. Res. 46, 201–206. doi: 10.4067/S0716-97602013000200012
62. Gao, Z., Meng, C., Zhang, X., Xu, D., Miao, X., Wang, Y., et al. (2012a). Induction of salicylic acid (SA) on transcriptional expression of eight carotenoid genes and astaxanthin accumulation in Haematococcus pluvialis. Enzyme Microb. Technol. 51, 225–230. doi: 10.1016/j.enzmictec.2012.07.001
63. Gao, Z., Meng, C., Zhang, X., Xu, D., Zhao, Y., Wang, Y., et al. (2012b). Differential expression of carotenogenic genes, associated changes on astaxanthin production and photosynthesis features induced by JA in H. pluvialis. PLoS ONE 7:e42243. doi: 10.1371/journal.pone.0042243
64. García-Malea, M. C., Acién, F. G., Del Río, E., Fernández, J. M., Cerón, M. C., Guerrero, M. G., et al. (2009). Production of

astaxanthin by Haematococcus pluvialis: taking the one-step system outdoors. Biotechnol. Bioeng. 102, 651–657. doi: 10.1002/bit.22076

65. Gomez, P. I., Inostroza, I., Pizarro, M., and Perez, J. (2013). From genetic improvement to commercial-scale mass culture of a Chilean strain of the green microalga Haematococcus pluvialis with enhanced productivity of the red ketocarotenoid astaxanthin. AoB Plants5:plt026. doi: 10.1093/aobpla/plt026

66. Greenwell, H. C., Laurens, L. M. L., Shields, R. J., Lovitt, R. W., and Flynn, K. J. (2010). Placing microalgae on the biofuels priority list: a review of the technological challenges. J. Royal Soc. Int. 7, 703–726. doi: 10.1098/rsif.2009.0322

67. Grewe, C. B., and Griehl, C. (2012). "The carotenoid astaxanthin from Haematococcus pluvialis," in Microalgal Biotechnology: Integration and Economy eds C. Posten and C. Walter (Berlin; Boston, MA: De Gruyter), 129–144. doi: 10.1515/9783110298321.129

68. Guedes, A. C., Amaro, H. M., and Malcata, F. X. (2011). Microalgae as sources of carotenoids. Mar. Drugs 9, 625–644. doi: 10.3390/md9040625

69. Guerin, M., Huntley, M. E., and Olaizola, M. (2003). Haematococcus astaxanthin: applications for human health and nutrition. Trends Biotechnol. 21, 210–216. doi: 10.1016/S0167-7799(03)00078-7

70. Gutiérrez, C. L., Gimpel, J., Escobar, C., Marshall, S. H., and Henríquez, V. (2012). Chloroplast genetic tool for the green microalgae Haematococcus pluvialis (Chlorophyceae, Volvocales)1. J. Phycol. 48, 976–983. doi: 10.1111/j.1529-8817.2012.01178.x

71. Gwak, Y., Hwang, Y. S., Wang, B., Kim, M., Jeong, J., Lee, C. G., et al. (2014). Comparative analyses of lipidomes and transcriptomes reveal a concerted action of multiple defensive systems against photooxidative stress in Haematococcus pluvialis. J. Exp. Bot. 65, 4317–4334. doi: 10.1093/jxb/eru206

72. Hagen, C., Siegmund, S., and Braune, W. (2002). Ultrastructural and chemical changes in the cell wall of Haematococcus pluvialis(Volvocales, Chlorophyta) during aplanospore formation. Eur. J. Phycol. 37, 217–226. doi: 10.1017/S0967026202003669

73. Han, D., Li, Y., and Hu, Q. (2013). "Biology and commercial aspects of Haematococcus pluvialis," in Handbook of Microalgal

Culture: Applied Phycology and Biotechnology, 2nd Edn., eds A. Richmond and Q. Hu (Hoboken, NJ: Blackwell), 388–405. doi: 10.1002/9781118567166.ch20

74. Harker, M., Tsavalos, A. J., and Young, A. J. (1996a). Factors responsible for astaxanthin formation in the chlorophyte Haematococcus pluvialis. Bioresour. Technol. 55, 207–214. doi: 10.1016/0960-8524(95)00002-X
75. Harker, M., Tsavalos, A. J., and Young, A. J. (1996b). Autotrophic growth and carotenoid production of Haematococcus pluvialis in a 30 liter air-lift photobioreactor. J. Ferment. Bioeng. 82, 113–118. doi: 10.1016/0922-338X(96)85031-8
76. Hashimoto, H., Arai, K., Hayashi, S., Okamoto, H., Takahashi, J., Chikuda, M., et al. (2013). Effects of astaxanthin on antioxidation in human aqueous humor. J. Clin. Biochem. Nutr. 53, 1–7. doi: 10.3164/jcbn.13-6
77. Hata, N., Ogbonna, J. C., Hasegawa, Y., Taroda, H., and Tanaka, H. (2001). Production of astaxanthin by Haematococcus pluvialis in a sequential heterotrophic-photoautotrophic culture. J. Appl. Phycol. 13, 395–402. doi: 10.1023/A:1011921329568
78. Hazen, T. E. (1899). The life history of Sphaerella lacustris. Mem. Torrey Bot. Club 6, 211–244.
79. He, P., Duncan, J., and Barber, J. (2007). Astaxanthin accumulation in the green alga Haematococcus pluvialis: effects of cultivation parameters. J. Integr. Plant. Biol. 49, 447–451. doi: 10.1111/j.1744-7909.2007.00468.x
80. Higuera-Ciapara, I., Felix-Valenzuela, L., and Goycoolea, F. M. (2006). Astaxanthin: a review of its chemistry and applications. Crit. Rev. Food Sci. Nutr. 46, 185–196. doi: 10.1080/10408690590957188
81. Hoeffler, J. F., Hemmerlin, A., Grosdemange-Billiard, C., Bach, T. J., and Rohmer, M. (2002). Isoprenoid biosynthesis in higher plants and in Escherichia coli: on the branching in the methylerythritol phosphate pathway and the independent biosynthesis of isopentenyl diphosphate and dimethylallyl diphosphate. Biochem. J. 366, 573–583. doi: 10.1042/bj20020337
82. Hoffman, Y., Aflalo, C., Zarka, A., Gutman, J., James, T. Y., and Boussiba, S. (2008). Isolation and characterization of a novel

chytrid species (phylum Blastocladiomycota), parasitic on the green alga Haematococcus. Mycol. Res. 112, 70–81. doi: 10.1016/j.mycres.2007.09.002

83. Hong, M.-E., Choi, S. P., Park, Y.-I., Kim, Y.-K., Chang, W. S., Kim, B. W., et al. (2012). Astaxanthin production by a highly photosensitive Haematococcus mutant. Process. Biochem. 47, 1972–1979. doi: 10.1016/j.procbio.2012.07.007
84. Hu, Z., Li, Y., Sommerfeld, M., Chen, F., and Hu, Q. (2008). Enhanced protection against oxidative stress in an astaxanthin-overproduction Haematococcus mutant (Chlorophyceae). Eur. J. Phycol. 43, 365–376. doi: 10.1080/09670260802227736
85. Hussein, G., Sankawa, U., Goto, H., Matsumoto, K., and Watanabe, H. (2006). Astaxanthin, a carotenoid with potential in human health and nutrition. J. Nat. Prod. 69, 443–449. doi: 10.1021/np050354+
86. Ikeuchi, M., Koyama, T., Takahashi, J., and Yazawa, K. (2007). Effects of astaxanthin in obese mice fed a high-fat diet. Biosci. Biotechnol. Biochem. 71, 893–899. doi: 10.1271/bbb.60521
87. In, M. (2009). Effect of enzyme treatments on the extraction efficacy and antioxidant activity of haematococcus extract from Haematococcus pluvialis. J. Korea Acad. Indus. Cooperat. Soc. 10, 194–199. doi: 10.5762/KAIS.2009.10.1.194
88. Inbbor, J. (1998). Haematococcus, the poultry pigmentor. Feed Mix. 6, 31–34.
89. Inborr, J., and Lignell, Å. (1997). "Effect of feeding astaxanthin-rich algae meal (Haematococcus pluvialis) on performance and carotenoid concentration of different tissues of broiler chickens," in Proceedings of the XIII WPSA Conference on Poultry Meat Quality in Poznan (Poland: Session M1), 39–43.
90. Industry Experts (2015). Global Astaxanthin Market – Sources, Technologies and Applications. Available online at: http://industry-experts.com/verticals/healthcare-and-pharma/global-astaxanthin-market-sources-technologies-and-applications
91. Issarapayup, K., Powtongsook, S., and Pavasant, P. (2009). Flat panel airlift photobioreactors for cultivation of vegetative cells of microalga Haematococcus pluvialis. J. Biotechnol. 142, 227–232. doi: 10.1016/j.jbiotec.2009.04.014

92. Iwamoto, T., Hosoda, K., Hirano, R., Kurata, H., Matsumoto, A., Miki, W., et al. (2000). Inhibition of low-density lipoprotein oxidation by astaxanthin. J. Atheroscler. Thromb. 7, 216–222. doi: 10.5551/jat1994.7.216

93. Iwasaki, T., and Tawara, A. (2006). Effects of astaxanthin on eyestrain induced by accommodative dysfunction. J. Eye (Atarashii Ganka). 6, 829–834.

94. Jaime, L., RodrIguez-Meizoso, I., Cifuentes, A., Santoyo, S., Suarez, S., and Ibánez, E. (2010). Pressurized liquids as an alternative process to antioxidant carotenoids extraction from Haematococcus pluvialis microalgae. LWT Food Sci. Tech. 43, 105–112. doi: 10.1016/j.lwt.2009.06.023

95. Jia, Z., Liu, Y., Daroch, M., Geng, S., and Cheng, J. J. (2014). Screening, growth medium optimisation and heterotrophic cultivation of microalgae for biodiesel production. Appl. Biochem. Biotechnol. 173, 1667–1679. doi: 10.1007/s12010-014-0954-7

96. Kaewpintong, K., Shotipruk, A., Powtongsook, S., and Pavasant, P. (2007). Photoautotrophic high-density cultivation of vegetative cells of Haematococcus pluvialis in airlift bioreactor. Bioresour. Technol. 98, 288–295. doi: 10.1016/j.biortech.2006.01.011

97. Kakizono, T., Kobayashi, M., and Nagai, S. (1992). Effect of carbon/nitrogen ratio on encystment accompanied with astaxanthin formation in a green alga Haematococcus pluvialis. J. Ferment. Bioeng. 74, 403–405.

98. Kamath, B. S., Srikanta, B. M., Dharmesh, S. M., Sarada, R., and Ravishankar, G. A. (2008). Ulcer preventive and antioxidative properties of astaxanthin from Haematococcus pluvialis. Eur. J. Pharmacol. 590, 387–395. doi: 10.1016/j.ejphar.2008.06.042

99. Kang, C. D., and Sim, S. J. (2008). Direct extraction of astaxanthin from Haematococcus culture using vegetable oils. Biotechnol. Lett.30, 441–444. doi: 10.1007/s10529-007-9578-0

100. Kang, C. D., Han, S. J., Choi, S. P., and Sim, S. J. (2010). Fed-batch culture of astaxanthin-rich Haematococcus pluvialis by exponential nutrient feeding and stepwise light supplementation. Bioproc. Biosys. Eng. 33, 133–139. doi: 10.1007/s00449-009-0362-5

101. Kang, C. D., Lee, J. S., Park, T. H., and Sim, S. J. (2005). Comparison

of heterotrophic and photoautotrophic induction on astaxanthin production by Haematococcus pluvialis. Appl. Microbiol. Biotechnol. 68, 237–241. doi: 10.1007/s00253-005-1889-2

102. Karppi, J., Rissanen, T. H., Nyyssönen, K., Kaikkonen, J., Olsson, A. G., Voutilainen, S., et al. (2007). Effects of astaxanthin supplementation on lipid peroxidation. Int. J. Vitam. Nutr. Res. 77, 3–11. doi: 10.1024/0300-9831.77.1.3

103. Katagiri, M., Satoh, A., Tsuji, S., and Shirasawa, T. (2012). Effects of astaxanthin-rich Haematococcus pluvialis extract on cognitive function: a randomised, double-blind, placebo-controlled study. J. Clinic. Biochem. Nutr. 51, 102–107. doi: 10.3164/jcbn.D-11-00017

104. Kenji, S., Kazuhiro, O., Takuya, N., Yasuhiro, S., Shinki, C., Kazuhiko, Y., et al. (2005). Effect of astaxanthin on accommodation and asthenopia-efficacy -identification study in healthy volunteers. J. Clin. Ther. Med. 21, 637–650.

105. Kim, J. H., Affan, M. A., Jang, J., Kang, M. H., Ko, A. R., Jeon, S. M., et al. (2015). Morphological, molecular, and biochemical characterization of astaxanthin-producin g green microalga Haematococcus sp. KORDI03 Haematococcaceae, Chlorophyta) isolated from Korea. J. Microbiol. Biotechnol. 25, 238–246. doi: 10.4014/jmb.1410.10032

106. Klochkova, T. A., Kwak, M. S., Han, J. W., Motomura, T., Nagasato, C., and Kim, G. H. (2013). Cold-tolerant strain of Haematococcus pluvialis (Haematococcaceae, Chlorophyta) from Blomstrandhalvoya (Svalbard). Algae 28, 185–192. doi: 10.4490/algae.2013.28.2.185

107. Kobayashi, M., Hirai, N., Kurimura, Y., Ohigashi, H., and Tsuji, Y. (1997b). Abscisic acid-dependent algal morphogenesis in the unicellular green alga Haematococcus pluvialis. Plant Growth Regul. 22, 79–85. doi: 10.1023/A:1005862809711

108. Kobayashi, M., Kakizono, T., and Nagai, S. (1993). Enhanced carotenoid biosynthesis by oxidative stress in acetateinduced cyst cells of a green unicellular alga, Haematococcus pluvialis. Appl. Environ. Microbiol. 59, 867–873.

109. Kobayashi, M., Kakizono, T., and Nagai, S. (1991). Astaxanthin production by a green alga, Haematococcus pluvialis accompanied

with morphological changes in acetate media. J. Ferment. Bioeng. 71, 335–339. doi: 10.1016/0922-338X(91)90346-I

110. Kobayashi, M., Kakizono, T., Yamaguchi, K., Nishio, N., and Nagai, S. (1992). Growth and astaxanthin formation of Haematococcus pluvialis in heterotrophic and mixotrophic conditions. J. Ferment. Bioeng. 74, 17–20. doi: 10.1016/0922-338X(92)90261-R

111. Kobayashi, M., Kurimura, Y., and Tsuji, Y. (1997a). Light independent, astaxanthin production by the green microalga Haematococcus pluvialis under salt stress. Biotechnol. Lett. 19, 507–509. doi: 10.1023/A:1018372900649

112. Kobayashi, M., Kurimura, Y., Sakamoto, Y., and Tsuji, Y. (1997c). Selective extraction of astaxanthin and chlorophyll from the green alga Haematococcus pluvialis. Biotechnol. Technol. 11, 657–660. doi: 10.1023/A:1018455209445

113. Koller, M., Muhr, A., and Braunegg, G. (2014). Microalgae as versatile cellular factories for valued products. Algal Res. 6, 52–63. doi: 10.1016/j.algal.2014.09.002

114. Krause, W., Henrich, K., Paust, J., and Ernst, H. (1997). Preparation of Astaxanthin. Available online at: https://www.google.com/patents/US5654488

115. Krichnavaruk, S., Shotipruk, A., Goto, M., and Pavasant, P. (2008). Supercritical carbon dioxide extraction of astaxanthin from Haematococcus pluvialis with vegetable oils as co-solvent. Bioresour. Technol. 99, 5556–5560. doi: 10.1016/j.biortech.2007.10.049

116. Kupcinskas, L., Lafolie, P., Lignell, A., Kiudelis, G., Jonaitis, L., Adamonis, K., et al. (2008). Efficacy of the natural antioxidant astaxanthin in the treatment of functional dyspepsia in patients with or without Helicobacter pylori infection: a prospective, randomized, double blind, and placebo-controlled study. Phytomedicine 15, 391–399. doi: 10.1016/j.phymed.2008.04.004

117. Lababpour, A., Shimahara, K., Hada, K., Kyoui, Y., Katsuda, T., and Katoh, S. (2005). Fedbatch culture under illumination with blue light emitting diodes (LEDs) for astaxanthin production by Haematococcus pluvialis. J. Biosci. Bioeng. 100, 339–342. doi: 10.1263/jbb.100.339

118. Leach, M., Hamilton, L. C., Olbrich, A., Wray, G. M., and

Thiemermann, C. (1998). Effects of inhibitors of the activity of cyclooxygenase-2 on the hypotension and multiple organ dysfunction caused by endotoxin: a comparison with dexa-methasone. Br. J. Pharmacol. 124, 586–592. doi: 10.1038/sj.bjp.0701869

119. Lei, A., Chen, H., Shen, G., Hu, Z., Chen, L., and Wang, J. (2012). Expression of fatty acid synthesis genes and fatty acid accumulation in Haematococcus pluvialis under different stressors. Biotechnol. Biofuels 5:18. doi: 10.1186/1754-6834-5-18
120. Leonardi, P. I., Popovich, C. A., and Damiani, M. C. (2011). "Feedstocks for second-generation biodiesel: microalgae's biology and oil composition," in Economic Effects of Biofuel Production, ed M. A. d. S. Bernardes (InTech), 317–347. doi: 10.5772/23125
121. Li, J., Liu, Y., Cheng, J. J., Mos, M., and Daroch, M. (2015). Biological potential of microalgae in China for biorefinery-based production of biofuels and high value compounds. N. Biotechnol. 32, 588–596. doi: 10.1016/j.nbt.2015.02.001
122. Li, J., Zhu, D. L., Niu, J., Shen, S. D., and Wang, G. (2011). An economic assessment of astaxanthin production by large scale cultivation of Haematococcus pluvialis. Biotechnol. Adv. 29, 568–574. doi: 10.1016/j.biotechadv.2011.04.001
123. Li, M., Wu, W., Zhou, P., Xie, F., Zhou, Q., and Kangsen Mai, K. (2014). Comparison effect of dietary astaxanthin and Haematococcus pluvialis on growth performance, antioxidant status and immune response of large yellow croaker Pseudosciaena crocea. Aquaculture434, 227–232. doi: 10.1016/j.aquaculture.2014.08.022
124. Li, Y., Sommerfeld, M., Chen, F., and Hu, Q. (2010). Effect of photon flux densities on regulation of carotenogenesis and cell viability of Haematococcus pluvialis (Chlorophyceae). J. Appl. Phycol. 22, 253–263. doi: 10.1007/s10811-009-9453-6
125. Lichtenthaler, H. K. (1999). the 1-Deoxy-D-Xylulose-5-Phosphate pathway of isoprenoid biosynthesis in plants. Annu. Rev. Plant Physiol. Plant Mol. Biol. 50, 47–65. doi: 10.1146/annurev.arplant.50.1.47
126. Lichtenthaler, H. K., Rohmer, M., and Schwender, J. (1997). Two independent biochemical pathways for isopentenyl diphosphate and isoprenoid biosynthesis in higher plants. Plant Physiol. 101, 643–

652. doi: 10.1111/j.1399-3054.1997.tb01049.x

127. Lignell, A. (2001). Medicament for Improvement of Duration of Muscle Function or Treatment of Muscle Disorders or Diseased. Varmdo: US patent # 6,245,818 Astacarotene AB.
128. Lignell, A., and Inborr, J. (1999). Agent for Increasing the Production of/in Breeding and Production Mammals. Varmdo: European patent. EP0912106.
129. Lignell, A. N. G. K., and Inborr, J. (2000). Agent for Increasing the Production of/in Breeding and Production Mammals. Varmdo: United States patent and trademark office granted patent. WO97/35491.
130. Linden, H. (1999). Carotenoid hydroxylase from Haematococcus pluvialis: cDNA sequence, regulation and functional complementation. Biochim. Biophys. Acta Gene Struct. Expr. 1446, 203–212. doi: 10.1016/S0167-4781(99)00088-3
131. Liu, B. H., and Lee, Y. K. (2003). Effect of total secondary carotenoids extracts from Chlorococcum sp. on Helicobacter pylori-infected BALB/c mice. Int. Immunopharmacol. 3, 979–986. doi: 10.1016/S1567-5769(03)00096-1
132. Liu, X., and Osawa, T. (2007). Cis astaxanthin and especially 9-cis astaxant hin exhibits a higher antioxidant activity in vitro compared to the all-trans isomer. Biochem. Biophys. Res. Commun. 357, 187–193. doi: 10.1016/j.bbrc.2007.03.120
133. López, M. C. G.-M., Sanchez, E. D. R., López, J. L. C., Fernández, F. G. A., Sevilla, J. M. F., Rivas, J., et al. (2006). Comparative analysis of the outdoor culture of Haematococcus pluvialis in tubular and bubble column photobioreactors. J. Biotechnol. 123, 329–342. doi: 10.1016/j.jbiotec.2005.11.010
134. Lorenz, R. T. (1999). A Technical Review of Haematococcus Algae. NatuRose™ Technical Bulletin #060 (Kailua-Kona, HI: Cyanotech Corporation).
135. Lorenz, R. T., and Cysewski, G. R. (2000). Commercial potential for Haematococcus microalgae as a natural source of astaxanthin. Trends Biotechnol. 18, 160–167. doi: 10.1016/S0167-7799(00)01433-5
136. Lotan, T., and Hirschberg, J. (1995). Cloning and expression in Escherichia coli of the gene encoding beta-C-4-oxygenase,

that converts beta-carotene to the ketocarotenoid canthaxanthin in Haematococcus pluvialis. FEBS Lett. 364, 125–128. doi: 10.1016/0014-5793(95)00368-J

137. Machmudah, S., Shotipruk, A., Goto, M., Sasaki, M., and Hirose, T. (2006). Extraction of astaxanthin from Haematococcus pluvialisusing supercritical CO2 and ethanol as entrainer. Ind. Eng. Chem. Res. Dev. 45, 3652–3657. doi: 10.1021/ie051357k

138. Mata, T. M., Martins, A. A., and Caetano, N. S. (2010). Microalgae for biodiesel production and other applications: a review. Renew. Sust. Energ. Rev. 14, 217–232. doi: 10.1016/j.rser.2009.07.020

139. McBride, R., Behnke, C., Botsch, K., Heaps, N., and Meenach, C. (2013). Use of Fungicides in Liquid Systems. Merryfield Row San Diego, CA: PCT Patent Application WO2013056166 A1.

140. Mendes-Pinto, M. M., Raposo, M. F. J., Bowen, J., Young, A. J., and Morais, R. (2001). Evaluation of different cell disruption processes on encysted cells of Haematococcus pluvialis: effects on astaxanthin recovery and implications for bioavailability. J. Appl. Phycol. 13, 19–24. doi: 10.1023/A:1008183429747

141. Mercer, P., and Armenta, R. E. (2011). Developments in oil extraction from microalgae. Eur. J. Lipid Sci. Tech. 113, 539–547. doi: 10.1002/ejlt.201000455

142. Miao, X., and Wu, Q. (2006). Biodiesel production from heterotrophic microalgal oil. Bioresour. Technol. 97, 841–846. doi: 10.1016/j.biortech.2005.04.008

143. Miki, W. (1991). Biological functions and activities of animal carotenoids. Pure Appl. Chem. 63, 141–146. doi: 10.1351/pac199163010141

144. Milledge, J. J. (2013). Energy Balance and Techno-economic Assessment of Algal Biofuel Production Systems. Doctoral Thesis, University of Southampton, Faculty of Engineering and the Environment.

145. Miyawaki, H., Takahashi, J., Tsukahara, H., and Takehara, I. (2008). Effects of astaxanthin on human blood rheology. J. Clin. Biochem. Nutr. 43, 69–74. doi: 10.3164/jcbn.2008048

146. Molina Grima, E., Belarbi, E. H., Acién Fernández, F. G., Robles Medina, A., and Chisti, Y. (2003). Recovery of microalgal biomass

and metabolites: process options and economics. Biotechnol Adv. 20, 491–515. doi: 10.1016/S0734-9750(02)00050-2

147. Moya, M. J., Sánchez-Guardamino, M. L., Vilavella, A., and Barbera, E. (1997). Growth of Haematococcus lacustris: a contribution to kinetic modelling. J. Chem. Technol. Biotechnol. 68, 303–309.
148. Nagaki, Y., Hayasaka, S., Yamada, T., Hayasaka, Y., Sanada, M., and Uonomi, T. (2002). Effects of astaxanthin on accommodation, critical flicker fusion, and pattern visual evoked potential in visual display terminal workers. J. Trad. Med. 19, 170–173.
149. Nagaki, Y., Mihara, M., Tsukuhara, H., and Ohno, S. (2006). The supplementation effect of astaxanthin on accommodation and asthenopia. J. Clin. Therap. Med. 22, 41–54.
150. Nagata, A., Tajima, T., and Takahashi, J. (2006). Effect of astaxanthin 5 mg on anti fatigue and task performance of human. Carotenoid Sci. 10, 102–106.
151. Naguib, Y. M. A. (2000). Antioxidant activities of astaxanthin and related carotenoids. J. Agric. Food Chem. 48, 1150–1154. doi: 10.1021/jf991106k
152. Nakamura, A., Isobe, R., Otaka, Y., Abematsu, Y., Nakata, D., Honma, C., et al. (2004). Changes in visual function following peroral astaxanthin. Jpn. J. Clin. Opthal. 58, 1051–1054.
153. Nakao, R., Nelson, O. L., Park, J. S., Mathison, B. D., Thompson, P. A., and Chew, B. P. (2010). Effect of astaxanthin supplementation on inflammation and cardiac function in BALB/c mice. Anticancer Res. 30, 2721–2725.
154. Nawrocki, W. J., Tourasse, N. J., Taly, A., Rappaport, F., and Wollman, F. A. (2015). The plastid terminal oxidase: its elusive function points to multiple contributions to plastid physiology. Annu. Rev. Plant Biol. 66, 49–74. doi: 10.1146/annurev-arplant-043014-114744
155. Nishikawa, Y., Minenaka, Y., Ichimura, M., Tatsumi, K., Nadamoto, T., and Urabe, K. (2005). Effects of astaxanthin and vitamin C on the prevention of gastric ulcerations in stressed rats. J. Nutr. Sci. Vit. 51, 135–141. doi: 10.3177/jnsv.51.135
156. Nobre, B., Marcelo, F., Passos, R., Beirao, L., Palavra, A., Gouveia, L., et al. (2006). Supercritical carbon dioxide extraction of astaxanthin and other carotenoids from the microalga Haematococcus pluvialis.

Eur. Food Res. Technol. 223, 787–790. doi: 10.1007/s00217-006-0270-8

157. Olaizola, M. (2000). Commercial production of astaxanthin from Haematococcuspluvialisusing25,000-literoutdoorphotobioreactors. J. Appl. Phycol. 12, 499–506. doi: 10.1023/A:1008159127672

158. Olaizola, M. (2003). Commercial development of microalgal biotechnology: from the test tube to the marketplace. Biomol. Eng. 20, 459–466. doi: 10.1016/S1389-0344(03)00076-5

159. Olguín, E. J., Giuliano, G., Porro, D., Tuberosa, R., and Salamin, F. (2012). Biotechnology for a more sustainable world. Biotechnol. Adv.30, 931–932. doi: 10.1016/j.biotechadv.2012.06.001

160. Orosa, M., Franqueira, D., Cid, A., and Abalde, J. (2005). Analysis and enhancement of astaxanthin accumulation in Haematococcus pluvialis. Bioresour. Technol. 96, 373–378. doi: 10.1016/j.biortech.2004.04.006

161. Orosa, M., Torres, E., Fidalgo, P., and Abalde, J. (2000). Production and analysis of secondary carotenoids in green algae. J. Appl. Phycol. 12, 553–556. doi: 10.1023/A:1008173807143

162. Palozza, P., Torelli, C., Boninsegna, A., Simone, R., Catalano, A., Mele, M. C., et al. (2009). Growth-inhibitory effects of the astaxanthin-rich alga Haematococcus pluvialis in human colon cancer cells. Cancer Lett. 283, 108–117. doi: 10.1016/j.canlet.2009.03.031

163. Pan, J. L., Wang, H. M., Chen, C. Y., and Chang, J. S. (2012). Extraction ofastaxanthin from Haematococcus pluvialis by supercritical carbon dioxide fluid with ethanol modifier. Eng. Life Sci. 12, 638–647.

164. Panis, G. (2015). Commercial Astaxanthin Production Derived by Green Alga Haematococcus pluvialis: A Microalgae Process Model and a Techno-Economic Assessment All Through Production Line. Master Thesis (45 EC), Utrecht University, Netherlands.

165. Parisenti, J., Beirao, L. H., Maraschin, M., Mourino, J. L., Nascimento, V.iera F., Do Nascimento Vieira, F., Bedin, L. H., et al. (2011). Pigmentation and carotenoid content of shrimp fed with Haematococcus pluvialis and soy lecithin. Aquacult. Nutr. 17, 530–535. doi: 10.1111/j.1365-2095.2010.00794.x

166. Park, J. C., Choi, S. P., Hong, M. E., and Sim, S. J. (2014). Enhanced

astaxanthin production from microalga, Haematococcus pluvialisby two-stage perfusion culture with stepwise light irradiation. Bioprocess Biosyst. Eng. 37, 2039–2047. doi: 10.1007/s00449-014-1180-y

167. Park, J. S., Chyun, J. H., Kim, Y. K., Line, L. L., and Chew, B. P. (2010). Astaxanthin decreased oxidative stress and inflammation and enhanced immune response in humans. Nutr. Metab. (Lond). 7:18. doi: 10.1186/1743-7075-7-18
168. Pérez-López, P., González-García, S., Jeffryes, C., Agathos, S. N., McHugh, E., Walsh, D., et al. (2014). Life-cycle assessment of the production of the red antioxidant carotenoid astaxanthin by microalgae: from lab to pilot scale. J. Clea. Prod. 64, 332–344. doi: 10.1016/j.jclepro.2013.07.011
169. Praveenkumar, R., Lee, K., Lee, J., and Oh, Y.-K. (2015). Breaking dormancy: an energy- efficient means of recovering astaxanthin from microalgae. Green Chem. 17, 1226–1234. doi: 10.1039/C4GC01413H
170. Pringsheim, E. G. (1966). Nutritional requirements of Haematococcus pluvialis and related species. J. Phycol. 2, 1–7. doi: 10.1111/j.1529-8817.1966.tb04584.x
171. Proctor, V. W. (1957). Some controlling factors in the distribution of Haematococcus pluvialis. Ecology 38, 457–462. doi: 10.2307/1929890
172. Radakovits, R., Jinkerson, R. E., Darzins, A., and Posewitz, M. C. (2010). Genetic engineering of algae for enhanced biofuel production. Eukaryot. Cell 9, 486–501. doi: 10.1128/EC.00364-09
173. Ranga Rao, A., Harshvardhan Reddy, A., and Aradhya, S. M. (2010). Antibacterial properties of Spirulina platensis, Haematococcus pluvialis, Botryococcus braunii micro algal extracts. Curr. Trends Biotechnol. Pharm. 4, 809–819.
174. Ranga Rao, A., Sarada, R., Shylaja, M. D., and Ravishankar, G. A. (2015). Evaluation of hepatoprotective and antioxidant activity of astaxanthin and astaxanthin esters from microalga-Haematococcus pluvialis. J. Food Sci. Tech. 52, 6703–6710. doi: 10.1007/s13197-015-1775-6
175. Ranga Rao, A., Siew Moi, P., Ravi, S., and Aswathanarayana, R.

G. (2014). Astaxanthin: sources, extraction, stability, biological activities and its commercial applications—a review. Mar. Drugs 12, 128–152. doi: 10.3390/md12010128

176. Ranga Rao, A., Sindhuja, H. N., Dharmesh, S. M., Sankar, K. U., Sarada, R., and Ravishankar, G. A. (2013). Effective inhibition of skin cancer, tyrosinase, and antioxidative properties by astaxanthin and astaxanthin esters from the green alga Haematococcus pluvialis. J. Agric. Food Chem. 61, 3842–3851. doi: 10.1021/jf304609j

177. Ranjbar, R., Inoue, R., Shiraishi, H., Katsuda, T., and Katoh, S. (2008). High efficiency production of astaxanthin by autotrophic cultivation of Haematococcus pluvialis in a bubble column photobioreactor. Biochem. Eng. J. 39, 575–580. doi: 10.1016/j.bej.2007.11.010

178. Razon, L. F., and Tan, R. R. (2011). Net energy analysis of the production of biodiesel and biogas from the microalgae: Haematococcus pluvialis and Nannochloropsis. Appl. Energ. 88, 3507–3514. doi: 10.1016/j.apenergy.2010.12.052

179. Recht, L., Zarka, A., and Boussiba, S. (2012). Patterns of carbohydrate and fatty acid changes under nitrogen starvation in the microalgae Haematococcus pluvialis and Nannochloropsis sp. Appl. Microbiol. Biotechnol. 94, 1495–1503. doi: 10.1007/s00253-012-3940-4

180. Renstrøm, B., Borch, G., Skulberg, O. M., and Liaaen-Jensen, S. (1981). Optical purity of (3S,3′S)-astaxanthin from Haematococcus pluvialis. Phytochemistry 20, 2561–2564. doi: 10.1016/0031-9422(81)83094-4

181. Reyes, F. A., Mendiola, J. A., Ibanez, E., and del Valle, J. M. (2014). Astaxanthin extraction from Haematococcus pluvialis using CO2-expanded ethanol. J. Supercrit. Fluids 92, 75–83. doi: 10.1016/j.supflu.2014.05.013

182. Rippka, R., Deruelles, J., Waterbury, M., Herdman, M., and Stanier, R. (1979). Generic assignments, strain histories and properties of pure culture of cyanobacteria. J. Gen. Microbiol. 111, 1–61. doi: 10.1099/00221287-111-1-1

183. Rohdich, F., Hecht, S., Gärtner, K., Adam, P., Krieger, C., Amslinger, S., et al. (2002). Studies on the nonmevalonate terpene biosynthetic pathway: metabolic role of IspH (LytB) protein. Proc. Natl. Acad.

Sci. U.S.A. 99, 1158–1163. doi: 10.1073/pnas.032658999

184. Rosenberg, J. N., Oyler, G. A., Wilkinson, L., and Betenbaugh, M. J. (2008). A green light for engineered algae: redirecting metabolism to fuel a biotechnology revolution. Curr. Opin. Biotechnol. 19, 430–436. doi: 10.1016/j.copbio.2008.07.008
185. Saha, S. K., McHugh, E., Hayes, J., Moane, S., Walsh, D., and Murray, P. (2013). Effect of various stressregulatory factors on biomass and lipid production in microalga Haematococcus pluvialis. Bioresour. Technol. 128, 118–124. doi: 10.1016/j.biortech.2012.10.049
186. Santos, M. F., and Mesquita, J. F. (1984). Ultrastructural study of Haematococcus lacustris (Girod.) Rostafinski (Volvocales) I. Some aspects of carotenogenesis. Cytologia 49, 215–228. doi: 10.1508/cytologia.49.215
187. Sarada, R., Bhattacharya, S., and Ravishankar, G. A. (2002a). Optimization of culture conditions for growth of the green alga Haematococcus pluvialis. World J. Microbiol. Biotechnol. 18, 517–521. doi: 10.1023/A:1016349828310
188. Sarada, R., Tripathi, U., and Ravishankar, G. A. (2002b). Influence of stress on astaxanthin production in Haematococcus pluvialisgrown under different culture conditions. Process Biochem. 37, 623–627. doi: 10.1016/S0032-9592(01)00246-1
189. Sarada, R., Vidhyavathi, R., Usha, D., and Ravishankar, G. A. (2006). An efficient method for extraction of astaxanthin from green alga Haematococcus pluvialis. J. Agric. Food Chem. 54, 7585–7588. doi: 10.1021/jf060737t
190. Satoh, A., Tsuji, S., Okada, Y., Murakami, N., Urami, M., Nakagawa, K., et al. (2009). Preliminary clinical evaluation of toxicity and efficacy of a new astaxanthin-rich Haematococcus pluvialis extract. J. Clin. Biochem. Nutr. 44, 280–284. doi: 10.3164/jcbn.08-238
191. Sawaki, K., Yoshigi, H., Aoki, K., Koikawa, N., Azum ane, A., Kaneko, K., et al. (2002). Sports performance benefits from taking natural astaxanthin characterized by visual acuity and muscle fatigue improvements in humans. J. Clinic. Therap. Med. 18, 73–88.
192. Seki, T., Sueki, H., Kono, H., Suganuma, K., and Yamashita, E. (2001). Effects of astaxanthin from Haematococcus pluvialis on human skin-patch test; skin repeated application test; effect on

wrinkle reduction. Fragrance J. 12, 98–103.

193. Sharon-Gojman, R., Maimon, E., Leu, S., Zarka, A., and Boussiba, S. (2015). Advanced methods for genetic engineering of Haematococcus pluvialis (Chlorophyceae, Volvocales). Algal Res. 10, 8–15. doi: 10.1016/j.algal.2015.03.022

194. Sheikhzadeh, N., Panchah, I. K., Asadpour, R., Tayefi-Nasrabadi, H., and Mahmoudi, H. (2012a). Effects of Haematococcus pluvialis in maternal diet on reproductive performance and egg quality in rainbow trout (Oncorhynchus mykiss). Anim. Reprod. Sci. 130, 119–123. doi: 10.1016/j.anireprosci.2011.12.010

195. Sheikhzadeh, N., Tayefi-Nasrabadi, H., Oushani, A. K., and Najafi Enferadi, M. H. (2012b). Effects of Haematococcus pluvialissupplementation on antioxidant system and metabolism in rain- bow trout (Oncorhynchus mykiss). Fish Physiol. Biochem. 38, 413–419. doi: 10.1007/s10695-011-9519-7

196. Solovchenko, A. E. (2015). Recent breakthroughs in the biology of astaxanthin accumulation by microalgal cell. Photosynth. Res. 125, 437–449. doi: 10.1007/s11120-015-0156-3

197. Sommer, T. R., Potts, W. T., and Morrissy, N. M. (1991). Utilization of microalgal astaxanthin by rainbow trout (Oncorhynchus mykiss). Aquaculture 94, 79–88.

198. Steinbrenner, J., and Linden, H. (2001). Regulation of two carotenoid biosynthesis genes coding for phytoene synthase and carotenoid hydroxylase during stress-induced astaxanthin formation in the green alga Haematococcus pluvialis. Plant Physiol. 125, 810–817. doi: 10.1104/pp.125.2.810

199. Strittmatter, M., Guerra, T., Silva, J., and Gachon, C. M. M. (2015). A new flagellated dispersion stage in Paraphysoderma sedebokerense, a pathogen of Haematococcus pluvialis. J. Appl. Phycol. 27, 1–6. doi: 10.1007/s10811-015-0700-8

200. Su, Y., Wang, J., Shi, M., Niu, X., Yu, X., Gao, L., et al. (2014). Metabolomic and network analysis of astaxanthin-producing Haematococcus pluvialis under various stress conditions. Bioresour. Technol. 170, 522–529. doi: 10.1016/j.biortech.2014.08.018

201. Sun, H., Kong, Q., Geng, Z., Duan, L., Yang, M., and Guan, B. (2015). Enhancement of cell biomass and cell activity of astaxanthin-

rich Haematococcus pluvialis. Bioresour. Technol. 186, 67–73. doi: 10.1016/j.biortech.2015.02.101

202. Sun, Z., Cunningham, F. X. Jr., and Gantt, E. (1998). Differential expression of two isopentenyl pyrophosphate isomerases and enhanced carotenoid accumulation in a unicellular chlorophyte. Proc. Natl. Acad. Sci. U.S.A. 95, 11482–11488. doi: 10.1073/pnas.95.19.11482
203. Suseela, M. R., and Toppo, K. (2006). Haematococcus pluvialis—a green alga, richest natural source of astaxanthin. Curr. Sci. 90, 1602–1603.
204. Thana, P., Machmudah, S., Goto, M., Sasaki, M., Pavasant, P., and Shotipruk, A. (2008). Response surface methodology to supercritical carbon dioxide extraction of astaxanthin from Haematococcus pluvialis. Bioresour. Technol. 99, 3110–3115. doi: 10.1016/j.biortech.2007.05.062
205. Tjahjono, A. E., Hayama, Y., Kakizono, T., Terada, Y., Nishio, N., and Nagai, S. (1994b). Hyper accumulation of astaxanthin in a green alga Haematococcus pluvialis at elevated temperatures. Biotechnol. Lett. 16, 133–138. doi: 10.1007/BF01021659
206. Tjahjono, A. E., Kakizono, T., Hayama, Y., Nishio, N., and Nagai, S. (1994a). Isolation of resistant mutants against carotenoid biosynthesis inhibitors for a green alga Haematococcus pluvialis, and their hybrid formation by protoplast fusion for breeding of higher astaxanthin producers. J. Ferment. Bioeng. 77, 352–357. doi: 10.1016/0922-338X(94)90003-5
207. Tolasa, S., Cakli, S., and Ostermeyer, U. (2005). Determination of astaxanthin and canthaxanthin in salmonid. Eur. Food. Res. Technol.221, 787–791. doi: 10.1007/s00217-005-0071-5
208. Tominaga, K., Hongo, N., Karato, M., and Yamashita, E. (2012). Cosmetic benefits of astaxanthin on humans subjects. Acta Biochim Pol.59, 43–47.
209. Torrissen, O. J., and Naevdal, G. (1984). Pigmentation of salmonids — Genetical variation in carotenoid deposition in rainbow trout. Aquaculture 38, 59–66. doi: 10.1016/0044-8486(84)90137-6
210. Torzillo, G., Goksan, T., Faraloni, C., Kopecky, J., and Masojidek, J. (2003). Interplay between photochemical activities and pigment

composition in an outdoor culture of Haematococcus pluvialis during the shift from the green to red stage. J. Appl. Phycol. 15, 127–136. doi: 10.1023/A:1023854904163

211. Triki, A., Maillard, P., and Gudin, C. (1997). Gametogenesis in Haematococcus pluvialis Flotow (Volvocales, Chlorophyta). Phycologia36, 190–194. doi: 10.2216/i0031-8884-36-3-190.1

212. Tripathi, U., Sarada, R., Ramachandra Rao, S., and Ravishankar, G. A. (1999). Production of astaxanthin in Haematococcus pluvialiscultured in various media. Bioresour. Technol. 68, 197–199. doi: 10.1016/S0960-8524(98)00143-6

213. Tripathi, U., Venkateshwaran, G., Sarada, R., and Ravishankar, G. A. (2001). Studies on Haematococcus pluvialis for improved production of astaxanthin by mutagenesis. World J. Microbiol. Biotechnol. 17, 143–148. doi: 10.1023/A:1016609815405

214. Valderrama, J. O., Perrut, M., Majewski, W., and Serena, L. (2003). Extraction of Astaxantine and Phycocyanine from Microalgae with Supercritical Carbon Dioxide, J. Chem. Eng. Data 48, 827–830. doi: 10.1021/je020128r

215. Vidhyavathi, R., Venkatachalam, L., Sarada, R., and Ravishankar, G. A. (2008). Regulation of carotenoid biosynthetic genes expression and carotenoid accumulation in the green alga Haematococcus pluvialis under nutrient stress conditions. J. Exp. Bot. 59, 1409–1418. doi: 10.1093/jxb/ern048

216. Wan, M., Hou, D., Li, Y., Fan, J., Huang, J., Liang, S., et al. (2014b). The effective photoinduction of Haematococcus pluvialis for accumulating astaxanthin with attached cultivation. Bioresour. Technol. 163, 26–32. doi: 10.1016/j.biortech.2014.04.017

217. Wan, M., Zhang, J., Hou, D., Fan, J., Li, Y., Huang, J., et al. (2014a). The effect of temperature on cell growth and astaxanthin accumulation of Haematococcus pluvialis during a light–dark cyclic cultivation. Bioresour. Technol. 167, 276–283. doi: 10.1016/j.biortech.2014.06.030

218. Wang, B., Zarka, A., Trebst, A., and Boussiba, S. (2003). Astaxanthin accumulation in Haematococcus pluvialis (Chlorophyceae) as an active photoprotective process under high irradiance. J. Phycol. 39, 1116–1124. doi: 10.1111/j.0022-3646.2003.03-043.x

219. Wang, J. F., Han, D. X., Sommerfeld, M. R., Lu, C. M., and Hu, Q. (2013a). Effect of initial biomass density on growth and astaxanthin production of Haematococcus pluvialis in an outdoor photobioreactor. J. Appl. Phycol. 25, 253–260. doi: 10.1007/s10811-012-9859-4
220. Wang, J. F., Sommerfeld, M. R., Lu, C. M., and Hu, Q. (2013b). Combined effect of initial biomass density and nitrogen concentration on growth and astaxanthin production of Haematococcus pluvialis (Chlorophyta) in outdoor cultivation. Algae 28, 193–202. doi: 10.4490/algae.2013.28.2.193
221. Wang, J., Sommerfeld, M., and Hu, Q. (2009). Occurrence and environmental stress responses of two plastid terminal oxidases in Haematococcus pluvialis (Chlorophyceae). Planta 230, 191–203. doi: 10.1007/s00425-009-0932-4
222. Wang, L., Yang, B., Yan, B., and Yao, X. (2012). Supercritical fluid extraction of astaxanthin from Haematococcus pluvialis and its antioxidant potential in sunflower oil. Innovative Food Sci. Emerg. Technol. 13, 120–127. doi: 10.1016/j.ifset.2011.09.004
223. Wang, P. (2014). Culture Medium and Culture Method for Culturing Haematococcus pluvialis by using Brewery Wastewater. Available online at: https://www.google.com/patents/CN103966103A?cl=en
224. Wayama, M., Ota, S., Matsuura, H., Nango, N., Hirata, A., and Kawano, S. (2013). Three-dimensional ultrastructural study of oil and astaxanthin accumulation during encystment in the green alga Haematococcus pluvialis. PLoS ONE 8:e53618. doi: 10.1371/journal.pone.0053618
225. Wu, Y. H., Yang, J., Hu, H. Y., and Yu, Y. (2013). Lipid-rich microalgal biomass production and nutrient removal by Haematococcus pluvialis in domestic secondary effluent. Ecol. Eng. 60, 155–159. doi: 10.1016/j.ecoleng.2013.07.066
226. Yamashita, E. (2002). Cosmetic benefi of dietary supplements containing astaxanthin and tocotrienol on human skin. Food Style 21, 112–117.
227. Yang, Y., Kim, B., and Lee, J. Y. (2013). Astaxanthin structure, metabolism, and health benefits. J. Hum. Nutr. Food Sci. 1, 1003:1–1003:11.

228. Yin, S., Wang, J., Chen, L., and Liu, T. (2015). The water footprint of biofilm cultivation of Haematococcus pluvialis is greatly decreased by using sealed narrow chambers combined with slow aeration rate. Biotechnol. Lett. 37, 1819–1827. doi: 10.1007/s10529-015-1864-7

229. Yoo, J. J., Choi, S. P., Kim, B. W., and Sim, S. J. (2012). Optimal design of scalable photobioreactor for phototropic culturing of Haematococcus pluvialis. Bioprocess Biosyst. Eng. 35, 309–315. doi: 10.1007/s00449-011-0616-x

230. Yu, X., Chen, L., and Zhang, W. (2015). Chemicals to enhance microalgal growth and accumulation of high-value bioproducts. Front. Microbiol. 6:56. doi: 10.3389/fmicb.2015.00056

231. Yuan, J. P., Peng, J., Yin, K., and Wang, J. H. (2011). Potential health promoting effects of astaxanthin: a high-value carotenoid mostly from microalgae. Mol. Nutr. Food Res. 55, 150–165. doi: 10.1002/mnfr.201000414

232. Zhang, B., Geng, Y., Li, Z., Hu, H., and Li, Y. (2009). Production of astaxanthin from Haematococcus in open pond by two-stage growth one-step process. Aquaculture 295, 275–281. doi: 10.1016/j.aquaculture.2009.06.043

233. Zhang, J., Luo, Z., Li, J., and Liu, Z. (2013). Production method and device for preventing and treating contamination of Paraphysoderma sedebokerensis in Haematococcus pluvialis. PCT Patent Application WO2013127280 A1.

234. Zhang, W., Wang, J., Wang, J., and Liu, T. (2014). Bioresource Technology. Attached cultivation of Haematococcus pluvialis for astaxanthin production. Bioresour. Technol. 158, 329–335. doi: 10.1016/j.biortech.2014.02.044

235. Zhang, X. W., Gong, X. D., and Chen, F. (1999). Kinetic models for astaxanthin production by high cell density mixotrophic culture of the microalga Haematococcus pluvialis. J. Ind. Microbiol. Biotechnol. 23, 691–696. doi: 10.1038/sj.jim.2900685

236. Zhang, X., Pan, L., Wei, X., Gao, H., and Liu, J. (2007). Impact of astaxanthin-enriched algal powder of Haematococcus pluvialis on memory improvement in BALB/c mice. Environ. Geochem. Health 29, 483–489. doi: 10.1007/s10653-007-9117-x

237. Zhekisheva, M., Boussiba, S., Khozin-Goldberg, I., Zarka, A., and

Cohen, Z. (2002). Accumulation of oleic acid in Haematococcus pluvialis (Chlorophyceae) under nitrogen starvation or high light is correlated with that of astaxanthin esters. J. Phycol. 38, 325–331. doi: 10.1046/j.1529-8817.2002.01107.

238. Zhekisheva, M., Zarka, A., Khozin-Goldberg, I., Cohen, Z., and Boussiba, S. (2005). Inhibition of astaxanthin synthesis under high irradiance does not abolish triacylglycerol accumulation in the green alga Haematococcus pluvialis (Chlorophyceae). J. Phycol. 41, 819–826. doi: 10.1111/j.0022-3646.2005.05015.x

239. Zou, T. B., Jia, Q., Li, H. W., Wang, C.-X., and Wu, H.-F. (2013). Response surface methodology for ultrasound-assisted extraction of astaxanthin from Haematococcus pluvialis. Mar. Drugs 11, 1644–1655. doi: 10.3390/md11051644

CHAPTER

7

EFFICIENCY OF VARIOUS FLOCCULANTS IN HARVESTING THE GREEN MICROALGAE TETRASELMIS TETRAHELE (CHLORODENDROPHYCEAE: CHLORODENDRACEAE)

Victor Marco Emmanuel N. Ferriols, Riza O. Aguilar

Institute of Aquaculture, College of Fisheries and Ocean Sciences, University of the Philippines Visayas, Miagao, Iloilo, Philippines

ABSTRACT

This study showed the efficiency of various compounds in flocculating cultures of the microalgae Tetraselmis tetrahele (G. S. West, 1916) for use as potential biodiesel feedstock. Of the five flocculants tested, NaOH and $Al_2(SO_4)_3$ at concentrations of 200 mgL-1 showed the highest ($p< 0.05$) flocculating efficiencies at 96.15% and 98.65%, respectively. In terms settling time, $Al_2(SO_4)_3$ at concentrations of 150 and 200 mgL-1 showed the fastest reactions reaching stable algal flocs only 1 hour after the addition of flocculants. Microscopic examination of resulting flocs showed no signs of cell plasmolysis or structural damage to cell walls for all treatments. Although significant ($p < 0.05$) changes in pH of the resulting supernatant were observed for all treatments, these still fell within the suitable range for algal culture indicating the possibility of reusing the waste water for further cultures.

Key Words: Tetraselmis tetrahele, flocculation, microalgae, harvesting

INTRODUCTION

The green microalgae Tetraselmis tetrahele (G. S. West, 1916) is a widely used microalgal species in aquaculture. It is mainly used in hatcheries as natural food for several aquatic species such as shrimps, bivalves, abalone, and rotifers (Estacion et al 1986; De Pauw & Persoone 1988; Ronquillo et al 1997; Madrones-Ladja et al 2002). Aside from its role in aquaculture, it is also a potential biodiesel feedstock due to its high lipid content and fast growth rate (Ferriols & Aguilar 2008).

In order to use T. tetrahele or any other microalgae species as feedstock for biodiesel production, efficient methods for separating and harvesting the algal biomass from the culture media must be employed. Separation of algal biomass from the culture media can be achieved using several methods. Physical or mechanical means of separation include sedimentation, filtration, and centrifugation. The two latter methods entail high energy costs making the production of algal paste for drying expensive (Shelef et al 1984). Sedimentation, although relatively cheaper takes a long time due to the colloidal nature of algal suspensions. The electric repulsion interactions between algal cells and cell interactions with the surrounding water both contribute to the colloidal nature of microalgal cultures (Tenney et al 1969). In order to increase the rate of sedimentation, chemical means of separating the algal biomass from the culture media are employed. This involves the use of various compounds such as inorganic salts containing polyvalent metal ions and cationic polymers to induce cell flocculation and coagulation. Upon addition of flocculants, microalgal cells form clumps that easily settle to the bottom and can be further separated by simpler filtration methods due to increased particle size of the algal biomass or by decanting due to the settling of the algal biomass out of the culture medium (Shelef et al 1984). The harvesting of algal cells by flocculation is a more convenient process than conventional methods such as centrifugation and gravity filtration because it allows the treatment of large quantities of culture (Lee et al 1998). In using flocculants for harvesting algal cells, considerations such as cost, efficiency, and overall effect on the algal biomass should also be noted (Grima et al 2003).

Inorganic salts such as aluminum sulfate ($Al_2(SO_4)_3$), ferrous sulfate (Fe_SO_4), and calcium chloride (CaCl2) are examples of compounds containing polyvalent metal ions. These inorganic salts have been used as clarifying and flocculating agents in other processes such as wastewater treatment in order to remove particulate matter and other dissolved compounds such as phosphorous (Dao & Daniel 2002; Al Mubaddal et al 2008).

Flocculation of algal cells is also sensitive to adjustments in pH of the culture media. The addition of chemicals such as sodium hydroxide (NaOH) and calcium hydroxide ($Ca(OH)_2$) are sometimes used in wastewater treatment to raise pH levels to the point where magnesium hydroxide ($Mg(OH)_2$) is formed and acts as the primary flocculant (Friedman et al 1977). Aside from the simple adjustment of pH, striking a balanced addition of ions such as Mg^{2+} at optimum pH ranges allows for the formation of stable flocs with faster settling rates (Smith & Davis 2012). Although this method of treating water has been found to be effective, the resulting medium would have to be retreated to neutralize its pH in order to be reused.

Moringa oleifera, considered as a tree vegetable, originates from India and Pakistan but has already been widely introduced in the tropical and subtropical regions of Africa and Asia. Almost all parts of the plant are used for various purposes due to high levels of proteins, vitamins, and minerals (Bosch 2004). Of its many uses, M. oleifera seeds have also been employed as clarifying agents in the treatment of water for the removal of particulate matter including algal cells (Katayon et al 2004; Shehata et al 2008). M. oleifera seeds have been found to have high concentrations of dimeric cationic polymers which act as coagulating agents through the adsorption and neutralization of colloidal charges (Ndbigengesere et al 1995). The use of organic cationic polymers as primary coagulants such as those present in M. oleifera seeds have been considered as low cost water purification methods with potential large scale applications in developing countries.

Although many organic and inorganic flocculants have already been tested for their efficiency in treating wastewater for the removal of particulate matter (including algal cells), the flocculation of algal cells can vary due to several factors. The variable flocculation potential of various algal species depends on the composition and properties of the cell wall, the extent and type of excretions, physiological conditions, age, and other factors (Avnimelech et al 1982). Flocculation reactions of algal biomass are also particularly sensitive to pH, properties of the cellular surface, the concentration of flocculants and divalent cations, and the ionic strength of culture solution (Bilanovic et al 1988). In this study, the flocculating efficiencies of various compounds on T. tetrahele were studied in relation to their concentration, settling time, and effect on pH of the resulting media/supernatant. Microscopic examination of algal cells was also carried out to determine if the various flocculants caused algal cells to rupture.

MATERIAL AND METHOD

Algal culture

This study was conducted at the Multispecies Hatchery of the Institute of Aquaculture, University of the Philippines Visayas, Miagao, and Iloilo, Philippines during the month of May, 2011. T. tetrahele starters were obtained from the Phycology Laboratory of the same institute. Test organisms were grown under a batch culture method using TMRL (Tungkang Marine Research Laboratory) media under constant illumination and aeration at ambient temperatures. T. tetrahele starters were first cultured in three 1 L vessels for 3 days then inoculated and scaled-up into a single 15 L vessel and allowed to grow for another 3 days. This was done to ensure homogenous samples to be used for the flocculation experiments.

Flocculation methods

To determine their flocculating efficiency on T. tetrahele cultures, four inorganic compounds (aluminum sulfate, ferrous sulfate, calcium chloride, and sodium hydroxide) and one organic flocculant (aqueous extract of M. oleifera) were used in jar tests at various concentrations. These compounds were selected based on previous studies on their use in either waste water treatment or harvesting other species of algae. Stock solutions for the four inorganic compounds were prepared at a concentration of 10,000 mgL^{-1} while the aqueous extract of M. oleifera was prepared following the methods of Katayon et al (2004).

Aliquots of algal culture (100 ml) were placed in 150 ml beakers. The test flocculants were then added accordingly to reach final concentrations of 100, 150, and 200 mgL^{-1}. Each concentration for every flocculant was carried out in triplicate. Upon addition of flocculants, beakers were stirred for 1 minute using a magnetic stirrer and left to settle over a 4 hour period. A control group with no flocculants added was also prepared to serve as a reference.

Determination of flocculation efficiency

Flocculation efficiency was determined by measuring the optical density of the cell cultures before flocculation and the residual optical density of the supernatant liquid every hour. Optical density (OD) was read using a spectrophotometer at 680 nm (Lee et al 1998). To determine the flocculating efficiency of the different compounds tested, the following formula was used:

Flocculating Efficiency = (Initial OD – Residual OD)/Initial OD x 100

Microscopic examination of algal cells and flocs

To determine whether the various flocculants used had any effect on the integrity of the cell walls of T. tetrahele, samples of settled material from each treatment after the 4 hour settling period were pipetted out and observed using a compound microscope under 100x magnification.

pH

The initial pH of the culture medium and the final supernatant were also monitored using a standard laboratory pH meter.

Statistical analysis

The data for flocculating efficiency (%) were arcsin transformed. The transformed data for flocculating efficiency and pH were analyzed by one-way ANOVA followed by Duncan's Multiple Range Test at a confidence level of 95%.

RESULTS AND DISCUSSION

Flocculating efficiency. Figure 1 shows the different flocculating efficiencies of the test flocculants at different concentrations after a settling period of 4 hours. Highest flocculating efficiencies for all test flocculants were observed at concentrations of 200 mgL-1 except for that of $FeSO_4$ which had the highest flocculating efficiency at 100 mgL^{-1}. At concentrations of 200 mgL^{-1}, there was no significant difference ($p>0.05$) between flocculating efficiency of NaOH and $Al_2(SO_4)3$ (98.7% and 96.2%, respectively) and both treatments were significantly higher ($p>$ 0.05) between flocculating efficiency of NaOH and $Al_2(SO_4)_3$ (98.7% and 96.2%, respectively) and both treatments were significantly higher ($p< 0.01$) than the other flocculants tested.

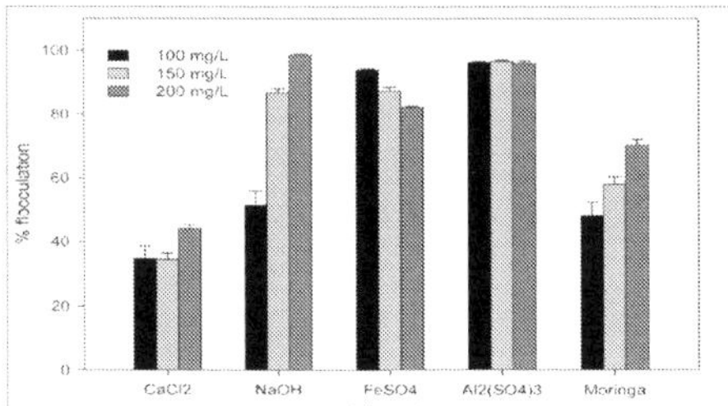

Figure 1: Flocculating efficiency (%) of various flocculants at different levels of concentration (mgL^{-1}) after the 4 hour settling period.

Comparing the flocculating efficiency of each flocculant at different concentrations, increasing efficiency was observed with increasing concentrations for NaOH, $CaCl_2$, and M. oleifera extracts. There was no significant difference ($p>0.05$) between the concentrations for $Al_2 (SO_4)_3$. Treatments with $FeSO_4$ on the other hand showed lower flocculating efficiency as concentrations increased. Settling time. Changes in the optical density of the algal culture under different flocculants throughout the 4 hour settling period are shown in Figure 2. Among the flocculants tested, treatments with $Al_2(SO_4)_3$ at 150 and 200 mgL^{-1} had the fastest settling times displaying stable optical densities 1 hour after the addition of flocculants. This was followed by Al_2 $(SO_4)_3$ at 100 mgL^{-1}, $FeSO_4$ at 200 mgL^{-1}, and M. oleifera extracts at 100 and 150 mgL^{-1} which displayed stable optical densities after 2 hours. All the other treatments, including NaOH at all concentrations, reached stable flocculating efficiencies 3 hours after the addition of flocculants.

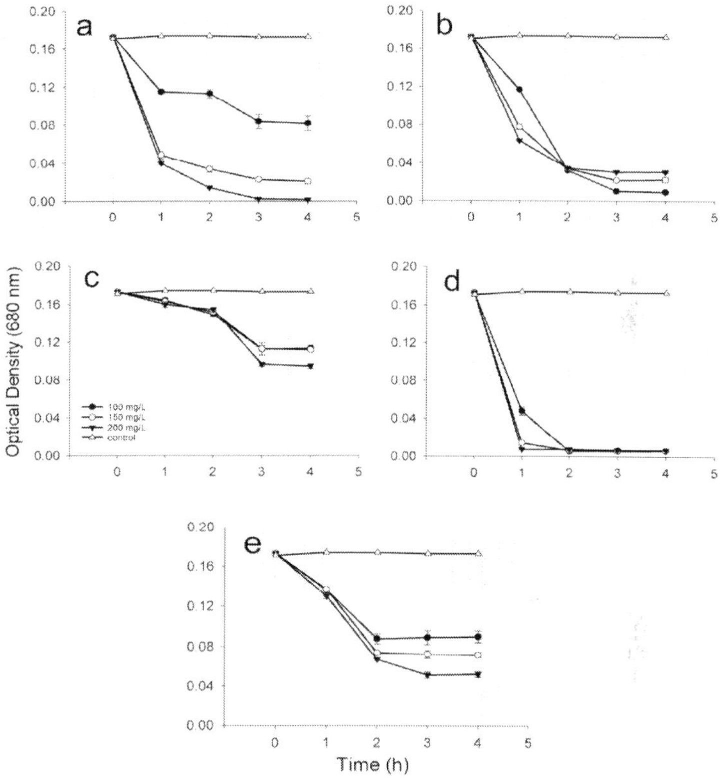

Figure 2: Changes in optical density of the algal culture with the addition of (a) NaOH, (b) $FeSO_4$, (c) $CaCl_2$, (d) $Al_2(SO_4)_3$, (e) M. oleifera extract at different concentrations over 4 hours.

Microscopic examination

Microscopic examination of cells treated with various flocculants showed that cells were in good physical condition. No signs of cell structural damage or plasmolysis were observed under all treatments (Figure

3). The formation of flocs were clearly observed for cells treated with NaOH, $FeSO_4$, and $Al_2(SO_4)_3$ (Figures 3a, 3b, 3c) while those treated with $CaCl_2$ and M. oleifera extracts, although indicating reductions in optical densities, displayed no prominent aggregations of algal cells (Figures 3d, 3e). Compared to untreated cells (Figure 3f), flocculated cells exhibited no distinct changes in size and shape.

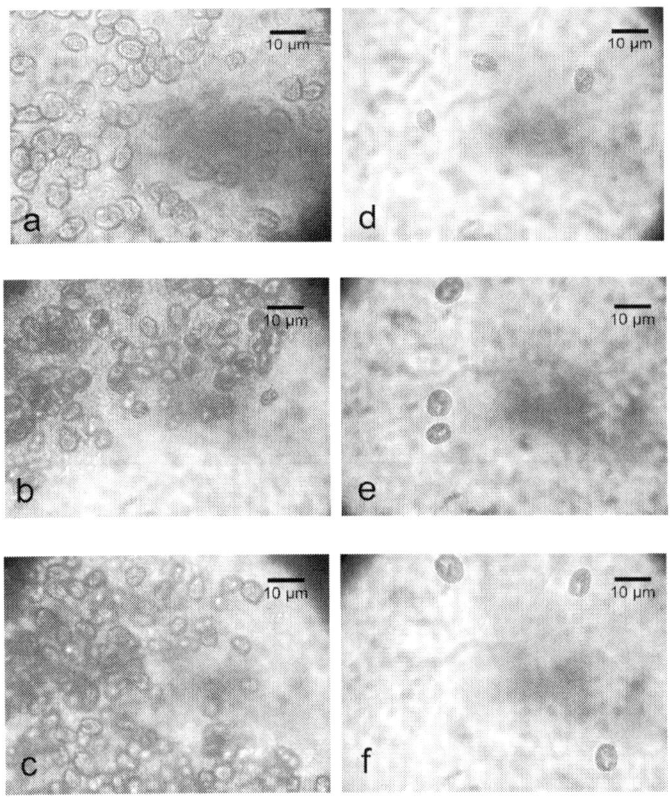

Figure 3: Microscopic examination of flocculated cells of T. tetrahele treated with (a) NaOH, (b) $FeSO_4$, (c) $Al_2(SO_4)_3$, (d) $CaCl_2$ and (e) M. oleifera showing no signs of cell plasmolysis. Comparison with untreated cells (f) shows similar cellular shape and size for all treatments.

pH

Mean pH levels of the resulting supernatants are shown in Table 1. Treatments with $FeSO_4$, $Al_2(SO_4)_3$, $CaCl_2$ and *M. oleifera* showed significantly lower ($p<0.01$) pH levels while treatments with NaOH showed significantly higher ($p<0.01$) pH levels when compared to the pH of the initial culture media. Although pH levels for the former flocculants were significantly lower, mean pH readings at all concentrations still showed that the resulting supernatants were neutral. Treatments using NaOH however resulted in supernatants that were slightly alkaline with mean pH levels ranging from 8.02 (for 100 mgL^{-1} NaOH) to 8.42 (for 200 mgL^{-1} NaOH). It was also observed that increasing concentrations of $FeSO_4$, $Al_2 (SO_4)_3$, and *M. oleifera* had an inverse relationship with pH levels of the resulting supernatant while the opposite was observed for treatments with NaOH. No significant difference was observed in the pH levels of samples treated with $CaCl_2$ at various concentrations.

Table 1: Mean pH of resulting supernatants after 4 hour settling period

Flocculant concentration (mgL^{-1})	*Flocculant*				
	$CaCl_2$	*NaOH*	*$FeSO_4$*	*$Al_2(SO_4)_3$*	*Moringa*
100	7.71±0.01	8.02±0.03	7.51±0.02	7.46±0.03	7.62±0.02
150	7.71±0.02	8.17±0.03	7.30±0.02	7.11±0.01	7.53±0.03
200	7.72±0.01	8.42±0.03	6.98±0.02	6.99±0.02	7.42±0.02

Note: mean pH of initial culture media is 7.79.

This study showed that the most effective flocculants for the separation of *T. tetrahele* cells from culture media were NaOH and $Al_2(SO_4)_3$. Knuckey et al (2006) reported that adjustment of pH using NaOH with subsequent addition of a non-ionic polymer was effectively applied to harvest cells of *Chaetoceros calcitrans, C. muelleri, Thalassiosira pseudonana, Attheya septentrionalis, Nitzschia closterium, Skeletonema sp., Tetraselmis suecica* and *Rhodomonas salina*, with efficiencies ≥80%. Similarly, a study by Lee et al (1998) also showed that pH adjustment using NaOH was more effective in harvesting *Botryococcus braunii* cells compared to aluminum sulfate and a cationic polyelectrolyte. It was also reported that aluminum sulfate was more superior in comparison with other inorganic salts in terms of optimal dose, pH, and the quality of the resulting water and algal slurry (Bare et al 1975; Moraine et al 1980). Lee et al (1998) also demonstrated that aluminum sulfate was more efficient in harvesting *Chlorella sp.* cells compared to pH adjustment using NaOH. Although the efficiency of one particular flocculant may vary from species to species depending on several factors (Avnimelech et al 1982; Bilanovic et al 1988), this study showed the efficiency of both aluminum sulfate and NaOH in separating *T. tetrahele* cells from the culture media.

This study also shows the direct relationship observed between concentration and flocculating efficiency for NaOH, $Al_2(SO_4)_3$, $CaCl_2$, and *M. oleifera*. Bilanovic et al (1988) and Wyatt et al (2012) reported flocculation reactions can be sensitive to a number of factors, one of which is flocculant concentration. As the trends in this study show, optimal doses for each flocculant could be determined in future studies. Although NaOH and $Al_2(SO_4)_3$ were shown to be more efficient flocculants, the use of aqueous extracts of *M. oleifera* could still be a cheaper alternative to chemical flocculants given higher concentrations. The use of aqueous extracts of *M. oleifera* as flocculants have been reported by Katayon et al (2004) with flocculating efficiencies of up to 94% at concentrations of 400 mgL^{-1}. Shehata et al (2008) also reported efficiencies of up to 97% in flocculating a mixture of algal cells using *M. oleifera* extracts. The decreasing flocculating efficiency of $Fe(SO_4)$ as concentrations increased was a notable observation since this digressed from the trend of all other tested flocculants. Similar observations by Aziz et al (2007) showed that using $Fe(SO_4)$ as a clarifying agent for landfill leachate at a controlled pH of 6 resulted in a notable reduction in the efficiency of $Fe(SO_4)$ as concentrations increased from 0 to 500 mgL^{-1}.

The maintenance of cell wall integrity during the harvesting process has been shown to improve the shelf life of harvested cells and the conservation of cell metabolites (Knuckey et al 2006; Knuckey 1998). In the process of extracting oil from algae, although the disruption of the cell wall is necessary to maximize lipid recovery, this is only done after cells have been sufficiently dewatered and dried. During the dewatering and drying stages of the algal biomass (and even during storage if extraction does not proceed directly), retention of cell wall integrity could help in minimizing the oxidation of lipids within the cells. This study showed that all flocculants used at varying concentrations did not exhibit any signs of cell wall disruption indicating the suitability of using these flocculants as a means of harvesting algal cells for oil extraction.

With regards to pH, this study showed that all flocculants used had a significant effect in changing the pH of the culture media after flocculation regardless of concentration. All flocculants lowered the pH of the resulting supernatant except for NaOH which had the opposite effect. Monitoring the pH of the resulting supernatant was done in order to determine if the resulting water after flocculation was suitable for reculture. A study by

Azov (1982) showed that lower pH ranges for culture media resulted in higher biomass concentrations. Although significant changes in pH were observed for all treatments, these levels were still within the tolerable range for algal culture of pH 7-9 (Coutteau 1996).

CONCLUSIONS

The present study showed the efficiency of various compounds in flocculating cultures of the microalgae *T. tetrahele*. Of the five flocculants tested, NaOH and $Al_2(SO_4)_3$ at concentrations of 200 mgL^{-1} showed the highest ($p<0.01$) flocculating efficiencies at 96.15% and 98.65%, respectively. In terms of settling time, $Al_2(SO_4)_3$ at concentrations of 150 and 200 mgL^{-1} showed the fastest reactions reaching stable algal flocs only 1 hour after the addition of flocculants.

All flocculants tested had no effect on the integrity of the cell walls of the test organism. Microscopic examination showed algal cells with no apparent structural damage or plasmolysis.

Measurement of pH for all treatments showed significant changes ($p<0.01$) in the pH of the culture media four hours after the addition of flocculants. Aqueous extract of *M. oleifera*, ferrous sulfate, aluminum sulfate, and calcium chloride all resulted in lower pH values for the resulting supernatant displaying an inverse relationship between flocculant concentration and pH. NaOH on the other hand had an opposite effect, raising pH levels of the resulting supernatant which was directly related with flocculant concentration. Despite significant changes in pH, all were still within the range suitable for algal culture indicating the possibility of reusing the supernatant for further cultures.

Further studies regarding optimum dosage of flocculants should still be carried out for *T. tetrahele*. This is of particular interest with regards to the use of *M. oleifera* since it has the potential to become a cheaper alternative to inorganic salts and other synthetic polymers. Although pH of the culture media plays a major role in the culture of algae, other parameters such as nutrient concentration (i.e. ammonia, nitrite, nitrate, and phosphorous) should also be included in future studies regarding the possibility of reusing waste water after flocculation.

ACKNOWLEDGEMENTS

This study was financed by the Department of Science and Technology – Philippine Council for Aquatic and Marine Research and Development (DOST-PCAMRD) within the program Biodiesel from Philippine Algae.

The authors would like to thank Prof. Ida Pahila, Sharon Rose dela Rama, Roland Joseph Calawigan, Mellie June Paulines for their valuable input during the conduct of this study.

REFERENCES

1. Al Mubaddal F., Al Rumaihi K., Ajbar A., 2008 Performance optimization of coagulation/flocculation in the treatment of wastewater from a polyvinyl chloride plant. Journal of Hazardous Materials 161:431-438.
2. Avnimelech Y., Troeger B. W., Reed L. W., 1982 Mutual flocculation of algae and clay: evidence and implications. Science 216:63–65.
3. Aziz H. A., Alias S., Assari F., Adlan M. N., 2007 The use of alum, ferric chloride and ferrous sulphate as coagulants in removing suspended solids, colour and COD from semi-aerobic landfill leachate at controlled pH. Waste Manag Res 25:556-565.
4. Azov Y., 1982 Effect of pH on inorganic carbon uptake in algal cultures. Applied and Environmental Microbiology 43:1300-1306.
5. Bare W. F. R., Jones N. B., Middlebrooks E. J., 1975 Algae removal using dissolved air flotation. J Water Poll Control Fed 47:153-169.
6. Bilanovic D., Shelef G., Sukenik A., 1988 Flocculation of microalgae with cationic polymers: effects of medium salinity. Biomass 17:65–76.
7. Bosch C. H., 2004 Moringa oleífera. In: Plant Resources of Tropical Africa: Vegetables. Grubben G. J. H., Denton O. A. (eds), PROTA Foundation, Wageningen, Netherlands/Backhuys Publishers, Leiden, Netherlands/CTA, Wageningen, Netherlands, pp. 392-395.
8. Coutteau P., 1996 Micro algae. In: Manual on the production and use of live food for aquaculture. Lavens P., Sorgeloos P. (eds), FAO Fisheries Technical Paper. No. 361. Rome, FAO, pp. 7-78.
9. Dao T. H., Daniel T. C., 2002 Particulate and dissolved phosphorus chemical separation and phosphorus release from treated dairy

manure. J Environ Qual 31:1388–1398.

10. De Pauw N., Persoone G., 1988 Micro-algae for aquaculture. In: Micro-algal Biotechnology. Borowitzka M. A., Borowitzka L. J. (eds), Cambridge University Press, Cambridge, U. K., pp. 197-221.
11. Estacion J., Solis E., Fabro L., 1986 A preliminary study of the effect of supplementary feeding on the growth of Tridacna maxima (Roding). Siliman Journal 33:111-116.
12. Ferriols V. M. E. N., Aguilar R. O., 2008 Analysis of the lipid content of micro algae species used in Philippine aquaculture for their potential in biodiesel production. Proceedings of the 5th World Fisheries Congress. Yokohama, Japan, October 20-24, 2008.
13. Friedman A. A., Peaks D. A., Nichols R. L., 1977 Algae separation from oxidation pond effluents. J Water Poll Control Fed 49:111-119.
14. Grima E. M., Belarbi E. H., Fernandez F. G. A., Medina A. R., Chisti Y., 2003 Recovery of microalgal biomass and metabolites: process options and economics. Biotechnology Advances 20:491–515.
15. Katayon S., Megat Mohd Noor M. J., Asma M., Thamer A. M., Liew Abdullah A. G., Idris A., Suleyman A. M., Aminuddin M. B., Khor B. C., 2004 Effects of storage duration and temperature of Moringa oleifera stock solution on its performance in coagulation. International Journal of Engineering and Technology 1:146–151.
16. Knuckey R. M., Brown M. R., Robert R., Frampton D. M. F., 2006 Production of microalgal concentrates by flocculation and their assessment as aquaculture feeds. Aquacultural Engineering 35:300-313.
17. Knuckey R. M., 1998 Australian microalgae and microalgal concentrates for use as aquaculture feeds. Ph. D. thesis, University of Tasmania, 254 pp.
18. Lee S. J., Kim S. B., Kwon G. S., Yoon B. D., Oh H. M., 1998 Effects of harvesting method and growth stage on the flocculation of the green alga Botryococcus braunii. Letters in Applied Microbiology 27:14-18.
19. Madrones-Ladja J., dela Peña M., Parami N., 2002 The effect of micro algal diet and rearing condition on gonad maturity, fecundity, and embryonic development of the window pane shell, Placuna placenta Linnaeus. Aquaculture 206:313-321.

20. Moraine R., Shelef G., Sandbank E., Bar Moshe Z., Schwarbard L., 1980 Recovery of sewage born algae: flocculation and centrifugation techniques. In: Algae Biomass. Shelef G., Solder C. J. (eds), Elsevier/North Holland, pp. 531-546.
21. Ndbigengesere A., Narasiah K. S., Talbot B. G., 1995 Active agent and mechanism of coagulation of turbid waters using Moringa oleifera. Wat Resources 2:703-710.
22. Ronquillo J. D., Saisho T., Yamasaki S., 1997 Culture of Tetraselmis tetrahele and its utilization in the hatchery production of different penaeid shrimps in Asia. Hydrobiologia 358:237-244.
23. Shelef G., Sukenik A., Green M., 1984 Microalgae harvesting and processing: a literature review. SERI Technical Paper. Golden, Colorado, USA, 57 pp.
24. Shehata S. A., Ali G. H., Wahba S. Z., 2008 Distribution pattern of Nile water algae with reference to its treatability in drinking water. Journal of Applied Sciences Research 4:722-730.
25. Smith B. T., Davis R. H., 2012 Sedimentation of algae flocculated using naturallyavailable, magnesium-based flocculants. Algae Research 1(1):32-39.
26. Tenney M. W., Echelberger W. F. Jr., Schuessler R. G., Pavoni J. L., 1969 Algae flocculation with synthetic organic polyelectrolytes. Appl Microb 18:965-971.
27. Wyatt N. B., Gloe L. M., Brady P. V., Hewson J. C., Grillet A. M., Hankins M. G., Pohl P. I., 2012 Critical conditions for ferric chloride-induced flocculation of freshwater algae. Biotechnol Bioeng 109(2):493-501.

CHAPTER

8

HARVESTING OF FRESHWATER MICROALGAE WITH MICROBIAL BIOFLOCCULANT: A PILOT-SCALE STUDY

Theoneste Ndikubwimana[1], Xianhai Zeng[2,3], Theophile Murwanashyaka[1], Emmanuel Manirafasha[1], Ning He[1,3], Wenyao Shao[1] and Yinghua Lu[1,3]

[1] Department of Chemical and Biochemical Engineering, College of Chemistry and Chemical Engineering, Xiamen University, Xiamen, China

[2] College of Energy, Xiamen University, Xiamen, China

[3]The Key Laboratory for Synthetic Biotechnology of Xiamen City, Xiamen University, Xiamen, China

ABSTRACT

Background

Nowadays, bioflocculation is considered as a potential technology that could be able to alleviate microalgae dewatering cost regarded as the cornerstone hindrance of their full-scale application. However, most bioflocculation studies reported are laboratory scales. This study examined a pilot-scale and in situ flocculation of freshwater microalgae *Desmodesmus brasiliensis* by microbial bioflocculant. Biochemical composition of microalgal biomass was analyzed to evaluate the applicability of bioflocculation for microalgae-based biofuel production.

Results

The flocculation efficiency >98 % was achieved at both pilot-scale and in situ treatment. Bioflocculation is simple, effective, economic, and environmentally friendly. Even though total proteins recovered from biomass harvested by centrifugation and that harvested by bioflocculation were significantly different, there was no significant difference in total carbohydrates and total lipids recovered from either biomass harvested by centrifugation or biomass harvested by bioflocculation.

Conclusion

The results herein presented, doubtlessly demonstrated that the γ-PGA bioflocculant produced by *Bacillus licheniformis* CGMCC 2876 is applicable for commercial-scale microalgae harvesting. In addition, bioflocculation process cost could greatly be reduced by in situ operation as no investment cost is needed for a separate flocculation tank and mixing device. Furthermore, bioflocculation method developed is a worthy microalgae harvesting method for algal-based biofuel production.

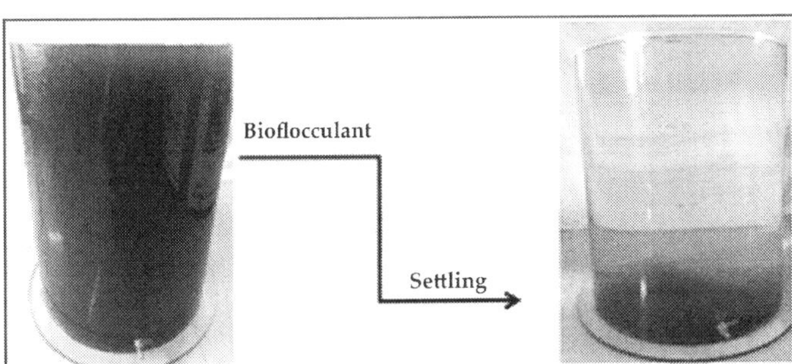

Graphical abstract

The addition of bioflocculant to microalgae cultures followed by mixing elicits, the formation of heavy flocs which settle out by gravity sedimentation in a relatively short settling time.

Keywords: Microalgae Pilot scale In situ flocculation Bioflocculant Biochemical composition

BACKGROUND

Algal biomass is considered as the most assuring raw material to counterbalance the unremitting global demand for food, feed, and biofuel and chemical production [1, 2]. Microalgae have appreciable growth rate, high lipids, and carbohydrate yield and other biochemicals such as proteins and vitamins, also microalgae cultivation can be incorporated in different environmental bioremediation schemes [3, 4, 5]. However, regardless of these advantages, the major challenge lies in the dewatering of the microalgae cultures due to their high dilution rate, minor cell dimensions, and electronegative cell surface charge [6].

The concentrating reactions of algal biomass are particularly sensitive to pH, properties of the cellular surface, concentrations of the flocculants, and ionic strength of the culture solution [1, 7, 8], thus the most dewatering methods currently available are obstructed by either economic or technical drawbacks. It is known that by varying the initial microalgae culture pH acts upon the membrane surface charge of the microalgal cells and the ionic forms of dissolved salts available in the culture suspension will be modified [9]. For example, flocculation induced by pH increase is ascribable to precipitation of $CaCO_3$, $Mg(OH)_2$, and calcium phosphate [9], while the flocculation at decreased pH is due to charge neutralization as the carboxylate ions of organic matters attached to microalgal cells accept protons as a result of pH reduction [10].

Most of the solid–liquid separation methods applied for microalgae suspensions dewatering are likely applied to lab-scale conditions, and would issue severe challenges such as high energy consumption, long processing times, low recovery, and high greenhouse gas emissions, once applied at large-scale conditions [11].

Lately, naturally occurring microbial flocculants have been used to harvest microalgae for aquaculture and biodiesel production because of their high harvesting efficiency, and biodegradability [12, 13].

Bioflocculation is believed to address substantially dewatering cost since little or no energy consumption is required compared to centrifugation mostly applied in industry [14]. Moreover, much less capital and maintenance costs are incurred [7], and microalgae dewatering by bioflocculation has achieved significant efficiencies [7, 15, 16, 17, 18]. Furthermore, bioflocculation is an innovative dewatering method, and environmentally friendly. Bioflocculation is a natural flocculation process hastened with biomolecules from microbial cells [14].

Recently, an innovative, economic, and environmentally friendly microalgae dewatering applying bacterial broth bioflocculant produced by *Bacillus licheniformis* CGMCC 2876 (containing active constituent of poly γ-glutamic acid, γ-PGA) was reported [7], with the mechanism governing this bioflocculation process [19]. High flocculation of 95 % for marine microalgae *Chlorella vulgaris* and freshwater microalgae *C. protothecoides* was reported by inducing flocculation with commercial γ-PGA bioflocculant produced by *B. subtilis* [18]. Furthermore, more than 98 % microalgae cells of *C. vulgaris* were entrapped in the fungal clumps as a result of co-cultivation of microalgae *C. vulgaris* with filamentous fungi [16, 17]. Although this harvesting method seemed promising, the increase in biomass was not proportional to lipid yield due to the decrease in culture pH. However, to our knowledge, no report was published about the microalgae dewatering by bioflocculation using bacterial bioflocculant at large scale.

This study is aimed to evaluate the scalability of the bioflocculation method for microalgae harvesting using bacterial broth bioflocculant produced by *B. licheniformis* CGMCC 2876, and the effectiveness of in situ treatment by the same method. The effect of bioflocculation technology on biochemical composition of microalgae biomass is investigated and presented.

RESULTS

Flocculation character of microalgae cultures

Significant flocculation efficiencies were achieved by only changing the initial culture pH [9, 10]. In addition, higher or lower bioflocculant dosage may decrease the flocculation efficiency [7, 18]. Moreover, flocculation process is significantly affected by mixing. Effective mixing is necessary to alleviate the dispersion of the flocculant within the particles in as short period as possible, thus granting to obtain a uniform mixture of the flocculant and suspended cells to maximize effective destabilization of colloidal particles in order to initiate flocculation. After flash mixing, it is important to averagely mix the microalgae cultures so as to step-up the contact between flocculating particles and to facilitate the growth of large flocs [7]. The settling time depends on the floc size and it has been pointed out that when the flocs settle faster, the quality of particle removal is better. Experimental results have demonstrated that the addition of the flocculating agents followed by flash mixing and slow mixing allowed better interactions and flocs growth thus enhancing the flocculation performance [7, 20]. Therefore, it is imperative to optimize the operating parameters prior to scale up the flocculation process.

In an attempt to scale up bioflocculation process for the freshwater microalgae *Desmodesmus brasiliensis*, firstly 5 L of microalgal cultures (biomass concentration: 0.5 and 1 g/L) were flocculated at operating conditions of 2.5 mL/L bioflocculant, 200 rpm mixing rate for 2 min at pH 3. The results are presented in Fig. 1.

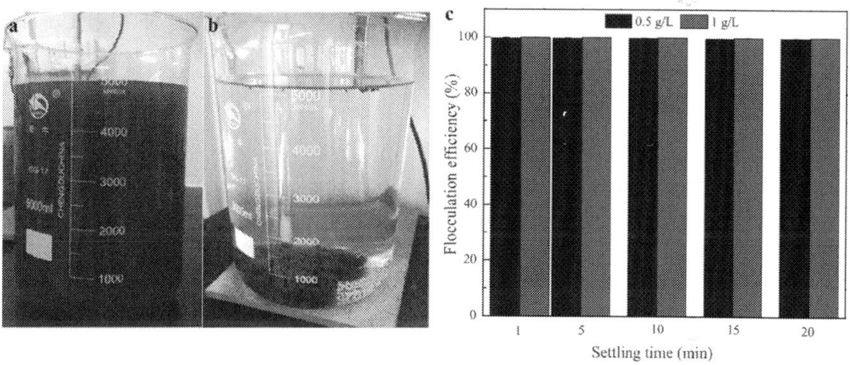

Figure 1: Bioflocculation of 5 L cultures of *D. brasiliensis* with the broth of *B. licheniformis*CGMCC 2876. **a** Microalgae cultures; **b** Flocculated microalgae; **c** Flocculation efficiency at different sedimentation times; *data* presented are the mean values of two independent replicates, and the *bars* represent the standard error of two replicates

The microalgal suspensions are seen with a dark-green color before flocculation (Fig. 1a) while flocs can be visualized settled at the bottom of the flocculation vessel after flocculation (Fig. 1b). The flocculation efficiency above 99 % was achieved after only 1 min of settling time either at 0.5 or 1 g/L of biomass concentration (Fig. 1c), the same flocculation rate was maintained after 20 min of settling time. These results are in agreement with small-scale results at which no significant differences were observed in flocculation efficiencies within the range of biomass concentration of 0.5–1.5 g/L [7]. Therefore, in further experiments, 0.5 g/L of biomass concentration was used.

After bioflocculation of 5 L of microalgae cultures, 50 L was also flocculated following the same procedures as described for flocculation of 5 L microalgae cultures. It can be seen from the results depicted in Fig. 2a that microalgal suspensions were evenly distributed in flocculation vessel while a deposit of dark green (flocs) is discernible at the bottom of the flocculation vessel (Fig. 2b). A flocculation efficiency of 99.2 ± 0.2 % was achieved after only 1 min of settling time (Fig. 2c), and more than 98 % flocculation efficiency was maintained after a long period of sedimentation time.

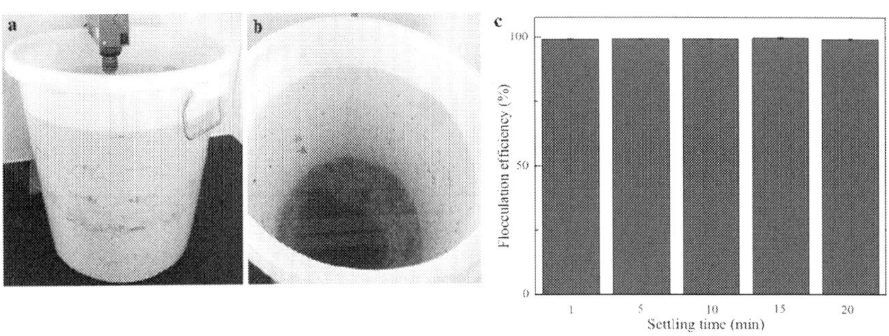

Figure 2: Bioflocculation of 50 L cultures of *D. brasiliensis* with the broth of *B. licheniformis*CGMCC 2876. **a** Microalgae cultures; **b** Flocculated microalgae; **c** Flocculation efficiency at different sedimentation times; *data* presented are the mean values of two independent replicates, and the *bars* represent the standard error of two replicates)

Further scale up was conducted with 200 L of microalgae cultures, and the operating parameters were the same as described above. The results presented in Fig. 3a, b show the visual characteristics of the cultures before and after flocculation, respectively. The dark-green color observable in Fig. 3a disappeared in Fig. 3b with dense algal biomass (flocs) at the bottom of the flocculation tank. From this view, high flocculation efficiency is expected, and this hypothesis is evidenced by the experimental results depicted in Fig. 3c. The flocculation efficiency of 97.5 ± 0.4 % was achieved after only 1 min of settling time, and after

10 min of settling time, flocculation efficiency above 99 % was achieved. These results demonstrate that this novel technology of harvesting freshwater microalgae by γ-PGA broth bioflocculant is effective.

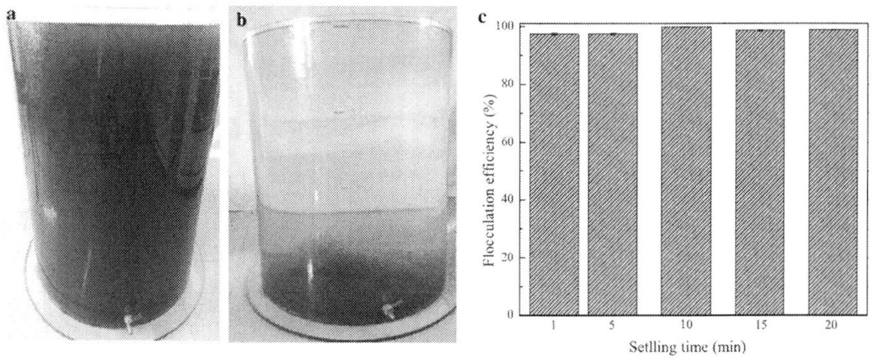

Figure 3: Bioflocculation of 200 L cultures of *D. brasiliensis* with the broth of *B. licheniformis*CGMCC 2876. **a** Microalgae cultures; **b** Flocculated microalgae; **c** Flocculation efficiency at different sedimentation times; *data* presented are the mean values of two independent replicates, and the *bars* represent the standard error of two replicates

In situ flocculation experiments produced the results presented in Fig. 4. The evenly dispersed microalgal cultures with dark-green color observable in Fig. 4a, c formed an algal biomass paste (flocs) at the bottom of the photobioreactor (PBR) (Fig. 4b, d). After only one min of settling time, a flocculation efficiency >98 % was achieved and the flocculation efficiency above 99 % was achieved after 20 min of settling time for both indoor and outdoor cultures treatment (Fig. 4e, f). These results demonstrate that in situ bioflocculation of freshwater microalgae *D. brasiliensis*is is feasible, however further studies are necessary to find optimum aeration rate required for effective mixing.

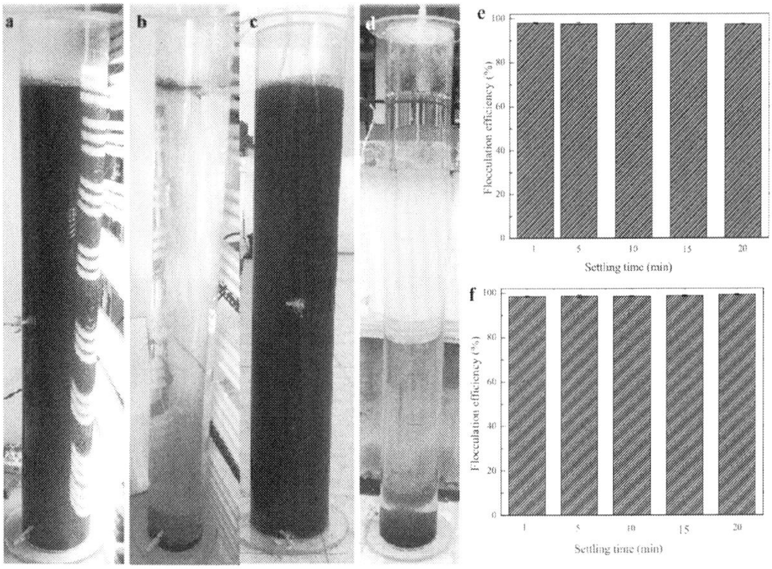

Figure 4: In situ bioflocculation of *D. brasiliensis* with the broth of *B. licheniformis* CGMCC 2876. **a**, **c** Microalgae cultures; **b**, **d** Flocculated microalgae; **e**, **f** Flocculation efficiency at different sedimentation times; *data* presented are the mean values of two independent replicates, and the *bars* represent the standard error of two replicates

Microalgae biomass characterization

Quantification of total carbohydrates, total proteins, and total lipids conducted as described in "Methods" generated the results presented in Table 1.

Table 1: Microalgal biomass biochemical composition

Harvesting method	Total carbohydrates (%)	Total proteins (%)	Total lipids (%)
Centrifugation	48.9 ± 3.4	18.4 ± 0	24.1 ± 2.5
Bioflocculation	44.25 ± 2.3	21.1 ± 0.42	24.2 ± 0.7

The table presents the composition in total percentage of carbohydrates, proteins, and lipids of microalgae *D. brasiliensis*

One can see from Table 1 that there is no significant difference in total carbohydrates and total lipids recovered from either biomass harvested by centrifugation or those recovered from biomass harvested by bioflocculation ($p = 0.251$, $p = 0.981$ for total carbohydrates and total lipids, respectively, $\alpha = 0.05$). However, total proteins recovered from biomass harvested by centrifugation and that harvested by bioflocculation were significantly different ($p = 0.001$, $\alpha = 0.05$).

As for the quantity of biochemicals recovered, there may be some variations due to the extraction method used. For example, the Phenol–Sulfuric Acid method used for total carbohydrates is notoriously variable, also it is known that all sugars do not exhibit a similar colorimetric response, therefore there could be over or underestimation of total carbohydrates if the calibration curve is conducted based on one neutral sugar [21]. The completeness of extraction and composition of total lipids depends largely on the biology of the algal species, the compatibility of the solvent polarity with the lipid molecules polarity and extraction conditions applied. Unavoidably the extractable oil fraction will contain amounts of chlorophyll, pigments, proteins, or soluble carbohydrates. Although gravimetric is the fast way to obtain total lipid data, the solvent system utilized will influence the gravimetric yield and composition of total lipids [22, 23]. If accurate data are required due to specific desired product or the objective of the microalgae study, alternative quantification methods are recommended. High-performance liquid chromatography (HPLC), Anion exchange chromatography (HPAEC) and Gas chromatography (GC) are recommended for identification and quantification of monomers after sequent hydrolysis of carbohydrate polymers in microalgae [21, 24]. The GC is recommended for quantification of the fatty acid methyl esters after an acid-catalyzed transesterification of extracted oils [22, 25, 26]. Furthermore, the most reliable method to quantify the total proteins recommended is the determination of amino acid composition by HPLC analysis [21].

Bioflocculation process utilizing the bioflocculant γ-PGA produced by *B. licheniformis* CGMCC 2876 herein presented achieve higher efficiency compared to other pilot or large-scale microalgae harvesting methods reported (Table 2). Furthermore, no significant differences found in biochemicals (e.g. carbohydrates and lipids) recovered from biomass harvested by either bioflocculation or centrifugation.

Table 2: Microalgae harvesting capableness of different technologies at pilot or large-scale conditions

Harvesting techniques	Microalgal species	Treated volume (L)	Harvesting efficiency (%)	References
Flocculation with *B. licheniformis*-broth	*Desmodesmus brasiliensis*	50–200	≥98	This study
pH increment	*Tetraselmis suecica, Chaetoceros calcitrans, Chlorella muelleri, Skeletonema* sp., *Rhodomonas salina, Attheya septentrionalis, Nitzschia closterium, Chlorella muelleri, Thalassiosira pseudonana*	10–1000	≥80	[28]
Foam fractionation	*Chaetoceros* sp.	220	90	[41]
Foam flotation	*Chlorella* sp.	10–10.2	≥92	[27, 54]
Flocculation	*Scenedesmus* sp.	1000	>96	[38]
Gravity sedimentation coupled with filtration	*Staurosira* sp., *Desmodesmus*sp.	200	80	[37]
Centrifugation	*Chaetoceros muelleri*	550	90	[40]
Flotation under vacuum	*Not precised*	2000	49.5	[39]

The table presents reported data about the efficiency of different technologies applied in microalgae harvesting at pilot or large-scale conditions in order to compare with the experimental results described in the present article

DISCUSSION

Effectiveness of different dewatering technologies at small-scale level have been extensively reported, however their scale up is questionable due to the severe problems they may elicit once applied at large-scale conditions [11]. Moreover, different methods currently applied for microalgae dewatering, although they could achieve significant efficiency, still are not economically workable especially for the low value products such as biofuels [11, 27]. Centrifugation is fast and can be applied successfully for microalgae dewatering, however cells are exposed to high risk of cell disruption and structural damage due to high gravitational and shear forces, also the method is energy consuming, thus the application of centrifugation for dewatering large quantities of microalgae culture is practically considered unsuitable [28, 29]. Different means of flocculation have been reported for their effective microalgae dewatering [28, 30, 31, 32, 33], however dosages of chemical flocculants which affect the product quality and that require additional separation steps from the final product, construct the hindrance for industrial application of flocculation in microalgae dewatering. The addition of surfactants or collectors to improve flotation processes result in the same challenges as chemical flocculants. Electrolytic methods are believed to be cost effective and environmentally friendly microalgae dewatering methods and different experimental reports have confirmed the assumption [34, 35], nevertheless regular replacement of electrode materials and fouling of cathodes constitute the main drawbacks associated with electrolytic processes. On the contrary, bioflocculation is an environmentally friendly dewatering method [12, 13], less energy consuming [14, 36], requires few minutes as running time, and can achieve high efficiency not only at lab scale as presented herein. Additionally, no greenhouse gases emitted during bioflocculation process, therefore it is worthy to be applied for harvesting algal biomass at industrial scale which has been recognized as the cornerstone challenge to microalgae industry referring to the high cost associated with microalgae biomass dewatering.

Different pilot-scale studies about microalgae harvesting have been reported [28, 37, 38, 39, 40, 41]. Pfeiffer and Rusch [40] have reported a harvesting efficiency of 90 % for *Chaetoceros muelleri* employing centrifugation, however it is well known that centrifugation is costing due to operational energy and capital cost thus centrifugation process is not suitable for the production of low value products such as biofuel. Knuckey et al. [28] have reported a flocculation efficiency ≥80 % while harvesting 10–1000 L cultures of different microalgae species by pH increment in the presence of a non-ionic polymer Magnafloc LT-25. Csordas and Wang [41] have reported a harvesting efficiency of 90 % while treating *Chlorella* sp. by foam fractionation. However the technology requires the use of chemicals (such as cetyl trimethylammonium bromide, CTAB) which may interfere with subsequent downstream processing of algal biomass harvested.

Innovative and effective microalgae harvesting/dewatering by various means of bioflocculation have been reported [7, 15, 16, 17, 18]. Recently the mechanism governing microalgae bioflocculation utilizing bacterial bioflocculant was developed [19]. Lee et al. [36] have estimated the mixing energy and process cost for algal microbial flocculation during their study on energy requirements and economic analysis of a full-scale microbial flocculation system of microalgal harvesting. Authors proposed a full-scale microalgae harvesting system based on laboratory data, and found that mixing energy required for the bioflocculation was estimated to be 0.893 kWh/10^3 kg of dry mass flocculated, with estimated overall process cost of \$0.12/$m^3$of the culture medium flocculated, a process cost relatively low compared to the self-cleaning centrifuge, conventional flocculation, flotation with flocculant, and electro-flocculation [36]. The energy consumption during harvesting step for any technology used was estimated previously between 8.2 and 32 kWh/kg dry mass [42], hence the commercialization of algal biomass would be suitable only for high value products. If algal biomass is to be utilized for biofuel production, the harvesting energy should be less than 1.8 kWh/kg dry mass [43].

In attempt to harvest algal biomass at low cost (low energy consumption), Berrut et al. [39] studied the separation efficiency of a vacuum gas lift for microalgae harvesting. Authors found that the harvesting cost could be greatly reduced as the harvesting system would require energy between 0.16 and 3.37 kWh/kg dry mass, therefore algal biomass would be suitable for production and commercialization of low value products such as biofuel. Even though the vacuum gas lift was found effective for microalgae harvesting, large volumes of the harvest would need additional cost for maintenance such as refrigerators and possibly affect further downstream processes. The same observations were remarked by Knuckey et al. [28] during their study on production of microalgal concentrates by flocculation and their assessment as aquaculture feeds. Authors realized that large volumes about 40 L remained after induced pH flocculation of 1000 L. Nevertheless, based on laboratory data, Lee et al. [44] estimated the mixing energy and harvesting cost for a designed electro-flocculation plant for harvesting marine microalgae. Authors achieved an overall energy consumption of 0.092 kWh/m^3 of culture medium treated and a total harvesting cost of $0.19/kg of the ash-free dry mass. However, only energy consumption, based on electro-flocculation separation and mechanical mixing, electrode dissolution and construction costs were considered in the study, while the electrode material cost and their regular replacement were neglected, which cause the most hindrance to electro-flocculation [45, 46, 47]. Recently, Selesu et al. [38] introduced a promising method for harvesting microalgae biomass by flocculation with a cheap and nontoxic flocculant (Tanfloc). Authors compared the cost of harvesting microalgae *Scenedesmus* sp. by flocculation utilizing chitosan and Tanfloc. The harvesting cost was relatively low for Tanfloc ($1.10/kg of dry mass) compared to chitosan ($10.00/kg of dry mass). However, mixing as one of the key factors for the success of flocculation was not considered in the study, apparently the harvesting cost could increase once the mixing energy is incorporated into the study.

Table 3 presents a summary of comparative energy requirement and cost analysis of different reported microalgae harvesting technologies. Electro-flocculation, microbial flocculation, flotation with vacuum gas lift, and TFF are in the same range of process cost (Table 3); however the long mixing time incurred during microbial flocculation increased the process cost, thus with the newly fast-developed bioflocculation herein presented, the process cost could be further reduced, thus placing bioflocculation at the forefront of the cheapest microalgae harvesting methods.

Table 3: Comparative energy and cost analysis of different microalgae harvesting technologies

Harvesting method	**Algae species**	**Energy input (kWh/m^3)**	**Cost ($/kg)**	**References**
Microbial induced flocculation	*Pleurochrysis carterae*	0.45×10^{-3}	0.24	[36]
Centrifugation	*Scenedesmus, Coelastrum proboscideum*	0.3–8	1.44–18	[55]
Flotation with vacuum gas lift	Not precised	0.8–1.69	0.022–0.44	[39]

Filtration with vacuum filter	*Scenedesmus, Chlorella proboscideum*	0.1–5.9	0.96–9	[55]
Electro-flocculation	*Tetraselmis* sp.	0.092	0.19	[44]
Flocculation with Tanfloc	*Scenedesmus* sp.	ND[a]	1.10	[38]
Flocculation with Zetag 7650 and $Al_2(SO_4)_3$	*Tetraselmis suecica*	14.81	~1.8	[56]
Filtration with pressure filters	*Chlorella proboscideum*	0.5–0.88	0.8–2.2	[55]
Flocculation with chitosan	*Scenedesmus* sp.	ND	10.00	[38]
Tangential flow filtration (TFF)	*Tetraselmis suecica*	2.06	~0.26	[56]

The energy consumption and cost data of microalgae harvesting are presented based on the culture medium (m^3) treated and dry mass (kg) produced, respectively

[a]Not determined

Considering experimental data previously reported [7, 15, 16, 17, 18], pilot-scale data herein presented and mixing energy requirement and process cost as estimated by Lee et al. [36], with many other advantages [12, 13], doubtlessly bioflocculation is a preferable choice for microalgae harvesting either for biofuel production or for feed and food purposes.

The results presented by Lee et al. [36, 44] are estimates for designed large-scale microalgae harvesting systems, therefore a full life-cycle assessment (LCA) study based on a pilot-scale or a full-scale data considering all input materials incurred for microalgae harvesting unit would be appropriate to provide more accurate information.

CONCLUSION

Bioflocculation is an efficient, environmentally friendly, and cost-effective microalgae dewatering technique. This work examined the pilot-scale and in situ flocculation of the freshwater microalgae *D. brasiliensis* induced by the bioflocculant γ-PGA produced by *B. licheniformis* CGMCC 2876. Advantages of bioflocculation in situ include reduced process cost as no mixing device and separate flocculation tank are required. Moreover, the process time is shortened as there is no need to move the cultures from cultivation area to flocculation area. The method was found scalable achieving over 98 % efficiency at the scale of 200 L of microalgae cultures and at in situ flocculation of 50 L of cultures. Doubtlessly, the process can be industrialized. Furthermore, biochemical composition analysis of microalgae biomass revealed that bioflocculation technology is worthy to be applied for harvesting algal biomass to be used for biofuel production. Even though the process was found scalable, the design of the mixing apparatus and air flow rate are the key parameters for effective mixing, therefore further studies are necessary to develop a most favorable and well-designed mixing device as well as optimizing the air flow rate for in situ treatment. Furthermore, it is essential to carry out a life-cycle assessment based on pilot-scale or full-scale data generated by flocculation process applying microbial flocculant in order to have more accurate information on energy consumption and process cost. This will contribute more to further development and commercialization of microalgal-based products.

METHODS

Microalgal strains and cultivation medium

The freshwater microalgae *D. brasiliensis* (collection number FACHB-1495) used in the current study was purchased from the Freshwater Algae Culture Collection at the Institute of Hydrobiology, CAS, Wuhan, China. Modified Bold 3 N medium was used for microalgae cultivation, consisting of (g/L): $NaNO_3$, 0.750; $CaCl_2{\cdot}2H_2O$, 0.025; $MgSO_4{\cdot}7H_2O$, 0.057; K_2HPO_4, 0.0383; KH_2PO_4, 0.088; and NaCl, 0.025. Metal and vitamin solutions are as described previously by Berges and Franklin [48].

Microalgae culture and bioflocculant production

The proposed pilot-scale scheme for microalgae cultivation and harvesting is described in Fig. 5. The culture of microalgae for primary and secondary seed cultures was operated according to the procedures described in our previous report [7]. The secondary pre-cultured microalgae (5 L) were inoculated into the 50 L photobioreactor (PBR) containing 45 L of the cultivation medium prepared with the tap water without sterilization; however, the culture growth rate was low compared to the cultures conducted with sterilized water possibly due to other microbes present in the unsterilized water culture. Cultivation conditions: 2.5 % CO_2, aeration rate 0.06 vvm. The light (light intensity ~ 2400 μmol/m²·s) was continuously (24 h) provided by external light sources (14 W TL5 tungsten filament lamps, Philips Co., China) mounted on one side of the PBR for indoor cultivation, while the sun light was the sole light source for outdoor cultivation. The microalgae were harvested at the end of the exponential growth phase.

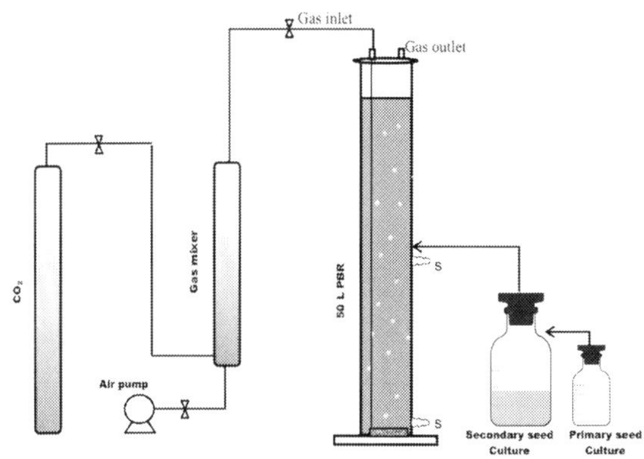

Figure 5: Microalgae cultivation system. The figure depicts a cultivation system applied for the 50 L column PBR (the primary seed culture was utilized to inoculate the second seed culture which was also utilized to inoculate the PBR culture, gas mixer helped to mix $CO_2$and air before they are feed into the PBR culture, *S* sample port)

The bacteria *B. licheniformis* CGMCC 2876 isolated and identified by the Department of Chemical and Biochemical Engineering (Xiamen University) was used to produce the γ-PGA broth bioflocculant and the production process was operated as previously described by Xiong et al. [49]. Analysis of the flocculating activity of the bioflocculant and determination of the bioflocculant concentration were conducted following the procedures described in our previous report [19]. The flocculation cost could be greatly reduced by in situ treatment as no investment cost is needed for a separate flocculation tank and mixing device. Therefore it was imperative to examine in situ flocculability of the freshwater microalgae *D. brasiliensis*. After the cultivation period, the supply of CO_2 was stopped; the pH value of microalgae cultures was adjusted to three using 1 M HCl. The bioflocculant (2.5 mL/L) was added to the cultures followed by mixing by air at aeration rate of 0.1 L/min for 2 min, and the flocs formed were allowed to settle under gravity. Samples were collected after 1, 5, 10, 15, and 20 min for flocculation efficiency analysis. The broth of *B. licheniformis* CGMCC 2876 was used as the bioflocculant without further purification processes.

Determination of algal biomass concentration

The calibration curve expressed in Eq. (1) relating the dry cell weight (DCW) to the optical density values at the wavelength of 685 nm (OD_{685nm}) was used to estimate the microalgae biomass concentration. The optical density values were determined utilizing the UV Spectrophotometer (UV-1780, SHIMADZU, Kyoto, Japan). The DCW was obtained by weighting the microalgae cells after washing two times with deionized water, followed by subsequent overnight drying till a constant weight was achieved in an oven at 80 °C.

$$y=1.526x-0.078 \quad (1)$$

with y the DCW (g/L) and x the $OD685_{nm}$.

Microalgae flocculation experiments

Flocculation experiments were carried out with 5, 50, and 200 L of microalgae cultures. After the distribution of microalgae cultures into flocculation vessels, the pH value of the culture was gradually adjusted to pH 3 using 1 M HCl, thereafter the bioflocculant was added (2.5 mL/L), followed by mixing at 200 rpm for 2 min, then the mixing was stopped to allow the flocs formed to settle at room temperature. The operating parameters for the pilot-scale were supposed to be the same as the optimum operating parameters reported at lab-scale [7], however the mixing capacity of the device used could not be adjusted below 200 rpm, thus only mixing at 200 rpm was possible and no slow mixing was applied. In situ flocculation was conducted in the same photobioreactor (50 L) used for microalgae cultivation. After the cultivation time, the CO_2 was stopped and air was used for mixing. The flocculation efficiency (FE) was calculated according to Eq. (2):

$$F.E\ (\%) = \frac{OD_0 - OD_1}{OD_0} \times 100 \qquad (2)$$

where OD0 and OD_1 are the $OD685_{nm}$ values of the microalgal suspension before and after flocculation, respectively.

Microalgae biomass characterization

Biomass characterization was conducted in order to examine the applicability of bioflocculation as a dewatering method for the recovery of different microalgal biomass biochemicals. The results will provide insight into the scalability of bioflocculation for the recovery of different microalgal biomass constituents vs. centrifugation widely applied in industry. Microalgae biomass harvested by either centrifugation at 6000 rpm for 10 min (Centrifuge, D-78532 Tuttlingen, ZENTRIFUGEN, Germany) or bioflocculation was freeze dried for 24 h at 3 kPa (Freeze dryer, FD-1000, EYELA, Japan) and stored at −4 °C before analysis. The biomass biochemicals recovered from both harvesting methods were compared through One-way ANOVA, using Fisher's protected least significant difference (PLSD) test for pair-wise comparisons (OriginPro 8.6, OriginLab Corporation, USA).

Determination of total carbohydrates

The colorimetric method (Phenol–Sulfuric Acid method) widely applied for the determination of total carbohydrate content in liquid solutions was employed in this study. Experiments were conducted according to the procedures described by Albalasmeh et al. [50] with modifications. Briefly, 10 mg of freeze-dried biomass was reconstituted in 10 mL of deionized water. 2 mL of reconstituted solution (carbohydrate solution) was mixed with 1 mL of phenol (5 % w/v) in a test tube. Subsequently, the mixture was rapidly reacted with 5 mL of concentrated sulfuric acid. After 10 min standing at room temperature, the tubes were vortexed for 1 min and then placed in water bath for 20 min at 30 °C for color development. Blanks were prepared in the same way as described above, and 2 mL of dH_2O was used instead of carbohydrate solutions. After the color development, light absorption at the wavelength of 490 nm was recorded using UVS pectrophotometer (UV-1780, SHIMADZU, Kyoto, Japan). The total carbohydrates were determined referring to the standard curve (Eq.(3)) based on glucose (Additional file 1: Fig. S1a), phenol (5 % w/v in water), and glucose solution for the standard curve were prepared freshly before the experiments. All samples were prepared in triplicates.

$$y=12.82x\pm0.14 \quad (3)$$

with y the $OD490_{nm}$ and *x* the glucose concentration (mg/L).

Determination of total proteins

In order to estimate the total proteins, Bradford method invented by Bradford [51] was utilized in the present study. Bradford method is the most convenient, simple, faster, and more preferred method for protein quantification in solutions [52]. This method is known to be less subjected to interference by common reagents and non-protein components of samples like Lowry method [52]. Bradford protein assay protocol [53] (http://www.bio-protocol.org/e45) was followed with modifications. The protein solution was prepared by solubilizing 10 mg of freeze-dried biomass in 10 mL of dH_2O. An aliquot of 0.1 mL of protein solution was reacted with 1 mL of the Bradford reagent, vortexed and incubated for 5 min at room temperature for color development. The blank was prepared by replacing the protein sample by dH_2O. All samples were prepared in triplicates. The absorbance was measured at 595 nm wavelength utilizing a UV Spectrophotometer (UV-1780, SHIMADZU, Kyoto, Japan) and the total protein concentration was determined referring to the Bovine Serum Albumin (BSA) standard curve (Eq. (4)) (Additional file 1: Fig. S1b):

$$y=4.45x\pm0.11 \quad (4)$$

with y the $OD595_{nm}$ and *x* the BSA concentration (mg/L).

Determination of total lipids

In the current study, total lipids extraction was performed according to the method developed by Axelsson and Gentili [25] with modifications. Freeze-dried biomass (20 mg) was placed in pre-weighted centrifuge tube, and then 8 mL of chloroform–methanol (2:1 v/v) mixture was added. Algal biomass was suspended manually by shaking vigorously the centrifuge tubes for 1 min, and then 2 mL of 0.73 % NaCl aqueous solution was added. The mixture was centrifuged at 120×10^3 rpm for 10 min (Centrifuge, himac CT15, HITACH, Japan) to allow phase separation.

The upper phase was siphoned and the solvent was evaporated under a nitrogen stream. After the solvent was completely evaporated, lipid fractions were quantified gravimetrically by reweighting the centrifuge tube. The percentage of total lipid content was determined by dividing the weight of recovered lipids by the weight of the dried biomass. All experiments were conducted in triplicates.

NOTES

Authors' contributions

NT carried out algae cultivation and flocculation experiments, participated in data analysis, and drafted the manuscript. ZX participated in the design and coordination of the study, and revised the manuscript. MT and ME participated in microalgae biomass characterization and data analysis. HN and SW carried out production and analysis of the bioflocculant experiments, and helped to revise the manuscript. LY conceived the study, participated in its design and coordination, and helped to revise and finalize the manuscript. All authors read and approved the final manuscript.

Acknowledgements

This research project was financially supported by National Natural Science Foundation of China (No. 21406185) and the Research Program from the Science and Technology Bureau of Xiamen City in China (No. 3502Z20151254). Theoneste Ndikubwimana appreciatively acknowledges the Xiamen University, China Scholarship Council, and the Government of Rwanda through Rwanda Education Board for the support.

Competing interests

The authors declare that they have no competing interests.

REFERENCES

1. Vandamme D, Foubert I, Muylaert K. Flocculation as a low-cost method for harvesting microalgae for bulk biomass production. Trends Biotechnol. 2013;31(4):233–9.

2. Foley JA, Ramankutty N, Brauman KA, Cassidy ES, Gerber JS, Johnston M, Mueller ND, O'Connell C, Ray DK, West PC. Solutions for a cultivated planet. Nature. 2011;478(7369):337–42.
3. Wijffels RH, Barbosa MJ. An outlook on microalgal biofuels. Science. 2010;329(5993):796–9. Posten C, Schaub G. Microalgae and terrestrial biomass as source for fuels-a process view. J Biotechnol. 2009;142(1):64–9.
4. Chisti Y. Biodiesel from microalgae beats bioethanol. Trends Biotechnol. 2008;26(3):126–31.
5. Garzon-Sanabria AJ, Davis RT, Nikolov ZL. Harvesting Nannochloris oculata by inorganic electrolyte flocculation: effect of initial cell density, ionic strength, coagulant dosage, and media pH. Bioresour Technol. 2012;118:418–24.
6. Ndikubwimana T, Zeng X, Liu Y, Chang JS, Lu Y. Harvesting of microalgae Desmodesmussp. F51 by bioflocculation with bacterial bioflocculant. Algal Res. 2014;6:186–93.
7. Cui Y, Yuan W, Cheng J. Understanding pH and ionic strength effects on aluminum sulfate-induced microalgae flocculation. Appl Biochem Biotechnol. 2014;173(7):1692–702.
8. Spilling K, Seppälä J, Tamminen T. Inducing autoflocculation in the diatom Phaeodactylum tricornutum through CO2 regulation. J Phycol. 2011;23(6):959–66.
9. Liu J, Zhu Y, Tao Y, Zhang Y, Li A, Li T, Sang M, Zhang C. Freshwater microalgae harvested via flocculation induced by pH decrease. Biotechnol Biofuels. 2013;6(1):1–11.
10. Uduman N, Qi Y, Danquah MK, Forde GM, Hoadley A. Dewatering of microalgal cultures: a major bottleneck to algae-based fuels. J Renew Sust Energ. 2010;2(1):012701–15.
11. Manheim D, Nelson Y. Settling and bioflocculation of two species of algae used in wastewater treatment and algae biomass production. Environ Progr Sust Energ. 2013;32(4):946–54.
12. Oh HM, Lee SJ, Park MH, Kim HS, Kim HC, Yoon JH, Kwon GS, Yoon BD. Harvesting of Chlorella vulgaris using a bioflocculant from Paenibacillus sp. AM49. Biotechnol Lett. 2001;23(15):1229–34.

13. Christenson L, Sims R. Production and harvesting of microalgae for wastewater treatment, biofuels, and bioproducts. Biotechnol Adv. 2011;29(6):686–702.
14. Salim S, Kosterink N, Wacka NT, Vermuë M, Wijffels R. Mechanism behind autoflocculation of unicellular green microalgae Ettlia texensis. J Biotechnol. 2014;174:34–8.
15. Zhang J, Hu B. A novel method to harvest microalgae via co-culture of filamentous fungi to form cell pellets. Bioresour Technol. 2012;114:529–35.
16. Xie S, Sun S, Dai SY, Yuan JS. Efficient coagulation of microalgae in cultures with filamentous fungi. Algal Res. 2013;2(1):28–33.
17. Zheng H, Gao Z, Yin J, Tang X, Ji X, Huang H. Harvesting of microalgae by flocculation with poly (γ-glutamic acid). Bioresour Technol. 2012;112:212–20.
18. Ndikubwimana T, Zeng X, He N, Xiao Z, Xie Y, Chang JS, Lin L, Lu Y. Microalgae biomass harvesting by bioflocculation-interpretation by classical DLVO theory. BiochemEng J. 2015;101:160–7.
19. Jang M, Lee HJ, Shim Y. Rapid removal of fine particles from mine water using sequential processes of coagulation and flocculation. Environ Technol. 2010;31(4):423–32.
20. Laurens LM, Dempster TA, Jones HD, Wolfrum EJ, Van Wychen S, McAllister JS, Rencenberger M, Parchert KJ, Gloe LM. Algal biomass constituent analysis: method uncertainties and investigation of the underlying measuring chemistries. Anal Chem. 2012;84(4):1879–87.
21. Laurens LM, Quinn M, Van Wychen S, Templeton DW, Wolfrum EJ. Accurate and reliable quantification of total microalgal fuel potential as fatty acid methyl esters by in situ transesterification. Anal Bioanal Chem. 2012;403(1):167–78.
22. Iverson SJ, Lang SL, Cooper MH. Comparison of the Bligh and Dyer and Folch methods for total lipid determination in a broad range of marine tissue. Lipids. 2001;36(11):1283–7.
23. Templeton DW, Quinn M, Van Wychen S, Hyman D, Laurens LM. Separation and quantification of microalgal carbohydrates. J Chromatogr A. 2012;1270:225–34.

24. Axelsson M, Gentili F. A single-step method for rapid extraction of total lipids from green microalgae. PLoS One. 2014;9(2):e89643. doi:10.1371/journal.pone.0089643.CrossRefGoogle Scholar
25. Li Y, Naghdi FG, Garg S, Adarme-Vega TC, Thurecht KJ, Ghafor WA, Tannock S, Schenk PM. A comparative study: the impact of different lipid extraction methods on current microalgal lipid research. Microb Cell Fact. 2014;13(14):1–9.
26. Coward T, Lee JG, Caldwell GS. Development of a foam flotation system for harvesting microalgae biomass. Algal Res. 2013;2(2):135–44.
27. Knuckey RM, Brown MR, Robert R, Frampton DM. Production of microalgal concentrates by flocculation and their assessment as aquaculture feeds. Aquacul Eng. 2006;35(3):300–13.
28. Rawat I, Kumar RR, Mutanda T, Bux F. Biodiesel from microalgae: a critical evaluation from laboratory to large scale production. Appl Energ. 2013;103:444–67.
29. Wyatt NB, Gloe LM, Brady PV, Hewson JC, Grillet AM, Hankins MG, Pohl PI. Critical conditions for ferric chloride-induced flocculation of freshwater algae. Biotechnol Bioeng. 2012;109(2):493–501.
30. Chen JJ, Yeh HH. The mechanisms of potassium permanganate on algae removal. Water Res. 2005;39(18):4420–8.
31. Henderson R, Parsons SA, Jefferson B. The impact of algal properties and pre-oxidation on solid-liquid separation of algae. Water Res. 2008;42(8):1827–45.
32. Vandamme D, Foubert I, Meesschaert B, Muylaert K. Flocculation of microalgae using cationic starch. J Phycol. 2010;22(4):525–30.
33. Weschler MK, Barr WJ, Harper WF, Landis AE. Process energy comparison for the production and harvesting of algal biomass as a biofuel feedstock. Bioresour Technol. 2014;153:108–15.
34. de Carvalho Neto RG, do Nascimento JGdS, Costa MC, Lopes AC, Neto EFA, Mota Filho CR, Dos Santos AB. Microalgae harvesting and cell disruption: a preliminary evaluation of the technology electroflotation by alternating current. Water Sci Technol. 2014. doi:10.2166/wst.2014.220.Google Scholar
35. Lee AK, Lewis DM, Ashman PJ. Energy requirements and economic

analysis of a full-scale microbial flocculation system for microalgal harvesting. Chem Eng Res Des. 2010;88(8):988–96.

36. Huntley ME, Johnson ZI, Brown SL, Sills DL, Gerber L, Archibald I, Machesky SC, Granados J, Beal C, Greene CH. Demonstrated large-scale production of marine microalgae for fuels and feed. Algal Res. 2015;10:249–65.
37. Selesu NF, de Oliveira TV, Corrêa DO, Miyawaki B, Mariano AB, Vargas JV, Vieira RB. Maximum microalgae biomass harvesting via flocculation in large scale photobioreactor cultivation. Can J Chem Eng. 2015. doi:10.1002/cjce.22391.
38. Barrut B, Blancheton J-P, Muller-Feuga A, René F, Narváez C, Champagne JY, Grasmick A. Separation efficiency of a vacuum gas lift for microalgae harvesting. Bioresour Technol. 2013;128:235–40.
39. Pfeiffer TJ, Rusch KA. An integrated system for microalgal and nursery seed clam culture. Aquacul Eng. 2000;24(1):15–31.
40. Csordas A, Wang JK. An integrated photobioreactor and foam fractionation unit for the growth and harvest of Chaetoceros spp. in open systems. Aquacul Eng. 2004;30(1):15–30.
41. Cadoret J-P, Bernard O. La production de biocarburant lipidique avec des microalgues: promesses et défis. J Soc Biol. 2008;202(3):201–11.
42. Borowitzka MA. Marine and halophilic algae for the production of biofuels. J Biotechnol. 2008;136:S7.
43. Andrew Lee K, David Lewis M, Ashman PJ. Harvesting of marine microalgae by electroflocculation: the energetics, plant design, and economics. Appl Energ. 2013;108:45–53.
44. Uduman N, Bourniquel V, Danquah MK, Hoadley AF. A parametric study of electrocoagulation as a recovery process of marine microalgae for biodiesel production. Chem Eng J. 2011;174(1):249–57.
45. Vandamme D, Pontes SCV, Goiris K, Foubert I, Pinoy LJJ, Muylaert K. Evaluation of electro-coagulation-flocculation for harvesting marine and freshwater microalgae. Biotechnol Bioeng. 2011;108(10):2320–9.

46. Granados M, Acien F, Gomez C, Fernández-Sevilla J, Molina Grima E. Evaluation of flocculants for the recovery of freshwater microalgae. Bioresour Technol. 2012;118:102–10.
47. Berges JA, Franklin DJ, Harrison PJ. Evolution of an artificial seawater medium: improvements in enriched seawater, artificial water over the last two decades. J Phycol. 2001;37(6):1138–45.
48. Xiong Y, Wang Y, Yu Y, Li Q, Wang H, Chen R, He N. Production and characterization of a novel bioflocculant from Bacillus licheniformis. Appl Environ Microbiol. 2010;76(9):2778–82.
49. Albalasmeh AA, Berhe AA, Ghezzehei TA. A new method for rapid determination of carbohydrate and total carbon concentrations using UV spectrophotometry. Carbohyd Polym. 2013;97(2):253–61.
50. Bradford MM. A rapid and sensitive method for the quantitation of microgram quantities of protein utilizing the principle of protein-dye binding. Anal Biochem. 1976;72(1):248–54.
51. Kruger NJ. The Bradford method for protein quantitation. In: The protein protocols handbook. Springer; 2009. p. 17–24.
52. He F. Bradford protein assay. Bio-protocol Bio101: e45 2011.
53. Coward T, Lee JG, Caldwell GS. Harvesting microalgae by CTAB-aided foam flotation increases lipid recovery and improves fatty acid methyl ester characteristics. Biomass Bioenerg. 2014;67:354–62.
54. Grima EM, Belarbi EH, Fernández FA, Medina AR, Chisti Y. Recovery of microalgal biomass and metabolites: process options and economics. Biotechnol Adv. 2003;20(7):491–515.
55. Danquah MK, Ang L, Uduman N, Moheimani N, Forde GM. Dewatering of microalgal culture for biodiesel production: exploring polymer flocculation and tangential flow filtration. J Chem Technol Biot. 2009;84(7):1078–83

CHAPTER

9

MINIMISING LOSSES TO PREDATION DURING MICROALGAE CULTIVATION

Kevin J. Flynn[1], Philip Kenny[1], Aditee Mitra[1]

Swansea University, Swansea, UK

ABSTRACT

We explore approaches to minimise impacts of zooplanktonic pests upon commercial microalgal crops using system dynamics models to describe algal growth controlled by light and nutrient availability and zooplankton growth controlled by crop abundance and nutritional quality. Losses of microalgal crops are minimised when their growth is fastest and, in contrast, also when growing slowly under conditions of nutrient exhaustion. In many culture systems, however, dwindling light availability due to self-shading in dense suspensions favours slow growth under nutrient sufficiency. Such a situation improves microalgal quality as prey, enhancing zooplankton growth, and leads to rapid crop collapse. Timing of pest entry is important; crop losses are least likely in established, nutrient-exhausted microalgal communities grown for high C-content (e.g. for biofuels). A potentially useful approach is to promote a low level of P-stress that does not adversely affect microalgal growth but which produces a crop that is suboptimal for zooplankton growth.

Keywords: Microalgae Production Pest control Zooplankton Predator-prey Optimisation Biofuels Stoichiometric ecology

INTRODUCTION

Microalgae have been proposed as a commercially important crop in support of a variety of products, ranging from biofuels to pharmaceuticals and feed-stocks for aquaculture (Greenwell et al. 2010; Milledge 2011; Borowitzka 2013). As with all crops, there is potential for production to be decreased or spoilt by the activity or presence of pests. In closed bioreactors, under laboratory conditions, there is the opportunity to exclude or control pests by using clean techniques, but in open pond systems and in large commercial bioreactor operations, there is enhanced scope for entry of pests. For microalgal crops, there are three potential pest types: contamination by other microalgae (Smith et al. 2005), infections caused by viruses which can destroy algal growth very rapidly (as noted sometimes in nature; Schroeder et al. 2003) and fungal attack such as those by chytrid fungi (Gutman et al. 2009; Strittmatter et al. 2016) and the presence of predators (Day et al. 2012). Predators may be present at very low levels in stock cultures and, depending on the growth conditions, may not normally be apparent at all (especially likely with protistan pests). In other instances, and notably in open ponds, predators of various types (protists, rotifers and crustacea such as *Daphnia* and copepods), may be introduced from the wider environment.

To date, there is no single established method successfully used to maximise microalgal production with simultaneous minimisation of crop loss through zooplanktonic predation. Cultivators of microalgae have resorted to a wide variety of strategies to control contamination. Most approaches rely on the culturing of extremophiles under highly selective growth conditions (Borowitzka 2005) with regards to pH or, in the case of marine species, salinity, with the latter having the potential double benefit of stimulating productivity while suppressing increases in invader populations (Bartley et al. 2013). Other methods can include filtration and the use of chemical pesticides (Bacellar Mendes and Vermelho 2013; Wang et al. 2013; McBride et al. 2014), although the former can only work when predators are relatively large, while imprudent use of pesticides can destroy the microalgae along with the predator (Méndez and Uribe 2012). Pulsed electric fields, intended to cause structural and functional damage to predators while leaving the microalgal cells intact, have also been suggested (Rego et al. 2015).

Other workers have proposed a more "top-down" approach to the problem by turning the hunter into the hunted with the introduction of zooplanktivorous fish into the system (Smith et al. 2010). The rationale behind such suggestions stems from the belief that monoculture states are naturally unstable; so, it is better to manage the inevitable increase in diversity by the creation of a "synthetic community" (Kazamia et al. 2012; Smith and Crews 2014). It has also been suggested that such a top-down bio-manipulation of trophic cascades may pay dividends through increased lipid production (Sturm et al. 2012). However, to preserve biochemical consistency within the crop at the point of harvest (which is usually a commercial imperative), uni-algal cultivation will most likely remain the favoured approach except perhaps for the formulation of aquaculture feeds. In view of the difficulty in applying effective predation mitigation strategies in an industrial setting, it is unsurprising that progress in this area remains slow (Chisti 2013b); this was the motivator for the current work.

The growth rate of microalgae, and the form of their biomass in biochemical terms (most basically, as indicated by their C/N/P elemental stoichiometry), is of paramount importance for commercial viability, crop production and also for the growth of grazing pests. Traditionally, a biomass C/N/P stoichiometry in accordance with the work of Redfield (1934), termed the Redfield ratio, is deemed to be optimal for microalgal growth and health (Geider and LaRoche 2002). Often, light limitation developing through self-shading affects the scope for nutrient limitation within dense microalgal populations, and in consequence, the so-called optimal N/P nutrient supply ratio does not simply, nor necessarily at all, drive balanced growth (Flynn 2010). There is a broadly linear relationship between cellular N/C and N-limiting growth rate and a strongly curvilinear relationship for cellular P/C under P-limiting growth (Elrifi and Turpin 1985; Flynn 2008); in consequence, some level of P-limitation can be incurred by microalgae without a significant impact on growth rate, nor significantly affect biochemical quality (Mayers et al. 2014).

Research on the relationship between microalgal C/N/P and growth of its natural predator, the zooplankton, indicates that prey C/N/P aligning closely with Redfield ratios best supports predator growth (Sterner and Elser 2002). Furthermore, this interaction is self-reinforcing; a predator feeding on poor quality algae (i.e. low N/C and/or low P/C) releases (regenerates) less nutrients to be re-assimilated by the remaining algae, and hence, the nutrient status of the microalgal population as prey for the grazer can deteriorate further (Mitra and Flynn 2006). For commercial exploitation, the relationship between the microalgal C/N/P and its value as a crop is most obviously divisible on whether the crop is intended for use as feedstocks that are either protein-rich (high N/C) or alternatively C-rich (high C/N, having accumulated extra C as carbohydrate and/or fatty acids). The options for optimising the rate of production for bulk biomass either for high-value chemicals, such as carotenoids and phycobilins (Borowitzka 2013), or for PUFAs and biofuel feedstocks (Fon Sing et al. 2013) are thus largely mutually exclusive. As a high microalgal C/N is associated with low growth rates and hence low biomass productivity, maximising production of C-rich products requires careful selection of microalgal physiology (Flynn et al. 2010, 2012) and also careful management of bioreactors with respect to their design, lighting and nutrient loading and harvesting or bioreactor dilution rates (Kenny and Flynn 2015).

For commercial growth of microalgae, the organisms are grown to high densities (with dry weight biomass concentration typically between 0.1 and 0.5 kg m^{-3}; Chisti 2007; Ozkan et al. 2012) and supplied with high nutrient loads. Under these conditions, light limitation of growth is likely (Richmond 2013) so that growth reactors contain dense suspensions of potentially high-quality prey for zooplankton. Further, in contrast to the situation in nature, prey availability is not limiting for zooplankton growth. If the reactor is operated in a continuous dilution (chemostat-like) mode, then conditions may further conspire to favour the growth of the zooplankton over the microalgae, as biomass-specific grazing rates typically exceed biomass-specific growth rates of the crop (Hansen et al. 1997). However, if the crop is grown for a high C-content (high C/N), then there is scope to minimise losses through the formation of a crop that is intrinsically of low nutrient quality for predators (Sterner and Elser 2002; Mitra and Flynn 2005). This work uses computational stoichiometric ecology to identify approaches (forgoing genetic modification, GM, of the crop) for the control of predation in commercial microalgal cultivation systems through manipulation of factors such as nutrient regimes and culture harvesting.

METHODS

System configuration

The investigation described here exploited the use of dynamic variable stoichiometric models (i.e. for algae, variable C/N/P/Chl). This is in recognition of the importance of simulating variable stoichiometry not only for describing biomass and carbohydrate/lipid production by microalgae (Kenny and Flynn 2015) but also for predator-prey interactions (Sterner and Elser 2002; Mitra and Flynn 2005). A schematic of the model is given in Fig. 1; a full description, with equations and examples of prior usage to establish the models providence, is given in the Supplementary material.

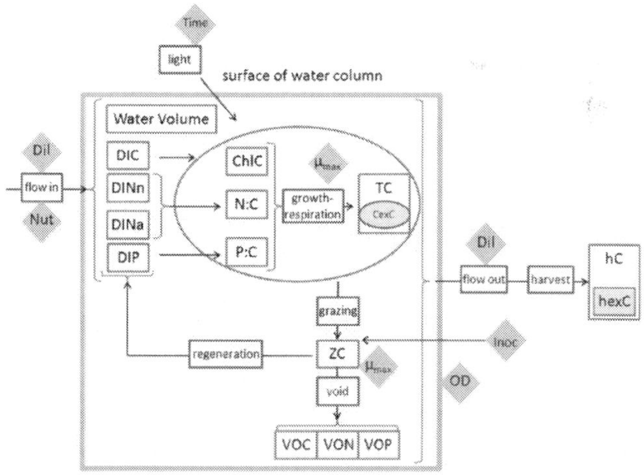

Figure 1: Schematic of the model. Items within *dark blue boxes* are state variables defined in the model.

Nutrients for algal consumption include dissolved inorganic C (DIC), ammonium (DINa) and nitrate (DINn) and phosphate (DIP). The increase in microalgal C-biomass (TC) is a function of the nutrient status (N/C and P/C) of the cells, the status of the photosystems (ChlC, which is itself a function of nutrient status and photoacclimation) and light availability. Light reaching the microalgae depends on surface irradiance, light absorbance by water and the pigmented microalgae and operation depth (OD; pond depth or diameter of a bioreactor tube). A proportion of algal biomass (TC) is C-rich storage products (carbohydrate + lipid; CexC). Zooplankton predator C-biomass (ZC) increases through grazing, with part of the ingestion biomass being voided as organic C/N/P, and part regenerated as nutrients for re-assimilation by the microalgae. The efficiency of grazing (conversion of TC to ZC) depends on algal N/C and P/C, according to stoichiometric rules. Harvesting contributes to harvested biomass (hC), including harvested material for biofuels (hexC). Indicated by *diamonds* are the key parameters explored in the simulations: *Dil* dilution rate and harvesting frequency, *Inoc* inoculation of the system with zooplankton, *Nut* nutrient concentration, *OD* operational depth, *Time* day and hour, μ_{max} maximum growth rate of the algae and of the zooplankton predator. Light available for microalgal photosynthesis is a function of that surface irradiance over the day-night cycle, absorbance by the algal suspension with reference to OD, TC and ChlC (the latter a function of NC and of light availability via photoacclimation). The value of hexC depends on TC and CexC, which in turn relates to Dil and Nut, such that growth rate is optimised while N/C is low (noting that N/C is linearly related to N-limited growth potential)

The microalgal sub-model was similar to that used in our studies on optimisation of microalgal production (Flynn et al. 2012; Kenny and Flynn 2015, 2016). Thus, microalgal growth was described using a mechanistic, acclimative, variable stoichiometric model of microalgal physiology. Changes in microalgal C/N/P/Chl occur with acclimation in response to changes in nutrient and light availability. With nutrient exhaustion, especially of nitrogen (N), excess carbon (C) is accumulated (i.e. cellular C/N increases; N/C declines). Whether that excess C accumulates in reality as carbohydrate and/or fatty acids depends on the taxonomic characteristics of the organism. This particular microalgal model has been used to describe the growth dynamics of many different species under different situations. A demonstration of the model operating against a published data set for a real reactor system (Quinn et al. 2012) is shown in Kenny and Flynn (2016).

To this original microalgal-bioreactor model, we added a model describing the growth dynamics of zooplankton (Mitra 2006). That zooplankton model has been previously configured to simulate the growth of both microzooplankton (protists) and mesozooplankton (crustacea) and has been used under various scenarios (Mitra and Flynn 2006; Mitra et al. 2007, 2014 and references therein).

The physiological descriptions of the microalgae and zooplankton were as used in Flynn et al. (2012). The maximum growth rate of the microalgae enabled a division per day (i.e. $\mu = 0.693\ day^{-1}$) under the 12:12-h light/dark cycle employed. This growth rate is consistent with the enzyme characteristics and cellular activity of RuBisCO (the enzyme that fixes CO_2), which may enable a maximum growth rate in continuous illumination approaching two divisions per day (Flynn and Raven 2017). The zooplankton were simulated with either a growth rate equivalent to a doubling per day, or two doublings per day. The maximum zooplankton assimilation efficiency was set at 0.75, but this decreased with deterioration in prey quality with reference to the prey and predator N/C and P/C ratios, using which ever was the lower of preyNC/predNC or preyPC/predPC (Mitra 2006). The result of this linkage is an increasingly poor trophic transfer from the prey to the predator as the prey quality declines (i.e. as microalgal N/C and/or P/C falls); this description accords with empirical evidence (e.g. Sterner et al. 1993; Flynn et al. 1996; Young et al. 1997).

The whole model, describing microalgae and zooplankton growth within a description of the physico-chemical environment of a bioreactor or pond, was run with an integration step size of 11.25 min. The growth environment included descriptions of the supply of inorganic dissolved N and P nutrients (and their recycling with any zooplankton activity), of light (changing in its availability to the microalgae as a function of the self-shading activity of the plankton biomass, as well as over a 12:12-h light/dark cycle), of bioreactor depth (assuming the typical homogenous mixing of the contents) and of harvesting rates (see below). The default primary limiting nutrient concentration was 440 μM inorganic N (which is half the concentration of the classic f/2 marine algal growth medium; Guillard 1975). We have previously shown that this nutrient provision allows potential nutrient exhaustion under non-light limiting conditions over a range of the shallower optical paths ("depths") typically used in photobioreactors (Kenny and Flynn 2015). In addition to this, P was supplied at either the Redfield ratio with N (mole ratio for N/P of 16) or at twice that ratio (N/P of 32) to enable a potentially moderate level of co-P stress in the microalgae. Growth at a nutrient N/P of 32 appears not to be deleterious to phytoplankton physiology or fatty acid production (Mayers et al. 2014), consistent with the distinct curvi-linear (quota) relationship between cellular P/C and growth rate (Flynn 2008).

Harvesting techniques

We explored different crop harvesting approaches. Harvesting can be continuous (akin to a chemostat operation, i.e. described by a dilution rate) or discontinuous (as in a batch growth system) with some fraction of the culture harvested every few days. However, in reality, in preparation for developing systems for either approach, a culture must first be grown up in what amounts to a batch growth system. Accordingly, here, growth was simulated from a low inoculum into a low-dilution continuous dilution system; this gave a period of nutrient-replete exponential growth (batch-culture like) prior to steady-state (chemostat-like) growth but without risk of an abrupt nutrient stress than may damage the crop.

Introducing the zooplankton

For the simulations presented here, the introduction of the zooplankton predator was considered to occur either concurrent with the microalgae at the start of the growth cycle (i.e. as a contamination in the original inoculum), or after establishment of steady-state conditions (here contamination occurred at day 20 of the simulation, into a well-developed algal system). In both instances, it was assumed that the initial zooplankton biomass was only 0.05% of that of the microalgae. If one considers a predator with a size (volume) of at least 10 times that of its prey (as would typically be the case—Hansen et al. 1997), then numerically, this level of zooplankton contamination would equate to only 0.005% (1 in 20,000). Such contamination levels would likely go unnoticed in routine microscope sample examination of cell counts of the microalgae (Day et al. 2012).

Presentation of results

Results are given in terms of the rates of areal biomass production (AP; gC m^{-2} day^{-1}) and areal production of C-rich components (fatty acids and/or carbohydrate; AXP; gC m^{-2} day^{-1}), as well as volumetric production (VP; gC m^{-3} day^{-1}). Areal productivity takes into account VP and bioreactor depth (m). Table 1 gives a summary of the simulated test conditions and their associated figures. Here, we highlight specific examples of simulations; additional results (as indicated in Table 1) are provided in the Supplementary Material and referenced in the format Fig. S*x*. It should be noted that maximum (standing stock) biomass levels reflect a balance of growth rate set against dilution rate. Emphasis here is placed on production rate (the commercial imperative) and not upon terminal biomass concentration.

Table 1: Summary of simulation conditions and plot locations

Nutrient mole N/P	Dilution rate (day^{-1})	Areal biomass: algal and zooplankton	Volumetric biomass: algal and zooplankton	Algal N/C and N/P	Algal AP and AXP	Zooplankton growth and ingestion rates	Zooplankton AE and GGE
16	0.1	Fig. S1					
32	0.1	Fig. S2					
16	0.3	Fig. 2	Fig. S5	Fig. 4	Fig. S7	Fig. S9	Fig. S11
32	0.3	Fig. 3	Fig. S6	Fig. 5	Fig. S8	Fig. S10	Fig. S12
16	0.5	Fig. S3					
32	0.5	Fig. S4					

Plots given in the Supplementary material are indicated by Fig. S*x*. In all instances, the simulations were performed over a range of reactor depths from 0.025 to 0.5 m

AP areal biomass production, AXP areal production of C-rich components (fatty acids and/or carbohydrate), AE assimilation efficiency (proportion of material ingested by zooplankton that is not voided), GGE zooplankton gross growth efficiency (ratio of growth rate to ingestion rate)

RESULTS

Control systems, with no pest introduction

Figures 2a and 3a show microalgal growth in systems of different depths (0.025 to 0.5 m), with different N/P nutrient supply ratios and different dilution rates (panels (a) in Figs. S1–S4). There are no substantial differences in the microalgal growth rates between systems supplied with nutrients at the higher N/P (N/P = 32) versus the lower, "optimal", N/P of 16. Peak areal biomass (i.e. as gC m^{-2}) decreases with increasing dilution rate (Figs. 2a and 3a versus Figs. S1–S4). The volumetric biomass density of microalgae exceeded 100 g C m^{-3} in shallow systems operating at a dilution rate of 0.3 day^{-1} (Figs. S5a and S6a).

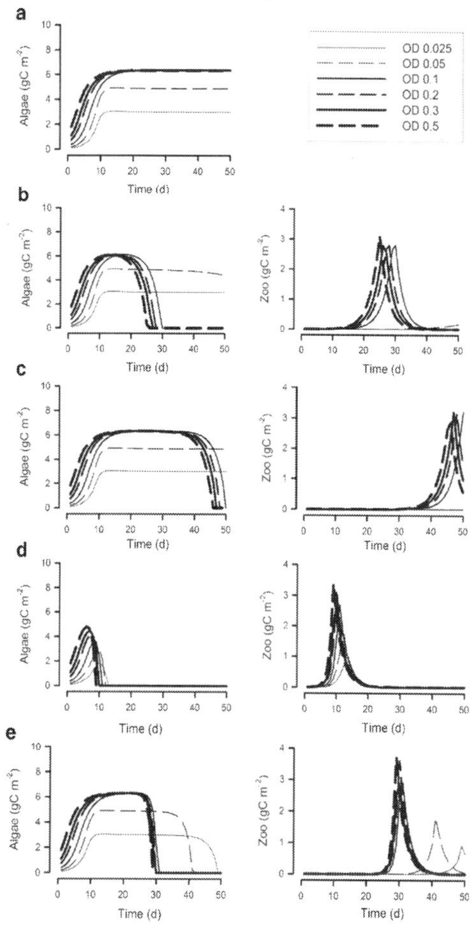

Figure 2: Areal biomass of algae and of the zooplankton contaminant when grown at 6 different reactor operation depths (OD; 0.025–0.5 m), with the supply nutrient mole ratio N/P at 16 and a dilution rate of 0.3 day^{-1}. **a** No contamination. **b** Contamination at 0 day. **c**Contamination at 20 days. **d** Contamination at 0 day with fast growing zooplankton. **e**Contamination at 20 days with fast growing zooplankton

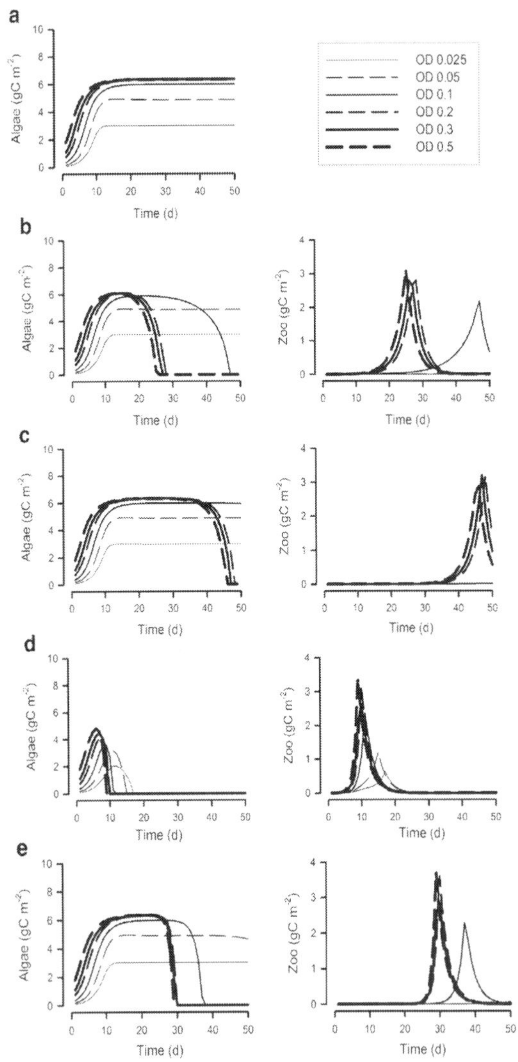

Figure 3: As Fig. 2 but with a supply nutrient mole ratio N/P of 32. **a** No contamination. **b**Contamination at 0 day. **c** Contamination at 20 days. **d** Contamination at 0 day with fast growing zooplankton. **e** Contamination at 20 days with fast growing zooplankton

Microalgal N/C was high (i.e. indicative of N-replete and/or light-limiting conditions) in all reactor systems except those with depths <0.1 m (Figs. 4 and 5). Accordingly, significant areal production of high-C metabolites, such as could be used for lipid nutrition or biofuels, only occurs in these shallow systems (AXP; Fig. S7a) where significant nutrient limitation can develop. Nutrient supply N/P does not greatly affect cellular N/C (Figs. 4 and 5) nor AXP (Figs. S7 and S8). Microalgal P/C in the N/P = 16 systems remains high at depths >0.2 m (Fig. 4), but in N/P = 32 systems (Fig. 5), only the deepest system (depth 0.5 m) enables an elevated microalgal P/C with shallower systems enabling the development of P-stress.

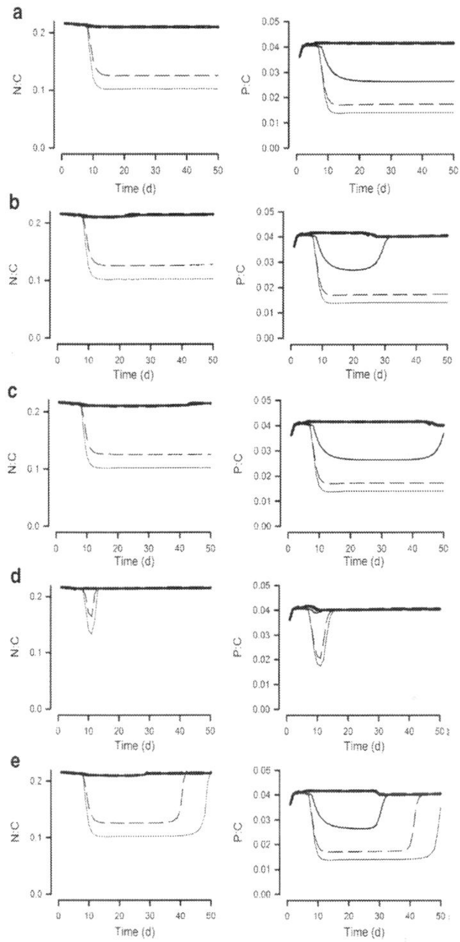

Figure 4: Algal mass ratios of N/C and P/C when grown at 6 different reactor operational depths (OD; 0.025–0.5 m), with the supply nutrient mole ratio N/P at 16 and a dilution rate of 0.3 day^{-1}. Cf. Fig. 2 for biomass and legend to line types. **a** No contamination. **B** Contamination at 0 day. **c** Contamination at 20 days. **d** Contamination at 0 day with fast growing zooplankton. **e** Contamination at 20 days with fast growing zooplankton

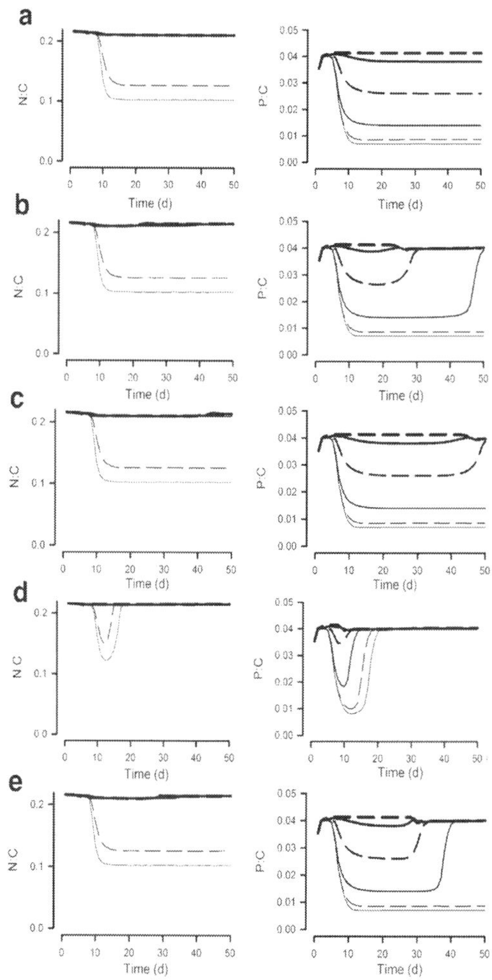

Figure 5: As Fig. 4, but with a supply nutrient mole N/P of 32. Cf. Fig. 3 for biomass and legend to line types. **a** No contamination. **b** Contamination at 0 day. **c** Contamination at 20 days. **d**Contamination at 0 day with fast growing zooplankton. **e** Contamination at 20 days with fast growing zooplankton

Systems contaminated with zooplanktonic pests

In the simulations, we considered two scenarios, one with contamination by zooplankton at the point of inoculation, and the other into an established crop. When contaminated with zooplankton with a maximum growth rate similar to that of the microalgae at the start of the system operation

(Figs. 2b and 3b; Figs. S1–S4), there was no effective grazing loss in the shallow systems (0.025 and 0.05 m deep) when supplied with a low nutrient N/P. This is because these systems contained microalgae with a C/N/P stoichiometry that makes them a poor food source for zooplankton growth. In the high N/P systems (Figs. 3, S2 and S4), which produce microalgae with low P/C (Fig. 5), microalgal losses in the 0.1-m-deep system were also slower, with low grazing pressure (Figs. S9 and S10) and poor zooplankton assimilation and growth efficiencies (Figs. S11 and S12).

Contamination of established systems is less likely to lead to crop losses than an initial contamination event (panels c versus panels(b in Figs. 2 and 3; Figs. S1–S4), provided that the system is shallow enough for nutrient limitation of microalgal growth to develop, and/or where the resultant zooplankton growth rates are slower than the system dilution rate (panels c in Figs. S3 and S4). The likelihood of some level of grazing resistance is enhanced in shallow depths by the P-depletion that developed in high N/P systems, which resulted in a depressed algal P/C (Fig. 3, S1, S2, S4 and S6); because of the resultant poor food quality, zooplankton in such systems have poor assimilation and growth efficiencies (Figs. S11 and S12).

Contamination with fast growing zooplankton leads to a more rapid demise of the microalgal crop; this is catastrophic when it occurs at the start of the culture process, when neither light nor nutrients are limiting and hence good food quality is assured (Figs. 2d and 3d; Figs. S1–S4). However, when contamination occurs into established shallow systems with a moderate system dilution rate (Figs. 2e and 3e; Figs. S1–S4), then microalgal losses are low. This is especially obvious with a high N/P medium (Fig. 3), where the poor P/C of the microalgae (i.e. poor quality food) restricts zooplankton growth to rates similar to, or below, the dilution rates of the systems (Figs. S9 and S10).

In our simulations, a high microalgal N/C ratio is maintained for depths >0.05 m and a high microalgal P/C for depths >0.2 m (Figs. 4 and 5), regardless of the timing of pest entry. These systems are more likely to support effective feeding by zooplankton, with higher assimilation and growth efficiencies (Figs. S11 and S12), and thence higher pest growth rates (Figs. S9 and S10).

DISCUSSION

Control systems, with no pest introduction

Our simulated microalgal biomass values are comparable to peak concentrations measured in commercial open ponds (Ozkan et al. 2012) assuming a conversion of C-biomass to dry weight, C/dw, between 0.3 and 0.5 (Heymans 2001; Geider and LaRoche 2002). Areal productivity levels, peaking at 2 gC m^{-2} day^{-1} and equating to 7.3 tC ha^{-1} $year^{-1}$, fall in the upper half of rates of the rate of 10–30 t dry weight ha^{-1} $year^{-1}$ reported for real systems (Garcia-Gonzalez et al. 2003; Jiménez et al. 2003; Crowe et al. 2012), again assuming a value for C/dw between 0.3 and 0.5. The simulations of microalgal growth thus give results consistent with expectations (see also Kenny and Flynn 2016).

Systems contaminated with zooplanktonic pests

The outcome of zooplankton-microalgal predator-prey interactions is dictated by the balance of growth and loss rates of both parties. Growth of the microalgal prey requires adequate nutrients and light, but as the formation of algal biomass increases at higher nutrient supply concentrations so does the likelihood of self-shading which leads to light limitation of photosynthesis. Loss of the microalgal crop is intrinsically related to the growth of the zooplanktonic predator which in turn depends on the nutritional quality and quantity of the available prey. In nature, both quantity and quality of prey affect zooplankton grazing. However, in artificial systems where microalgae are grown for biomass production, prey quality is the overriding issue. It is important to note that prey quality in this context need not equate to commercial crop quality; a microalgal crop grown for high lipid content will have a high C/N and thus constitute poor food for a predator (Sterner and Elser 2002; Mitra and Flynn 2005). In theory, genetic modification (GM) approaches could be used to configure microalgae that are unpalatable to zooplanktonic pest. Indeed, an analysis on the optimal configuration of GM-microalgae for biofuels and lipid production describes an organism that is coincidentally very poor feed for zooplankton (Flynn et al. 2012). While outwardly a win-win situation, the (inevitable) escape of such a GM organism from large-scale open ponds into the wild would carry a very real risk of generating a harmful algal species *par excellence*, being able to grow rapidly using little nutrient and be ungrazable (Flynn et al. 2012). For this reason, here we consider exploiting stoichiometric ecology to control pest growth.

The greatest risk from zooplankton pests is at the initial phases of crop growth, when the microalgal biomass density is relatively low (though far above levels likely to limit zooplankton grazing rates—Hansen et al. 1997) and hence neither light nor nutrients are limiting for algal growth. Microalgae in such situations are typically of good feed value for predators. Contamination of established systems is less likely to lead to crop losses if the cultivation system is shallow enough and the nutrient loading low enough to enable nutrient limitation of microalgal growth to develop. Minimising risks is improved further if a high N/P nutrient regime is operated allowing some level of P-depletion to develop, depressing algal P/C (Figs. 3, S1, S2, S4 and S6). High dilution systems populated by microalgae with growth rates that are higher relative than that of their predators are also less susceptible to zooplankton attack (Moheimani and Borowitzka 2006). However, contamination with fast growing zooplankton lessens the likelihood of crop survival whenever that contamination should occur.

To optimise production of biofuels or lipids, shallow culture systems populated with microalgae grown into nutrient limitation at a low fraction of their maximum possible growth rate (μ_{max}) are required (Kenny and Flynn 2015). The higher the microalgal μ_{max}, the better, as the system can also be operated at a higher absolute dilution rate and this then also washes out any zooplankton. Hence, to minimise crop losses through predation, a balance needs to be struck with bioreactor dilution rates. A slow diluted, light limiting system will favour predators. Conversely, a high dilution rate system configured for high production rates of a C-rich crop appears of lesser susceptibility to predators because the crop is of poor value as food and the elevated dilution rate then exceeds the zooplankton growth rate, washing out the pest. To illustrate this, compare the high peaks of slow-growing zooplankton contamination in Fig. S1b and S1c (where dilution = 0.1 day^{-1}) to the fast dilution (0.5 day^{-1}) situation in Fig. S3b and S3c; in the latter, zooplankton biomass only starts to rise very late in the simulation period.

Manipulating the systems to commercial advantage

Setting the best culture conditions to attain optimal chemical composition and elemental C/N/P for commercial growth of microalgae is challenging. For instance, any naturally lit system is susceptible to diel and weather-induced changes in irradiance levels that pose a serious risk to production. This is not solely due to a decrease in growth rate, but because during periods of lowered irradiance, there is an attendant improvement in food quality resulting from changes in microalgal C/N/P. Hence, decreased illumination over a significant period of a day (due to cloud cover, for example) may be expected to enhance the likelihood of grazer control due to a combination of a decreased growth rate of the phototroph, coupled with a concomitant improvement in prey nutritional quality as microalgal N/C and P/C increases. Thus, in studies by Sterner et al. (1998) and Urabe et al. (2002), the balance of light and nutrient limitation affected whether the stoichiometric quality of the microalgae remained low under high light, suppressing zooplankton growth, or was improved under low light conditions, where the zooplankton quickly dominated the microalga, *Scenedesmus*.

If prey are of good nutritional quality (i.e. microalgal N/C and P/C broadly similar to that of their predator), then grazing and assimilation efficiency are expected to be optimal (Sterner and Elser 2002). Through a positive feedback loop, the nutrients regenerated by the zooplankton enhance the nutrient status of their prey and the predator population grows rapidly; the microalgal crop is thence removed rapidly as the well-fed zooplankton proliferate. However, if microalgal N/C and/or P/C become decreased due to nutrient exhaustion, then trophic transfer to the grazer is less effective; nutrient regeneration by the zooplankton is decreased or even stalled and the crop remains of poor nutritional status. Such effects on grazer activity were explored by Hessen et al. (2002) under a variety of nutrient/light conditions. When zooplankton were introduced to a stable algae culture, an increase in P-availability combined with low light levels increased the algal P/C and the zooplankton soon dominated. Conversely, zooplankton growth was slowest under high light/low P conditions, and hence, algal loss was low due to poor grazing (Hessen et al. 2002). The effects of such positive feedback processes have been demonstrated using models (Mitra and Flynn 2006). The approach needed in commercial microalgal production is to exploit these ecological features to advantage.

That is relatively easy if the crop is grown for high-C products, as such microalgae are naturally poor prey. But what if the crop is grown for high protein content?

Recent evidence indicates that biomass and biochemical production by microalgal crops may be maintained when the microalgae are grown to N/P ratios higher than the Redfield N/P mole ratio of 16 (Mayers et al. 2014). This is of immediate benefit because it decreases the demand for phosphorus fertiliser—an expensive and dwindling resource (Elser and Bennett 2011; Chisti 2013a). However, there is an additional benefit because growth of zooplankton pests appears to be affected more by a lack of P (Hessen et al. 2002) than are the microalgae, providing a more grazer-resistant crop (Fig. 2 vs 3). In the simulations run here, the stoichiometric interactions between predator and prey were, in accordance with simple stoichiometric ecology (Sterner and Elser 2002), those directly and simply (linearly) related to differences in elemental C/N/P. In reality, deviations in C/N/P in phototrophs are often associated with other biochemical events such as the accumulation of secondary metabolites which are distasteful to grazers, if not potent toxins. Being stressed by P, rather than N, is most closely allied to accumulation of secondary metabolites that are noxious to zooplankton (Granéli and Flynn 2006). N-stress may also produce microalgae that are distasteful to zooplankton (Flynn et al. 1996) even though they may not be classed as noxious in a human nutrition context. Given that zooplankton grazing is most damaging when the prey (the microalgal crop) has a well-balanced C/N/P, it is fortuitous that many products of value from microalgae are derived from high-C metabolites (Greenwell et al. 2010) which are products of cells with elevated C/N and that minor changes in algal C/P may not be damaging to crop production (Mayers et al. 2014). Interestingly, P-limitation also limits chytrid parasitic infections of microalgae (Bruning 1991). Coupled with the projected increase in the cost of P-fertilisers, there appear sound reasons for minimising the addition of P in all microalgal culture systems.

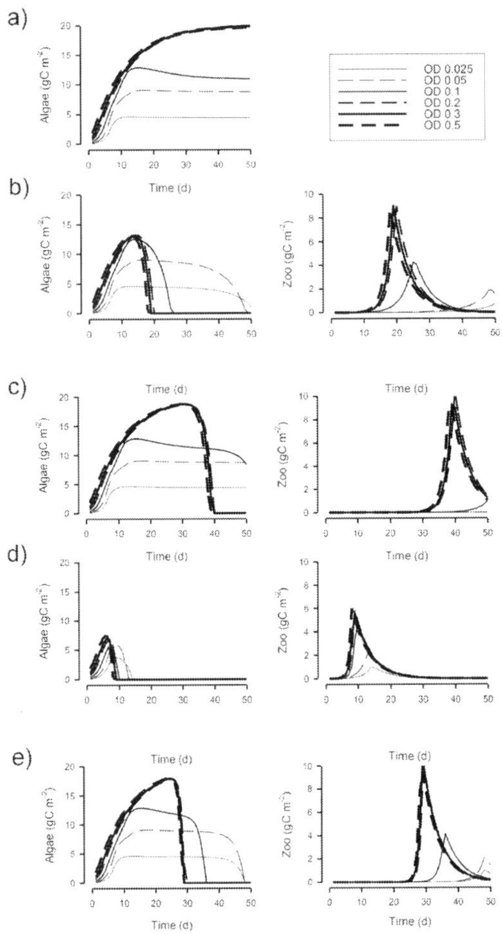

Figure S1: Areal biomass of algae and a zooplankton contaminant when grown at 6 different operation depths (OD; 0.025 – 0.5 m), with the supply nutrient mole ratio N:P at 16 and a dilution rate of $0.1d^{-1}$. a) No contamination. b) Contamination at 0d. c) Contamination at 20d. d) Contamination at 0d with fast growing zooplankton. e) Contamination at 20d with fast growing zooplankton.

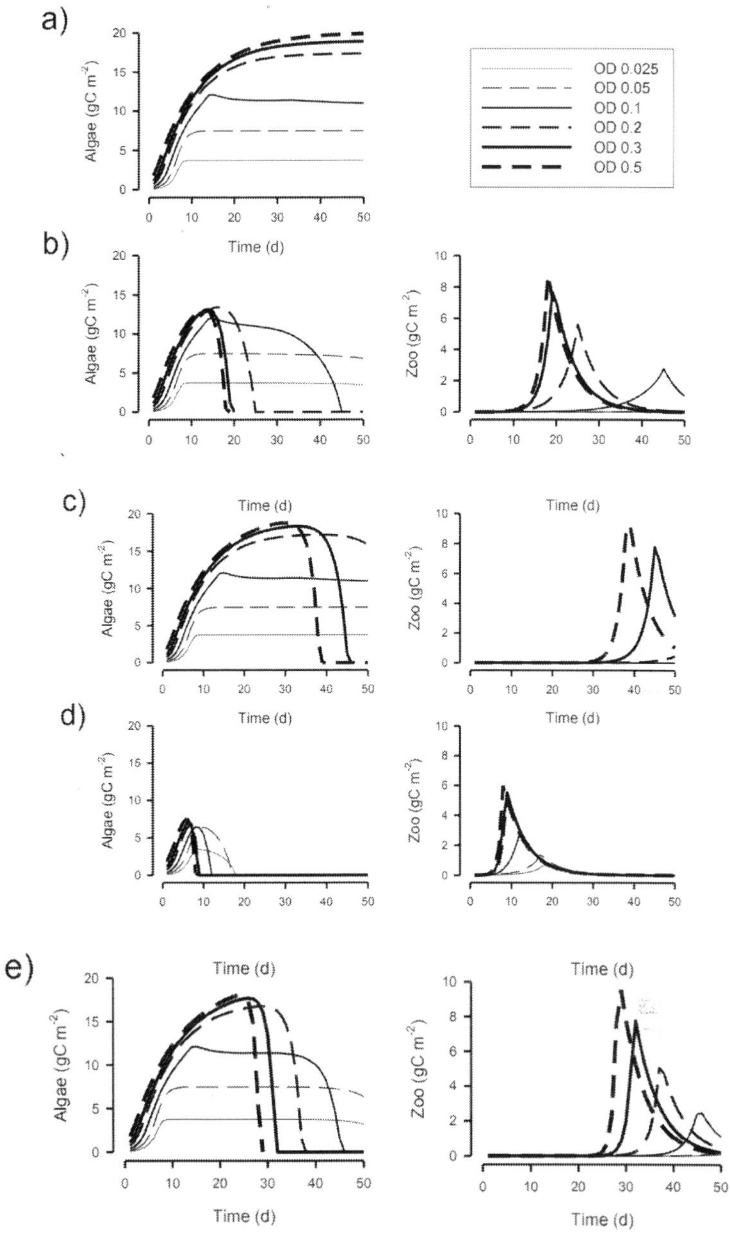

Figure S2: Areal biomass, as Fig.S1, at 6 different operation depths (OD; 0.025 – 0.5 m), with a dilution rate 0.1 d^{-1} , but with the supply nutrient mole ratio N:P at 32. a) No contamination. b) Contamination at 0d. c) Contamination at 20d. d) Contamination at 0d with fast growing zooplankton. e) Contamination at 20d with fast growing zooplankton.

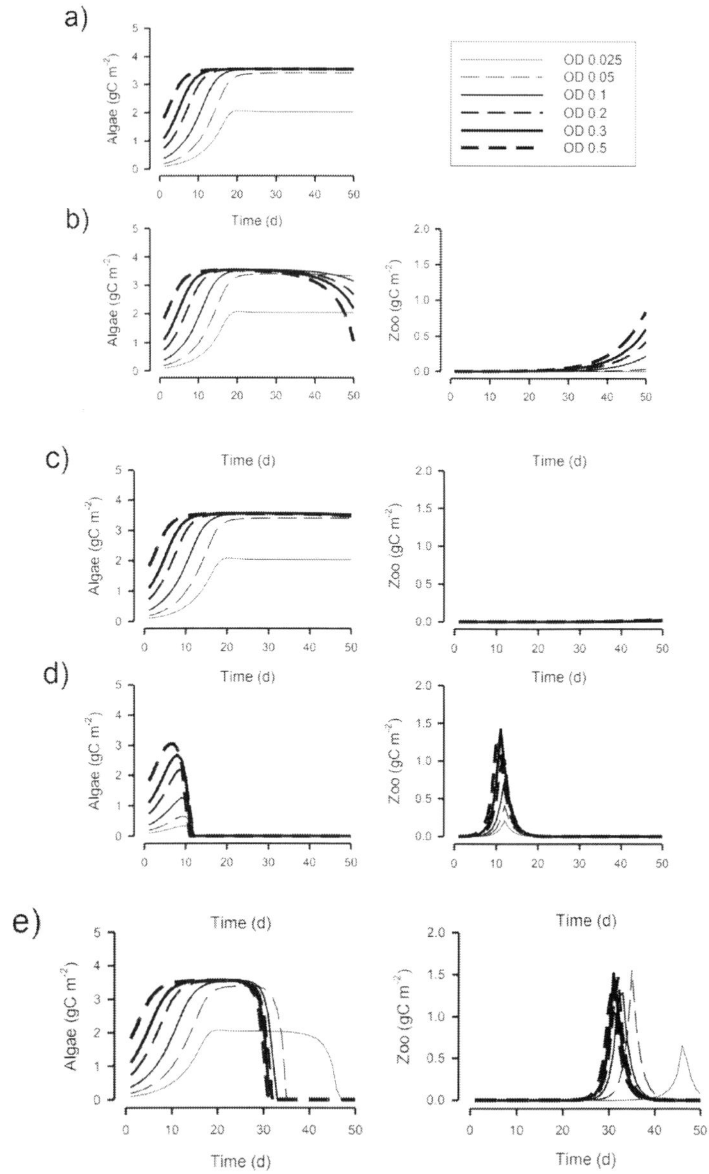

Figure S3: Areal biomass, as Fig.S1, at 6 different operation depths (OD; 0.025 – 0.5 m), with a supply nutrient mole ratio N:P of 16, but a dilution rate of 0.5 d-1 . a) No contamination. b) Contamination at 0d. c) Contamination at 20d. d) Contamination at 0d with fast growing zooplankton. e) Contamination at 20d with fast growing zooplankton.

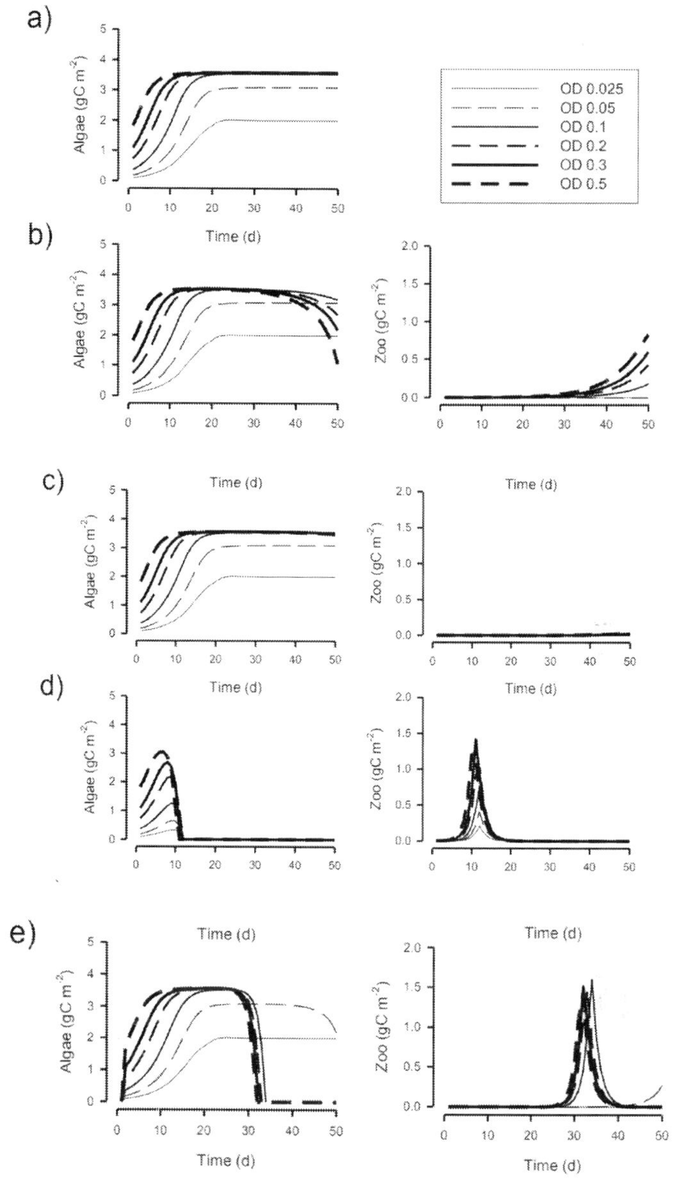

Figure S4: Areal biomass, as Fig.S1, at 6 different operation depths (OD; 0.025 – 0.5 m), but with a supply nutrient mole ratio N:P of 32, and a dilution rate of 0.5 d^{-1} . a) No contamination. b) Contamination at 0d. c) Contamination at 20d. d) Contamination at 0d with fast growing zooplankton. e) Contamination at 20d with fast growing zooplankton.

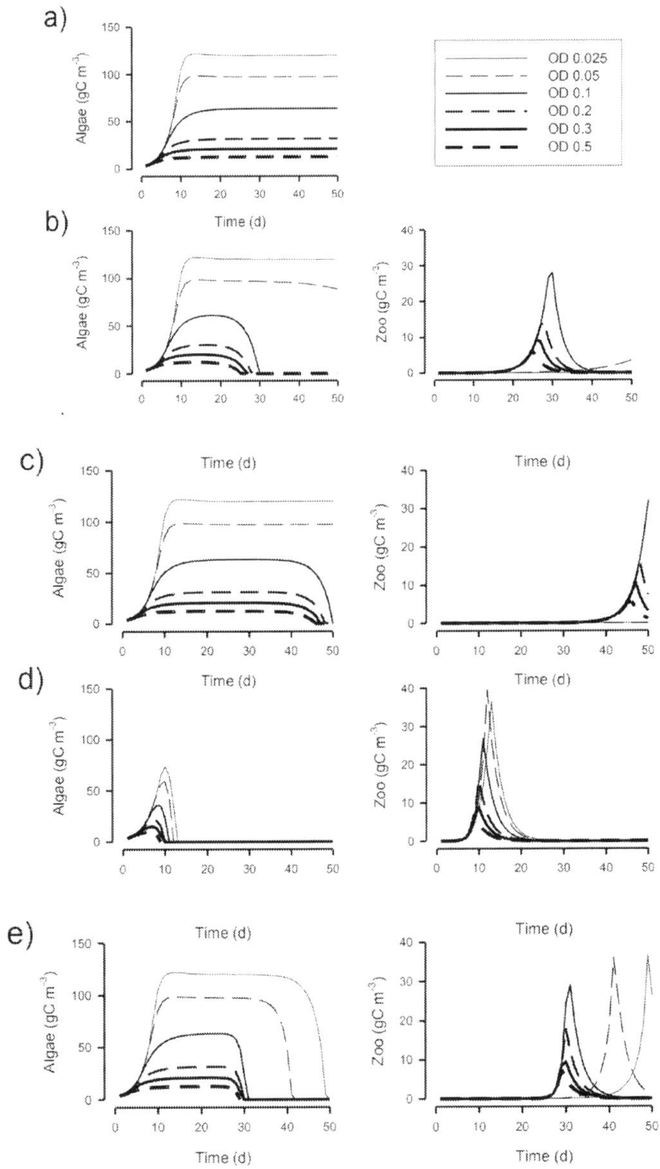

Figure S5: Volumetric biomass of algae and a zooplankton contaminant grown at 6 different operational depths (OD; 0.025 – 0.5 m), with the supply nutrient mole ratio N:P at 16 and a dilution rate of 0.3d-1 . Cf. Fig. 2 for biomass. a) No contamination. b) Contamination at 0d. c) Contamination at 20d. d) Contamination at 0d with fast growing zooplankton. e) Contamination at 20d with fast growing zooplankton.

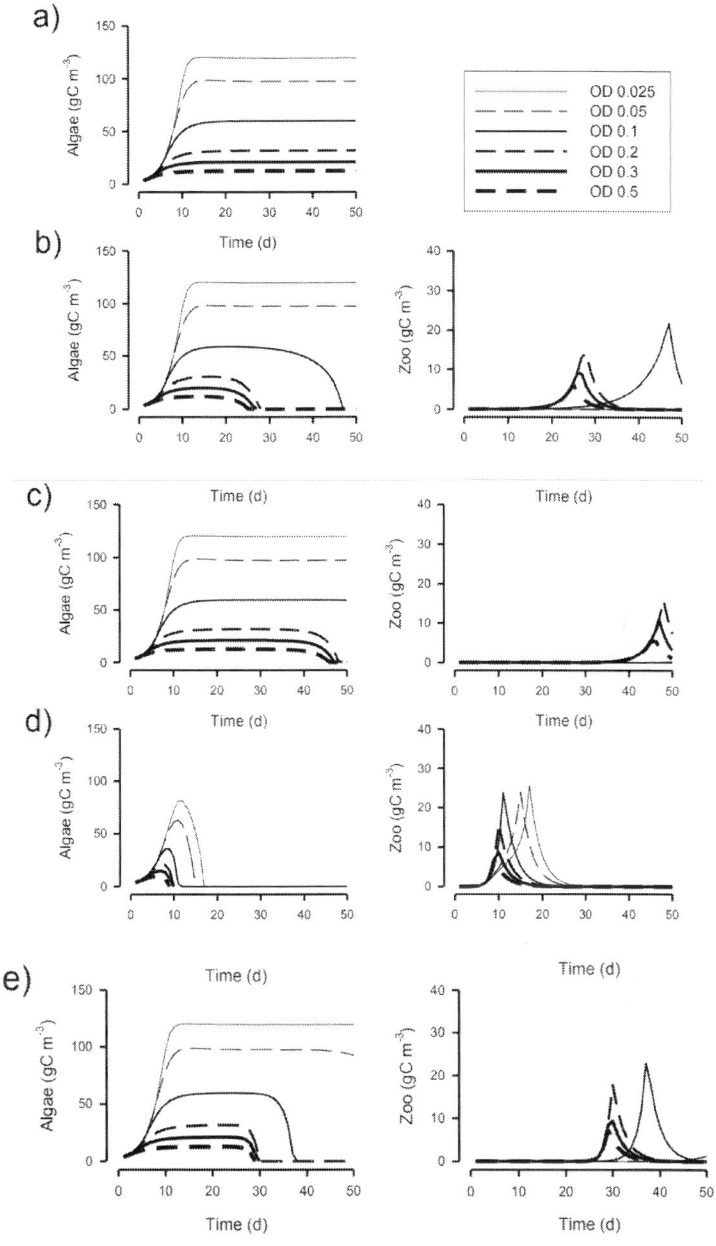

Figure S6: Volumetric biomass, as Fig.S5 at 6 different operation depths (OD; 0.025 – 0.5 m), with a dilution rate of 0.3 d^{-1} , but with a supply nutrient mole ratio N:P of 32. Cf. Fig. 3 for biomass. a) No contamination. b) Contamination at 0d. c) Contamination at 20d. d) Contamination at 0d with fast growing zooplankton. e) Contamination at 20d with fast growing zooplankton.

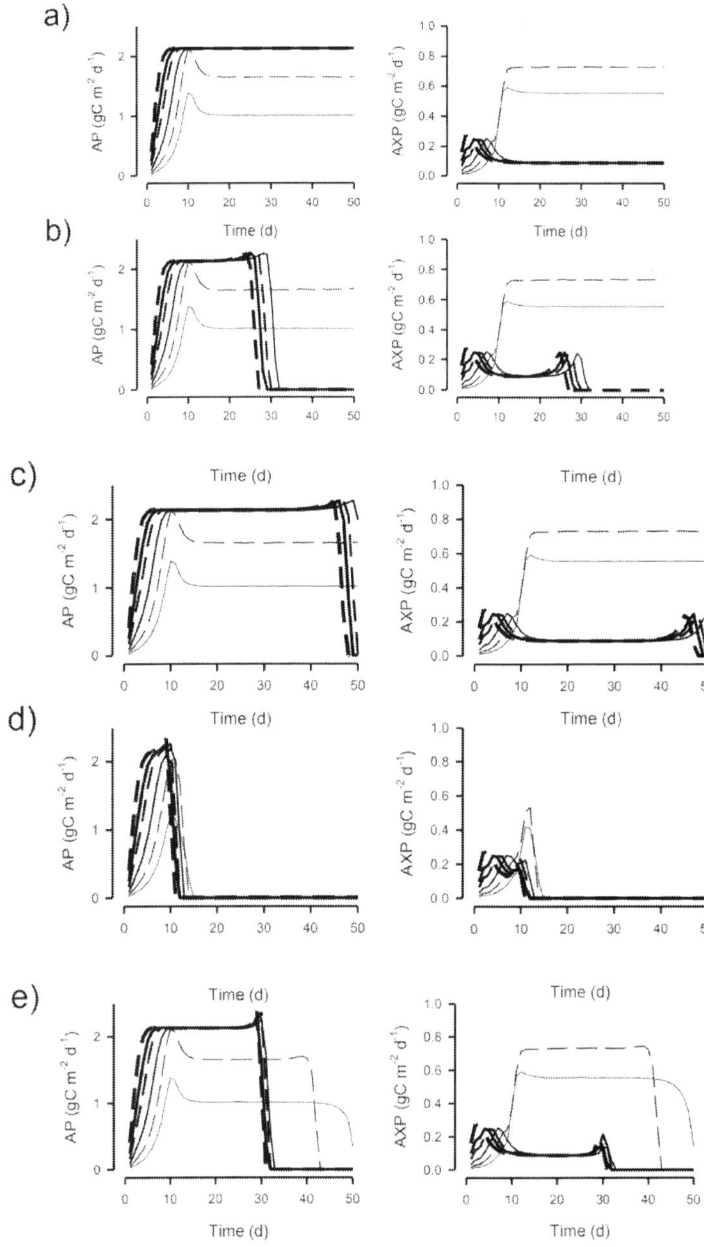

Figure S7: Areal production of biomass (AP) and biofuels (AXP) at 6 different operational depths (OD; 0.025 – 0.5 m; see line style definitions in Fig. S6), with the supply nutrient mole ratio N:P at 16 and a dilution rate of 0.3d-1 . Cf. Fig. 2 for biomass. a) No contamination. b) Contamination at 0d. c) Contamination at 20d. d) Contamination at 0d with fast growing zooplankton. e) Contamination at 20d with fast growing zooplankton.

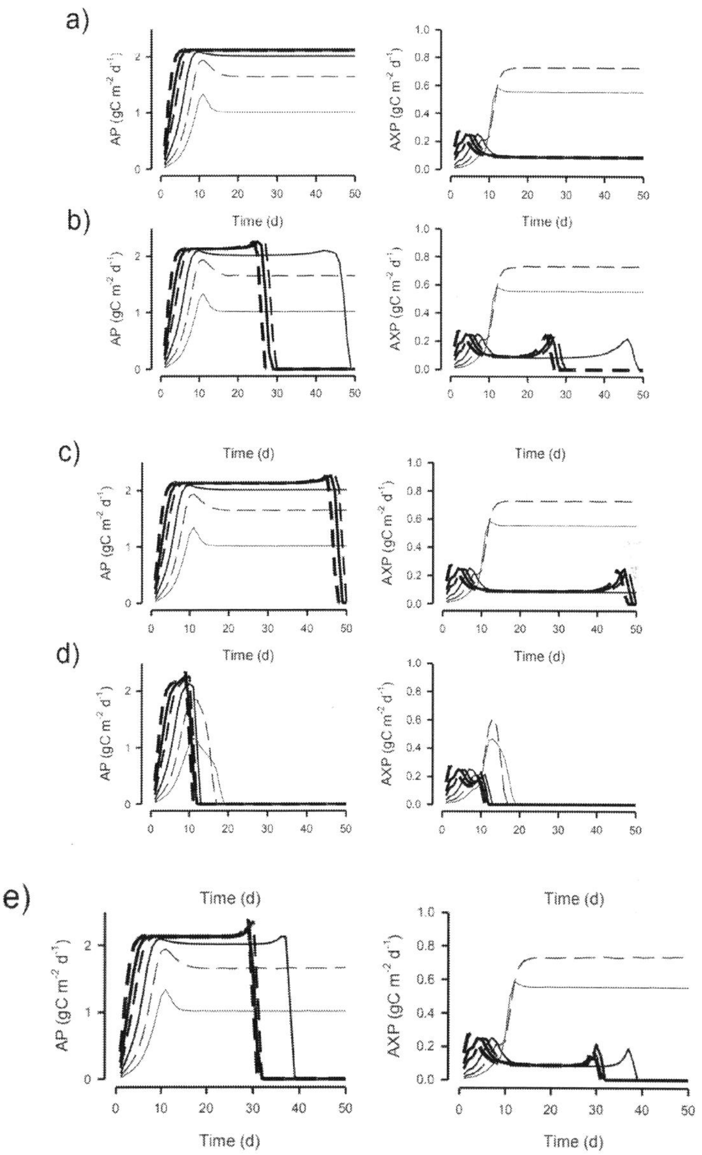

Figure S8: Areal production of biomass (AP) and biofuels (AXP), as Fig. S7, with a dilution rate of 0.3 d^{-1} , but with a supply nutrient mole N:P of 32. Cf. Fig. 3 for biomass; see line style definitions in Fig. S6. a) No contamination. b) Contamination at 0d. c) Contamination at 20d. d) Contamination at 0d with fast growing zooplankton. e) Contamination at 20d with fast growing zooplankton.

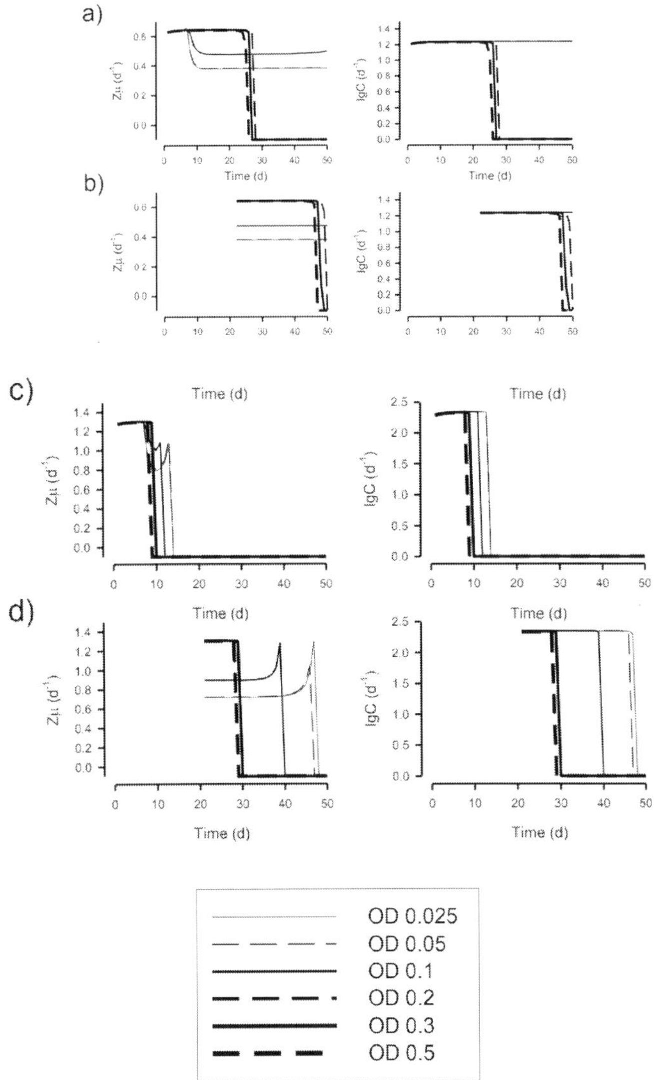

Figure S9: Zooplankton growth rate (Zμ) and ingestion rate (IgC) at 6 different operational depths (OD; 0.025 – 0.5 m), with the supply nutrient mole ratio N:P at 16 and a dilution rate of $0.3d^{-1}$. Cf. Fig. 2 for biomass. a) Contamination at 0d. b) Contamination at 20d. c) Contamination at 0d with fast growing zooplankton. d) Contamination at 20d with fast growing zooplankton.

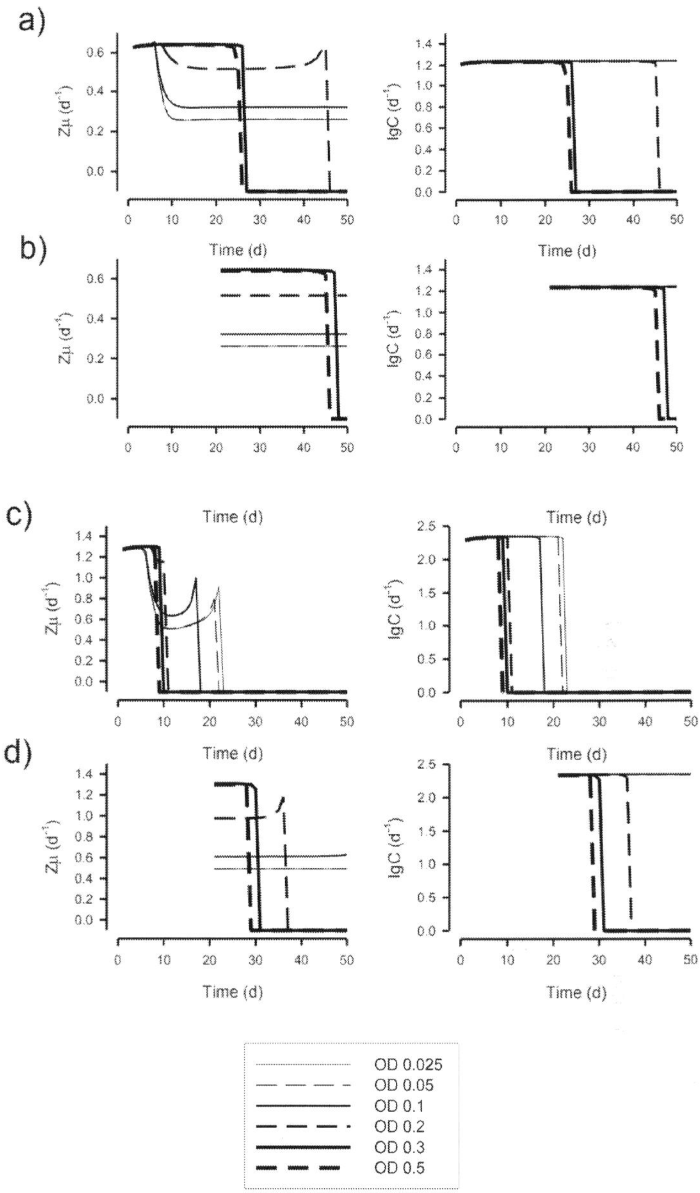

Figure S10: Zooplankton growth rate (Zμ) and ingestion rate (IgC), as Fig. S9 at 6 different operation depths (OD; 0.025 – 0.5 m), with a dilution rate of 0.3 d^{-1} , but with a supply nutrient mole N:P of 32. Cf. Fig. 3 for biomass. a) Contamination at 0d. b) Contamination at 20d. c) Contamination at 0d with fast growing zooplankton. d) Contamination at 20d with fast growing zooplankton

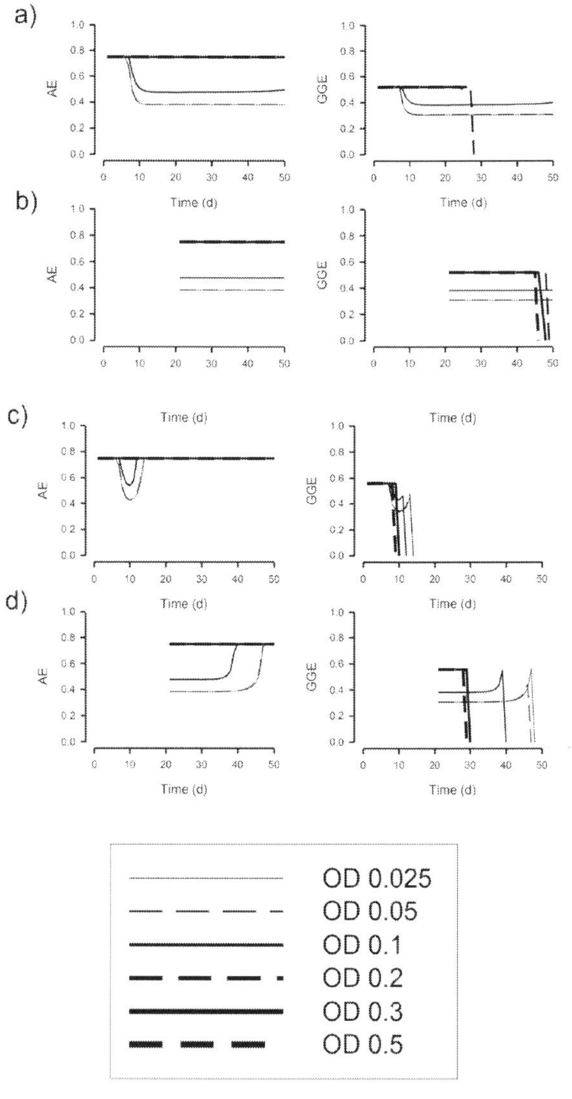

Figure S11: Zooplankton assimilation efficiency (AE) and gross growth efficiency (GGE) at 6 different operational depths (OD; 0.025 – 0.5 m), with the supply nutrient mole ratio N:P at 16 and a dilution rate of 0.3d-1 . Cf. Fig. 2 for biomass. a) Contamination at 0d. b) Contamination at 20d. c) Contamination at 0d with fast growing zooplankton. d) Contamination at 20d with fast growing zooplankton.

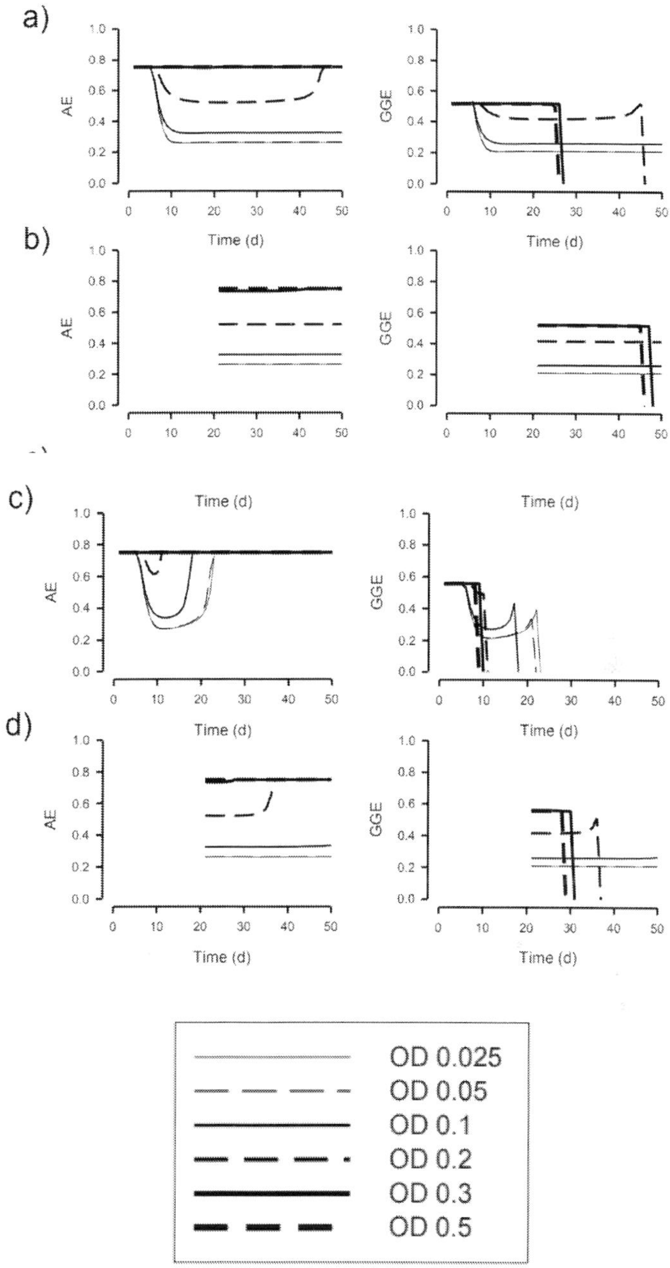

Figure S12: Zooplankton assimilation efficiency (AE) and gross growth efficiency (GGE), as Fig. S11 at 6 different operation depths (OD; 0.025 – 0.5 m), with a dilution rate of 0.3 d^{-1} , but with a supply nutrient mole N:P of 32. Cf.

Fig. 3 for biomass. a) Contamination at 0d. b) Contamination at 20d. c) Contamination at 0d with fast growing zooplankton. d) Contamination at 20d with fast growing zooplankton.

NOTES

Acknowledgements

This work was supported through the EnAlgae project (ref. 215G), which received ERDF funding through the INTERREG IVB North West Europe programme with co-financing from the Welsh Government, and Natural Environment Research Council (UK) grants NE/K001345/1 (via subcontract to A.Mitra) and NE/F003455/1.

REFERENCES

1. Bacellar Mendes LB, Vermelho AB (2013) Allelopathy as a potential strategy to improve microalgae cultivation. Biotechnol Biofuels 6:152. doi:10.1186/1754-6834-6-152
2. Bartley ML, Boeing WJ, Corcoran AA, Holguin FO, Schaub T (2013) Effects of salinity on growth and lipid accumulation of biofuel microalga Nannochloropsis salina and invading organisms. Biomass Bioenerg 54:83–88
3. Borowitzka MA (2005) Culturing microalgae in outdoor ponds. In: Andersen RA (ed) Algal Culturing Techniques. Elsevier, New York, pp 205–218
4. Borowitzka MA (2013) High-value products from microalgae—their development and commercialisation. J Appl Phycol 25:743–756
5. Bruning K (1991) Effects of phosphorus limitation on the epidemiology of a chytrid phytoplankton parasite. Freshw Biol 25:409–417
6. Chisti Y (2007) Biodiesel from microalgae. Biotechnol Adv 25:294–306
7. Chisti Y (2013a) Constraints to commercialization of algal fuels. J Biotechnol 167:201–214
8. Chisti Y (2013b) Raceways-based production of algal crude oil.

Green 3(3–4):195–216

9. Crowe B, Attalah S, Agrawal S, Waller P, Ryan R, Wagenen JV, Chavis A, Kyndt J, Kacira M, Ogden KL, Huesemann M (2012) A comparison of Nannochloropsis salina growth performance in two outdoor ponds. Int J Chem Eng 2012:920608. doi:10.1155/2012/920608
10. Day JG, Thomas NJ, Achilles-Day UEM, Leakey RJG (2012) Early detection of protozoan grazers in algal biofuel cultures. Bioresource Technol 114:715–719
11. Elrifi IR, Turpin DH (1985) Steady-state luxury consumption and the concept of optimum nutrient ratios: a study with phosphate and nitrate limited Selenastrum minutum(Chlorophyta). J Phycol 21:592–602
12. Elser J, Bennett E (2011) Phosphorus cycle: a broken biogeochemical cycle. Nature 478:29–31
13. Flynn KJ (2008) The importance of the form of the quota curve and control of non-limiting nutrient transport in phytoplankton models. J Plankton Res 30:423–438
14. Flynn KJ (2010) Do external resource ratios matter?: implications for modelling eutrophication events and controlling. J Mar Sys 83:170–180
15. Flynn KJ, Raven JA (2017) What is the limit for photoautotrophic plankton growth rates? J Plankton Res 39:13–22
16. Flynn KJ, Davidson K, Cunningham A (1996) Prey selection and rejection by a microflagellate; implications for the study and operation of microbial food webs. J Exp Mar Biol Ecol 196:357–372
17. Flynn KJ, Greenwell HC, Lovitt RW, Shields RJ (2010) Selection for fitness at the individual or population levels: modelling effects of genetic modifications in microalgae on productivity and environmental safety. J Theor Biol 263:269–280
18. Flynn KJ, Mitra A, Greenwell HC, Sui J (2012) Monster potential meets potential monster: pros and cons of deploying genetically modified microalgae for biofuels production. Interface Focus 3:20120037. doi:10.1098/rsfs.2012.0037
19. Fon Sing S, Isdepsky A, Borowitzka MA, Moheimani NR (2013)

Production of biofuels from microalgae. Mitig Adapt Strat Global Change 18:47–72

20. Garcia-Gonzalez M, Moreno J, Canavate JP, Anguis V, Prieto A, Manzano C, Florencio FJ, Guerrero MG (2003) Conditions for open-air outdoor culture of Dunaliella salina in southern Spain. J Appl Phycol 15:177–184
21. Geider RJ, LaRoche J (2002) Redfield revisited: variability of C:N:P in marine microalgae and its biochemical basis. Eur J Phycol 37:1–17
22. Granéli E, Flynn KJ (2006) Chemical and physical factors influencing toxin content. In: Granéli E, Turner JT (eds) Ecology of harmful algae. Springer-Verlag, Berlin, pp 229–241
23. Greenwell HC, Laurens LML, Shields RJ, Lovitt RW, Flynn KJ (2010) Placing microalgae on the biofuels priority list: a review of the technological challenges. J Roy Soc Interface 7:703–726
24. Guillard RRL (1975) Culture of phytoplankton for feeding marine invertebrates. In: Smith WL, Chanley MH (eds) Culture of marine invertebrate animals, pp 29–60 Springer USA
25. Gutman J, Zarka A, Boussiba S (2009) The host-range of Paraphysoderma sedebokerensis, a chytrid that infects Haematococcus pluvialis. Eur J Phycol 44:509–514
26. Hansen PJ, Bjornsen PK, Hansen BW (1997) Zooplankton grazing and growth: scaling within the 2–2,000-μm body size range. Limnol Oceanogr 42:687–704
27. Hessen DO, Færøvig PJ, Andersen T (2002) Light, nutrients, and P: C ratios in algae: grazer performance related to food quality and quantity. Ecology 83:1886–1898
28. Heymans JJ (2001) The Gulf of Maine, 1977-1986. Fisheries impacts on North Atlantic ecosystems: models and analyses. Fish Centre Res Rep 9:129–149
29. Jiménez C, Cossío BR, Labella D, Niell FX (2003) The feasibility of industrial production of Spirulina (Arthrospira) in southern Spain. Aquaculture 217:179–190
30. Kazamia E, Aldridge DC, Smith AG (2012) Synthetic ecology - a way forward for sustainable algal biofuel production? J Biotechnol

162:163–169

31. Kenny P, Flynn KJ (2015) In silico optimization for production of biomass and biofuel feedstocks from algae. J Appl Phycol 27:33–48
32. Kenny P, Flynn KJ (2016) Coupling a simple irradiance description to a mechanistic growth model to predict algal production in industrial-scale solar-powered photobioreactors. J Appl Phycol 28:3203–3212
33. Mayers JJ, Flynn KJ, Shields RJ (2014) Influence of the N:P supply ratio on biomass productivity and time-resolved changes in elemental and bulk. Bioresource Technol 169:588–595
34. McBride RC, Lopez S, Meenach C, Burnett M, Lee PA, Nohilly F, Behnke C (2014) Contamination management in low cost open algae ponds for biofuels production. Indust Biotechnol 10:221–227
35. Méndez C, Uribe E (2012) Control of Branchionus sp. and Amoeba sp. in cultures of Arthrospira sp. Lat Am J Aquat Res 40:553–561
36. Milledge JJ (2011) Commercial application of microalgae other than as biofuels: a brief review. Rev Environ Sci Biotechnol 10:31–41
37. Mitra A (2006) A multi-nutrient model for the description of stoichiometric modulation of predation in micro- and mesozooplankton. J Plankton Res 28:597–611
38. Mitra A, Flynn KJ (2005) Predator-prey interactions: is "ecological stoichiometry" sufficient when good food goes bad? J Plankton Res 27:393–399
39. Mitra A, Flynn KJ (2006) Promotion of harmful algal blooms by zooplankton predatory activity. Biol Lett 2:194–197
40. Mitra A, Flynn KJ, Fasham MJR (2007) Accounting correctly for grazing dynamics in nutrient-phytoplankton-zooplankton models. Limnol Oceanogr 52:649–661
41. Mitra A, Castellani C, Gentleman WC, Jónasdóttir SH, Flynn KJ, Bode A, Halsband C, Kuhn P, Licandro P, Agersted MD, Calbet A, Lindeque PK, Koppelmann R, Møller EF, Gislason A, Nielsen TG, St. John M (2014) Bridging the gap between marine biogeochemical and fisheries sciences; configuring the zooplankton link. Prog Oceanogr 129, Part B:176–199
42. Moheimani NR, Borowitzka MA (2006) The long-term culture of

the coccolithophore Pleurochrysis carterae (Haptophyta) in outdoor raceway ponds. J Appl Phycol 18:703–712

43. Ozkan A, Kinney K, Katz L, Berberoglu H (2012) Reduction of water and energy requirement of algae cultivation using an algae biofilm photobioreactor. Bioresource Technol 114:542–548
44. Quinn JC, Yates T, Douglas N, Weyer K, Butler J, Bradley TH, Lammers PJ (2012) Nannochloropsis production metrics in a scalable outdoor photobioreactor for commercial applications. Bioresource Technol 117:164–171
45. Redfield AC (1934) On the proportions of organic derivatives in sea water and their relation to the composition of plankton. In: Daniel RJ (ed) James Johnstone memorial volume. University Press of Liverpool, UK, pp 176–192
46. Rego D, Redondo LM, Geraldes V, Costa L, Navalho J, Pereira MT (2015) Control of predators in industrial scale microalgae cultures with pulsed electric fields. Bioelectrochemistry 103:60–64
47. Richmond A (2013) Biological principles of mass cultivation of photoautotrophic microalgae. In: Richmond A, Hu Q (eds) Handbook of microalgal culture: applied phycology and biotechnology, 2nd edn. John Wiley & Sons, Chichester, pp 171–204
48. Schroeder DC, Oke J, Hall M, Malin G, Wilson WH (2003) Virus succession observed during an Emiliania huxleyi bloom. Appl Environ Microbiol 69:2484–2490
49. Smith VH, Crews T (2014) Applying ecological principles of crop cultivation in large-scale algal biomass production. Algal Res 4:23–34
50. Smith VH, Foster BL, Grover JP, Holt RD, Leibold MA, deNoyelles F (2005) Phytoplankton species richness scales consistently from laboratory microcosms to the world's oceans. Proc Natl Acad Sci U S A 102:4393–4396
51. Smith VH, Sturm BSM, deNoyelles FJ, Billings SA (2010) The ecology of algal biodiesel production. Trends Ecol Evol 25:301–309
52. Sterner RW, Elser JJ (2002) Ecological stoichiometry: the biology of elements from molecules to the biosphere Princeton. Princeton University Press, NJ

53. Sterner RW, Hagemeier DD, Smith WL, Smith RF (1993) Phytoplankton nutrient limitation and food quality for Daphnia. Limnol Oceanogr 38:857–871
54. Sterner RW, Clasen J, Lampert W, Weisse T (1998) Carbon:phosphorus stoichiometry and food chain production. Ecol Lett 1:146–150
55. Strittmatter M, Guerra T, Silva J, Gachon CMM (2016) A new flagellated dispersion stage in Paraphysoderma sedebokerense, a pathogen of Haematococcus pluvialis. J Appl Phycol 28:1553–1558
56. Sturm BSM, Peltier E, Smith V, deNoyelles F (2012) Controls of microalgal biomass and lipid production in municipal wastewater-fed bioreactors. Environ Prog Sustain Energy 31:10–16
57. Urabe J, Kyle M, Makino W, Yoshida T, Andersen T, Elser J (2002) Reduced light increases herbivore production due to stoichiometric effects of light/nutrient balance. Ecology 83:619–627
58. Wang H, Zhang W, Chen L, Wang J, Liu T (2013) The contamination and control of biological pollutants in mass cultivation of microalgae. Bioresource Technol 128:745–750
59. Young S, Palm M, Grover JP, McKee D (1997) How Daphnia cope with algae selected for inedibility in longrunning microcosms. J Plankton Res 19:391–397

SECTION III

BIOFUEL PRODUCTION

CHAPTER

10

DUAL POTENTIAL OF MICROALGAE AS A SUSTAINABLE BIOFUEL FEEDSTOCK AND ANIMAL FEED

Krystal K Lum, Jonggun Kim and Xin Gen Lei

Department of Animal Science, Cornell University, Ithaca, NY 14850, USA

ABSTRACT

The rise in global population has led to explorations of alternative sources of energy and food. Because corn and soybean are staple food crops for humans, their common use as the main source of dietary energy and protein for food-producing animals directly competes with their allocation for human consumption. Alternatively, de-fatted marine microalgal biomass generated from the potential biofuel production may be a viable replacement of corn and soybean meal due to their high levels of protein, relatively well-balanced amino acid profiles, and rich contents of minerals and vitamins, along with unique bioactive compounds. Although the full-fatted (intact) microalgae represent the main source of omega-3 (n-3) polyunsaturated fatty acids including docohexaenoic acid (DHA) and eicosapentaenoic acid (EPA), the de-fatted microalgal biomass may still contain good amounts of these components for enriching DHA/EPA in eggs, meats, and milk. This review is written to highlight the necessity and potential of using the de-fatted microalgal biomass as a new generation of animal feed in helping address the global energy, food, and environmental issues. Nutritional feasibility and limitation of the biomass as the new feed ingredient for simple-stomached species are elaborated. Potential applications of the biomass for generating value-added animal products are also explored.

Keywords: Algae, Biofuel, Feed, Health, Nutrition

INTRODUCTION

The projected rise of global population from the current 7 billion to over 9 billion within the next several decades [1, 2] garners urgent needs for renewable energy and alternative foods. As global supplies of petroleum continue to decline, renewable nature fuels including biofuel are being explored. Biofuels are defined as the energy derived from raw biological materials, and show promise in harnessing adequate energy and reducing greenhouse gas emissions associated with fossil fuels. As corn and soybean are staple food crops for humans as well as the two main conventional feedstuffs that provide energy and protein for food-producing animals, the projected global population rise, along with the expansion of animal production, presents a serious threat to nutrition security for both humans and animals. However, the increasing use of these feedstocks for biodiesel and bioethanol production has driven up their global prices. From 2007 to 2011, the worldwide production of bioethanol nearly doubled from 49.6 to 84.6 billion liters [3]. In parallel, the price of corn was increased from \$163 to \$291 per metric ton over the four years [4]. In 2011, the United States produced approximately 12.4 billion bushels of corn, and 38% of which was used to produce bioethanol or to generate other co-products [3]. Apparently, current allocations of corn and (or) soybean for the biofuel and animal productions are unsustainable. Alternatives to these ingredients are required to maintain a harmonious infrastructure among the fuel, food, and feed industries.

Marine microalgae bear attractive properties for biofuel production [5, 6, 7, 8]. Subsequently, the fat-extracted microalgal biomass derived from the biofuel production may be a promising carbon-neutral animal feed supplement [9, 10, 11]. In fact, our laboratory has recently demonstrated that the de-fatted biomass of *Staurospira sp.* with 19% crude protein replaced 7.5% of corn and soybean meal without affecting the growth performance or health status of broiler chickens [12]. Notably, the dual application of microalgae as a new source of biofuel and animal feed will help alleviate the greenhouse effects associated with current energy and food production.

MICROALGAE FOR BIOFUEL PRODUCTION

Marine microalgae sequester carbon dioxide (CO_2) through photosynthesis, and may be used to produce biogas including methane and hydrogen via anaerobic processing [13, 14, 15]. While certain species of microalgae were recognized in the 1940s to yield high amounts of cellular lipids under selective growth pressures, it was not until the 1950's when algae were viewed as a potential energy source, and were tested for methane gas production via anaerobic digestion of their cell carbohydrates [16, 17]. The flexibility and (or) adaptivity of microalgal species to water and cultural conditions allows us to spare fresh water and arable land for crop production [18]. The land use efficiency of microalgae for biofuel production, grown with 30% oil content by weight, was 130 and 338 times greater than the conventional biodiesel feedstock soybean and corn, respectively [6]. While optimal growth conditions for microalgae are species-specific, photoautotrophic cultivation of these single cell species at large scales for biofuel and co-products depends on the technical and economic feasibility. At the present time, the photoautotrophic production of microalgae is marginally cost-effective only for generating value-added co-products or feed additives used in aquaculture [19, 20]. In such productions, microalgae are grown in the presence of light within constructions such as open raceway ponds. To extract the lipids, microalgae are first de-watered. The concentrated biomass is subsequently processed to optimize the solvent extraction through cell disruption, particle size reduction, and drying [21]. The remaining microalgae skeleton after lipid extraction is the so-called de-fatted microalgal biomass to be used as an animal feed. Without the presumed feed application, the commercial microalgae cultivation and processing for biofuel production [22, 23] remains largely cost-ineffective. Therefore, the feed application of the de-fatted biomass would not only create a new source of animal feed to mitigate the current competition with human food supply, but also help make the biofuel production of microalgae economically feasible.

MICROALGAE FOR ANIMAL NUTRITION

Feasibility of microalgae as animal feed

Although the use of whole microalgae in animal diets has long been studied, dating as far back as in the 1950's, only recent literature has shown attempts to supplement lipid-extracted microalgae in animal diets [12, 24]. Initially, researchers sought out methods to culture algae in ponds, and the developments of which were quickly followed by studies on algae supplementation into animal diets as a protein source [25, 26]. Different sources of cultivated algae were effective in maintaining animal growth performance, and in some cases improving daily body weight gain. Sewage-grown *Chlorella* and *Scenedesmus* sp. gained attention as potential nutrient sources due to their high crude protein and carotenoid contents [27, 28, 29, 30]. In 1952, Combs reported that supplementing 10% *Chlorella* sp. into a diet deficient in riboflavin and vitamin A improved feed efficiency and growth of chicks [27]. Later work showed no adverse response of growth by chicks to diets containing 20% sewage-grown, aluminum-free *Chlorella* and *Scenedesmus* sp., compared with those fed a corn-soybean meal based diet[28].

During the following decades, pond and tap water-grown algae were found to sustain fish growth in aquaculture, and 6-10% sewage-grown algae incorporated in a barley-based diet for growing-finishing pigs maintained their growth rate and feed conversion efficiency[31, 32]. Other algal variants, such as the blue-green algae *Spirulina* sp., were also investigated not only for their effects on overall growth performance, but also on organ health and reproductive characteristics of animals. Ross and Dominy supplemented diets for broiler chickens with up to 20% blue-green algae, and found the 3 wk-old broilers experienced depressed growth when algae inclusion levels were higher than 10% [33]. At the inclusion rate of 12%, broilers showed slower growth in comparison with those fed 0, 1.5, 3.0, or 6.0% algae. Toyomizu et al.[34] fed broilers with up to 8% *Spirulina* sp. and reported no differences in their body weight, liver weight, abdominal fat, or kidney weight at 16 d of age. When early-weaned piglets were fed a diet with *S. maxima* replacing up to 12% of the total protein from skim milk, apparent digestibilities of the diets were reduced, but the piglet growth remained insignificantly different from that of the control pigs [35]. Fevrier and Seve[36] supplemented *S. maxima* into the diets of sows, and showed that the algae-fed sows had lower net weight gains with more and heavier piglets at the first reproductive cycle. By the end of the second reproductive cycle, the growth and litter characteristics were similar between the algae-fed and control groups, although the algae-fed sows produced a higher culling rate of piglets. Another study showed that feeding weanling pigs with a mixture of *S. maxima* and *A. platensis*, or *Chlorella* sp. in replacing 33% of soybean meal did not affect their body weight gain, feed efficiency, or incidence of diarrhea compared with pigs fed a control diet. Apparent toxicity or gastro-intestinal lesion was absent in the algae-fed pigs[35].

In contrast, the de-fatted biomass of microalgal species, derived from the biofuel production research, has only recently been shown feasibility in replacing corn and soybean meal in diets for poultry, swine, and cattle. Studies supplementing the de-fatted biomass from *Staurospira*sp. to replace 7.5% corn and soybean meal in diets for weanling pigs did not affect their overall growth performance or plasma biochemical indicators[37, 38]. However, the pigs were incapable of tolerating a 15% replacement of corn and soybean meal [38]. Factors causing this intolerance might include amino acid imbalances, disruption of the acid:base balance, high ash content of the algal biomass, and(or) an overall reduction in the buffering capacity within the gastrointestinal tract of the weanling pigs. Austic et al [12] replaced 7.5% of soybean meal with the de-fatted biomass of the *Staurosira* sp. in the diets of broilers, and showed decreased body weight gain and feed efficiency during the first three weeks of experiment. During the following three weeks, these differences were no longer seen. In the same study, broiler chicks fed a diet containing essential amino acids (Met, Lys, Ile, Thr, Trp, and Val) co-supplemented with 7.5% of the de-fatted biomass did not show growth performance differences from the control group. A diet with 10% lipid-extracted *Nannochloropsis oculata* meal was well tolerated by adolescent male rabbits, which showed similar final body weight, serum urea nitrogen, blood glucose, and organ histology in comparison with rabbits fed a control diet[39, 40]. Finishing wethers maintained similar growth performance and carcass characteristics (*longissimus* muscle area, dressing percentage, marbling score, hot carcass weight, and subcutaneous adipose depth) when fed up to 20% de-fatted algal biomass on a dry matter basis as a protein replacement, in comparison with wethers fed a control diet[24]. In a study that examined effects of supplementing dietary dry matter with de-fatted *Lithothamnium calcareum* meal in Holstein cows, up to 1% algal meal in the diet mediated venous acid–base balance after acidosis induction, but did not improve the tract digestibility or growth performance of the cows[41].

Nutritional values of microalgae

While the nutritional profiles of microalgae vary considerably with the species used, a large majority are characterized by protein, carbohydrate, and lipid contents that are comparable, if not superior, to conventional feedstuffs (Table 1). Dietary soybean meal typically contains up to 48% crude protein, and lays claim to a relatively well balanced amino acid profile. The diversity of microalgae makes certain species amenable to cultivation for diet-specific needs of humans and animals. A commonly cultivated algae species for human consumption: *S. maxima*, contains high levels of vitamin B_1, vitamin B_2, and β-carotene, and up to 71% crude protein with sufficient concentrations of all essential amino acids except for the sulfur-containing ones (Table 2)[5, 42]. Since protein is considered to be the most expensive nutrient in animal feed[43], developing natural alternatives to soybean meal may be cost competitive.

Table 1: Nutrient composition of conventional feedstuffs and various algae (% dry matter)[a]

Source	Crude protein	Carbohydrates	Lipids
Soybean	37	30	20
Corn	10	85	4
Wheat	14	84	2
Anabaena cylindrical	43-56	25-30	4-7
Arthrospira maxima	60-71	13-16	6-7
Chlorella vulgaris	51-58	12-17	14-22

Spirogyra sp.	6-20	33-64	11-21
Synechococcus sp.	73	15	11

[a]Adapted from[11, 44, 45, 46].

Table 2: Amino acid profile of conventional protein sources and various algae (g/100 protein)[a]

Source	Ala	Arg	Asp	Cys	Glu	Gly	His	Ile	Leu	Lys	Met	Phe	Pro	Ser	Thr	Try	Tyr	Val
Egg	-	6.2	11.0	2.3	12.6	4.2	2.4	6.6	8.8	5.3	3.2	5.8	4.2	6.9	5.0	1.7	4.2	7.2
Soybean	5.0	7.4	1.3	1.9	19.0	4.5	2.6	5.3	7.7	6.4	1.3	5.0	5.3	5.8	4.0	1.4	3.7	5.3
Dried Whole Milk	0.9	0.9	2.0	0.2	5.5	0.5	0.7	1.6	2.6	2.1	0.6	1.3	2.5	1.4	1.2	0.4	1.2	1.7
C. vulgaris	9.4	6.9	9.3	-	13.7	6.3	2.0	3.2	9.5	6.4	1.3	5.5	5.0	5.8	5.3	-	2.8	7.0
D. bardawil	7.3	7.3	10.4	1.2	12.7	5.5	1.8	4.2	11.0	7.0	2.3	5.8	3.3	4.6	5.4	0.7	3.7	5.8
S. platensis	9.5	7.3	11.8	0.9	10.3	5.7	2.2	6.7	9.8	4.8	2.5	5.3	4.2	5.1	6.2	0.3	5.3	7.1
Aphanizomenon flos-aquae	4.7	3.8	4.7	0.2	7.8	2.9	0.9	2.9	5.2	3.2	0.7	2.5	2.9	2.9	3.3	0.7	-	3.2

[a]Based on reference[11].

Among all dietary amino acids, lysine and methionine are the first and second limiting amino acids. Many microalgal species contain relatively high amounts of lysine, but as previously described, are somewhat deficient in the sulfur-containing amino acids: cysteine and methionine[5]. To maximize amino acid utilization by animals, diets are typically formulated by mixing different feedstuffs to balance amino acid profiles and(or) by supplementing synthetic amino acids to meet their nutrient requirements. Austic et al.[12] reported that the decreased growth performance of broilers fed the de-fatted *Staurospira* sp. biomass in the first three weeks was prevented by the supplementation of essential amino acids. In a laying hen study, a 7.5% replacement of corn and soybean meal with the de-fatted microalgal biomass did not negatively affect hen production or health parameters, yet significantly increased the redness and decreased both the lightness and yellowness of the egg yolks[47]. However, inclusions of the biomass at 10% reduced feed intake, plasma uric acid, and egg albumen weight. Up to a third of soybean meal was successfully replaced in the diets of weanling pigs by algae biomass from the cyanobacteria *S. maxima*, *Arthrospira platensis*, and *Chlorella* sp.[35]. In another study conducted on laying hens, an inclusion of 10% *Porphyridium* sp. red algal biomass did not affect their body weight, egg production rate, or egg weight, but lowered egg yolk cholesterol levels by 24%[48].

MICROALGAE FOR VALUE-ADDED ANIMAL PRODUCTS

The research on microalgal biomass supplementation in food-producing animals has opened a new gateway to improve human health. A study conducted in the last decade showed not only 10% enhancement of growth but also potential in producing iodine-rich pork for human consumption when pigs were fed algae containing naturally high iodine[49]. Likewise, microalgae may be cultivated for their advantageous fatty acid profile (Table 3), with notable enrichment in omega-3 (n-3) polyunsaturated fatty acids (PUFA) including arachidonic acid, docohexaenoic acid (DHA), eicosapentaenoic acid (EPA), and γ–linoleic acid[21]. The enriched concentrations of these omega-3 fatty acids by a variety of microalgal species represent a largely untapped natural resource with multiple health implications for both humans and animals. Among many fat sources, fish products represent the major source of n-3 fatty acids [50]. However, marine fish species are incapable of synthesizing n-3 fatty acids by themselves; they may obtain n-3 fatty acids by consuming microalgae or other algae-consuming fish [51]. Most notably, dietary intakes of n-3 fatty acids are associated with decreased risks of chronic diseases [52]. However, the average fish consumption in the U.S. has not changed since 1983, at approximately 6.5 kg per year [53]. To increase the public consumption of n-3 fatty acids, studies have been conducted with full-fatted or de-fatted microalgae to generate n-3 fatty acid-fortified animal products such as milk, meats, and eggs. Apparently, the profile of PUFA in the de-fatted microalgal biomass is affected by the lipid extraction procedure. While a practical solvent extraction can remove 95% of lipids from oilseeds, it is ineffective to extract microalgal lipid due to the relatively high moisture content and cellular elasticity [21]. Grima et al.[54] reported that the lipid extraction yields of dried algal biomass range from 50 to 93% depending on the polarity of solvent systems. In fact, a higher proportion of long chain polyunsaturated fatty acids in the de-fatted biomass may bolster its appeal as an n-3 fatty acids source for developing value-added animal products.

Table 3: Generalized fatty acid profiles (%) of oil extracts from*Spirulina maxima*(SP),*Chlorella vulgaris*(Cv),*Scenedesmus obliquus*(Sc),*Dunaliella te rtiolecta*(Dt),*Nannochloropsis*sp, (Nanno), and*Neochloris oleabundans*(Neo)[a]

Fatty acid	Sp	Cv	Sc	Dt	Nanno	Neo
14:0	0.34	3.07	1.48	0.47	7.16	0.43
16:0	40.16	25.07	21.78	17.70	23.35	19.35
16:1	9.19	5.25	5.95	0.88	26.87	1.85
16:2	N.D.	N.D.	3.96	3.03	0.39	1.74
16:3	0.42	1.27	0.68	1.24	0.48	0.96
16:4	0.16	4.06	0.43	10.56	N.D.	7.24
18:0	1.18	0.63	0.45	N.D.	0.45	0.98
18:1	5.43	12.64	17.93	4.87	13.20	20.29
18:2	17.89	7.19	21.74	12.37	1.21	12.99
18:3	18.32	19.05	3.76	30.19	N.D.	17.43
18:4	0.08	N.D.	0.21	N.D.	N.D.	2.10
20:0	0.06	0.09	N.D.	N.D.	N.D.	N.D.

20:1	N.D.	0.93	N.D.	N.D.	N.D.	N.D.
20:2	0.48	N.D.	N.D.	N.D.	N.D.	N.D.
20:3	N.D.	0.83	N.D.	N.D.	N.D.	N.D.
20:4	N.D.	0.23	N.D.	N.D.	2.74	N.D.
20:5	N.D.	0.46	N.D.	N.D.	14.31	N.D.
SFA	41.74	28.86	23.71	18.17	30.96	20.76
MUFA	14.62	18.82	23.88	5.75	40.07	22.14
PUFA	37.35	33.09	30.78	57.39	19.13	42.46

[a]Based on reference[55].

Microalgal biomass or oil may be supplied in the feed of ruminants to manipulate their milk fatty acid composition. The content of n-3 fatty acids in milk, especially DHA, has been shown to increase with the inclusion of algal biomass or oil, without affecting the milk yield[56, 57, 58]. However, a study with similar dietary algae incorporation reported a decrease in milk fat content[59]. Many studies in ruminants have focused on producing n-3 fatty acid-fortified milk. In such an attempt, Stamey et al.[51] supplied either algal biomass or oil to dairy cows. The overall milk fat and yield were not affected by either supplement, but the milk DHA (C22:6) content was increased by both supplementations. Glover et al.[57] reported that feeding microalgae and fresh forage decreased total milk fat content, but elevated its DHA concentration. A parallel study was conducted with sheep to determine if algae and (or) their co-supplementation with sunflower oil in the diet could enhance the nutrient profile of milk[59]. While the milk yield was unaffected by the dietary treatments, the milk fat content was decreased and the milk DHA concentration was increased as dietary algae concentration rose. Franklin et al.[58] determined that feeding dairy cows with either algae protected against ruminal biohydrogenation or unprotected algae elevated the milk DHA concentration without affecting the milk fat content. Collectively, the supplementation of microalgae to the diets of ruminants increased the concentration of n-3 PUFA in milk, with a mixed effect on the milk fat content.

Likewise, fatty acid compositions in meats and eggs are also affected by dietary supplementations of microalgae. In lambs and horses, dietary microalgae increased the n-3 fatty acid content in meat and blood, respectively [60, 61]. In pigs, dietary microalgae increased

DHA concentrations in the loin and subcutaneous fat [62]. However, this increment did not manifest as a dose-dependent response [63]. To enrich eggs with n-3 fatty acids, flaxseed is usually supplemented in diets of laying hens [64]. Alternatively, dietary algae have been shown to increase n-3 and decrease n-6 fatty acids in eggs [64, 65]. Supplementing flaxseed at 10% or higher usually decreases egg production and the body weight of hens. Moreover, such high flaxseed supplementations may induce fatty acid oxidation, resulting in off-odor eggs [66]. Herber and Van Elswyk[65] supplemented up to 4.8% of golden marine algae into the diet of hens, and reported increased levels of DHA with a concomitant reduction of n-6 fatty acids. The inclusion of algae in the diet affected neither the egg production nor egg quality in the 56 wk-old hens. When up to 4.3% of fermented *Schizochytrum* sp. was included in the diets of hens, their egg production and feed conversion ratio were elevated [67]. The 0.86% and 4.3% supplementation of the microalgae elevated the DHA content to 134 and 220 mg/egg, respectively.

SPECIAL CONSIDERATIONS OF MICROALGAE

Despite their high nutritional values and health implications, caution must be taken to avoid possible toxicity of the microalgal biomass [8]. Among over 200,000 existing algal species, approximately 35,000 species are characterized [68] without apparent toxicological property [5]. However, certain species contain biogenic toxins including purines and non-biogenic toxins such as heavy metals. Some algal species are known to rapidly accumulate heavy metals at concentrations higher than their surroundings [5], while others generate pathological metabolites that cause neuro-degenerative disorders [69]. Such species require establishment and evaluation prior to their commercialization as feed supplements. Currently, no official regulations for heavy metals in microalgal products exist [5].

In spite of the relatively high crude protein contents, many microalgae species may still show limited biological values for the proteinaceous biomass due to the presence of non-protein nitrogen that consists of nucleic acids, nitrogen-containing cell walls, and amines. As single cell proteins, microalgae are photosynthetic unicellular eukaryotes containing nucleic acids that may represent approximately 10% of the crude protein [5]. Ruminal bacteria are effective in utilizing non-protein nitrogen to synthesize protein [70]. Simple-stomached animals lack ruminal microbes and thus are incapable of utilizing the relatively high levels of non-protein nitrogen in the microalgal biomass. Because the individual amino acid content and digestibility are important factors for formulating diets to support the maximum growth of animals without excreting excess nitrogen, it is necessary to optimize dietary amino acid profile and digestibility when the microalgal biomass is used to replace conventional feedstuffs.

Notably, the microalgal cell wall is largely indigestible by simple-stomached animals, yet contains immunostimulating properties that may provide added value to the biomass [71, 72]. Approaches to improve the absorption of microalgal nutrients include employing enzymatic treatments to enhance cell wall digestibility [71], as well as developing genetically-altered microalgae that lack cell walls [73]. Microalgal solubility is high at alkaline conditions [74]. Comprising approximately 10% cellular dry matter [71], the cell wall of *Chlorella* sp. contains high levels of β-1, 3-glucan, a bio-active immunostimulating free-radical scavenger [75]. Other species, such as those of cyanobacterial origin, have well-recognized anti-tumor, anti-bacterial, and anti-viral properties [76]. Microalgal cell wall-derived sulfated polysaccharides and carrageenans show strong T and B cell mitogenic effects [77]. Animals fed the whole *Spirulina*sp. Samples accumulated greater degrees of cytotoxic lymphocytes that were critical to innate immunity, and had elevated phagocytic activity and antibody production [78].

CONCLUSIONS

The use of microalgal biomass in animal feed will not only improve human and animal food security, but also facilitate cost-effective biofuel production and reduces greenhouse gas production of agriculture[6, 79, 80, 81]. Recent estimates indicate that 30% of the global algal production is used by the animal feed industry [5], amounting to a fast-growing $300 million in retail value[82]. Pragmatic species-specific large-scale algae refinery techniques must be devised to reduce the cost of the biomass production. It is necessary to determine limiting factors of the microalgal biomass that hinder its digestion and utilization by animals. Novel technology should be explored to improve microalgal nutrient utilization, and their long term nutritional and metabolic effects should be assessed.

In addition, the production of DHA from microalgal biomass alone is a rapidly growing market. The last estimate is that approximately 300 tons is produced to generate $1.5 billion market value each year [72, 82]. However, the tremendous potential of using the microalgal biomass in producing DHA/EPA enriched eggs, meats, and milk for improving human health is yet to be fully explored.

NOTES

Acknowledgement

Research on this subject in the authors' laboratory has been supported in part by a USDA/DOE Biomass R&D Initiative grant.

Competing interests

The authors declare that they have no competing interests.

REFERENCES

1. United States Census Bureau: International Database: World population information.http://www.census.gov/population/international/data/idb/informationGateway.php.
2. Population Reference Bureau: 2012 World Population Data Sheet. http://www.prb.org/pdf12/2012-population-data-sheet_eng.pdf.
3. Accelerating Industry Innovation: 2012 Ethanol Industry Outlook. http://ethanolrfa.3cdn.net/d4ad995ffb7ae8fbfe_1vm62ypzd.pdf.
4. Monthly commodity futures price chart.http://futures.tradingcharts.com/chart/CN/M.
5. Becker W: Microalgae in human and animal nutrition. Handbook of Microalgal Culture. Edited by: Richmond A. 2004, Oxford, UK: Blackwell Publishing Ltd, 312-351.
6. Chisti Y: Biodiesel from microalgae. Biotechnol Adv. 2007, 25: 294-306. 10.1016/j.biotechadv.2007.02.001.
7. Alzate ME, Muñoz R, Rogalla F, Fdz-Polanco F, SI P-E: Biochemical methane potential of microalgae biomass after lipid extraction. Chem Eng J. in press
8. Kovač DJ, Simeunović JB, Babić OB, Mišan AČ, Milovanović IL:

Algae in food and feed. Food Feed Res. 2013, 40: 21-32.

9. Brune DE, Lundquist TJ, Benemann JR: Microalgal biomass for greenhouse gas reductions: potential for replacement of fossil fuels and animal feeds. J Environ Eng. 2009, 135 (11): 1136-10.1061/(ASCE)EE.1943-7870.0000100.
10. Shields RJ, Lupatsch I: Algae for aquaculture and animal feeds. J Anim Sci. 2012, 21: 23-37.
11. Becker EW: Microalgae: Biotechnology and Microbiology. 1994, Cambridge: Cambridge University Press
12. Austic RE, Mustafa A, Jung B, Gatrell S, Lei XG: Potential and limitation of a new defatted diatom microalgal biomass in replacing soybean meal and corn in diets for broiler chickens. J Agric Food Chem. 2013, 61: 7341-7348. 10.1021/jf401957z.
13. Mostafa S: Microalgal biotechnology: prospects and applications. Plant Science. Edited by: Nabin Kumar D. INTech,http://www.intechopen.com/books/plant-science/microalgal-biotechnology-prospects-and-applications.
14. Li Y, Horsman M, Wu N, Lan CQ, Dubois-Calero N: Biofuels from microalgae. Biotechnol Prog. 2008, 24 (4): 815-820.
15. Hughes AD, Kelly MS, Black KD, Stanley MS: Biogas from microalgae: is it time to revisit the idea?. Biotechnol Biofuel. 2012, 5: 86-10.1186/1754-6834-5-86.
16. Meier RL: Biological cycles in the transformation of solar energy into useful fuels. J Anim Sci. 1955, 14: 179-183.
17. Oswald WJ, Golueke CG: Biological transformation of solar energy. Adv Appl Microbiol. 1960, 2: 223-262.
18. Mata TM, Martins AA, Caetano NS: Microalgae for biodiesel production and other applications: a review. Renew Sust Energ Rev. 2010, 14: 217-232. 10.1016/j.rser.2009.07.020.
19. Borowitzka M: Microalgae for aquaculture: opportunities and constraints. J Appl Physiol. 1997, 9: 393-401.
20. Slade R, Bauen A: Micro-algae cultivation for biofuels: Cost, energy balance, environmental impacts and future prospects. Biomass Bioenerg. 2013, 53: 29-38.
21. U.S. Department of Energy: National algal biofuels technology

roadmap: 9–10 December 2008. 2010, Maryland,http://www1.eere.energy.gov/biomass/pdfs/algal_biofuels_roadmap.pdf.

22. Chisti Y: Constraints to commercialization of algal fuels. J Biotechnol. 2013, 167: 201-214. 10.1016/j.jbiotec.2013.07.020.
23. Saharan BS, Sharma D, Sahu R, Sahin O, Warren A: Towards algal biofuel production: a concept of green bio energy development. Innovat Rom Food Biotechnol. 2013, 12: 1-21.
24. Dib MD, Engle TE, Han H, Roman-Muniz IN, Archibeque SL: Effects of algal meal supplementation to finishing wethers on performance and carcass characteristics [abstract]. J Anim Sci. 2012, 90 (Suppl 3): 148.
25. Gotaas HB, Oswald WJ, Ludwig HF: Photosynthetic reclamation of organic wastes. Sci Monthly. 1954, 79: 368-378.
26. Oswald WJ, Gotaas HB: Photosynthesis in sewage treatment. Trans Am Soc Civ Eng. 1957, 122: 73-105.
27. Combs GF: Algae (chlorella) as a source of nutrients for the chick. Science. 1952, 116: 453-454. 10.1126/science.116.3017.453.
28. Grau CR, Klein NW: Sewage-grown algae as a feedstuff for chicks. Poult Sci. 1957, 36: 1046-1051. 10.3382/ps.0361046.
29. Hundley JM, Ing RB: Algae as sources of lysine and threonine in supplementing wheat and bread diets. Science. 1956, 124: 536-537. 10.1126/science.124.3221.536.
30. Fink H, Herold E: Biological value of the protein of unicellular algae. On the protein value of unicellular algae and of young green leaves of higher plants/A contribution on alimentary liver necrosis of the rat and the treatment of protein deficiency in the tropics. Naturwissenschaften. 1955, 42: 516-517.
31. Ahmad MR: Observations on the effect of feeding Labeo rohita(Ham.) with microcystis aeruginosa Mycrocystis aeroginosa. Hydrobiologia. 1967, 29: 388-388. 10.1007/BF00189903.
32. Hintz HF, Heitman H: Sewage-grown algae as a protein supplement for swine. Anim Prod. 1967, 9: 135-140. 10.1017/S0003356100038393.
33. Ross E, Dominy W: The nutritional value of dehydrated, blue-green algae (spirulina plantensis) for poultry. Poult Sci. 1990, 69: 794-

800. 10.3382/ps.0690794.

34. Toyomizu M, Sato K, Taroda H, Kato T, Akiba Y: Effects of dietary spirulina on meat colour in muscle of broiler chickens. Br Poult Sci. 2001, 42: 197-202. 10.1080/00071660120048447.
35. Yap TN, Wu JF, Pond WG, Krook L: Feasibility of feeding spirulina-maxima, arthrospira-platensis or chlorella sp to pigs weaned to a dry diet at 4 to 8 days of age. Nutr Rep Int. 1982, 25: 543-552.
36. Fevrier C, Seve B: Incorporation of a spiruline (spirulina maxima) in swine food. Ann Nutr Aliment. 1975, 29: 625-650.
37. Isaacs R, Roneker KR, Huntley M, Lei XG: A partial replacement of soybean meal by whole or defatted algal meal in diet for weanling pigs does not affect their plasma biochemical indicators [abstract]. J Anim Sci. 2011, 89 (Suppl 1): 723.
38. Lum KK, Roneker KR, Lei XG: Effects of various replacements of corn and soy by defatted microalgal meal on growth performance and biochemical status of weanling pigs [abatract]. J Anim Sci. 2012, 90 (Suppl 3): 701.
39. Howe BA, Roman-Muniz IN, Willson BD, Archibeque SL: Evaluation of algae meal from Nannochloropsis oculata as a protein source for non-ruminant animals [abstract]. J Anim Sci. 2010, 88 (Suppl 2): 101-102.
40. Howe BA, Roman-Muniz IN, Willson BD, Archibeque SL: Nannochloropsis oculata meal did not alter nutrient usage and had no adverse health effects when fed to rabbits as a protein source [abstract]. J Anim Sci. 2011, 89 (Suppl. 2): 724.
41. Lopex NM, Pereira RAN, Pereira MN: Intake, milk yield, and blood acid–base balance of cows in response to marine algae meal [abstract]. J Anim Sci. 2013, 91 (Suppl. 2): 29.
42. Clement G, Giddey C, Menzi R: Amino acid composition and nutritive valve of the alga Spirulina maxima. J Sci Food Agric. 1967, 18: 497-501. 10.1002/jsfa.2740181101.
43. Rezaei R, Wang W, Wu Z, Dai Z, Wang J, Wu G: Biochemical and physiological bases for utilization of dietary amino acids by young Pigs. J Anim Sci Biotechnol. 2013, 4: 7-10.1186/2049-1891-4-7.
44. Aaronson S, Dubinsky Z: Mass production of microalgae.

Experientia. 1982, 38: 36-40. 10.1007/BF01944523.

45. Fabregas J, Herrero C: Marine microalgae as a potential source of single cell protein (SCP). Appl Microbiol Biotechnol. 1985, 23: 110-113. 10.1007/BF00938962.
46. Miles R, Chapman R: The benefits of fish meal in aquaculture diets. 2009,https://edis.ifas.ufl.edu/pdffiles/FA/FA12200.pdf.
47. Leng XJ, Hsu HN, Austic RE, Lei XG: Defatted algae biomass may replace one-third of soybean meal in diets for laying hens [abstract]. J Anim Sci. 2012, 90 (Suppl 3): 701.
48. Ginzberg A, Cohen M, Sod-Moriah U, Shany S, Rosenshtrauch A, Arad S: Chickens fed with biomass of the red microalga Porphyridium sp have reduced blood cholesterol level and modified fatty acid composition in egg yolk. J Appl Phycol. 2000, 12: 325-330. 10.1023/A:1008102622276.
49. He ML, Hollwich W, Rambeck WA: Supplementation of algae to the diet of pigs: a new possibility to improve the iodine content in the meat. J Anim Physiol Anim Nutr (Berl). 2002, 86: 97-104. 10.1046/j.1439-0396.2002.00363.x.
50. Daviglus ML, Stamler J, Orencia AJ, Dyer AR, Liu K, Greenland P, Walsh MK, Morris D, Shekelle RB: Fish consumption and the 30-year risk of fatal myocardial infarction. N Engl J Med. 1997, 336: 1046-1053. 10.1056/NEJM199704103361502.
51. Stamey JA, Shepherd DM, de Veth MJ, Corl BA: Use of algae or algal oil rich in n-3 fatty acids as a feed supplement for dairy cattle. J Dairy Sci. 2012, 95: 5269-5275. 10.3168/jds.2012-5412.
52. Albert CM, Hennekens CH, O'Donnell CJ, Ajani UA, Carey VJ, Willett WC, Ruskin JN, Manson JE: Fish consumption and risk of sudden cardiac death. JAMA. 1998, 279: 23-28. 10.1001/jama.279.1.23.
53. United States Department of Agriculture: Food Availability (Per Capita) Data System.http://www.ers.usda.gov/data-products/food-availability-(per-capita)-data-system.aspx#.Ua6jU0DrzVU.
54. Grima EM, González MJI, Giménez AG: Solvent extraction for microalgae lipids. Algae for biofuels and energy. Edited by: Borowitzka MA, Moheimani NR. 2013, New York, London: Springer, 187-205.

55. Gouveia L, Oliveira AC: Microalgae as a raw material for biofuels production. J Ind Microbiol Biotechnol. 2009, 36: 269-274. 10.1007/s10295-008-0495-6.
56. Papadopoulos G, Goulas C, Apostolaki E, Abril R: Effects of dietary supplements of algae, containing polyunsaturated fatty acids, on milk yield and the composition of milk products in dairy ewes. J Dairy Res. 2002, 69: 357-365.
57. Glover KE, Budge S, Rose M, Rupasinghe HP, Maclaren L, Green-Johnson J, Fredeen AH: Effect of feeding fresh forage and marine algae on the fatty acid composition and oxidation of milk and butter. J Dairy Sci. 2012, 95: 2797-2809. 10.3168/jds.2011-4736.
58. Franklin ST, Martin KR, Baer RJ, Schingoethe DJ, Hippen AR: Dietary marine algae (Schizochytrium sp.) increases concentrations of conjugated linoleic, docosahexaenoic and transvaccenic acids in milk of dairy cows. J Nutr. 1999, 129: 2048-2054.
59. Toral PG, Hervas G, Gomez-Cortes P, Frutos P, Juarez M, de la Fuente MA: Milk fatty acid profile and dairy sheep performance in response to diet supplementation with sunflower oil plus incremental levels of marine algae. J Dairy Sci. 2010, 93: 1655-1667. 10.3168/jds.2009-2769.
60. Cooper SL, Sinclair LA, Wilkinson RG, Hallett KG, Enser M, Wood JD: Manipulation of the n-3 polyunsaturated fatty acid content of muscle and adipose tissue in lambs. J Anim Sci. 2004, 82: 1461-1470.
61. Hess TM, Rexford JK, Hansen DK, Harris M, Schauermann N, Ross T, Engle TE, Allen KG, Mulligan CM: Effects of two different dietary sources of long chain omega-3, highly unsaturated fatty acids on incorporation into the plasma, red blood cell, and skeletal muscle in horses. J Anim Sci. 2012, 90: 3023-3031. 10.2527/jas.2011-4412.
62. Sardi L, Martelli G, Lambertini L, Parisini P, Mordenti A: Effects of a dietary supplement of DHA-rich marine algae on Italian heavy pig production parameters. Livest Sci. 2006, 103: 95-103. 10.1016/j.livsci.2006.01.009.
63. Marriott NG, Garrett JE, Sims MD, Abril JR: Performance characteristics and fatty acid composition of pigs fed a diet with docosahexaenoic acid. J Muscle Foods. 2002, 13: 265-277. 10.1111/

j.1745-4573.2002.tb00335.x.

64. Sim JS, Qi G: Designing poultry products using flaxseed. Flaxseed in Human Nutrition. Edited by: Thompson LU, Cunnane LU. 1995, Champaign, IL: American Oil Chemists' Society Press, 315-333.
65. Herber SM, Van Elswyk ME: Dietary marine algae promotes efficient deposition of n-3 fatty acids for the production of enriched shell eggs. Poult Sci. 1996, 75: 1501-1507. 10.3382/ps.0751501.
66. Leeson S, Caston L, MacLaurin T: Organoleptic evaluation of eggs produced by laying hens fed diets containing graded levels of flaxseed and vitamin E. Poult Sci. 1998, 77: 1436-1440.
67. Abril JR, Barclay WR, Abril PG: Safe use of microalgae (DHA GOLD) in laying hen feed for the production of DHA-enriched eggs. Egg nutrition and biotechnology. Edited by: Sim JS, Nakai S, Guenter W. 1999, New York: CABI Publishing, 197-202.
68. Munir N, Sharif N, Shagufta N, Saleem F, Manzoor F: Harvesting and processing of microalgae biomass fractions for biodiesel production (a review). Sci Tech and Dev. 2013, 32: 235-243.
69. Jin DQ, Lim CS, Sung JY, Choi HG, Ha I, Han JS: Ulva conglobata, a marine algae, has neuroprotective and anti-inflammatory effects in murine hippocampal and microglial cells. Neurosci Lett. 2006, 402: 154-158. 10.1016/j.neulet.2006.03.068.
70. McNaught ML, Owen EC: The utilization of non-protein nitrogen in the bovine rumen. 8. The nutritive value of the proteins of preparations of dried rumen bacteria, rumen protozoa and brewer's yeast for rats. Biochem J. 1954, 56: 151-156.
71. Becker EW: Micro-algae as a source of protein. Biotechnol Adv. 2007, 25: 207-210. 10.1016/j.biotechadv.2006.11.002.
72. Spolaore P, Joannis-Cassan C, Duran E, Isambert A: Commercial applications of microalgae. J Biosci Bioeng. 2006, 101: 87-96. 10.1263/jbb.101.87.
73. Davies DR: Single cell protein and the exploitation of mutant algae lacking cell walls. Nature. 1971, 233: 143-144. 10.1038/233143a0.
74. Pulz O, Gross W: Valuable products from biotechnology of microalgae. Appl Microbiol Biotechnol. 2004, 65: 635-648. 10.1007/s00253-004-1647-x.

75. Iwamoto H: Industrial production of microalgal cell-mass and secondary products - major industrial species: Chlorella. Handbook of microalgal culture: biotechnology and applied phycology. Edited by: Richmond A. 2004, New Jersey: Willey-Blackwell Publishing, 255-263.
76. Singh S, Kate BN, Banerjee UC: Bioactive compounds form cyanobacteria and microalgae: an overview. Crit Rev Biotechnol. 2005, 25: 73-95. 10.1080/07388550500248498.
77. Schaeffer DJ, Grylov VS: Anti-HIV activity of extracts and compounds form algae and cyanobacteria. Ectoxicol Environ Saf. 2000, 45: 208-227. 10.1006/eesa.1999.1862.
78. Luescher-Mattli M: Algae, a possible source for new drugs in the treatment of HIV and other viral diseases. Curr Med Chem Anti Infect Agents. 2003, 7: 219-225.
79. Vijffels RA, Barbosa M: An outlook of microalgal biofuels. Science. 2010, 329: 796-799. 10.1126/science.1189003.
80. Darzins A, Pienkos P, Edye L: Current status and potential for algal biofuels production.http://www.fao.org/uploads/media/1008_IEA_Bioenergy_-_Current_status_and_potential_for_algal_biofuels_production.pdf.
81. Singh J, Gu S: Commercialization potential of microalgae for biofuels production. Renew Sust Energ Rev. 2010, 14: 2596-2610. 10.1016/j.rser.2010.06.014.
82. Guil-Guerrero JL, Navarro-Juarez R, Lopez-Martinez JC, Campra-Madrid P, Rebolloso-Fuentes MM: Functional properties of the biomass of three microalgal species. J Food Eng. 2004, 65: 511-517. 10.1016/j.jfoodeng.2004.02.014.

CITATION

Krystal K Lum, Jonggun Kim and Xin Gen Lei, Dual potential of microalgae as a sustainable biofuel feedstock and animal feed

CHAPTER

11

ENERGY ASPECTS OF MICROALGAL BIODIESEL PRODUCTION

Edith Martinez-Guerra and Veera Gnaneswar Gude

Department of Civil and Environmental Engineering, Mississippi State University, Mississippi State, MS 39762, USA

INTRODUCTION

Depleting fossil fuel reserves and escalating environmental pollution associated with their consumption have created an urge for researchers around the world to investigate into renewable and sustainable energy and fuel supplies such as biofuels. The stimulus for research in biofuel production comes from their additional benefits of high energy density and ease in process utilization. Current biodiesel technologies are not economically feasible since they require government subsidies to be profitable to the producers and affordable to the users. This is mainly due to: 1) high feedstock costs; and 2) energy-intensive process steps involved in their production. Moreover, the feedstock should be derived from non-food related renewable materials to be sustainable and have less environmental impacts when compared with fossil fuels. Microalgae, a high oil-yielding feedstock, can be used to reduce production costs and make biodiesel competitive with petroleum diesel. Microalgae as a feedstock do not compete with any of the current human interests and offer many environmental benefits that make them an attractive feedstock for biodiesel production. Microalgae have much higher growth rates and productivity when compared with conventional forestry, agricultural crops, and other aquatic plants, requiring much less land area than other biodiesel feedstock of agricultural origin, i.e., up to 49-132 times less when compared to rapeseed or soybean crops (basis: 30% w/w of oil content in microalgae biomass [1]. Therefore, the competition for arable land with other crops, in particular for human consumption is greatly reduced by using microalgae as biodiesel feedstock [2]. Microalgae show great promise as a potential future energy source due to their environmental-friendliness and high oil yielding capacity per given area. Although microalgae can be grown in most of the tropical climates, dry and wet weather conditions and even in marginal lands where solar insolation is high, the lipid yield from algal biomass varies around the world [3]. In addition, current algal biodiesel production methods are not efficient since the process steps involved from algal biomass cultivation to final biodiesel separation/purification are all energy-intensive and cost-prohibitive. Therefore, rigorous research is being pursued all over the world to develop novel and energy-efficient process techniques for sustainable algal biodiesel production.

NET ENERGY BALANCE—MICROALGAE CASE STUDY

Algal biodiesel production consists primarily of five steps: a) microalgae biomass production; b) biomass harvesting; c) oil and lipid extraction; d) trans esterification or chemical treatment; and e) separation and purification [4]. All steps involved in algal biodiesel production are both energy- and cost-intensive. Currently, major hurdles for the algal biodiesel production are dewatering the algal biomass (algal cell concentration) and drying and oil extraction [4, 5]. The algal culture is usually concentrated to 15-20% from its original concentration of 0.02-0.05% through various physical and chemical processing techniques. Apart from this, extraction of algal oil is not as simple as that would be from other crop seeds (which are usually done by mechanical pressing and solvent extraction methods) due to their rigid cell wall structure. As such, these three steps add significantly to the cost of the algal biodiesel product.

Algal biodiesel production can be sustainable only if the net energy gain from entire process is higher than one. The net energy ratio is defined as the ratio of the energy available from the end product (algal biodiesel) to the energy invested in its production cycle. Microalgae have an energy content of 5-8 kWh/kg (18,000-28,800 kJ/kg) of dry weight depending on the species and lipid content [4]. Therefore, in order for algal biodiesel production to be feasible, the amount of energy required to produce the microalgae and to process it into a useable fuel must be less than this amount. Therefore, the Net Energy Ratio can be written as:

$$E_{NER} = \frac{E_{out}}{E_{in}} = \frac{\text{Energy in algal biofule}}{\text{Energy invested}}$$

Figure 1 shows the energy requirements for algal biodiesel production. Energy consumption for cultivation and algal biomass production depends on the cultivation methods. Low specific energy consumption is often reported for open raceway ponds. Photobioreactors are currently energy-intensive.

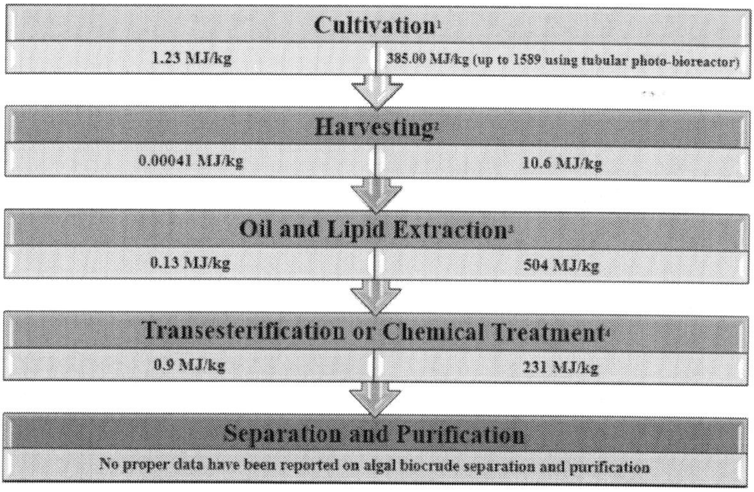

Figure 1. Energy consumption in algal biodiesel production (low and high values of energy consumption in each step).

Major energy consumption (60-70%) occurring in harvesting, drying, and extraction steps is unavoidable to prepare lipids suitable for the transesterification reaction. High energy consumption for lipid extraction is often combined with drying the biomass. This clearly suggests the need for alternative methods to dry microalgae biomass, or eliminating the need for drying entirely by hydrothermal liquefaction processes.

[1] Cultivation (freshwater or brackish-to-saline water): open ponds (raceway) and photobioreactors (i.e. flat-plate, tubular, and biofilm PBR) [4,9,10,11,12,13]

[2] Harvesting: centrifugation, gravity sedimentation, filtration (natural), filtration (pressurized), tangential flow filtration, vacuum filtration, polymer flocculation, electro-coagulation, electro-flotation, electro-flocculation [4,9,14,15,16]

[3] Oil extraction or lipid extraction (dry or wet): chemical extraction or direct extraction (i.e. oil mill and mechanical pressing), hydrothermal liquefaction [4,9,16]

[4] Transesterification (in-situ transesterification often combines the extraction and transesterification step into a single step): conventional (sub-critical water, solvent-conventional heating), non- conventional (microwave, ultrasound, ultrasonic bath, hydrodynamic cavitation, pyrolysis) heating technologies, supercritical alcohol transesterification (supercritical methanol, supercritical ethanol) [4,17,18,19,20,21,22,23,24,25,26,27,28,29,30]

For alternative methods to be effective, the goal will be to extract all of the lipid content from the biomass and transesterify in the steps that follow. Microalgae have a very rigid cell wall that is hard to break or penetrate through mechanical or chemical extraction methods. Mechanical pressing proved to be inefficient for algal oil extraction for this reason. Chemical extraction using solvents can be expensive as well as creating other byproducts and waste products and separation/recovery problems. From the data reported by several researchers, a minimum energy consumption of 1.7 kWh/kg of dry algal biomass with a net energy gain ratio of up to 4.7 may be possible if all of the steps involved in algal biodiesel production are achieved with minimum energy consumption [4,6,7,8]. Energy requirements higher than 8 kWh/kg of dry algal biomass are not beneficial and sustainable. With drying operation taking up to 60% of the total energy, it provides an opportunity to save energy by eliminating the step or by finding other energy saving alternative. As such, hydrothermal processes that maximize recovery of carbon content in the final product in the form of biocrude are highly desirable.

A CLOSER LOOK AT THE PROCESS STEPS AND ENERGY CONSUMPTION

A closer look at the energy requirements for biodiesel production from two different forms of algal feed stock (dry and wet microalgae) is shown below [4]. It can be noted that drying and extraction steps consume about 84.2% and 72.8% of the total energy requirements for dry and wet algal biomasses respectively (Table 1). Microalgae cell harvesting and oil extraction steps also make significant contributions to the energy balance. The following sections discuss the energy requirements for algal cell harvesting and extraction and transesterification steps in algal biodiesel production.

Table 1: Energy requirements for different steps in algal biodiesel production [9]

Biodiesel production step (basis: 1 kg of algal biodiesel)	Dry algal biomass (MJ)	Wet algal biomass (MJ)
Microalgae culture and harvesting	7.5	10.6
Drying	90.3	0
Extraction	8.6	30.8
Oil transesterification	0.9	0.9
Total	107.3	42.3

Harvesting methods

Microalgae can be harvested from the culture medium (water) by chemical (coagulation-flocculation); and physical (membrane filtration, hydrocyclones and centrifugation) methods. Microalgal cells carry a negative charge that preventsaggregation of cells in suspension. The commonly used salts to separate algal suspensions include ferric chloride ($FeCl_3$), aluminum sulfate ($Al_2(SO_4)_3$, alum) and ferric sulfate ($Fe_2(SO_4)_3$) [31, 32,33,34,35]. These mineral coagulants (alum and ferric chloride) produce large volumes of sludge which are toxic to animals when consumed due to high concentration of residual aluminum and iron in the biomass harvested [36,37]. The main problem with membrane filtration is membrane fouling and clogging due to the small size of the microalga. Membrane processes operate at high pressure which means high energy requirements and high capital costs [38,39,40]. Centrifugation is the application of centripetal acceleration to separate the algal growth medium into regions of greater and less densities. Once separated, the microalgae can be removed from the culture by simply draining the excess medium [41]. However, shear forces experienced during spinni ng can disrupt cells, thus limiting the speed of centrifugation [11]. Macro-filtration is widely used for larger microalgae species like *Arthrospira*. Belt filters are able to filter up to 20% with an energy consumption of 0.5 kWh m^{-3} (2.04 kJ/kg) if the feed is pre-concentrated at 4%. Micro-filtration with appropriate pore size can retain the majority of common species. However, micro-filtration could be even less economical than centrifugation for the recovery of microalgal cells on a large scale [11]. Recent studies showed that harvesting by submerged filtration in combination with centrifugation could achieve concentration up to 22% and reduce energy needs to under 1 kWh m^{-3} (4.09 kJ/kg) [42]. Ultrafiltration is a possible alternative in particular for very fragile cells, but has not generally been used for recovery of microalgae since operating and maintenance costs are high [11,43]. Energy consumption is believed to be between 1 and 3 kWh m^{-3} (4.09 and 12.27 kJ/kg). Disc stack centrifuges are suited for separating particles of the size (3-30 μm) and concentration (0.02-0.05%) of microalgae cultures up to 15% solids while consuming 0.7-1.3 kWh m^{-3} (2.86-5.32 kJ/kg) [43]. The unit conversion was done by using the average biodiesel density of 0.88 g/mL [44,45]. Table 2 shows the comparison of several algal cell harvesting techniques [15]. It should be noted that the final product concentration is dependent on the separation mechanism.

Table 2: Comparison of algal cell harvesting techniques [15]

Method	Advantages	Drawbacks	Dry solids (%)	Energy Requirements (kWhm^{-3})
Centrifugation	rapid, efficient, suitable for most microalgae species	high capital and operating costs	10–22	0.7–8 (2.86–32.73 kJ/kg)
Filtration	high system variety	species specific, fouling	2–27	0.5–3 (2.04–12.7 kJ/kg)
Flotation	faster than sedimentation	species specific, high capital costs	2.5–7	0.015–1.5 (0.06–6.14 kJ/kg)
Sedimentation	low capital and operating costs	species specific, low final concentration	0.5–3	0.1–0.3 (0.41–1.23 kJ/kg)

Oil extraction methods

Microalgal oil extraction alone may cost up to $15 per gallon of oil produced involving use of non-renewable energy and/or high quality electrical energy with yet another energy-intensive step of drying [46]. It is crucial to consider the costs for large scale production feasibility of the algal biodiesel. Well known methods for oil extraction are namely mechanical pressing, milling, solvent extraction, supercritical fluid extraction, and enzymatic extraction. These methods require high volumes of solvents, long extraction times, and mechanical or thermal energy resulting in environmental pollution and hazardous byproduct formation/disposal. Once the oils are extracted from microalgae, biodiesel can be produced thorough widely known technique "transesterification". Transesterification process is simply the replacement of one group of ester with another to make the carbon chain less complex. Transesterification of vegetable-, waste cooking-, non-edible-oils (jatropha, kharanja, and animal fats) and other feedstocks under microwave and ultrasound irradiations were reported by many researchers [47,48].

Dry vs wet routes in biodiesel production

Biodiesel production using dry algal biomass as feedstock can be less energy intensive as it was reported earlier [19,20]. Because, biomass drying can be performed using solar energy which is available "free". However, the details and requirements for such process are yet unknown. Table 3 shows the energy requirements for single-pot extractive-transesterification of algal biomass (*Nannochloropsis sp.*) [19,20]. It can be shown that the dry microalgae processing under microwaves for biodiesel production is less intensive. While the supercritical methanol process is a non-catalytic process which requires less separation and purification steps, the process time was longer and the feed sample was large due to the water content in the paste.

Table 3: Energy requirements for dry vs. wet routes

Species	Feed type	Process	Optimum parameters	FAME (%)	Feed (g)	MeOH (mL)	Energy consumption (kJ)	Reference
Nannochloropsis (CCMP1776)	Wet	SCM	250 °C 8 (wt/vol) 25 min	84.15	4	32	thermal 525 mechanical 75 Total 600	[21]
	Dry	MW	65 °C Cat. 2 wt% 9 (wt/vol) 6 min	80.13	2	18	thermal 240 mechanical 15 Total 255	[21]

SCM: supercritical methanol, MW: microwaves.

Microwave and ultrasound for extraction and transesterification

It is critical to develop energy-efficient methods that would reduce the chemical and energy consumption and processing time of the overall algal biodiesel production. Ultrasonic-assisted extraction combined with microwaves could be an attractive option for algal oil extraction since it does not require excess solvents or mechanical energy [49]. Individually, microwaves and ultrasound have received considerable attention in recent years due to their unique process enhancing effects. Numerous studies accounted for their process intensification benefits in pharmaceutical, chemical, and industrial applications. Microwaves and ultrasound have also been extensively used and equally investigated for their benefits in biofuel synthesis ever since their discovery, although it is fairly recent for biodiesel production. Microwaves enhance the reaction rates by ionic conduction and dipolar polarization mechanisms. Due to these effects localized superheating of the reactantsresults in hotspots increasing the rate of reaction tremendously. Advantages of microwave technique can be short extraction time and higher oil recovery since the microwaves have the ability to penetrate through algal cell walls to heat the lipid pockets and force them to be excreted out of biological matrix into extraction medium [47,48,50,51,52]. Ultrasound technology was employed in various stages of biodiesel production. Stavarache et al. [53] used low frequency ultrasound energy for biodiesel production and compared the results with conventional biodiesel production processes. They used three different types of alcohols and NaOH as a catalyst. The study showed that ultrasonication had a positive effect on transesterification process and reduced the process time and saved energy in the biodiesel production [53]. Santos et al. [54] studied the effect of ultrasonication in biodiesel production from soybean oil and showed the positive effect of ultrasound on biodiesel yield enhancement. Cintas et al. [55] used high power ultrasound in a continuous system for biodiesel production from soybeans. They used ultrasound after heating the oil and premixing with a mechanical stirrer. Their results showed considerable improvement on reaction time and energy savings.

Table 4 summarizes the specific energy consumption reported in various studies. These processes include supercritical and subcritical processes, and microwave and ultrasound techniques for extraction and transesterification of algal lipids including traditional methods such as bead mills. It can be noted that a wide range of specific energy consumption from 1.1 MJ/kg to 504 MJ/kg was reported for various processes. From this data, it is clear that traditional bead mill is an inefficient method. However, the data reported from across the studies is not consistent. For example, some studies reported the data for pre-treatment only while others reported the specific energy consumption for the entire process. Some studies report the data based on the enthalpy of the reaction contents which does not represent the actual energy supplied to the process. There are several variables that can be identified in these results. The feedstock is different across the studies. The algal biomass physical cell characteristics are species-specific meaning that optimized conditions for one algal species may not work for another algal species. Some studies were performed at gram-scale while the other reported at kilogram-scale which may introduce some process and reaction specific differences. In microwave and ultrasound mediated extraction and transesterification processes, different power levels and different frequencies were identified to be effective. Apart from that a wide range of extraction solvents, purification agents are utilized in the final steps. All these parameters introduce experimental errors and variations.

Table 4: requirements for algal biodiesel production reported in literature

Process Conditions	Method	Energy (MJ/kg)	Reference	Comments
Nannochloropsis salina: 220 °C, 7.5% biomass (80 mL) loading, 25 min, 24.2 bar, 15 mL hexane. 4 kg of hexane are required to separate crude extract.	conventional heating subcritical water (C-SCW) extraction	12.2	[17]	Calculations are based on specific enthalpy of water and latent heat of evaporation of hexane
Nannochloropsis salina: 220 °C, 7.5% biomass (60 mL) loading, 25 min, 24.2 bar, 15 mL hexane. 4 kg of hexane are required to separate crude extract.	Microwave assisted subcritical water (MW-SCW)	11.26	[17]	Calculations are based on water's enthalpy and latent heat of evaporation of hexane
4.3 g wt algal biomass (1 g dry), 5 mL DI water, 1021 W, 5min, 15 mL of ethanol	Microwave	54.6	[18]	Based on 70% solvent recovery with distillation setup. Includes 8.77 MJ/kg from harvesting
4.3 g wt algal biomass (1 g dry), 5 mL DI water, 635 W, 5min, 15 mL of ethanol	Microwave	54.5	[18]	Based on 70% solvent recovery with distillation setup. Includes 8.77 MJ/kg from harvesting
1 kg of wet microalgae, methanol 114 g, KOH	Solvent/conventional heating extraction	106.4	[4]	Balance −2.6 MJ/kg. Includes harvesting (7.5 MJ/kg) and drying (81.8 MJ/kg)
1 kg of dry microalgae, methanol 114 g KOH	Solvent/conventional heating extraction	41.4	[4]	Balance: 105 MJ/kg. Includes harvesting (10.6 MJ/kg)
100 g of *Chlamydomonas sp.*, 100 mL ethanol, 350 W, 10 min	Microwave	2.1	[22]	Cell disruption step (pre-treatment) only
Nannochloropsis sp. (18 mL, 6 min, 400 W, 2 g)	Microwave	72	[21]	Accounts for power output in extraction-transesterification only
Chlorella pyrenoidosa (1 g, 400 W, 40 s)	Microwave	16	[23]	Accounts for power output during extraction and transesterification only
Chlorella sp. (4 g, 48 mL, 5 min, 350 W)	Microwave	26.3	[19]	Accounts for power output during extraction and transesterification only
Chlorella sp. (4 g, 48 mL, 5 min, 490 W)	Ultrasound	44.1	[20]	Accounts for power output during extraction and transesterification only
Nannochloropsis (5 min, 770 W, 1 g)	Microwave	231	[24]	Accounts for power output during extraction and transesterification only
(5 g, 42.5 mL, 40 min, 29.7 W/L; 25 mL, 20 min)	Ultrasonic bath	1.1	[25]	Accounts for power output during extraction and transesterification only
Nannochloropsis (10g, 30 min, 150 W)	Ultrasonic bath	27	[26]	Accounts for power output during extraction and transesterification only
Scenedesmus sp. (2 g, 30 min, 100 W)	Ultrasound	90	[27]	Accounts for power output during extraction and transesterification only
Nannochloropsis sp. (4 g, 32 mL, 25 min, 350 W)	Supercritical Methanol	131.25	[21]	Accounts for power output during extraction and transesterification only
Nannochloropsis oculata (100 g (30% DW), 1000 W, 30 min)	Ultrasound	60	[28]	Accounts for power output during extraction and transesterification
Botryococcus, Chlorella, Scenedesmus (100 mL, 5 kg/m^3, 840 W, 5 min)	Bead Mills	504	[29]	Accounts for power output during extraction and transesterification only
Saccharomyces cerevisiae (50 L, 10 kg/m^3, 5.5 kW, 50 min)	Hydrodynamic Cavitation	33	[30]	Accounts for power output during extraction and transesterification only

AIMS Energy Volume 4, Issue 2, 347-362.

RESEARCH NEEDS

Chisti (2013) suggested that a net energy ratio (profit) of 7 is desirable for algal biodiesel production to be cost-competitive with other conventional fuel supply options [56]. However, as shown in Figure 2, net energy rations of less than 1 as well as higher than 4 are reported in the literature [4,6,7,12,57,58,59,60,61,62,63,64,65,66,67]. To achieve the goal of net energy gain of 7, algal biodiesel production process should be improved in several aspects and process steps [68]. Energy extraction has to be maximized not only from the biodiesel production, but also from the spent biomass through biogas production or incineration. This may add to greater overall energy gains.

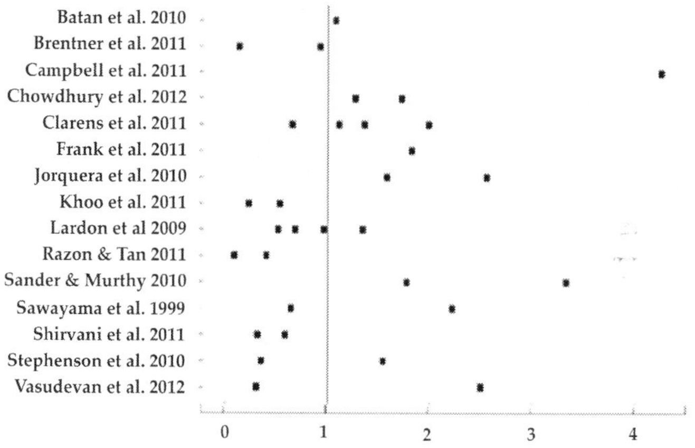

Figure 2: Net energy ratios for algal biodiesel production [4,6,7,12,57,58,59,60,61,62,63,64,65,66,67].

Microalgae growth and lipid yield estimations

Many reports focusing on algal biomass growth and lipid yields are available. But the estimations show large variations in microalgae productivity levels. This is in terms of land requirements, energy needs, water footprints, climate parameters and geographical locations. As this is the case, it is challenging to perform accurate techno-economic or life cycle impact analysis for algal biofuel production. In addition, these disparities are even worse in the case of process pathways reported for microalgae biomass processing and biodiesel conversion. For algal growth systems, the main differences occur between the raceway ponds and photobioreactors. To address this issue, Moody et al. [3] modeled various growth systems in the different assessments, including open raceway ponds and photobioreactors. Their results represented a promising production scenario based on cultivation in closed photobioreactors, which have been demonstrated to be robust culture platforms albeit cost-intensive.

Microalgae harvesting techniques

Among the harvesting methods, centrifugation and membrane filtration are not suitable for large scale application due to high energy consumption by these processes. Bio-flocculation, traditional chemical coagulation-flocculation-sedimentation methods with natural coagulants such as extracellular polymeric substances and chitosan should be studied more aggressively to provide a cost-effective solution. Novel techniques such as ultrasonic wave treatment combined with biopolymers and other energy-efficient methods should be considered [69,70,71,72,73].

Microalgae extraction and transesterification methods

A wide range of procedures have been followed to extract the algal lipids. There are many altered and customized procedures deviating from the well-established Bligh and Dyer [74] and Folch et al. [75] methods. In addition, transesterification step was performed using different experimental protocols. Extraction of lipids followed by transesterification or extractive-transesterification (also known as "in situ") are the most commonly reported methods in recent years [71,72,73]. These methods will need to be standardized soon. Energy consumption should be reported in commonly acceptable expressions such as MJ or kWh per unit biomass. A large disparity remains in the results reported to date.

Specific energy consumption calculations

As much as possible, specific energy consumption for the entire algal biodiesel production should be reported. These calculations should consider theoretical limitations but should be formed around the practical feasibility. For example, studies based on theoretical estimates tend to be more unrealistic either reporting ambitious values or too pessimistic scenarios. This should be given proper attention. From the discussion presented in other sections, it is clear that the process schemes utilizing centrifugal or mechanical separation/extraction for algal cell harvesting and drying or supercritical conditions for extraction and transesterification steps will not result in desirable energy gains. In addition, nutrient demands in algal cultivation stage should be given proper attention. Because production of these nutrients results in non-renewable energy consumption. Resource recovery, recycling and reuse of these chemical compounds should be considered a priority. Water footprint for algal cultivation is another area of concern. Adequate amounts of water supplies should be made available. Water extraction, conveyance, treatment (if any) and transport requires significant quantities of energy. This is also location and climate specific and species-specific. For example, *Chlorella vulgaris* cultivation has less water footprint when compared with other algal species identified to be suitable for biofuel production [76]. Above all, algal biorefinery schemes for multiple product recovery should be developed. This will improve overall process economics and achieve higher energy gains [77].

Separation and purification methods

There is no single standard procedure for separation and purification of the final product (biocrude). This introduces a large disparity in the results as well. Very high or low estimates of oil yield and product conversion can be reported if proper protocols are not developed. Efficient (in terms of energy, chemical and costs) methods should be identified and used in all studies regardless of the differences in the extraction and transesterification schemes. This will address the reproducibility and reliability issues.

CONCLUSION

Algal biodiesel production has yet to overcome several hurdles before it can be commercialized. All steps involved in the feedstock preparation to the final product purification steps need significant improvements. It is not possible to achieve energy-positive algal biodiesel production with single product recovery since the process steps are energy- and capital-intensive. Biorefinery schemes that increase the valuable bio-product recovery as well as utilization of spent biomass should be developed to make the overall process affordable and energy-positive. Conventional mechanical separation, harvesting and oil extractions steps should be avoided as much as possible. Theoretical studies should include more realistic assumptions to estimate the lipid yields and process benefits for future scenarios. Novel process intensification techniques such as microwaves and ultrasound should be developed further with more uniform specific energy consumption data to address the reliability issues. Finally, biocrude separation and purification steps are labor intensive consuming significant amounts of chemicals. Proper protocols and procedures should be developed for this step as well. All these efforts will provide exciting opportunities for research and development for scientists, engineers and practitioners all over the world.

ACKNOWLEDGMENTS

This work was supported by the Department of Civil and Environmental Engineering, the Bagley College of Engineering, and The Office of Research and Economic Development at Mississippi State University.

CONFLICT OF INTEREST

No conflicts ofinterest.

REFERENCES

1. Chisti Y (2007) Biodiesel from microalgae. Biotechnol Adv 25: 294-306.
2. Mata TM, Martins AA, Caetano NS (2010) Microalgae for biodiesel production and other applications: A review. Renew Sust Energ Rev 14: 217-232.
3. Moody JW, McGinty CM, Quinn JC (2014) Global evaluation of biofuel potential from microalgae. Proc Natl Acad Sci 111: 8691-8696.
4. Lardon L, Helias A, Sialve B, et al. (2009) Life-cycle assessment of biodiesel production from microalgae. Environ Sci Technol 43: 6475-6481.
5. Uduman N, Qi Y, Danquah MK, et al. (2010) Dewatering of microalgal cultures: a major bottleneck to algae-based fuels. J Renew Sustain Energy 20: 12701-12715.
6. Batan L, Quinn J, Willson B, et al. (2010) Net Energy and Greenhouse Gas Emission Evaluation of Biodiesel Derived from Microalgae. Environ Sci Technol 44: 7975-7980.

30. Balasundaram B, Skills SC, Llewellyn CA (2012) A low energy process for the recovery of bioproducts from cyanobacteria using a ball mill. Biochem Eng J 69: 48-56.

31. McGarry MG (1970) Alga flocculation with aluminium sulfate and polyelectrolytes. J Water Pollut Control Fed 42: R191-201.

32. Benemann JR, Kopman BL, Weissman DE, et al. Development of microalgae harvesting and high rate pond technologies in California. In: Shelef G, Soeder CJ, editors. Algal biomass. Amsterdam: Elsevier; 1980, 457.

33. Moraine R, Shelef G, Sandbank F, Bar-Moshe Z, Shvartzbard I. Recovery of sewage borne algae: flocculation and centrifugation technique. In: Shelef G, Soeder CJ, editors. Algae biomass. Amsterdam: Elsevier; 1980, 531-546.

34. Koopman B, Lincoln EP (1983) Autoflotation harvesting of algae from high rate pond effluents. Agric Wastes 5: 231-246.

35. Lincoln EP (1985) Resource recovery with microalgae. Arch Hydrobiol 20: 25-34.

36. Buelna G, Bhattarai KK, de la Nuoe J, et al. (1990) Evaluation of various flocculants for the recovery of algal biomass grown on pig-Waste. Biol Waste 31: 211-222.

37. Harith ZT, Yusoff FM, Mohamed MS, et al. (2009) Effect of different flocculants on the flocculation performance of microalgae, Chaetoceroscalcitrans, cells. Afr J Biotechnol 8: 5971-5978.

38. Bosma R, van Spronsen WA, Tramper J, et al. (2003) Ultrasound, a new separation technique to harvest microalgae. J Appl Phycol 15: 143-153.

39. Zhang X, Hu Q, Sommerfeld M, et al. (2010) Harvesting algal biomass for biofuelsusing ultrafiltration membranes. Bioresour Technol 101: 5297-5304.

40. Zou S, Gu Y, Xiao D, et al. (2011) The role of physical and chemical parameters on forwardosmosis membrane fouling during algae separation. J Membr Sci 366: 356-362.

41. Mohn F (1980) Experiences and strategies in the recovery of biomass from mass cultures of microalgae. Algae Biomass: 547-571.

42. Bilad MR, Vandamme D, Foubert I, et al. (2012) Harvesting

microalgal biomass using submerged microfiltration membranes. Bioresource Technol 111: 343-352.

43. Milledge JJ, Heaven S (2013) A review of the harvesting of microalgae for biofuel production. Rev Environ Sci Bio 12: 165-178.
44. Supple D. MIT Energy Club (https://mitei.mit.edu/research). Accessed February 5, 2016.
45. University of Washington (http://www.ocean.washington.edu). Accessed February 5, 2016.
46. Singh J, Gu S (2010) Commercialization potential of microalgae for biofuels production. Renew Sust Energ Rev 14: 2596-2610.
47. Gude VG, Patil PD, Deng S, et al. (2011) Microwave enhanced methods for biodiesel production and other environmental applications. Green chemistry for environmental remediation. New York: Wiley Interscience, 209-249.
48. Gude VG, Patil P, Martinez-Guerra E, et al. (2013). Microwave energy potential for biodiesel production. Sustainable Chemical Processes 1: 1-31.
49. Ma G, Hu W, Pei H, et al. (2016) In situ heterogeneous transesterification of microalgae using combined ultrasound and microwave irradiation.Energ Convers Manage 90: 41-46.
50. Kanitkar AV, Master's Thesis. Lousiana State University, Baton Rougue, LA.2010
51. Kanitkar A, Balasubramanian S, Lima M, et al. (2011) A critical comparison of methyl and ethyl esters production from soybean and rice bran oil in the presence of microwaves. Bioresource Technol 102: 7896-7902.
52. Balasubramanian S, Allen JD, Kanitkar A, et al. (2011) Oil extraction from Scenedesmusobliquus using a continuous microwave system--design, optimization, and quality characterization. Bioresource Technol 102: 3396-3403.
53. Stavarache C, Vinatoru M, Maeda Y, et al. (2007) Ultrasonically driven continuous process for vegetable oil transesterification. Ultrason Sonochem 14: 413-417.
54. Santos FFP, Rodrigues S, Fernandes FAN (2009) Optimization of the production of biodiesel from soybean oil by ultrasound assisted

methanolysis. Fuel Process Technol 90: 312-316.

55. Cintas P, Mantegna S, Gaudino EC, et al. (2010) A new pilot flow reactor for high intensity ultrasound irradiation. Application to the synthesis of biodiesel. Ultrason Sonochem 17: 985-989.
56. Chisti Y (2013) Constraints to commercialization of algal fuels. J biotechnol 167: 201-214.
57. Brentner LB, Eckelman MJ, Zimmerman JB (2011) Combinatorial Life Cycle Assessment to Inform Process Design of Industrial Production of Algal Biodiesel. Environ Sci Technol 45: 7060-7067.
58. Campbell PK, Beer T, Batten D (2011) Life cycle assessment of biodiesel production from microalgae in ponds. Bioresour Technol 102: 50-56.
59. Chowdhury R, Viamajala S, Gerlach R (2012) Reduction of environmental and energy footprint of microalgal biodiesel production through material and energy integration. Bioresour Technol 108: 102-111.
60. Clarens AF, Nassau H, Resurreccion EP, et al. (2011) Environmental Impacts of Algae-Derived Biodiesel and Bioelectricity for Transportation. Environ Sci Technol 45: 7554-7560.
61. Frank ED, Palou-Rivera I, Elgowainy A, et al. (2011) Life- Cycle Analysis of Algal Lipid Fuels with the GREET Model; Argonne National Laboratory: Argonne, IL, 2011.
62. Khoo HH, Sharratt PN, Das P, et al. (2011) Life cycle energy and CO2 analysis of microalgae-to-biodiesel: Preliminary results and comparisons. Bioresour Technol 102: 5800-5807.
63. Razon LF, Tan RR (2011) Net energy analysis of the production of biodiesel and biogas from the microalgae Haematococcus pluvialis and Nannochloropsis. Appl Energy 88: 3507-3514.
64. Sander K, Murthy G (2010) Life cycle analysis of algae biodiesel. Int J Life Cycle Assess 15: 704-714.
65. Sawayama S, Minowa T, Yokoyama SY (1999) Possibility of renewable energy production and CO2 mitigation by thermochemical liquefaction of microalgae. Biomass Bioenergy 17: 33-39.
66. Shirvani T, Yan X, Inderwildi OR, et al. (2011) Life cycle energy and greenhouse gas analysis for algae-derived biodiesel. Energy

Environ Sci4: 3773-3778.

67. Vasudevan V, Stratton RW, Pearlson MN, et al. (2012) Environmental Performance of Algal Biofuel Technology Options. Environ Sci Technol46: 2451-2459.
68. Chisti Y (2008) Biodiesel from microalgae beats bioethanol. Trends in Biotechnology 26: 126-131.
69. Fast SA, Kokabian B, Gude VG (2014) Chitosan enhanced coagulation of algal turbid waters–Comparison between rapid mix and ultrasound coagulation methods. Chem Eng J 244: 403-410.
70. Fast SA, Gude VG (2015) Ultrasound-chitosan enhanced flocculation of low algal turbid waters. J Ind Eng Chem 24: 153-160.
71. Patil PD, Gude VG, Mannarswamy A, et al. (2011) Optimization of direct conversion of wet algae to biodiesel under supercritical methanol conditions. Bioresource Technol 102: 118-122.
72. Patil PD, Gude VG, Mannarswamy A, et al. (2011) Optimization of microwave-assisted transesterification of dry algal biomass using response surface methodology. Bioresource Technol 102: 1399-1405.
73. Martinez-Guerra E, Gude VG (2016) Energy analysis of extractive transesterification of algal lipids for biocrude production. Biofuels, in press.
74. Yang J, Xu M, Zhang X, et al. (2011) Life-cycle analysis on biodiesel production from microalgae: water footprint and nutrients balance. Bioresource Technol 102: 159-165.
75. Bligh EG, Dyer WJ (1959) A rapid method of total lipid extraction and purification. Biochem Cell Bio 37: 911-917
76. Folch J, Lees M, Sloane Stanley GH (1957) A simple method for the isolation and purification of total lipides from animal tissues. J Biol Chem 226: 497-509.
77. Gude VG (2015) Synergism of microwaves and ultrasound for advanced biorefineries. Resource-Efficient Technologies 1: 116-125.

CHAPTER

12

DYNAMIC MODELING OF THE MICROALGAE CULTIVATION PHASE FOR ENERGY PRODUCTION IN OPEN RACEWAY PONDS AND FLAT PANEL PHOTOBIOREACTORS

Matteo Marsullo[1],Alberto Mian[2], Adriano Viana Ensinas[2,3], Giovanni Manente[1], Andrea Lazzaretto[1] and François Marechal[2]

[1]Department of Industrial Engineering, University of Padova, Padova, Italy

[2]Industrial Process and Energy System Engineering Group (IPESE), École Polytechnique Fédérale de Lausanne, Lausanne, Switzerland

[3]Universidade Federal do ABC, Santo Andre, Brazil

Keywords: Microalgae cultivation phase, open raceway pond, flat panel photobioreactor, repeated batch cultivation, carbon dioxide

A dynamic model of microalgae cultivation phase is presented in this work. Two cultivation technologies are taken into account: the open raceway pond and the flat panel photobioreactor. For each technology, the model is able to evaluate the microalgae areal and volumetric productivity and the energy production and consumption. Differently from the most common existing models in literature, which deal with a specific part of the overall cultivation process, the model presented here includes all physical and chemical quantities that mostly affect microalgae growth: the equation of the specific growth rate for the microalgae is influenced by CO_2 and nutrients concentration in the water, light intensity, temperature of the water in the reactor, and by the microalgae species being considered. All these input parameters can be tuned to obtain reliable predictions. A comparison with experimental data taken from the literature shows that the predictions are consistent and slightly overestimating the productivity in the case of closed photobioreactor. The results obtained by the simulation runs are consistent with those found in literature, being the areal productivity for the open raceway pond between 50 and 70 t/(ha × year) in Southern Spain (Sevilla) and Brazil (Petrolina) and between 250 and 350 t/(ha × year) for the flat panel photobioreactor in the same locations.

INTRODUCTION

Microalgae are single cell organisms, which can be found in colonies or individual cells. Their most interesting characteristic is the ability of realizing a photosynthetic reaction in a single cell. They are extremely resistant and may grow in many different environments, from fresh water to marine and hyper-saline water (Le et al., 2010; Mata et al., 2010). Microalgae can be seen as an interesting alternative to more typical biomass as a source for biofuels production. Studies on microalgae have been carried out since the 1980s, but in recent years their importance has grown fast: as Chisti (2007) claims, the reason is that microalgae appear to be the only source of biodiesel that has the potential to completely displace fossil diesel. Moreover, compared to other biofuels sources, such as traditional crops and wood, microalgae have several advantages: they grow extremely fast, reaching high areal and volumetric productivities; they do not require arable land; they ask for less freshwater than normal crops, being able to use wastewater; they can directly capture CO_2 released by industries and they overtake the food vs. fuel debate; and their ability to grow in a wide range of conditions, resisting to severe temperatures, pH, and salinity (NREL, 0000), makes them even more attractive. In 1978, the National Renewable Energy Laboratory of the United States of America started a 20-year program (Aquatic Species Program) to develop renewable transportation fuels from algae. The researches explored both the genetic engineering for manipulating the metabolism of microalgae and the engineering of microalgae production systems (NREL, 0000). In the last decades of the twentieth century, many other academic studies have been carried out all over the world, focusing on different aspects: the biological studies try to identify the best strain for a specific purpose, the engineering work tries to define open or closed systems to optimize the growth process and the system analysis works aim at defining the possible impact of the microalgae as a source of biomass for different applications, adopting a holistic approach.

Microalgae can be cultivated with the main purpose of producing lipids that can substitute biodiesel (Rodolfi et al., 2009) or produce higher value products that can be used as base chemicals for the production of biobased chemical products (Hempel et al., 2012). Cultivation of microalgae can be realized in open ponds systems where hydrodynamics (Hadiyanto et al., 2013) and CO_2 supply (Yang, 2011) are key drivers of the growth efficiency. Pollution and contamination related to mass transfer between open systems and environment together with the use of mechanical methods to obtain a proper mixing in the reactor make it impossible for a large number of microalgae to grow: only the most resistant strains might be used in open systems (Borowitzka, 1999). An alternative is the use of closed systems named photobioreactors that have the potential of maximizing the growth rate by controlling the growth conditions (pH, nutrients) and the gas supply at the expense of a much higher (factor ten) energy consumption. Examples are flat panel reactors or tubular reactors (Cuaresma et al., 2011). Studies have been conducted regarding the geometric characterization of flat panel photobioreactors, being the vertical positioning and east–west orientation the most suitable for microalgae proliferation (Sierra et al., 2008). Laboratory scale prototypes have been tested and compared to evaluate the relation between light availability, composition and growth rate of microalgae finding that high light availability leads to a change in the microalgae composition more than in a growth rate increase (Münkel et al., 2013). Tubular reactors need a much higher energy consumption (factor ten) than flat panel photobioreactors to maintain a proper mixing and circulation of the solution in the reactor and to ensure the oxygen removal, since there might be a decrease of the growth rate if oxygen saturation is reached (Molina et al., 2001). Mathematical models for light distribution have shown that photo-inhibition is a key problem for outdoor closed photobioreactors, both for flat panel technology and tubular reactors, being 100–200 W/m^2 the saturation light intensity for most of microalgae species (Fernández et al., 1997).

The land use and the solar efficiency of the microalgae system require adopting a holistic approach for the microalgae production. On the one hand, it is important to assess the overall conversion chain to define the substitution potential of the microalgae with respect of the extracted products. It is, therefore, important to adopt a holistic vision considering the complete life-cycle assessment for both the biobased products and the fossil one that are substituted (Azadi et al., 2014). Recent studies analyze the microalgae cultivation and transformation technologies evaluating their environmental impact in terms of GHG emissions and non-renewable energy consumption (Azadi et al., 2014) through LCA methodology, resulting that microalgae drying phase and water (Stephenson et al., 2010) and nutrient (Collet et al., 2011) reuse are the crucial aspects from an environmental point of view. From techno-economic analysis, irradiation conditions, mixing, medium, and CO_2 costs, together with dewatering technology, are the most important cost factors (Norsker et al., 2011). The opportunity to use microalgae strains capable of sustaining high growth rates and high lipid content give to costs the potential for significant improvement in the future (Davis et al., 2011).

The goal of this paper is to develop a comprehensive modeling framework to assess and compare different microalgae cultivation technologies in a consistent and holistic manner. A dynamic model of microalgae cultivation is developed. It aims at representing the influence of the most significant parameters on the system productivity and consumptions. The model is able to evaluate the variation of the most significant parameters, which influence microalgae growth, and therefore to estimate the microalgae biomass productivity and the energy productivity and consumption. The cultivation technologies taken into account are the open raceway pond and the flat panel photobioreactor: they are considered the most suitable technologies for extensive microalgae cultivation for biofuel production (Jorquera et al., 2010). Microalgae growth rate depends on several factors, such as light accessibility and nutrient availability, temperature, pH and salinity of the water, and CO_2 and O_2 concentration in the water. As all these terms are not constant in time, the model has to consider the dynamic behavior of the system and integrate the possible control strategies to be adopted to optimize the system performance.

The paper aims at including the effect of all the physical and chemical parameters that mostly affect microalgae growth, giving a consistent and robust modeling framework in which the contributions of recent publications are improved to interact together. Repeated batch operating strategy by Radmann et al. (2007) has been applied to microalgae growth model and mass balances by Yang (2011) together with thermal balances by Slegers et al. (2013) and pH control strategy by Sills (2013); reactor geometry has been taken into account for both the flat panel photobioreactor and the open pond to define the effect of solar radiation to microalgae growth, using the Beer–Lambert equations, and geometric evaluations by Duffie and Beckman (2013), included in Slegers et al. (2011).

MATERIALS AND METHODS

Reactor Geometry

Two photo bioreactor technologies are considered in this work, which are likely to be the most promising technologies for extensive microalgae production.

Open raceway pond is the most used artificial system for microalgae cultivation. The geometry used in this work is a closed oval-shaped channel: the most common values for pond depth are between 0.2 and 0.5 m [0.3 m being the most used value (Chisti, 2007)]. Circulation around the oval ring and mixing are guaranteed with the use of a mechanical rotating device, usually a paddlewheel (Brennan and Owende, 2010). The paddlewheel is in continuous operation to prevent sedimentation, giving the water a speed between 0.15 and 0.25 m/s, the most commonly used velocity range for open pond microalgae cultivations (Doucha and Lívanský, 2006; Hadiyanto et al., 2013). To satisfy the microalgae CO_2 requirement, the open pond is equipped with submerged aerators. The main characteristics of the raceway pond used in the model have been taken from the literature (Borowitzka, 1999; James and Boriah, 2010;

Norsker et al., 2011; Yang, 2011; Sompech et al., 2012) and presented in Table 1 and Figure 1. Open cultivation technologies have several problems, such as lower volumetric productivity compared to closed systems [0.1 compared to 1.5 kg m^{-3} day^{-1} (Chisti, 2007)], evaporation of the water through the open surface, contamination which causes a limitation in species sustainability and the need for large land area (Singh and Dhar, 2011).

Table 1: Open pond characteristics.

Pond surface	1 ha
Length to width ratio	10
Pond depth	30 cm
Speed of the water	0.20 m/s
CO_2 absorption system efficiency	0.9

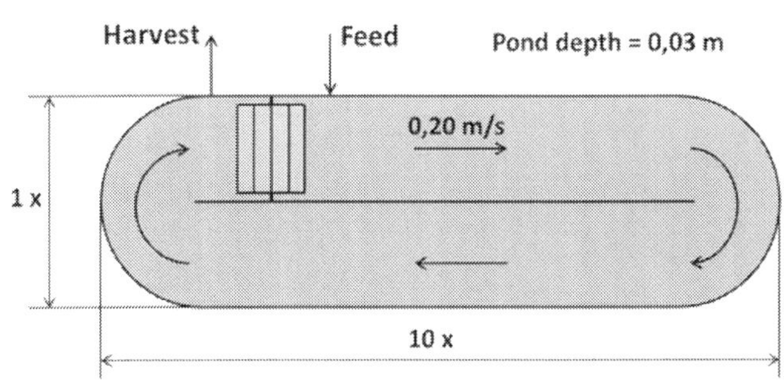

Figure 1: Raceway open pond.

Closed photobioreactors are designed to overcome the limitations of open systems. They have higher efficiency and biomass productivity, shorter harvesting times (2–3 days compared to 7–10 days for open pond), high surface-to-volume ratios, reduced contamination risks, and can be used to cultivate wider range of algal species than open systems (Bahadar and Bilal Khan, 2013). However, closed systems are more expensive to be constructed since they need high quality materials, and difficult to operate and scale up (Muñoz and Guieysse, 2006).

Flat panel photobioreactor is a flat, transparent vessel, made of glass, Plexiglas or plastic or other transparent material. The mixing of the water is carried out directly with air bubbling, which is introduced via a perforated tube at the bottom of the reactor. Flat panel PBRs are never thicker than 5–6 cm as the light entering the panel would not penetrate deeper in the culture. Only panels with both a height and width of <1 m have been studied (Janssen et al., 2003). The flat panel reactors are positioned vertically, closely spaced, to reach a higher photosynthetic efficiency through self-shading of the panels (Rawat et al., 2013). As for the open raceway pond, the main geometric characteristics have been taken from the literature (Janssen et al., 2003; Sierra et al., 2008; Pruvost et al., 2011; Ruiz et al., 2013; Kochem et al., 2014; Sugai-Guérios et al., 2014) and presented in Table 2 and Figure 2.

Table 2: Flat panel photobioreactor characteristics.

Height of the panels	1.5 m
Thickness of the panels	5 cm
Distance between the panels	1 m
Reactor surface	1 ha
Operating temperature	Constant

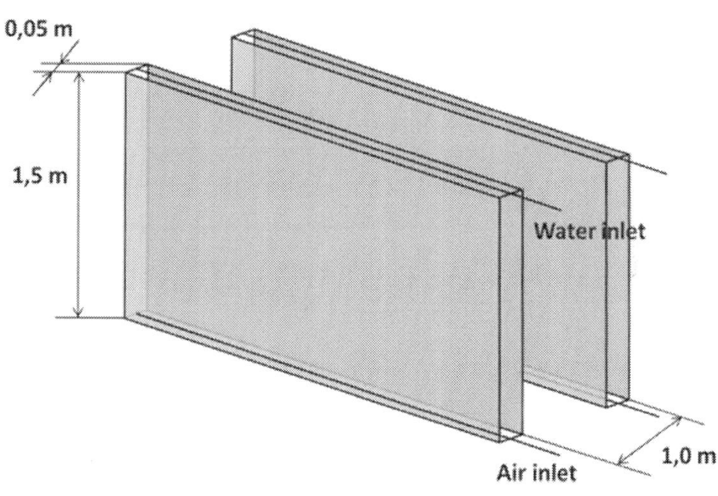

Figure 2: Flat panel photobioreactor.

Operating Strategy

The reactors can be operated either in continuous or in batch mode. The continuous operation requires a continuous extraction of the microalgae, while the batch operation extracts the produced microalgae after a certain time of production. An inoculum is, therefore, maintained in the reactor to serve as a basis for the new batch. The nutrient feeding strategy changes depending on the operating strategy: in continuous operation, two different reactors in series would be needed for lipid production, while in batch operation, a recipe for the batch feeding with nutrients has to be applied. As explained by Radmann et al. (2007), in the repeated batch cultivation after a certain period in the reactor, a specific culture volume is removed and replaced with an equal amount of fresh medium. Consequently, a part of cultivation medium is kept in the reactor as a starting inoculum. Repeated batch cultivation presents several operational advantages, the most important of which are the maintenance of a constant inoculum and high growth rates.

The operating strategy is also influenced by two considerations regarding the upstream and downstream processes. The downstream processing includes a settler tank where the extracted microalgae undergo a first concentration by sedimentation. The separated water is recycled while the products enter the downstream processing step. In this situation, the time required by the microalgae to grow can be decoupled from the time of the downstream technology. Second, in case of shortage of nutrients, the microalgae growth rate decreases as well as the production of new biomass, while the quantity of lipids in the biomass increases: this growth condition is particularly interesting if the downstream process aims at producing biodiesel. The models presented in this work do not consider the input condition of scarcity of nutrients, which instead are always supplied in excess.

MODEL

Cultivation Model Structure

The mass and energy produced by the cultivation systems in the form of biomass are calculated by the mathematical model through time-dependent mass and energy balances (Figure 3).

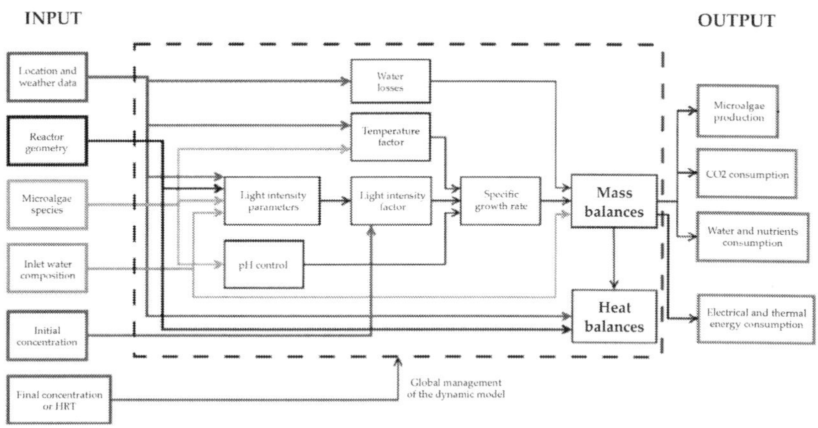

Figure 3: Model structure.

Before calculating the results of these balances, it is necessary to introduce some auxiliary equations. The auxiliary equations include the microalgae growth model, the pH control through CO_2 injection, and the equations to evaluate the radiation affecting the microalgae cells.

A set of differential equations is developed: it is solved by an explicit integration scheme with fixed time decomposition. To simplify the resolution of this system, the finite differences approximation was adopted. All the differential equations were treated as finite differences, and the backward difference formula was implemented

$$\frac{dx}{dt} = f(t) \rightarrow \frac{x(t) - x(t-1)}{\Delta t} = f(t) \quad (1)$$

This expression was preferred to the forward difference formula, being less vulnerable to instability.

If the backward difference formula is applied to matrices and vectors, the following expression is obtained:

$$\overline{x}(t) = \boldsymbol{F}(t) * \Delta t + \overline{x}(t-1) \quad (2)$$

The model is a differential algebraic equations system that is converted into a set of algebraic equations by defining the proper solving sequence: some of the differential equations depend on one another, making impossible to solve them in a sequential logic: it is, therefore, necessary to solve them altogether as a system. Therefore, all the equations were manipulated to separate the known terms from the unknown variables and to obtain an easily solvable system of algebraic equations.

A dynamic approach to the resolution of energy and mass balances has been preferred to a steady state analysis because the laws describing the microalgae growth are time dependant; moreover, a dynamic model gives the opportunity of a deeper and more accurate evaluation of microalgae production and energy consumption over the time period.

Inputs to the Model

The following inputs are required by the model.

- *Location data*: latitude and longitude of the reactor are needed together with the hour difference between the location of the bioreactor and Greenwich.
- *Weather data* are also required by the model both to calculate the irradiation factor, which influences the microalgae growth and to evaluate the thermal balance. The open raceway pond model requires the global irradiation over a horizontal surface, whereas the flat panel model requires the direct and the diffuse radiation over a horizontal surface as two distinct data. Moreover, both the two models require other weather data, such as the atmospheric temperature, wind speed, and relative humidity. Hourly data are used in all models.
- *Microalgae species characterization*: many parameters of the model depend on this input, such as the maximum growth rate, the optimal growth temperature, and the saturation light irradiation. All the characteristics assume different values depending on the microalgae species. Table 3 presents the model parameters considered for two different microalgae strains. The model assumes a constant pH that is kept at the optimal value for microalgae species using CO_2 injection: to evaluate the correct quantities of CO_2 to be injected, the pH optimal value is represented as an optimal value of alkalinity of the reactor. LHV is the lower heating value for wet microalgae biomass. I_S is the saturation light intensity, μ_{max} is the maximum microalgae growth rate, T_{let} is the lethal temperature, T_{opt} is the optimal temperature for microalgae growth (maximum growth

rate), and β is a curve modulating constant for the temperature factor. The decay rate is a coefficient that is used to evaluate the microalgae mass losses. K_C and K_{NA} are the half-saturation constants, i.e., the compound concentration when the specific growth rate [day^{-1}] is $\mu_A = \hat{\mu}_A / 2$. From Yang (2011), K_C is set to 0.001 molCO2/m30.001 molCO2/m3 while K_{NA} = 0.001 mol_{NA}/m^3. When nitrogen and carbon dioxide concentrations are next to the saturation values, these terms assume a value close to 1 and they do not affect the microalgae growth.

- *The composition of the inlet water*, which is used to fill the reactor at the beginning of each cultivation: since each cultivation cycle works as a batch reaction, nutrients must be supplied in sufficient quantities to feed the microalgae for the entire duration of the batch process. The model gives as input a surplus of nutrients in the water to guarantee they are not the limiting factor to microalgae growth. Then, as output, the model gives the exact quantity of nutrients consumed by the microalgae.
- *The operating strategy*, which is used to predict microalgae production. Two different strategies can be adopted: if the objective consists in searching the hydraulic retention time (HRT) required to reach a certain microalgae concentration in the reactor, the initial and the final microalgae concentrations are to be supplied as input information to the model. If the HRT is fixed along the whole time period, then the input data will be the initial concentration and the HRT, leaving the final concentration in the reactor free to change.

Table 3: Microalgae species characteristics.

Typology	*P. tricornutum*	*T. pseudonana*
pH_opt	8.3	8.3
alk_opt [meq/l]	0.032	0.032
LHV [kJ/kg]	21,527	21,527
I_S [W/m^2]	37.118	21.834
μ_{max} [1/day]	1.392	3.288
T_{set} [°C]	30	31
T_{opt} [°C]	21	24
β	1.57	1.83
Decay rate [1/day]	0.048	0.048
K_{NA} [mol_{NA}/m^3]	0.001	0.001
K_C [mol_C/m^3]	0.001	0.001

Outputs of the Model

The mass production of microalgae is the most significant output of the cultivation model. In addition to this parameter, it is possible to define two other global performance parameters: the volumetric productivity {expressed in [g/(l × day)] or in [kg/(m^3 × day)]} and the areal productivity {expressed in [t/(ha × y)] or in [kg/m^2 × day)]}. These parameters refer to the volume of the reactor and to the area of the soil covered by the reactor. Moreover, the model is able to give as output the exact quantities of all the compounds consumed or produced in the time period considered by the simulation (usually 1 year). For example, the model gives the CO_2 that is needed to be bubbled in the reactor to maintain a constant pH level and to feed the microalgae. In this way, the model gives the possibility to analyze the carbon footprint of the technology or at least to understand the quantity of CO_2, which can be fixed by the biomass produced, being a part of the injected CO_2 lost to the atmosphere: this quantity can be relevant in the case of open raceway pond.

Fundamental outputs of the model are the total electrical and thermal energies required by the reactor: to evaluate these quantities, the energy balance is not sufficient, as it is necessary to include the electrical energy spent at the boundary of the system, to refill the reactor and to harvest the biomass from the reactor itself; the electrical energy also includes the quantities which are needed for mixing and for the air bubbling in the bioreactor. Finally, another output of the model is the quantity of water required by the reactor, without considering the possible water recirculation coming from the downstream process.

Model Equations

Growth model

Following Yang (2011), the growth rate is calculated by Eq. (3):

$$r_{gA} = \mu_A X_A \qquad (3)$$

where X_A is the mass concentration of microalgae [g/m^3] and μ_A is the specific growth rate [day^{-1}] that is calculated as:

$$\mu_A = \hat{\mu}_A \left(\frac{CO_{2D}}{K_C + CO_{2D}} \right) \left(\frac{N_T}{K_{NA} + N_T} \right) f_I f_T \qquad (4)$$

where the specific growth rate $\hat{\mu}_A$ depends on the microalgae species; CO_{2D}, N_T are the quantities of dissolved nutrients and CO_2 [mol/m^3] at time t. Nitrogen is the only nutrient that has been taken into account in the specific growth rate expression; other nutrients, such as phosphorus, are not explicitly considered, as suggested by Yang (2011), under the reasonable assumption that the metabolism of the microalgae is not limited or inhibited by these compounds. Although the growth rate may be affected by the presence of micro nutrients, they have not been introduced in the model to avoid the definition of a more complex reaction scheme with the corresponding experimental data to represent those influences. The specific growth rate can reach higher values thanks to the use of some micronutrients, such as iron and silicon (Chisti, 2007; Çelekli and Yavuzatmaca, 2009; Le et al., 2010; Ak, 2012; Bahadar and Bilal Khan, 2013). Nitrogen and CO_2 are considered limiting nutrients for microalgae growth and so microalgae growth rate dependence is expressed as saturation functions. K_C and K_{NA} are the half-saturation constants, as explained in the previous paragraphs. f_I is the light intensity factor representing the influence of the light on the microalgae production. Following Yang (Yang, 2011; Fernández et al., 2013), it is calculated by the following equation:

$$f_I = \frac{I_a}{I_s} \exp\left(1 - \frac{I_a}{I_s} \right) \qquad (5)$$

where I_a is the average light intensity in the volume of the bioreactor at a given time t, while I_s is the saturation light intensity, which depends on the microalgae species being considered; saturation light intensity is usually in the range between 30 and 100 W/m^2 (Kumar et al., 2011). The average light intensity I_a depends on the weather, the turbidity caused by the microalgae and other substances in the water and on the reactor geometry. The model includes the hypothesis of perfect mixing: the whole quantity of microalgae in the water is affected by the same quantity of light radiation. f_T is the temperature factor. This term is equal to one and does not influence microalgae growth when the reactor temperature corresponds to the optimal growth temperature for the given microalgae species: this is what happens in the closed reactor, where a heat exchanger is used to keep a constant optimal temperature. Below and above the optimal growth temperature, the growth is negatively influenced by the water temperature. Above the optimal temperature, the value of f_T decreases fast and reaches 0 for a certain temperature that again depends on the algae species: this value is called lethal temperature (T_{let}). The expression of the temperature factor is shown in the following equation, taken from Slegers et al. (2013):

$$C_T = \left[H_2CO_3^*\right] + \left[HCO_3^-\right] + \left[CO_3^{2-}\right] \quad (6)$$

β_{algae} depends on the algae species and it is a curve modulating constant.

Figure 4 shows the shape of the temperature factor f_T: it is possible to note that the equation used for the temperature factor suggested by Slegers et al. (2013) overestimates the growth for low temperature values, since the growth is not possible for low temperatures.

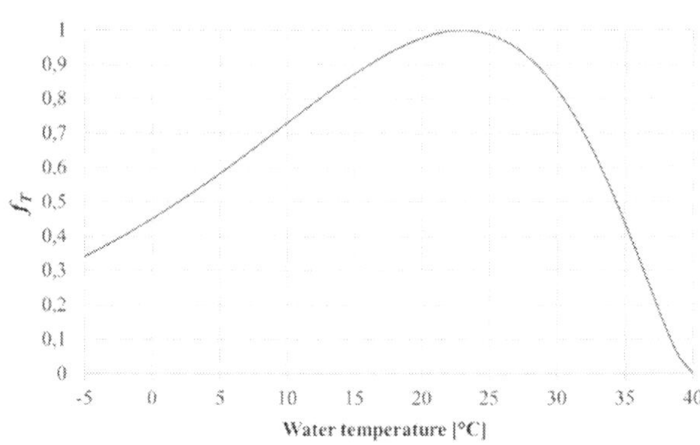

Figure 4: Temperature factor (Slegers et al., 2013).

pH control strategy. Each microalgae species has an optimal pH value: the model includes a control system to keep the optimal value of pH in the reactor through CO_2 injection. As shown in the following equations from Sills (2013), the pH can be controlled by keeping constant two parameters: dissolved CO_2concentration and alkalinity. Since alkalinity does not change with time in the model, to keep pH constant, it is necessary to calculate the quantity of dissolved CO_2 which has to be maintained constant during time.

The total concentration (C_T) of carbonate species in solution is defined as

$$C_T = \left[H_2CO_3^*\right] + \left[HCO_3^-\right] + \left[CO_3^{2-}\right] \quad (7)$$

Dissolved CO_2 concentration is $\left[CO_{2(aq)}\right] \cong \left[H_2CO_3^*\right]$. The molar concentration of each carbonate species (as a fraction of C_T) depends on pH, according to the following equations:

$$\left[H_2CO_3^*\right] = \frac{C_T}{1 + \frac{k_1}{\left[H^+\right]} + \frac{k_1 k_2}{\left[H^+\right]^2}} = \alpha_0 C_T \quad (8)$$

$$\left[HCO_3^-\right] = \frac{C_T}{1 + \frac{\left[H^+\right]}{k_1} + \frac{k_2}{\left[H^+\right]}} = \alpha_1 C_T \quad (9)$$

From Eqs (8) and (9), α_0 and α_1 are calculated.

$$\left[CO_3^{2-}\right] = \frac{C_T}{1 + \frac{\left[H^+\right]}{k_2} + \frac{\left[H^+\right]^2}{k_1 k_2}} = \alpha_2 C_T \quad (10)$$

Finally:

$$C_T = \frac{\text{alk} - OH^- + H^+}{\alpha_1 + 2\alpha_2} \quad (11)$$

Since pH is the concentration of dissolved [H^+] in the water and pH = 14 – pOH, all terms are known and the total concentration of carbonate species can be obtained by

$$C_T = \frac{\text{alk} - OH^- + H^+}{\alpha_1 + 2\alpha_2} \quad (12)$$

This term will be a part of the time-dependent CO_2 mass balance, which is used to calculate the CO_2injection at each time *t*.

Mean light intensity for open pond. The light intensity factor in the specific growth rate expression contains I_a, that is the average light intensity in the bioreactor at a given time. Starting from global horizontal radiation, the model evaluates I_a using the Beer–Lambert's law, which assumes an exponential decay of the light intensity from the external surface of the cultivation system:

$$I_a(s) = I_0 \exp(-\sigma X_A s) \tag{13}$$

As it is explained by Béchet et al. (2013), $I_a(s)$ is the local light intensity, s is the distance from the external surface of the system to the position under consideration, I_0 is the incident light intensity, σ is the extinction coefficient, and X_A the cell concentration. The Beer–Lambert's law can be applied only if two conditions are verified by the culture medium: first, it must be isotropic (i.e., the optical properties of the broth are independent from light direction); this condition is often verified in well-mixed systems. Second, algae cells should not scatter light: this second condition is not always met, but the model considers both the two requirements always verified. The equations to calculate I_a strictly depend on the geometry of the bioreactor: for the open pond, as suggested by Yang (2011), an integration through the pond depth of the Beer–Lambert's law has been used:

$$I_a = \frac{1}{Z}\int_0^Z I_0 \exp(-K_e z)\,dz \tag{14}$$

where I_0 is obtained directly from the global horizontal radiation I_{GHR}: $I_0 = PAR \times I_{GHR}$. PAR is the photosynthetic active radiation (PAR), i.e., the amount of solar radiation which is used by the microalgae for the photosynthesis and which corresponds to 45% of the total incoming light. K_e is related to algal concentration in the pond, and it is called extinction coefficient:

$$K_e = K_{e1} + K_{e2} X_A \tag{15}$$

Light intensity for flat panel. In case of flat panel, the complexity of the geometry requires the use of different equations from those applied for the open pond. The equations used to calculate the light intensity for flat panels are taken from Slegers et al. (2011). For this typology of reactors, the input data are the direct and diffuse radiation on the horizontal surface; moreover, the equations must take into account also the reflection of the radiation on the surface of the panel. As explained by Slegers et al. (2011), to evaluate the effect of direct light irradiation over a tilted surface (panels positioned vertically), it is necessary to introduce a geometrical parameter for direct radiation:

$$G_{\text{direct}}(t) = \frac{\cos\theta}{\cos\theta_z} \quad (16)$$

in which θ is the solar incidence angle, and θ_z is the solar zenith angle. The values assumed by these angles during each day of the year depend on the location, and have been calculated using the equations taken from Duffie and Beckman (2013).

For large scale cultivations, parallel positioned flat panels are used. Parallel placement causes shading, and consequently part of the panels no longer receive direct sky light. The shadow height on vertical reactor panels is given by

$$h_{\text{shadow}} = h - \frac{d\tan(90-\theta_z)}{\cos(\gamma)} \quad (17)$$

which is a function of the reactor height h [m], distance between the reactor panels d [m], solar elevation, which is equal to $90 - \theta_z$ and angle between solar rays and the azimuth of the surface. If $h_{\text{shadow}} > 0$, the flat panel is divided into two parts. The upper part receives direct and diffuse radiation, the lower part only diffuse light. The separation between the upper and the lower part varies with the solar position. Parallel placement of the reactors also influences the penetration of diffuse sky light into the space between panels; the light intensity decreases from the top to the bottom. A similar situation can be seen with the penetration of light in urban street canyons (Robinson and Stone, 2004). For these reasons, the geometrical factor for diffuse light depends on the position over the surface of the panel.

At height $y < h_{\text{shadow}}$:

$$G_{\text{diffuse}} = \frac{1+\cos(\beta+u)}{2} \tag{18}$$

where u = atan(y/d), β [deg] is the slope of the reactor, i.e., the angle that the surface makes with the surface of the earth.

At height $y > h_{\text{shadow}}$:

$$G_{\text{diffuse}} = \frac{1+\cos(\beta)}{2} \tag{19}$$

All panels are treated similarly in the calculations. Moreover, ground reflection is low for parallel placed panels, and is therefore not taken into account.

The total amount of light falling on each reactor surface at a given height y, of a given time t is:

$$I_0(y,t) = G_{\text{direct}}(t)I_{\text{direct}}(t) + G_{\text{diffuse}}(y)I_{\text{diffuse}}(t) \tag{20}$$

At this point, it is fundamental to consider the reflected fraction of the irradiation, which does not enter into the reactor and so does not contributes to the microalgae growth. The amount of reflected light on each interface is related to the differences in refractive indices and the incidence angle (Sukhatme and Sukhatme, 1996).

The incidence angle for diffuse radiation which is considered to evaluate the light reflection is assumed to be 60°, as it is suggested by Duffie and Beckman (2013).

Light reflection by the flat panel walls follows Fresnel equations:

$$R_s = \left[\frac{\eta_i \cos(\theta_i) - \eta_t \sqrt{1-\left(\frac{\eta_i}{\eta_t}\sin(\theta_i)\right)^2}}{\eta_i \cos(\theta_i) + \eta_t \sqrt{1-\left(\frac{\eta_i}{\eta_t}\sin(\theta_i)\right)^2}} \right]^2 \tag{21}$$

$$R_p = \left[\frac{\eta_i \sqrt{1 - \left(\frac{\eta_i}{\eta_t} \sin(\theta_i) \right)^2} - \eta_t \cos(\theta_i)}{\eta_i \sqrt{1 - \left(\frac{\eta_i}{\eta_t} \sin(\theta_i) \right)^2} + \eta_t \cos(\theta_i)} \right]^2 \quad (22)$$

where θ_i [°] is the incidence angle, η_i [−] is the refractive index of the material before the interface, and η_t[−] is the refractive index for the material after the interface. Normal sunlight is non-polarized, therefore the overall reflection coefficient equals the average of the reflection coefficients for *s*-polarized and *p*-polarized light:

$$R' = \frac{R_s + R_p}{2} \quad (23)$$

The light reflected within the reactor wall is completely transmitted to the air, introducing the hypothesis of a non-absorbing material for the walls of the reactor. The light transmitted to the culture, which has to be calculated separately for direct and diffuse radiation, is

$$I_i(t) = I_0(t)(1 - R_1'R_2')T_m \quad (24)$$

T_m [−] is a factor which takes into account a possible low transparency of the material.

The calculation is performed for both the two sides of the reactor and, for parallel positioned panels, for each height. R_1' and R_2' are the reflection coefficients for the air–reactor wall interface and the reactor wall–culture volume interface, respectively. Two light intensity gradients exist in the culture volume. First, as a function of height due to shading and the penetration of diffuse light between parallel positioned panels. Second, in the liquid between the two reactor walls. The second gradient runs from the reactor wall to the center of the reactor and is caused by the absorption of light by the medium and the algae (Slegers et al., 2011). Only the PAR of the spectrum is absorbed by the algae. This accounts for about 45% of the total light. The Lambert–Beer's law is used for the overall light gradient in the culture volume, as it was done for the open pond:

$$I(y,z,t) = I_{\text{front}} e^{-(K_{e1}+K_{e2}X_A)z} + I_{\text{back}} e^{-(K_{e1}+K_{e2}X_A)(s-z)} \quad (25)$$

This equation gives the light intensity at location z [m] inside the reactor thickness [m], at a given height y[m] in the reactor at time t.

At this point, to simplify the model, the values of light intensity $I(y,z,t)$ are integrated to find a mean value of irradiation for the whole culture inside the whole reactor at a given time t: with these integrations, a single value of radiation is obtained and used in the growth model, for the whole panel, at a given time t. The hypothesis of a perfect mixing inside the culture at each time t is necessary to integrate the equation in both height and depth.

Mass balances. Mass balances of nitrogen and oxygen inside the pond can be modeled in the same way, suggested by Yang (2011):

$$\frac{dM}{dt} = \mu_A X_A Y_{AM} - k_{Lg}\alpha\left(M - M^*\right) \quad (26)$$

where M is the concentration of the respective component in the water in the bioreactor, Y_{AM} is the mass of the respective component consumed or generated by the microalgae per unit mass of microalgae produced. The last term of the right-end side of Eq. (26) represents the mass transfer between the atmosphere and the pond, where $k_{Lg}\alpha$ is the mass transfer coefficient for a given element, M^* is the saturation concentration of the associated dissolved element.

For total inorganic carbon, the mass balance assumes a different formulation since the CO_2 is injected continuously in the pond during the growth phase to keep a constant concentration of dissolved CO_2 in the reactor, balancing the losses of CO_2 to the atmosphere and the consumption of CO_2 by the microalgae.

$$\frac{dCO_2}{dt} = \mu_A X_A Y_{ACO_2} + f_{CO_2} - k_{Lg}\alpha\left(CO_2 - CO_2^*\right) \quad (27)$$

where fCO2fCO2 represents the flux of CO_2 introduced by the supply of gas flow into the system. Since the quantity of dissolved CO_2 is kept constant ($dCO_2/dt = 0$), the mass balance can be written again as

$$f_{CO_2} = -\mu_A X_A Y_{ACO_2} + k_{Lg}\alpha\left(CO_2 - CO_2^*\right) \tag{28}$$

Finally, the mass balance for microalgae species can be written as

$$\frac{dX_A}{dt} = \mu_A X_A - k_{dA} X_A \tag{29}$$

As it can be seen, all these equations are time dependent, and all strictly depend on one another: this means that they form altogether a system of differential equations. The strategy to solve the equations with the finite difference methodology is used to solve this system in Matlab, and therefore to solve the system of differential equations as a system of algebraic equations.

Thermal balance for open pond. The model includes the thermal balance as it is suggested by Slegers et al. (2013):

$$V_R cp_w \rho_w \frac{dT_w}{dt} = Q_{irr} - Q_{algae} - Q_{rad} - Q_{evap} - Q_{conv} - Q_{cond} \tag{30}$$

where V_R is the volume of the pond, cp_w the heat capacity of the growth medium, ρ_w the density of the growth medium, T_w the temperature in the pond, Q_{irr} the heat flow rate to the pond by the sunlight, Q_{algae} the light energy flow rate to algae during growth, Q_{rad} the heat flow rate by emission of long-wave radiation in the infrared region, Q_{evap} [W] the heat flow rate caused either by evaporation or condensation, Q_{conv} the heat flow rate by convection, and Q_{cond} the heat flow rate between pond and ground via conduction. The water in the pond is heated by sunlight that enters the culture volume. Solar energy that is not used by algae for growth is considered as thermal energy that is dissipated through the different thermal exchange mechanisms (Q_{rad}, Q_{evap}, Q_{conv}, and Q_{cond}) or as a contribution to temperature growth in the pond. The total heat flow rate by the sunlight is given by

$$Q_{irr} = A_w I_{surface}(t) \tag{31}$$

where A_w [m^2] is the water surface area of the pond and $I_{surface}$ [W/m^2] is the total light arriving on the pond. Part of this light is absorbed by microalgae for growth:

$$Q_{algae} = h_{comb}\mu_A X_A V_R \tag{32}$$

which is a function of the lower heating value on a dry basis of algae biomass h_{comb} [J/kg], specific growth rate μ_A [s^{-1}] and the biomass concentration X_A [kg/m^3].

The water in the pond emits thermal energy by long-wave radiation. The overall long-wave radiation flow between water in the pond and sky is calculated using Duffie and Beckman (2013):

$$Q_{rad} = A_w \varepsilon_w \sigma_{SB}\left(\left(T_w + 273.15\right)^4 - T_{sky}^4\right) \tag{33}$$

where ε_w [−] is the emissivity of the water in the infrared region, σ_{SB} [W/(m^2K^4)] the Stefan–Boltzmann constant, and T_{sky} [K] the equivalent sky temperature for clear sky days, which is expressed by Duffie and Beckman (2013) as

$$T_{sky} = (T_a + 273.15)(0.711 + 0.0056T_{dew} + 0.000073T_{dew}^2 + 0.013\cos(15t_{solar}))^{0.25} \tag{34}$$

where T_a [°C] is the air temperature, T_{dew} [°C] the dew point temperature, and t_{solar} [−] the number of hours after solar midnight.

Evaporation has a large effect on the water temperature, especially in locations with low humidity and high wind velocities. The evaporation rate depends on the shape of the water area, wind velocity, and consequently also movement of the water. The evaporation flow is driven by the difference of water vapor pressures between ambient air the saturated water body. The evaporation energy flow is given by

$$Q_{evap} = A_w h_{evap}\left(p'_s - p'_a\right) \tag{35}$$

The evaporation flow depends on the heat transfer coefficient for evaporation h_{evap} [W/(m^2 Pa)], the saturated water pressure p'_s [Pa] at water temperature T_w, and the water pressure of air p'_a [Pa] at air temperature T_a. The evaporation rates have been calculated using the heat transfer coefficient h_{evap} found in Duffie and Beckman (2013):

$$h_{evap} = 0.036 + 0.025v \tag{36}$$

where v [m/s] is the wind speed. The Antoine equation is applied to calculate the saturated water pressure p'_s [Pa] at water temperature T_w and the water pressure of the air p'_a [Pa] at air temperature T_a:

$$p' = \text{RH}\ 10^{\left(8.07131+\log_{10}\left(\frac{101325}{760}\right)\right)-\frac{1730.63}{233.46+T_a}} \tag{37}$$

where RH [−] is the relative humidity and T_a [°C] the temperature.

Convection and evaporation are related processes, as it is shown in Eq. (37). The flow for passive and forced convection at the water surface mainly depends on the difference between water and air temperature. The convection flow is given by

$$Q_{conv} = C_{Bowen} \frac{p_a\left(T_w - T_a\right)}{p_{ref}\left(p'_s - p'_a\right)} Q_{evap} \tag{38}$$

where C_{Bowen} is the Bowen constant [Pa/°C], p_a is the ambient pressure [Pa] and p_{ref} the reference pressure [Pa], and p'_s and p'_a are derived using equation

$$p' = \text{RH}\ 10^{\left(8.07131+\log_{10}\left(\frac{101325}{760}\right)\right)-\frac{1730.63}{233.46+T_a}} \tag{39}$$

Conductive heat transfer takes place between the open pond and the soil. The soil is assumed to be an infinite source for heat transfer. This heat transfer calculation is derived from Fourier's law:

$$Q_{cond} = h_{soil} A_{soil}\left(T_w - T_{soil}\right) \tag{40}$$

where h_{soil} [W/(m^2°C)] is the heat transfer coefficient of the surrounding soil layer, A_{soil} [m^2] is the area of the pond that is embedded in the soil, and T_{soil} [°C] is the temperature of the soil surrounding the pond.

Thermal balance for flat panel. Due to the significant difference in the geometry of the reactor and the temperature strategy being considered, the thermal balance for the flat panel assumes a different form from that implemented for the open raceway pond:

$$V_R cp_w \rho_w \frac{dT_w}{dt} = Q_{irr} - Q_{algae} - Q_{exch} - Q_{conv+cond} \tag{41}$$

In a flat panel, a constant value of the water temperature T_w [°C] is desired, and obtained using a heat exchanger placed at the bottom of the reactor to remove or supply heat.

Thus, the thermal balance can be written as follows:

$$\frac{dT_w}{dt} = 0 \rightarrow Q_{irr} - Q_{algae} - Q_{exch} - Q_{conv+cond} = 0 \tag{42}$$

Q_{irr} [W] is given by the following expression:

$$Q_{irr} = 2A_{panel} I_{mean} \tag{43}$$

where A_{panel} is the surface of the reactor, which is multiplied by two, as it is necessary to consider both the front and the back surfaces; I_{mean} [W/m^2] is the radiation previously calculated taking into account both the direct and diffuse radiation and the reactor geometry that is the radiation which interacts with the water in the reactor and with the microalgae. As calculated for the open pond, a part of the incoming heat is used by the microalgae for growing:

$$Q_{algae} = h_{comb} \mu_A X_A V_R \tag{44}$$

which is a function of the combustion energy of algae biomass h_{comb} [J/kg], specific growth rate μ_A [s^{-1}], biomass concentration X_a [kg/m^3], and reactor volume V_R [m^3].

The model takes into account the natural convection over the panel surface caused by the wind and the glass conductivity:

$$Q_{conv+cond} = 2A_{panel}U_{tot}\left(T_w - T_{atm}\right) \quad (45)$$

where

$$U_{tot} = U_{cond} + U_{conv} \quad (46)$$

where U_{cond} is the glass conductivity [W/(m^{2}K)] and U_{conv} is the conductivity of the natural convection that is obtained by Duffie and Beckman (2013). Using the equations above, the quantity of heat that has to be removed or supplied by the heat exchanger at each time *t* can be calculated. Heat exchange through radiation can be omitted from this balance, since the water temperature is always kept between 20 and 30°C, depending on microalgae growth optimal temperature.

Electrical energy consumption for harvesting, refilling, mixing and bubbling. For both the open pond and flat panel, the harvesting of the water from the reactor and its refilling are carried out in 8 h, during one night: 3.5 h are required for harvesting and 3.5 h for refilling. These operations are performed using a pump. The electrical consumption has been calculated as follows:

$$E_{harv/refil} = P_{pump}t_{harv/refil} \quad (47)$$

where the power of the pump is

$$P_{pump} = \frac{\rho g Q h}{\eta_{pump}} \quad (48)$$

where ρ is the water density [kg/m^3], *g* is the acceleration of gravity [m/s^2], η_{pump} [–] is the efficiency of the pump, set at 0.85, *Q* [m^3/s] is the volumetric flow rate which has to be pumped, and *h* [m] is the height difference between the two basins before and after the pump: thanks to the fact that the model includes also the design of a settler positioned after the bioreactor, it is possible to know the exact *h* for the harvesting, which has been increased to consider the losses in the pipes, whereas for the refilling data have been taken from the literature, considering $h = 1$ m for the open pond and 3 m for the flat panel: the difference between these two values is again associated with the energy losses in pipes.

The power required for mixing in the open pond is

$$P_{mix} = \frac{\rho g Q h}{\eta_{paddlewheel}} \quad (49)$$

where Q [m^3/s] is obtained from the water speed that is desired in the reactor (0.20 m/s) and from the cross-section of the open pond (which depends on the geometry), h is the given height difference before and after the paddle wheel, taken from literature (0.05 m). $\eta_{paddlewheel}$ is lower than the efficiency of a normal pump and is assumed to be equal to 0.25.

Both for the open pond and flat panel, a bubbling system has to be taken into account: for the flat panel, this system should be able both to supply the CO_2 necessary for the photosynthesis of the microalgae and to guarantee an adequate mixing inside the reactor. For this reason, the amount of air bubbled in the flat panel is higher than the air supplied to the open pond. These quantities are controlled by the CO_2 molar fraction inside the injected air which is 0.04 in the case of open pond and 0.02 for the flat panel. The design and the energy consumption of the bubbling system have been calculated through Belsim Vali modeling software, using a compressor. The result is that it is necessary to supply 4 kJ for each kg of air injected in the reactor.

RESULTS

Open Pond

Tables 4 and 5 show the input and output data in the open pond modeling:

Table 4: Input data for open pond simulations.

HRT not fixed, X target fixed		HRT fixed, X target not fixed	
Location	Sevilla (SPA) Petrolina (BRA)	Location	Sevilla (SPA) Petrolina (BRA)
Typology	*P. tricurnutum* *T. pseudonana*	Typology	*P. tricumutum* *T. pseudonana*
Xa_init [g/m³]	100	Xa_init [g/m³]	100
Xa_target [g/m³]	490	HRT [day]	7
Tw_in [°C]	15	Tw_in [°C]	15
CO_2 rate [%]	0.04	CO_2 rate [%]	0.04
Z_pond [m]	0.3	Z_pond [m]	0.3
LW [–]	10	LW [–]	10

Table 5: Output from open pond simulations.

	HRT not fixed X target fixed		HRT fixed X target not fixed	
	Petrolina	Sevilla	Petrolina	Sevilla
Number of harvesting in 1 year	45	50	–	–
HRT [day]	–	–	7	7
Mass microalgae [t/(ha y)]	57.29	64.68	54.04	67.88
CO_2 captured [t/(ha y)]	140	156	133	163
CO_2 injected [t/(ha y)]	243	260	236	268
CO_2 lost to atmosphere [t/(ha y)]	78.62	77.35	78.14	77.27
CO_2 ratio losses	0.324	0.298	0.33	0.29
CO_2 ratio algae	0.57	0.6	0.56	0.61
N absorbed [t/(ha y)]	5.8	6.55	5.57	6.8
Water injected [t/(ha y)]	109,210	121,640	120,220	119,800
Water evaporation [t/(ha y)]	5.8	5.01	5.63	5.24
Energy microalgae [kWh/(ha y)]	342,610	386,790	323,180	405,930
Electrical energy [kWh/(ha y)]	18,497	19,024	18,522	19,111
Thermal energy [kWh/(ha y)]	–	–	–	–
Volumetric productivity [kg/(m³ × day)]	0.0523	0.0591	0.0494	0.062
Areal productivity [kg/(m² × day)]	0.0157	0.0177	0.0148	0.0186
NER	0.108	0.0603	0.1146	0.0942

The open pond productivity that is obtained by the model is consistent with the literature: Slegers et al. (2013) estimated an annual biomass production for a 1 ha surface in the Netherlands and Algeria equal to 41.5 and 63.7 t, respectively. Fernández et al. (2013) report a maximum microalgae areal productivity equal to 30 g/(m^2 day), higher than those resulting from the model calculations, but still of the same order of magnitude. Jiménez et al. (2003) obtained a volumetric productivity equal to 0.05 kg/(m^3 day) for a location in Southern Spain (Malaga), by cultivating the microalgae until a maximum concentration of 470 g/m^3: the cultivation conditions and the results from the dynamic model are consistent with the results of this work.

In this work, the Net Energy Ratio (NER)

$$\text{NER} = \text{Net Energy Ratio} = \frac{\text{total energy requirement for operation}}{\text{total energy production (biomass)}} \quad (50)$$

expresses the fraction of the energy produced in the cultivation system that is used by the system itself to generate the biomass. If this value is close to 1, the cultivation technology of microalgae is too energy intensive, requiring a big share of the energy produced. From Table 5, it appears that for both the locations analyzed by the model the NER is quite far from 1, showing that the energy demand for the operation of the cultivation system in Petrolina is 10.8% of the energy contained in the biomass produced, while for Sevilla is 6%. These values are quite promising for a potential production of microalgae in these locations, since the NER is far enough from 1: even if the energy content of open pond construction and materials are added to the total energy requirement, it seems reasonable to state that the plant would still be convenient from an energetic point of view. Another important output of the model from an environmental point of view is the quantity of CO_2 fixed by the microalgae during the growth process: the downstream process that transforms microalgae biomass into biofuel releases CO_2 to the atmosphere, with a negative environmental impact. Since the cultivation phase captures more CO_2 than the quantity released, using the data of CO_2 fixed in the algae it is possible to calculate a global CO_2 balance, for the entire biofuel production chain: this environmental analysis may lead to a comparison with other biofuel chain production and with traditional fossil fuels, to evaluate which product has a positive or negative environmental impact.

With the first operating strategy (meaning that the inputs are the initial and final concentration), one of the output would be the number of days that are needed to reach the final target concentration. The number of days is not the same along the year, being variable with weather conditions. During wintertime, the harvesting period in Sevilla is extremely long due to low irradiation and low temperatures, which make the microalgae growth in the pond slow and difficult. A possible strategy to overcome this limitation might be to interrupt the production during wintertime: in this case, the productivity during the whole year would decrease, since a part of the biomass would not be produced, but the NER would positively decrease, because of the absence of energy consumption during winter: the production lost is less than the energy saving obtained, and so the overall effect would be positive. For Petrolina, the HRT is more homogeneous along the year, but longer: the reason might be that weather conditions in Petrolina cause a strong increase of pond temperature that may cause a consequent reduction in microalgae productivity when pond temperature is higher than the optimal temperature for microalgae growth.

Results of dynamic simulation runs are reported in Figures 5–9, showing the effect of PAR and pond temperature over the specific growth rate and microalgae growth rate. Input parameters for simulation runs are shown in Table 6.

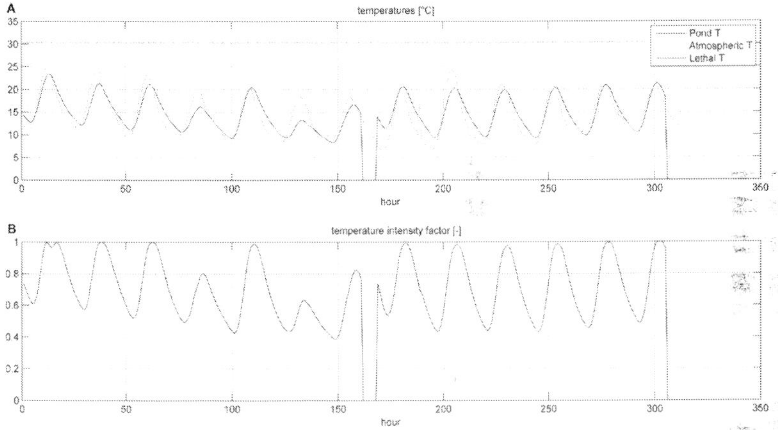

Figure 5: Temperatures (A) and temperature factor (B) for a simulation time equal to two HRTs.

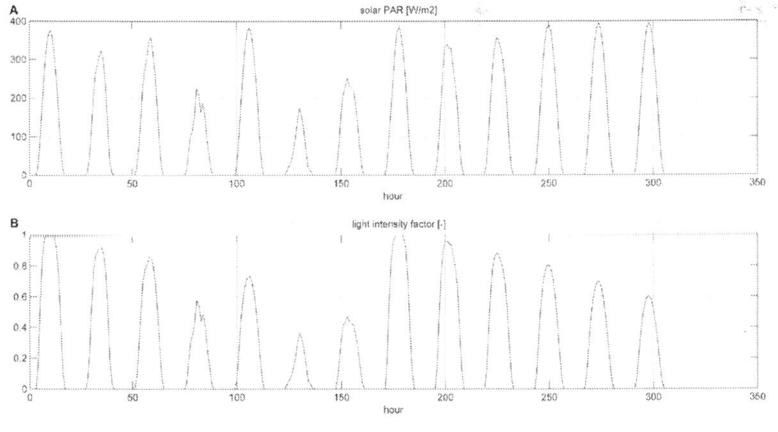

Figure 6: Photosynthetic active radiation (PAR) (A) and light intensity factor (B) for a simulation time equal to two HRTs.

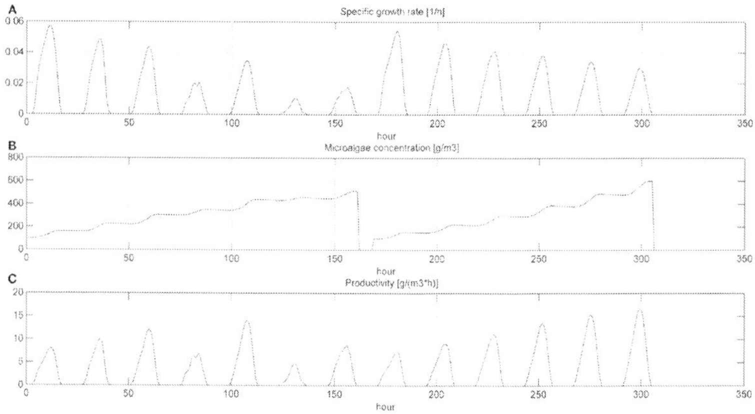

Figure 7: Specific growth rate (A), microalgae concentration (B) and microalgae volumetric productivity (C) in the open pond during two consequent batch cultivations (simulation time equal to two HRTs).

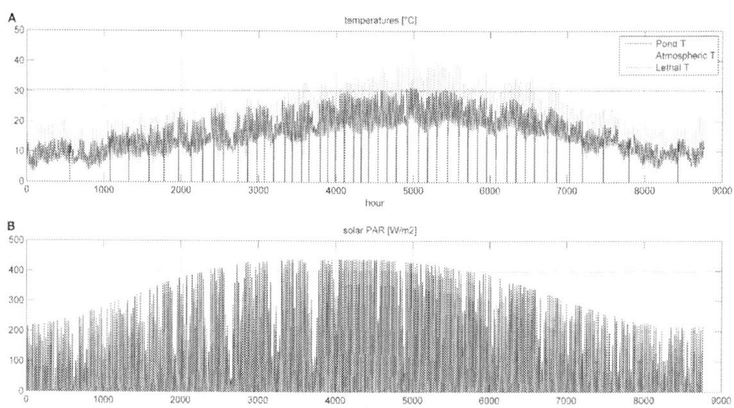

Figure 8: Temperatures (A) and photosynthetic active radiation (B) for a simulation time equal to 1 year.

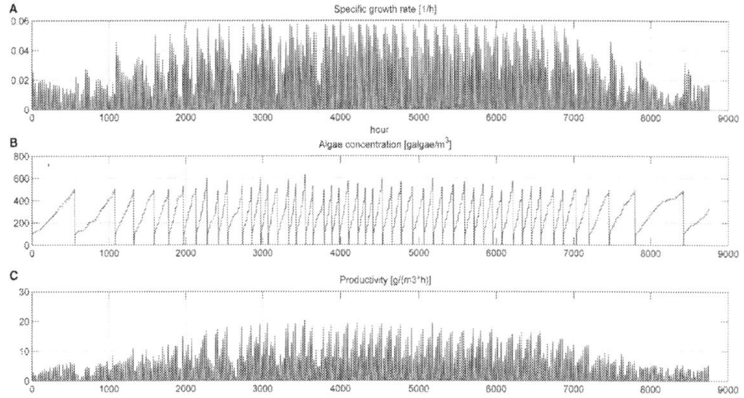

Figure 9: Specific growth rate (A), microalgae concentration in the pond (B) and microalgae growth rate (C) for a simulation time equal to 1 year.

Table 6: Input data for open pond simulation runs.

HRT not fixed, X target fixed	
Location	Sevilla (SPA)
Typology	*P. tricurnutum*
Xa_init [g/m³]	100
Xa_target [g/m³]	490
Tw_in [°C]	15
CO_2 rate [%]	0.04
Z_pond [m]	0.3
LW [–]	10

Even if the simulation has been run for the entire year, as it is shown in Figures 8 and 9, Figures 5–7report a small time range that includes two consequent batch cultivations, for an easier understanding of the light and temperature effect over the microalgae growth. Figure 5 includes the pond temperature values (A), calculated in each hour of the entire simulation period, and the consequent temperature factor (B). In Figure 5, the model does not evaluate the pond temperature during the hours when the pond is emptied. Figure 6 reports the PAR and the consequent light intensity factor (B). Figure 7 shows the specific growth rate (A) and the microalgae concentration (B) in the open pond during two consequent batch cultivations. Microalgae growth rate (C) is obtained, as explained in Eq. (3), by the product of the specific growth rate and the microalgae concentration. As shown in Figure 7B, each batch cultivation starts with the same initial concentration, while the final concentration depends on each batch: the model waits for the end of the solar radiation to start to empty the pond that takes almost the whole night.

In Figures 8 and 9, the time of the entire simulation run has been considered presenting input values (temperature in Figure 8A and PAR in Figure 8B) and output [specific growth rate in Figure 9A, microalgae concentration in the pond in Figure 9B and microalgae growth rate (productivity) in Figure 9C].

The time to reach the minimum microalgae target concentration to start the harvest in winter time in Sevilla is really long and it might be useful to interrupt the production for some months.

Flat Panel

Table 7 shows the input data for the flat panel model for both the two operating strategies.

Table 7: Input data for flat panel simulations.

HRT not fixed, X target fixed		HRT fixed, X target not fixed	
Location	Sevilla Petrolina	Location	Sevilla Petrolina
Typology	*P. tricurnutum* *T. pseudonana*	Typology	*P. tricurnuturn* *T. pseudonana*
Azimuth [°]	0 −90	Azimuth [°]	0 −90
Slope [°]	90	Slope [°]	90
Xa_init [g/m³]	3000	Xa_init [g/m³]	3000
Xa_target [g/m³]	6000	HRT [day]	4 or 5
CO_2 rate [%]	0.02	CO_2 rate [%]	0.02
h [m]	1.5	*h* [m]	1.5
s [m]	0.05	*s* [m]	0.05
d [m]	0.5	*d* [m]	0.5

The azimuth of the panel is an input, which may vary from 0°, when the panel faces south, to −90°, when the panel faces east. The model is not able to consider different slopes of the reactor, which may only be positioned vertically: this position is the most favorable for light distribution and dilution, leading to the highest values of productivity (Sierra et al., 2008). CO_2 molar concentration in the injected gases is lower than that of the open raceway pond, since the injected gases in the flat panel photobioreactor do not only have to supply the CO_2 needed by microalgae for photosynthesis, but also to generate the necessary mixing in the reactor: adequate mixing is essential to obtain good levels of productivity in every microalgae cultivation system. Due to the complex geometry of the flat panel photobioreactor, more geometrical data are needed as input if compared to the open raceway pond system. As explained by Fernández et al. (2013), heights <1.5 m and widths <0.10 m are preferred; following this indication and data from Slegers et al. (2011), the distance between the vertical panels has been set equal to 0.5 m, the height of each panel to 1.5 m and the thickness to 5 cm. Both height and distance between panels have been varied within a range of possible values in a further parametrical analysis. The initial and final concentrations have been suggested by Münkel et al. (2013); the values are higher than in the case of open raceway pond: the more sophisticate closed photobioreactor allows higher concentrations to be reached without compromising the productivity of the cultivation system. This is possible thanks to an optimal light distribution over the whole reactor for the entire operation time that is guaranteed also by an adequate mixing of the medium. The flat panel model, as well as the open raceway pond model, is able to produce the dynamic trend of all the time dependent physical quantities, which are included in the analysis. The global results for the flat panel reactor are shown in Tables 8 and 9: Table 8 reports the results for the first operating strategy, i.e., the case in which both input and output microalgae concentrations are known, whereas Table 9 corresponds to the second operating strategy, in which the input parameters are the input concentration and the HRT.

Table 8: Results from flat panel simulations using the first operating strategy, meaning that input and output microalgae concentrations are known.

	HRT not fixed South		HRT not fixed East	
	Petrolina	Sevilla	Petrolina	Sevilla
N harvesting	65	77	88	81
Mass microalgae [t]	289	345	393	366
CO_2 captured [t]	860	976	1083	1025
CO_2 injected [t]	956	1065	1204	1139
CO_2 ratio algae	0.9	0.9	0.9	0.9
N absorbed [t]	35.92	40.77	45.2	42.77
Water injected [t]	46,023	54.592	62,353	57,742
Energy microalgae [kWh]	1733.700	2063.700	2350,700	2191,100
Electrical energy [kWh]	35,634	40,476	44.906	42,473
Thermal energy [kWh]	1310,300	2124,000	1267,400	2093,900
Volumetric productivity [kg/(m³ × day)]	0.5851	0.6965	0.7934	0.7395
Areal productivity [kg/(m² × day)]	0.0794	0.0946	0.1077	0.1004
NER	1.55	2.09	1.1165	1.95

Table 9: Results from flat panel simulation using the second strategy, meaning that input microalgae concentration and HRT are known.

	HRT fixed South		HRT fixed East	
	Petrolina	Sevilla	Petrolina	Sevilla
HRT [day]	5	5	4	4
Mass microalgae [t]	284	367	390	380
CO_2 captured [t]	844	1041	1081	1052
CO_2 injected [t]	938	1157	1201	1168
CO_2 ratio algae	0.9	0.9	0.9	0.9
N absorbed [t]	35,023	43,044	45.11	43.88
Water injected [t]	46,691	52,620	62,422	57,201
Energy microalgae [kWh]	1701,300	2197,000	2336,600	2273.000
Electrical energy [kWh]	34,973	43,043	44,819	43,542
Thermal energy [kWh]	1292,400	2264,900	1262,700	2113.900
Volumetric productivity [kg/(m³ × day)]	0.5742	0.7415	0.7886	0.7672
Areal productivity [kg/(m² × day)]	0.078	0.1007	0.1071	0.1041
NER	1.56	2.101	1.1192	1.8983
Xa final mean [kg/m³]	5.87	6.707	6.016	6.077

For the two operating strategies, results are reported for both the locations (Sevilla and Petrolina) and for the two most significant orientations, that is when the flat panel is facing south (and north) and when it is facing east (and west). The east–west orientation appears to be preferable in both locations because it leads to a higher microalgae production: as explained by Sierra et al. (2008), if the orientation of the two faces of the reactor is east–west, the intercepted radiation is maximum during the first and last solar hours, because of the orientations toward sunrise and sunset. Thus, light availability during the daylight solar cycle is also more homogenous for this configuration. Areal and volumetric productivities are consistent with those found in literature: Chisti (2007) reports a volumetric productivity equal to 1.535 kg m^{-3} day^{-1}, which is higher than the values obtained from the model; if compared to some other works, the results of the dynamic model, in particular the areal productivity, seem to be optimistic: for example, the microalgae production coming from the model [~400 t/(ha × year)] is two times higher than the production obtained by Slegers et al. (2011) from a flat panel reactor located in Algeria which produces up to 200 t/(ha × year). Moreover, the volumetric productivity for a flat panel reported by Jorquera et al. (2010) is equal to 0.27 kg m^{-3} day^{-1}, whereas the model supplies values equal to 0.8 kg m^{-3} day^{-1}. From a recent work by Münkel et al. (2013), volumetric productivities equal to 1.25 kg m^{-3} day^{-1} have been reached in experimental analyzes. The difference from some values found in literature could be a consequence of a series of related factors: the microalgae species chosen may strongly influence the productivity of the reactor. Moreover, the model created in this work contains a temperature control, which fixes the temperature inside the reactor at the optimal level for microalgae growth: this means that the specific growth rate is not affected by the temperature factor (which is always equal to one), and consequently it is nearer to the maximum growth rate than in the real operating conditions. Furthermore, the locations chosen for the analysis present optimal values of irradiation, next to the saturation irradiation, where the light intensity factor affecting the growth rate is next to 1. From Tables 8 and 9 the high dependence on the orientation of the panel clearly appears in the productivity of Petrolina: if the two faces of the reactor are oriented toward east and west, the productivity is higher than in case of south and north orientation. A possible explanation might be related to radiation reflection by the panel: when the sun is high in the sky, the radiation hits the flat panel with an incidence angle close to 90°; if the incidence angle is too high, radiation could not enter the reactor, due to glass reflection. For this reason, if the panel is oriented toward south, the largest part of the radiation during summer is lost and does not contribute to microalgae growth: if the orientation if east–west, the radiation is collected during morning and afternoon with an incidence angle next to 0°. The same situation does not take place in Sevilla, because the sun does not reach high elevations during the whole year; the east–west orientation is preferable also in Sevilla, as suggested by Sierra et al. (2008).

In general, higher volumetric and areal productivity values have been reached both in Sevilla and in Petrolina, and the difference between the two locations is less remarkable than it was for the open pond; the flat panel photobioreactor growth model includes a temperature control, which maintains the temperature of the water at a constant level. In spite of the higher productivities, Sevilla appear to be less suitable for a flat panel photobioreactor than Petrolina, as shown by NER. The thermal energy requirement is extremely high in Sevilla, where winter time brings low atmospheric temperatures: the photobioreactor should operate only during summertime.

Microalgae mass production in flat panel reactor is much higher than in open raceway pond, CO_2absorption is more efficient, but the thermal energy requirement implies a NER value >1. There are different possible strategies to solve this problem: keeping the water in the reactor within a range of suboptimal temperatures where the productivity of microalgae is still high and the thermal energy requirement is lower; leaving the temperature in the reactor without any control during the night: this might be an interesting solution also to limit the microalgae losses due to dark respiration, which are higher for optimal temperature. If water temperature is lower than the optimal value for growth, the metabolic energy required by the microalgae for their maintenance during the night is lower.

DISCUSSION

Unlike the most common existing models in the literature, which deal with a specific part of the overall cultivation process, the models presented in this paper include all physical and chemical quantities that mostly affect microalgae growth. These features allow the model to correctly predict the overall behavior of microalgae cultivation plants for energy production. All input parameters can be tuned and varied to obtain reliable predictions.

This paper aims to present a model for microalgae cultivation where geometric and physical parameters, presented in recent works by Slegers et al. (2011, 2013), Duffie and Beckman (2013), Béchet et al. (2013)and more, receive a robust analysis together with the chemical and biological aspects, as presented by Yang (2011), Chisti (2007), Le et al. (2010), and more.

A comparison with experimental data taken from the literature shows that the predictions are consistent, only slightly overestimating the productivity in case of closed photobioreactor. The reason for the overestimation might be found in the temperature control strategy of the reactor, which is kept constant in the simulation runs. This strategy is currently applied in laboratory scale plants but it is expensive both from an economic and energetic point of you for a large scale cultivation plant.

The model is a part of a wider activity on microalgae cultivation plants for energy production which consider the additional directions of work:

- Further analysis could be conducted at the boundaries of the system taken into consideration by the model, evaluating and modeling the downstream microalgae transformation process and the upstream technologies that generate the input streams entering the cultivation phase.

- The cultivation phase model might be modified to include other less important, but still valuable parameters that influence microalgae growth, such as some other nutrients like phosphorus.
- The microalgae growth rate equation could be compared with other models in the literature and with results coming from real pilot plants to make predictions more reliable.
- The flat panel model should be tested with different temperature control strategies, which might lead to a lower productivity but also to a lower thermal energy consumption, making the technology more interesting from an economic point of view.

Conflict of Interest Statement

The authors declare that the research was conducted in the absence of any commercial or financial relationships that could be construed as a potential conflict of interest.

REFERENCES

1. Ak, İ (2012). Effect of an organic fertilizer on growth of blue-green alga *Spirulina platensis*. *Aquacult. Int.* 20, 413–422. doi: 10.1007/s10499-011-9473-5
2. Azadi, P., Brownbridge, G., Mosbach, S., Smallbone, A., Bhave, A., Inderwildi, O., et al. (2014). The carbon footprint and non-renewable energy demand of algae-derived biodiesel. *Appl. Energy* 113, 1632–1644. doi:10.1016/j.apenergy.2013.09.027
3. Bahadar, A., and Bilal Khan, M. (2013). Progress in energy from microalgae: a review. *Renew. Sustain. Energ. Rev.* 27, 128–148. doi:10.1016/j.rser.2013.06.029
4. Béchet, Q., Shilton, A., and Guieysse, B. (2013). Modeling the effects of light and temperature on algae growth: state of the art and critical assessment for productivity prediction during outdoor cultivation. *Biotechnol. Adv.* 31, 1648–1663. doi:10.1016/j.biotechadv.2013.08.014

5. Borowitzka, M. A. (1999). Commercial production of microalgae: ponds, tanks, tubes and fermenters. *J. Biotechnol.* 70, 313–321. doi:10.1016/S0168-1656(99)00083-8
6. Brennan, L., and Owende, P. (2010). Biofuels from microalgae – a review of technologies for production, processing, and extractions of biofuels and co-products. *Renew. Sustain. Energ. Rev.* 14, 557–577. doi:10.1016/j.rser.2009.10.009
7. Çelekli, A., and Yavuzatmaca, M. (2009). Predictive modeling of biomass production by *Spirulina platensis* as function of nitrate and NaCl concentrations. *Bioresour. Technol.* 100, 1847–1851. doi:10.1016/j.biortech.2008.09.042
8. Chisti, Y. (2007). Biodiesel from microalgae. *Biotechnol. Adv.* 25, 294–306. doi:10.1016/j.biotechadv.2007.02.001
9. Collet, P., Hélias, A., Lardon, L., Ras, M., Goy, R.-A., and Steyer, J.-P. (2011). Life-cycle assessment of microalgae culture coupled to biogas production. *Bioresour. Technol.* 102, 207–214. doi:10.1016/j.biortech.2010.06.154
10. Cuaresma, M., Janssen, M., Vílchez, C., and Wijffels, R. H. (2011). Horizontal or vertical photobioreactors? How to improve microalgae photosynthetic efficiency. *Bioresour. Technol.* 102, 5129–5137. doi:10.1016/j.biortech.2011.01.078
11. Davis, R., Aden, A., and Pienkos, P. T. (2011). Techno-economic analysis of autotrophic microalgae for fuel production. *Appl. Energy* 88, 3524–3531. doi:10.1016/j.apenergy.2011.04.018
12. Doucha, J., and Lívanský, K. (2006). Productivity, CO2/O2 exchange and hydraulics in outdoor open high density microalgal (*Chlorella sp.*) photobioreactors operated in a Middle and Southern European climate. *J. Appl. Phycol.* 18, 811–826. doi:10.1007/s10811-006-9100-4
13. Duffie, J. A., and Beckman, W. A. (2013). *Solar Engineering of Thermal Processes*, 4th edn. New York: A Wiley-Interscience Publication, John Wiley and Sons, Inc.
14. Fernández, F. G. A., Camacho, F. G., Pérez, J. A. S., Sevilla, J. M. F., and Grima, E. M. (1997). A model for light distribution and average solar irradiance inside outdoor tubular photobioreactors for the microalgal mass culture. *Biotechnol. Bioeng.* 55, 701–714.

doi:10.1002/(SICI)1097-0290(19970905)55:5

15. Fernández, F. G. A., Sevilla, J. M. F., and Grima, E. M. (2013). Photobioreactors for the production of microalgae. *Rev. Environ. Sci. Biotechnol.* 12, 131–151. doi:10.1007/s11157-012-9307-6
16. Hadiyanto, H., Elmore, S., Van Gerven, T., and Stankiewicz, A. (2013). Hydrodynamic evaluations in high rate algae pond (HRAP) design. *Chem. Eng. J.* 217, 231–239. doi:10.1016/j.cej.2012.12.015
17. Hempel, N., Petrick, I., and Behrendt, F. (2012). Biomass productivity and productivity of fatty acids and amino acids of microalgae strains as key characteristics of suitability for biodiesel production. *J. Appl. Phycol.* 24, 1407–1418. doi:10.1007/s10811-012-9795-3
18. James, S. C., and Boriah, V. (2010). Modeling algae growth in an open-channel raceway. *J. Comput. Biol.* 17, 895–906. doi:10.1089/cmb.2009.0078
19. Janssen, M., Tramper, J., Mur, L. R., and Wijffels, R. H. (2003). Enclosed outdoor photobioreactors: light regime, photosynthetic efficiency, scale-up, and future prospects. *Biotechnol. Bioeng.* 81, 193–210. doi:10.1002/bit.10468
20. Jiménez, C., Cossio, B. R., Labella, D., and Xavier Niell, F. (2003). The feasibility of industrial production of Spirulina (*Arthrospira*) in Southern Spain. *Aquaculture* 217, 179–190. doi:10.1016/S0044-8486(02)00118-7
21. Jorquera, O., Kiperstok, A., Sales, E. A., Embiruçu, M., and Ghirardi, M. L. (2010). Comparative energy life-cycle analyses of microalgal biomass production in open ponds and photobioreactors. *Bioresour. Technol.* 101, 1406–1413. doi:10.1016/j.biortech.2009.09.038
22. Kochem, L. H., Da Fré, N. C., Redaelli, C., Rech, R., and Marcílio, N. R. (2014). Characterization of a novel flat-panel airlift photobioreactor with an internal heat exchanger. *Chem. Eng. Technol.* 37, 59–64. doi:10.1002/ceat.201300420
23. Kumar, K., Dasgupta, C. N., Nayak, B., Lindblad, P., and Das, D. (2011). Development of suitable photobioreactors for CO2 sequestration addressing global warming using green algae and cyanobacteria. *Bioresour. Technol.* 102, 4945–4953. doi:10.1016/j.biortech.2011.01.054
24. Le, P. J., Williams, B., and Laurens, L. M. L. (2010). Microalgae

as biodiesel & biomass feedstocks: review & analysis of the biochemistry, energetics & economics. *Energy Environ. Sci.* 3, 554–590. doi:10.1039/b924978h

25. Mata, T. M., Martins, A. A., and Caetano, N. S. (2010). Microalgae for biodiesel production and other applications: a review. *Renew. Sustain. Energ. Rev.* 14, 217–232. doi:10.1016/j.rser.2009.07.020
26. Molina, E., Fernández, J., Acién, F. G., and Chisti, Y. (2001). Tubular photobioreactor design for algal cultures. *J. Biotechnol.* 92, 113–131. doi:10.1016/S0168-1656(01)00353-4
27. Münkel, R., Schmid-Staiger, U., Werner, A., and Hirth, T. (2013). Optimization of outdoor cultivation in flat panel airlift reactors for lipid production by *Chlorella vulgaris*. *Biotechnol. Bioeng.* 110, 2882–2893. doi:10.1002/bit.24948
28. Muñoz, R., and Guieysse, B. (2006). Algal-bacterial processes for the treatment of hazardous contaminants: a review. *Water Res.* 40, 2799–2815. doi:10.1016/j.watres.2006.06.011
29. Norsker, N.-H., Barbosa, M. J., Vermuë, M. H., and Wijffels, R. H. (2011). Microalgal production – a close look at the economics. *Biotechnol. Adv.* 29, 24–27. doi:10.1016/j.biotechadv.2010.08.005
30. NREL, . (0000). *Biomass Research – Publications*. Available at: http://www.nrel.gov/biomass/publications.html?print
31. Pruvost, J., Cornet, J. F., Goetz, V., and Legrand, J. (2011). Modeling dynamic functioning of rectangular photobioreactors in solar conditions. *AIChE J.* 57, 1947–1960. doi:10.1002/aic.12389
32. Radmann, E. M., Reinehr, C. O., and Costa, J. A. V. (2007). Optimization of the repeated batch cultivation of microalga *Spirulina platensis* in open raceway ponds. *Aquaculture* 265, 118–126. doi:10.1016/j.aquaculture.2007.02.001
33. Rawat, I., Ranjith Kumar, R., Mutanda, T., and Bux, F. (2013). Biodiesel from microalgae: a critical evaluation from laboratory to large scale production. *Appl. Energy* 103, 444–467. doi:10.1016/j.apenergy.2012.10.004
34. Robinson, D., and Stone, A. (2004). Solar radiation modelling in the urban context. *Solar Energy* 77, 295–309. doi:10.1016/j.solener.2004.05.010

35. Rodolfi, L., Chini Zittelli, G., Bassi, N., Padovani, G., Biondi, N., Bonini, G., et al. (2009). Microalgae for oil: strain selection, induction of lipid synthesis and outdoor mass cultivation in a low-cost photobioreactor. *Biotechnol. Bioeng.* 102, 100–112. doi:10.1002/bit.22033

36. Ruiz, J., Álvarez-Díaz, P. D., Arbib, Z., Garrido-Pérez, C., Barragán, J., and Perales, J. A. (2013). Performance of a flat panel reactor in the continuous culture of microalgae in urban wastewater: prediction from a batch experiment. *Bioresour. Technol.* 127, 456–463. doi:10.1016/j.biortech.2012.09.103

37. Sierra, E., Acién, F. G., Fernández, J. M., García, J. L., González, C., and Molina, E. (2008). Characterization of a flat plate photobioreactor for the production of microalgae. *Chem. Eng. J.* 138, 136–147. doi:10.1016/j.cej.2007.06.004

38. Sills, D. (2013). *Modeling CO2 Requirements for Cultivation of Microalgae in Open Raceway Pond.*

39. Singh, N. K., and Dhar, D. W. (2011). Microalgae as second generation biofuel. A review. *Agron. Sustain. Dev.* 31, 605–629. doi:10.1007/s13593-011-0018-0

40. Slegers, P. M., Lösing, M. B., Wijffels, R. H., van Straten, G., and van Boxtel, A. J. B. (2013). Scenario evaluation of open pond microalgae production. *Algal. Res.* 2, 358–368. doi:10.1016/j.biortech.2012.11.123

41. Slegers, P. M., Wijffels, R. H., van Straten, G., and van Boxtel, A. J. B. (2011). Design scenarios for flat panel photobioreactors. *Appl. Energy* 88, 3342–3353. doi:10.1016/j.apenergy.2010.12.037

42. Sompech, K., Chisti, Y., and Srinophakun, T. (2012). Design of raceway ponds for producing microalgae. *Biofuels* 3, 387–397. doi:10.4155/bfs.12.39

43. Stephenson, A. L., Kazamia, E., Dennis, J. S., Howe, C. J., Scott, S. A., and Smith, A. G. (2010). Life-cycle assessment of potential algal biodiesel production in the United Kingdom: a comparison of raceways and air-lift tubular bioreactors. *Energy Fuels* 24, 4062–4077. doi:10.1021/ef1003123

44. Sugai-Guérios, M. H., Mariano, A. B., Vargas, J. V. C., de Lima Luz, L. F., and Mitchell, D. A. (2014). Mathematical model of

the CO_2 solubilisation reaction rates developed for the study of photobioreactors. *Can. J. Chem. Eng.* 92, 787–795. doi:10.1002/cjce.21937

45. Sukhatme, K., and Sukhatme, S. P. (1996). *Solar Energy: Principles of Thermal Collection and Storage*. Noida, UP: Tata McGraw-Hill Education.
46. Yang, A. (2011). Modeling and evaluation of CO_2 supply and utilization in algal ponds. *Ind. Eng. Chem. Res.* 50, 11181–11192. doi:10.1021/ie200723w

CHAPTER

13

BIOMASS PRODUCTIVITY AND PRODUCTIVITY OF FATTY ACIDS AND AMINO ACIDS OF MICROALGAE STRAINS AS KEY CHARACTERISTICS OF SUITABILITY FOR BIODIESEL PRODUCTION

Niels Hempel[1] Ingolf Petrick[1] and Frank Behrendt[2]

[1]Faculty of Natural Sciences, Lausitz University of Applied Sciences, Senftenberg, Germany

[2] Department of Energy Engineering, Berlin Institute of Technology, Berlin, Germany

ABSTRACT

Microalgae are discussed as an alternative source for the production of biofuels. The lipid content compared to cultivation time of used species is the main reason for any choice of a special strain. This paper reviews more analytical data of 38 screened microalgae strains. After the cultivation period, total content of lipids was analysed. The extracted fatty acids were quantified as fatty acid methyl esters by GC analysis. The amino acids were analysed by HPLC. *Chlorella* sp., *Chlorella saccharophila*,

Chlorella minutissima and *Chlorella vulgaris* were identified as species with the highest productivity of fatty acids relevant to transesterification reactions. The components were mainly linoleic acid, palmitic acid and oleic acid. To increase productivity of highly saturated fatty acids, cultivation parameters light intensity and temperature were varied. In this manner, the ideal conditions for biodiesel production were defined in this publication.

Keywords: Amino acids Biodiesel *Chlorella* Fatty acid productivity Species selection

INTRODUCTION

For environmental and economic sustainability of the world, renewable and carbon-neutral biofuels are necessary (Patil et al. 2008). Compared to conventional energy sources, alternative possibilities have only a minor economic basis. Today, optimisation of production processes is the main reason for increasing research activities in this area. A multi-faceted approach for an energy generation that includes biofuels as well as nuclear, solar, hydrogen, wind and fossil fuels (from which carbon is sequestered) will provide the backbone for industries in any future world (Hoffart et al. 2002; Pacala and Socolow 2004; Patil 2007). Alternative routes will reduce greenhouse gases such as carbon dioxide (CO_2), sulphur dioxide (SO_2) and nitrogen oxides (NOx) and help to maintain our high living standards in a less negatively affected environment.

Besides economic reasons, consumption of vast swathes of farmland and native habitats with increasing food prices are a major criticism against biomass usage for large-scale production (Crutzen et al. 2007; Pacala and Socolow 2004; Righelato and Spracklen 2007), the usage of agricultural crops alone unable to satisfy the entire demand. Oil palm crops, with their high content of oil, used for 50% of US transport fuel, would need 24% of the total cropland area of the USA (Chisti 2007). The resulting 'food versus fuel' argument is already widely discussed. In contrast, serious studies postulate that biofuels are able to reach an amount of 30% of global demand in an environmentally responsible manner without affecting food production (Koonin 2006).

Biomass is principally able to be used for production of various energy storaging substrates such as biodiesel, bioethanol, biomethane and biohydrogen, with only biodiesel and bioethanol produced to industrial quantities at this time (Patil et al. 2008). First-generation biofuels have already been displaced by second-generation biofuels produced by non-food feedstocks. Lignocellulosic technologies and microalgae play an important role in this manner (Chisti 2007; Hankamer et al. 2007; Kruse et al. 2005; Schaub et al. 2007). A wide range of advantages shift the focus to microalgae biomass. Microalgae are miniature biochemical factories using similar photosynthetic processes in the same manner as higher plants for generation of energy and intermediates for metabolism. Photosynthetic efficiency of microalgae, however, is notably higher than in terrestrial plants (Pirt 1986; Shay 1993). Microalgae already reach up to 300 times more oil productivity for biodiesel production than traditional crops on an area basis (Chisti 2007; Schenk et al. 2008). Alongside these advantages, microalgae assimilate CO2 into their biomass resulting in a reduction of concentrations of greenhouse gasses (Brown and Zeiler 1993). Furthermore, microalgae can be harvested in continuous processes (1–10 days depending on culture volume and culture techniques) enabling a high efficiency of cultivation processes. In contrast, conventional crop plants usually can only be harvested once or twice a year. Growing of crops is only possible in vegetation periods dependent on climatic and geographic requirements. Cultivation of microalgae can potentially be carried out on marginal or non-arable land allowing the possibility of opening up new economic opportunities for arid, drought- or salinity-affected regions (Chisti 2007). Greater photosynthetic efficiency results in reduced fertiliser and nutrient inputs and less waste and pollution, with the usage of waste water instead of drinking water for algal cultivation also being a viable option (Hammouda et al. 1995; Mallick 2002; Muñoz and Guieysse 2006). Compared to conventional agricultural production, the evaporation loss of fresh water is reduced during algal cultivation in closed systems. In addition, water can be used in a cycling system, thus considerably increasing water balance. Cultivation systems for microalgae can consist of open ponds, bubble columns, flat-plate photobioreactors (PBRs) or tubular PBRs (Borowitzka 1997; Ugwu et al. 2008; Xu et al. 2009). For large-scale production, PBRs are the most promising systems because of low contamination risk, low space requirement, almost no CO_2 loss, high adaptability to many species and high biomass concentration (Pulz 2001; Richmond 2000). Growing of microalgae strains in closed systems under defined conditions allows usage of biomass for other applications such as production of foods. The utilisation of amino acids of algae biomass alongside fatty acids increases the economic basis of the biodiesel production process. However, for user acceptance, microalgal biodiesel needs to comply with existing standards, such as ASTM Biodiesel Standard D 6751 (USA) or Standard EN 14214 (European Union) meaning a high cetane number is therefore required. The cetane number is directly related to good performance characteristics of biofuel, showing ignition quality in an engine (Van Gerpen 2009). Highly saturated fatty acids give an excellent cetane number and oxidative stability to biodiesel (Chinnasamy et al. 2010). The content of such kinds of fatty acids is dependent on species-specific enzyme configuration of microalgae strains, with cultivation conditions also having a possible influence (Sato et al. 1979).

The first step in developing such an algal process is the screening of various strains to choose the most suitable species (Pulz and Gross 2004) with a high productivity of fatty acids. Relevant properties for specific culture conditions, as well as the additional identification of high valuable byproducts, play an important role. In this publication, the screening and characterisation results of 38 microalgae strains are given for the identification of potent biodiesel producers.

MATERIALS AND METHODS

Strains and culture media used in this work are listed in Table 1. The strains were obtained from UTEX Culture Collection of Algae, Austin; SAG Culture Collection of Algae, Göttingen; SAS Culture Collection of Algae, Senftenberg; MACC Algal Culture Collection, Mosonmagyaróvár; and IPPAS Culture Collection of Microalgae, Moscow. The following culture media were used:

Table 1: List of laboratory algal strains (screened strains and reference strain), cultivation media and culture collection

Medium[a]	Algal strain (strain numbers; culture collection)
T	Chlorella fusca strain 251 (UTEX, Austin)
T	Chlorella minutissima strains 444; 452; 494 (MACC, Mosonmagyaróvár)
T	Chlorella saccharophila strains 363; 477 (MACC, Mosonmagyaróvár)

T	Chlorella sp. strains 03; 04; 11; 17-1; 18 (SAS, Senftenberg)
T	Chlorella sp. strains 4; 313; 318; 391; 418; 459; 474; 550; 552; 572; 589; 732; 800 (MACC, Mosonmagyaróvár)
T	Chlorella vulgaris strains 125; 132[b] (SAS, Senftenberg)
T	Chlorella vulgaris strains 211-1b; 211-11f; 211-11j (SAG, Göttingen)
T	Chlorella vulgaris strain C1 (IPPAS, Moscow)
T	Chlorella zofingiensis strain 133 (SAS, Senftenberg)
B	Chlorococcum ellipsoideum strain 33 (MACC, Mosonmagyaróvár)
B	Cosmarium sp. strain 25 (SAS, Senftenberg)
T	Neochloris sp. strain 421 (MACC, Mosonmagyaróvár)
B	Pediastrum boryanum strain 39 (SAS, Senftenberg)
B	Phaeodactylum tricornutum strain 1090 (SAG, Göttingen)
T	Scenedesmus rubescens strain 5.95 (SAG, Göttingen)
S	Spirulina maxima strain 20 (SAS, Senftenberg)
S	Spirulina platensis strain 85.79 (SAG, Göttingen)

[a]Key to media: *B* modified BG11, *S* spirulina, *T* modified Tamiya

[b]Reference strain

Modified BG11 medium contained (g L^{-1}): $NaNO_3$, 1.5; $MgSO_4{\cdot}7H_2O$, 0.075; K_2HPO_4, 0.035; $CaCl_2{\cdot}2H_2O$, 0.036; Na_2CO_3, 0.02; ferric ammonium citrate, 0.006; citric acid, 0.006; Na_2EDTA, 0.001; and trace elements solution, 1 mL L^{-1}. The trace elements solution contained (g L^{-1}): H_3BO_3, 0.061; $MnSO_4{\cdot}H_2O$, 0.169; $ZnSO_4{\cdot}7H_2O$, 0.287; $CuSO_4{\cdot}5H_2O$, 0.0025; and $(NH_4)_6Mo_7O_{24}{\cdot}4H_2O$, 0.0125.

Spirulina medium contained (g L^{-1}): $NaHCO_3$, 6.8; Na_2CO_3, 2.0; K_2HPO_4, 0.25; $NaNO_3$, 1.25; K_2SO_4, 0.5; NaCl, 0.5; $MgSO_4{\cdot}7H_2O$, 0.1; $CaCl_2{\cdot}2H_2O$, 0.02; $FeSO_4{\cdot}7H_2O$, 0.005; Na_2EDTA, 0.04; cyanocobalamine, 0.000005; and trace elements solution, 5 mL L^{-1}. The trace elements solution contained (g L^{-1}): $ZnSO_4{\cdot}7H_2O$, 0.001; $MnSO_4{\cdot}4H_2O$, 0.002; H_3BO_3, 0.01; $Co(NO_3)_2{\cdot}6H_2O$, 0.001; $Na_2MoO_4{\cdot}2H_2O$, 0.001, $CuSO_4{\cdot}5H_2O$, 0.000001; $FeSO_4{\cdot}7H_2O$, 0.7; and Na_2EDTA, 0.8.

Modified Tamiya medium contained (g L^{-1}): KNO_3, 2.5; KH_2PO_4, 0.625; $MgSO_4{\cdot}7H_2O$, 1.25; $FeSO_4{\cdot}7H_2O$, 0.0045; Na_2EDTA, 0.0186; and trace elements solution, 0.5 mL L^{-1}. The trace elements solution contained (g L^{-1}): H_3BO_3, 2.86; $MnCl_2{\cdot}4H_2O$, 1.81; $ZnSO_4{\cdot}7H_2O$, 0.222; NH_4VO_3, 0.023; and MoO_3, 0.018.

Each microalgal strain was precultured in Erlenmeyer flasks: temperature 25°C; shaken at 80 rpm. For growth assay, 0.2 L culture suspension with dry weight (dw) of 2.0 $g L^{-1}$ was inoculated in 2 L glass cylinder photobioreactors containing 1.8 L liquid medium (Table 1). Sterile filtered air enriched with 2% (v/v) CO_2 was continuously pumped through special nozzles and glass tubes into the bottom of a photobioreactor. The aeration was performed at 0.33 $L min^{-1}$ for 24 h. Cultures were incubated at 25°C with continuous stirring. The photobioreactors were illuminated with eight horizontally fixed 40 W white lamps which were placed behind the photobioreactors. The light intensity was approximately that of 200 μmol photons m^{-2} s^{-1}. Each strain was cultivated in duplicate. To exclude a nitrate or phosphate limitation, the contents were determined every 2 or 3 days by using special test kits (Machery-Nagel, Germany). For a good comparability of results, all screened microalgae strains were harvested at 3 g L^{-1} by centrifugation at 4,050 × *g* for 10 min and washed with distilled water. Cell pellets were freeze-dried and stored in polypropylene cups until further analysis was conducted.

Biomass assay

Cell growth was monitored by gravimetric determination of algal biomass dry weight (dw) with each determination made in duplicate every 2 or 3 days. Aliquots of 10 mL algal suspension were centrifuged at 4,050 × *g* for 10 min, washed with distilled water to remove the salt, dried at 105°C for 24 h to a constant weight, cooled in a desiccator and weighed. The growth curves were determined from parallel cultures starting from the same inocula and were calculated from average algal biomass as function of time.

Volumetric biomass productivity $P_{Biomass}$ was calculated by $P_{Biomass}(g\,L^{-1}day^{-1}) = (X_2 - X_1) \cdot (t_2 - t_1)^{-1}$ where X_1 and X_2 were the biomass dw concentrations (g L^{-1}) on days t_1 (startpoint of cultivation) and t_2 (endpoint of cultivation), respectively, and was given as an average productivity.

Analysis of lipid content and fatty acid profile

Lipid content of algae probes was determined gravimetrically. One gram of freeze-dried biomass was resuspended in 6 mL of chloroform/methanol (2:1, v/v) and transferred to a grinding beaker with 4 g steel balls. The algae cells were disrupted in the ball mill (Retsch, Germany) at full activity (1,800 min^{-1}) for 25 min and afterwards centrifuged at $4{,}050 \times g$ for 10 min. Following centrifugation, the supernatant was filtered and collected. The rotary evaporator was then used to separate the chloroform and methanol from the dissolved oil fraction, with the residual oil being dried in an oven until constant weight was reached.

The amount of fatty acids was estimated as the amount of FAMEs, which was determined by direct transesterification. The dried lipid extract was dissolved in 10 mL *n*-heptane, and 10 mL 0.5 N methanolic KOH was added with the resulting mixture shaken for 30 min. Aliquots of upper phase with FAMEs were analysed using standard gas chromatography with a capillary column (100 m, 0.25 mm, 0.2 µm; SP™-2560; Supelco, USA) and a flame ionisation detector (7820 A, Agilent Technologies, Germany). H_2 at 35 mL min^{-1} was used as carrier gas. Temperature was programmed to increase from 50 to 65°C at 2°C min^{-1} then to 180°C at 30°C min^{-1} and thereafter ramped to 220°C at 15°C min^{-1}. The injector and detector were kept at 220°C and 250°C, respectively. The amount of FAMEs was quantified to comparison to a standard solution containing 37 FAMEs (Supelco, USA).

Determination of amino acid profile

Amino acids were quantified using HPLC System S433 (Sykam, Germany). Then 100 mg of freeze-dried biomass was resuspended in 5 mL 1 NHCl with 1% (w/v) phenol and was rinsed with nitrogen gas for 1 min. Solutions were dried at 110°C for 24 h and filtered. Probes were evaporated, dissolved in a sample solution buffer and injected on a cation separation column (4.6 × 150 mm, LCA K06/Na; Sykam, Germany) and were detected at 440 nm and at 570 nm. The analyses were run under the following conditions: analysis cycle time 60 min; flow rates 0.45 mL min^{-1} for buffer and 0.25 mL min^{-1} for ninhydrin. The amino acid content is given as the summed content of alanine, asparagine, aspartic acid, arginine, cysteine, glutamic acid, glutamine, glycine, histidine, isoleucine, leucine, lysine, methionine, phenylalanine, proline, serine, threonine, tyrosine and valine. Based on acidic hydrolysis, the content of asparagine as well as aspartic acid and the content of glutamine and glutamic acid are given as the summed content of both amino acids.

Analytics and calculation of productivities

All determinations were made in duplicate and were given as average content or average productivity. To exclude drift of analytical data, in addition to internal controls, the reference strain *Chlorella vulgaris* 132 was analysed every 3 months.

Productivity of lipids, fatty acids and amino acids were calculated by

$$P_{\text{lipids, fatty acids, amino acids}}(\text{mg L}^{-1}\text{day}^{-1})=P_{\text{Biomass}}\cdot Cf$$

where $P_{\text{lipids, fatty acids, amino acids}}$ is productivity of lipids, fatty acids or amino acids; P_{Biomass} is productivity of biomass; and C_f is the final content of lipids, fatty acids or amino acids and were given as percent dry weight.

For identification of productivities influencing factors in this publication, various analytical parameters were correlated. For linear trendlines, added to the data series, coefficients of determination (r^2) were determined and the levels of mutual influences were estimated.

RESULTS

Biomass productivity

To test the suitability of microalgae for biodiesel production, screening data were collected for 35 strains of the green algae (Chlorophyta), two strains of the blue-green algae (cyanobacteria) and one strain of diatoms (Bacillariophyta). For all strains, growth curves were determined. No lag phase was observed, possibly because exponentially growing cells of precultures were used as inocula. Because of stagnating growth, *Phaodactylum tricornutum*, *Spirulina maxima* and *Spirulina platensis* were harvested prematurely (end-dw ranged between 1.0 and 2.2 g L^{-1}) before reaching the end concentration of 3 g L^{-1}. The stationary growth phase was not reached in any case, indicating no limiting conditions. Because of linearity of all growth curves, the biomass productivity was calculated over the whole cultivation period.

Figure 1 shows biomass productivity of all tested strains. *Chlorella* sp. 800, *Chlorella* sp. 313 and *Chlorella minutissima* 494 were identified as fast-growing strains with the highest biomass productivity. For *Chlorella* sp. 800, biomass productivity of 0.495 g L^{-1} day^{-1} was calculated, for *Chlorella* sp. 313 of 0.451 g L^{-1} day^{-1} and for *C. minutissima* 494 of

0.396 g L^{-1} day^{-1}. Reference strain *C. vulgaris* 132 reached values of between 0.38 and 0.435 g L^{-1} day^{-1}, indicating a good validity of growth data.

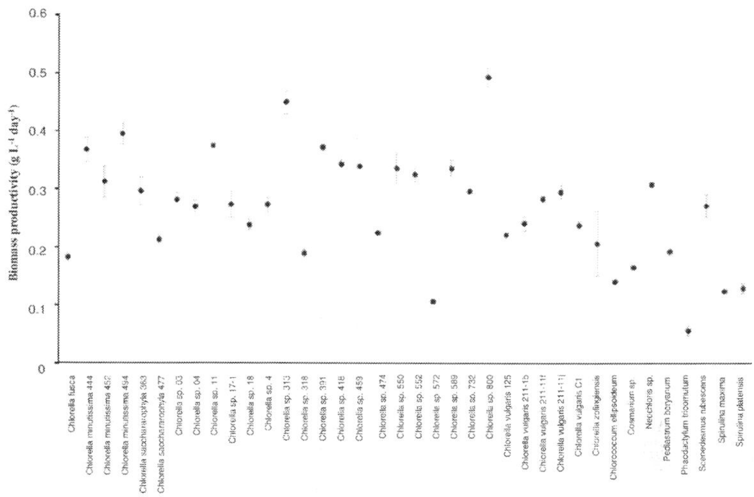

Figure 1: Average specific biomass productivity of 38 screened micro-algae strains. For reference strain *Chlorella vulgaris* 132, the biomass productivity ranged between 0.38 and 0.435 g L^{-1} day^{-1} over the whole test period. *Error bars* show the minimum and the maximum recorded for each species

To increase biomass productivity of *Chlorella* sp. 800, the influence of light intensity and cultivation temperature was examined. At varied light intensity, the biomass productivity was 0.423 g L^{-1} day^{-1} at 200 µmol photons m^{-2} s^{-1}, 0.475 g L^{-1} day^{-1} at 300 µmol photons m^{-2} s^{-1}, 0.62 g L^{-1} day^{-1} at 400 µmol photons m^{-2} s^{-1} and 0.707 g L^{-1} day^{-1} at 500 µmol photons m^{-2} s^{-1}. A continuous increase of biomass productivity was postulated, and therefore a saturation effect of light was not detected. In further studies, a cultivation temperature of 25°C was identified as optimum. The biomass productivity was 0.538 g L^{-1} day^{-1} compared to 0.435 g L^{-1} day^{-1} at 35°C and 0.34 g L^{-1} day^{-1} at 15°C.

Lipid productivity and productivity of fatty acids

Strains with the highest lipid content were *Chlorella* sp. 589 (30.2% dw), *Chlorella saccharophila* 477 (27.6% dw) and *Chlorella* sp. 800 (24.4%

dw). For reference strain, analysed lipid content ranged between 18.5% and 23.3% dw. The resulting lipid productivity (Fig. 2) was recorded as follows: Based on both high biomass productivity and high lipid content, *Chlorella* sp. 800 was identified as the strain with the highest lipid productivity of 121 mg L^{-1} day^{-1}. Strain *Chlorella* sp. 589 reached 102 mg L^{-1} day^{-1} and *Chlorella* sp. 459 72.5 mg L^{-1} day^{-1}. For reference strain *C. vulgaris* 132, over the whole test period, lipid productivity between 80.5 and 88.5 mg L^{-1} day^{-1} was measured.

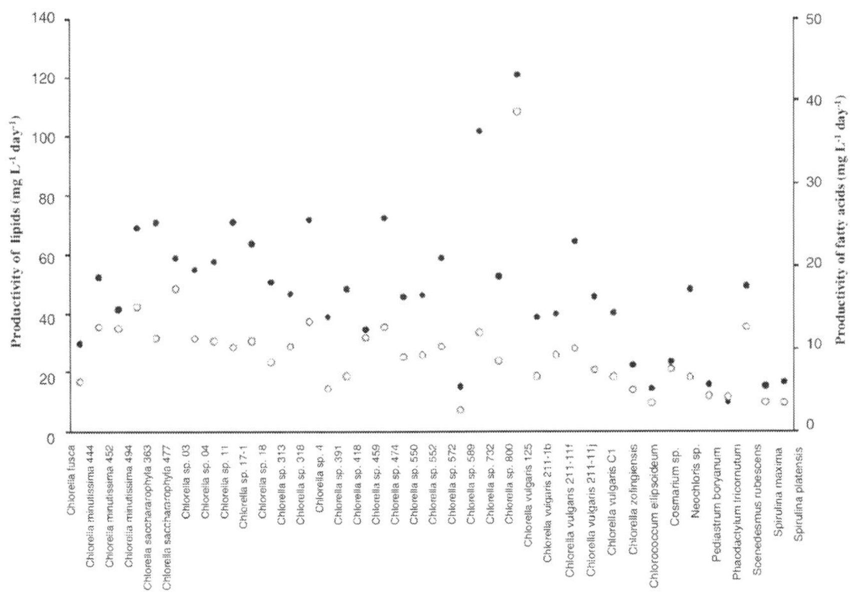

Figure 2: Specific lipid productivity (*filled circles*) and specific productivity of fatty acids (*open circles*) of 38 screened microalgae strains. For reference strain *Chlorella vulgaris* 132, the lipid productivity ranged between 80.5 and 88.5 mg L^{-1} day^{-1} over the whole test period and the productivity of fatty acids ranged between 12.6 and 15.6 mg L^{-1} day^{-1} over the whole test period

For estimation of content of fatty acids in screened microalgae strains, clear identifiable fatty acids were summarised. *C. saccharophila* 477 reached a content of 8.1% dw for fatty acids, *Chlorella* sp. 800 a content of 7.8% dw and *P. tricornutum* 7.2% dw. These strains were identified as species with the highest content of fatty acids of all screened strains. Despite 37 detectable fatty acids, only six fatty acids formed the main fraction: palmitic (C16:0), palmitoleic (C16:1), stearic (C18:0), oleic

(C18:1ω9), linoleic (C18:2ω6) and α-linolenic (C18:3ω3). Summarised content of these fatty acids give more than 90% of all fatty acids. For most productive strains, these contents and productivities are given in Table 2. Differences were detected in contents of linoleic acid and α-linolenic acid. Strain *Chlorella* sp. 800 contained high amount of tri-unsaturated α-linolenic acid compared to di-unsaturated linoleic acid, indicating a high desaturase activity. In contrast, the remaining most productive strains had low comparable desaturase activity based on quantified fatty acids. Besides main fraction of fatty acids in six strains, divergent spectra of fatty acids were assumed. The reason for this phenomenon was founded in abnormal content of special fatty acids. High content of trans-configured linolelaidic acid (C18:2ω6t) was measured in *Chlorella* sp. 391, in *Cosmarium* sp. and in *Chlorococcum ellipsoideum*. For γ-linolenic acid (C18:3ω6), high content was detected in *C. ellipsoideum*, of *Cosmarium* sp. and of *S. platensis*. Extreme low content of linoleic acid (C18:2ω6) was the reason for the discrepancy in main spectrum of fatty acids in *Chlorella* sp. 391 and in *Chlorella zofingiensis*. In *P. tricornutum* biomass, a high amount of eicosapentaenoic acid (C20:5ω3) was detected.

Table 2: Content and productivity of fatty acids palmitic, palmitoleic, stearic, oleic, linoleic and α-linolenic of five microalgal strains

Fatty acid	Strains with highest productivity of fatty acids				
	***Chlorella*sp. 800**	***Chlorella sacchar-ophila*477**	***Chlorella minutis-sima*494**	***Chlorella*sp. 313**	***Chlorella minutis-sima*444**
Palmitic (C16:0)					
Content (% dw)	1.86	1.56	1.18	1.08	1.00
Productivity (mg L^{-1} day^{-1})	9.21	3.33	4.67	4.87	3.68
Palmitoleic (C16:1)					
Content (% dw)	0.055	0.045	0.037	0.032	0.041
Productivity (mg L^{-1} day^{-1})	0.27	0.096	0.15	0.14	0.15
Stearic (C18:0)					
Content (% dw)	0.60	0.35	0.036	0.04	0.04
Productivity (mg L^{-1} day^{-1})	2.97	0.74	0.14	0.18	0.15
Oleic (C18:1ω9)					
Content (% dw)	2.59	1.86	0.18	0.16	0.14
Productivity (mg L^{-1} day^{-1})	12.80	3.96	0.71	0.72	0.52
Linoleic (C18:2ω6)					
Content (% dw)	1.97	4.04	2.05	1.41	1.95
Productivity (mg L^{-1} day^{-1})	9.80	8.61	8.09	6.36	7.17
α-Linolenic (C18:3ω3)					
Content (% dw)	0.67	0.19	0.25	0.19	0.23
Productivity (mg L^{-1} day^{-1})	3.31	0.40	0.99	0.86	0.85
Sum of fatty acids					
Content (% dw)	7.75	8.04	3.73	2.91	3.40
Productivity (mg L^{-1} day^{-1})	38.30	17.14	14.78	13.13	12.52

The determination of volumetric productivity of summarised fatty acids allows prediction of amount of synthesised fatty acids, both per day and per litre. Results for screening strains are given in Fig. 2. Strain *Chlorella* sp. 800 was identified as the top fatty acid producer. The productivity of fatty acids was 38.7 mg L^{-1} day^{-1}, based on both fast-growing and high content of fatty acids. The next most productive strains were *C. saccharophila* 477 (17.3 mg L^{-1} day^{-1}) and *C. minutissima* 494 (15.1 mg L^{-1} day^{-1}). Reference strain also reached high fatty acid productivity, 12.6–15.6 mg L^{-1} day^{-1} over the whole test period.

For increasing productivity of fatty acids, cultivation temperature and light intensity were varied for top fatty acid producer strain *Chlorella* sp. 800. Cultivation at other temperatures than at room temperature resulted in a decreased fatty acid productivity: Compared to 27.4 mg L^{-1} day^{-1}, value reached in these experiments for cultivation temperature of 25°C, at 15°C fatty acid productivity decreased by 54% and at 35°C by 64%. At increasing light intensity, an increase of fatty acid productivity was postulated: At 500 μmol photons m^{-2} s^{-1}, the fatty acid productivity increased by 26% compared to cultivation at 200 μmol photons m^{-2} s^{-1}.

Variation of cultivation parameters and influence to quality of biodiesel

To increase the suitability of a single microalgae strain for biodiesel production, varied cultivation temperature and light intensity were tested, with both high fatty acid productivity and a good quality of biodiesel being equally necessary. A chain length of fatty acids between 16 and 18 C atoms and a low content of unsaturated fatty acids are optimal. At varied cultivation temperature, the spectrum of synthesised fatty acids changed significantly: If growth temperature for *Chlorella* sp. 800 was lowered from 35 to 15°C increased relative content compared to total fatty acids of threefold unsaturated α-linolenic acid (C18:3ω3) from 6.2 to 23.4% was found (Fig. 3). The content of linoleic acid (C18:2ω6) was decreased in the same manner from 39.1 to 34.0% and of palmitic acid (C16:0) from 40.7 to 24.9%. The content of other fatty acids was nearly constant (Fig. 3).

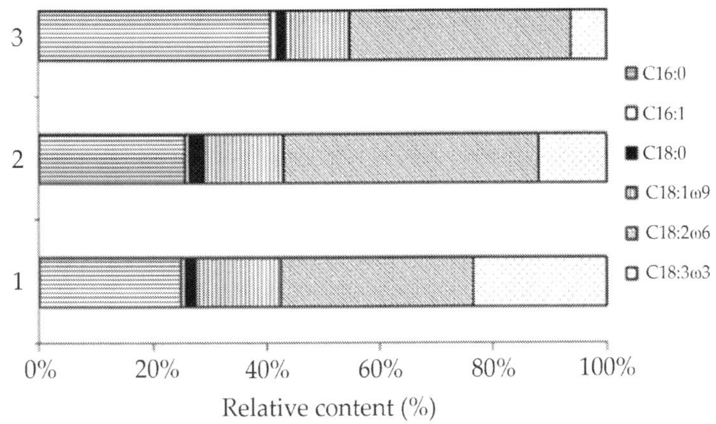

Figure 3: Spectrum of fatty acids palmitic (C16:0), palmitoleic (C16:1), stearic (C18:0), oleic (C18:1ω9), linoleic (C18:2ω6) and α-linolenic (C18:3ω3) given as relative amounts to 100% in *Chlorella* sp. 800 under different cultivation temperatures: 15°C (*1*), 25°C (*2*), 35°C (*3*)

Summarising these effects in *Chlorella* sp. 800, desaturation of fatty acids was low at a high cultivation temperature. A decrease of the average number of double bonds per fatty acid molecule (1.54 at 15°C, 1.41 at 25°C, 1.09 at 35°C, Fig. 4) was postulated. The chain length of fatty acids was lower at each increase in cultivation temperature. The quotient of summarised C16 fatty acids and summarised C18 fatty acids increased from 0.34 at 15°C to 0.36 at 25°C and to 0.72 at 35°C, mainly a result of increasing content of palmitic acid (Fig. 4). Despite the effect of cultivation temperature, no influence of light intensity on the number of double bonds per fatty acid molecule was detectable. Increasing light intensity tended to a slight shortening of the chain length of fatty acids. Alongside these effects, the relative amount of fatty acids was largely unchanged.

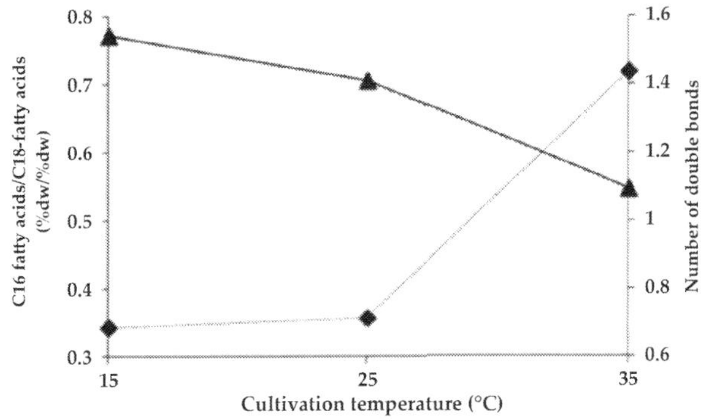

Figure 4: Effect of cultivation temperature on the number of double bonds per fatty acid molecule (*solid line*) and of quotient of summarised C16 fatty acids and summarised C18 fatty acids (*dashed line*) in strain *Chlorella* sp. 800

Productivity of amino acids

As a result of screening experiments, four strains with an amino acids content of more than 40% dw were identified. The blue-green alga *S. platensis* reached a protein content of 46.8% dw and *S. maxima* of 44.9% dw. For green algae *Chlorella* sp. 589, a protein content of 44.3% dw and for *C. saccharophila* 363 a protein content of 42.4% dw were found. Protein content of the reference strain reached values of between 35.7% and 41.9% dw. The main fraction of amino acids consisted of alanine, arginine, aspartic acid, glutamic acid, lysine and leucine. Relative amounts of single amino acids were nearly similar in all screened microalgae strains. The spectrum of main fraction amino acids in the five most productive strains. The results indicate that no differences, or only minimal differences, exist in anabolic pathways for amino acids. Results for productivity of amino acids of all screened strains are given in Fig. 5. Outstanding productivity was reached by the reference strain (155–159 mg L^{-1} day^{-1}). The next most productive strains were *Chlorella* sp. 589 (149 mg L^{-1} day^{-1}), *Chlorella* sp. 313 (145 mg L^{-1} day^{-1}) and *C. minutissima* 494 (137 mg L^{-1} day^{-1}).

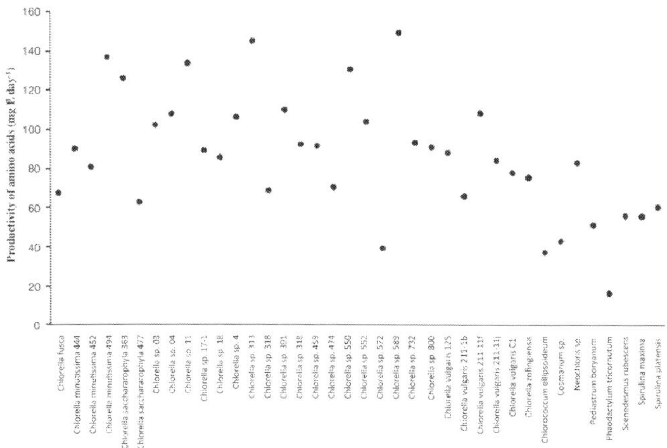

Figure 5: Specific amino acid productivity of 38 screened microalgae strains. For reference strain *Chlorella vulgaris* 132, the productivity of amino acids ranged between 155 and 159 mg L^{-1} day^{-1} over the whole test period

DISCUSSION

Growth data in this paper concur with already published data: Rodolfi et al. (2008) screened 30 microalgae strains and obtained biomass productivity between 0.04 and 0.37 g L^{-1} day^{-1}. Compared to biomass productivity in this study, ranging from 0.06 g L^{-1} day^{-1} (*P. tricornutum*) to 0.495 g L^{-1} day^{-1} (*Chlorella* sp. 800), a good accordance is given. For top biomass producers *Chlorella* sp. and *C. minutissima*, a high practicable suitability for biodiesel production was postulated in literature (Rasoul-Amini et al. 2011; Tang et al. 2011). In addition, *Chlorella*strains have a high fixation capacity for CO_2 from flue gas generated by combustion processes (Doucha et al. 2005). For *C. minutissima*, a high pollution tolerance has been reported (Bhatnagar et al. 2010) indicating a plausible role in wastewater remediation. The usage of wastewater and industrial fumes for cultivation processes will clearly reduce the production costs of biodiesel production.

In order to test if high biomass productivity excludes high amounts of lipids, fatty acids or amino acids, corresponding correlation analyses for all screened strains were calculated, with no relationship between these data detected. The top biomass producer strain *Chlorella* sp. 800, for example, also reached the highest content of fatty acids as well as the third highest lipid content. The r^2 for all tested strains, all tested in comparison to biomass productivity, was 0.08 for lipids, 0.00 for fatty acids and 0.05 for amino acids. In literature, sometimes a strong negative correlation between lipid content and biomass productivity has been postulated based on the high metabolic cost of lipid biosynthesis (Rodolfi et al. 2008). However, based on screening results of this study, no such effect was detectable.

Strains with highest lipid productivity

Absolute lipid content of screened microalgae strains ranged from 8% dw (*Pediastrum boryanum*) to 30% dw (*Chlorella* sp. 589) with an average lipid content of 17% dw. In a review paper, Griffiths and Harrison (2009) reported screened microalgae strains with a lipid content of between 5% and 55% dw with an average lipid content of 23% dw. This higher average content was reached due to nine strains (of the 55 tested) containing extremely high lipid contents of more than 30% dw. In this study, there was no possibility of testing these lipid-rich strains. Lipid productivity, of all tested strains, ranged from 10 mg L^{-1} day^{-1} (*P. tricornutum*) to 121 mg L^{-1} day^{-1} (*Chlorella* sp. 800). Studies of existing literature show comparable lipid productivities. Huerlimann et al. (2010) reported lipid productivities between 2 and 23 mg L^{-1} day^{-1} in tested microalgae strains and Rodolfi et al. (2008) between 17 and 61 mg L^{-1} day^{-1}. The results were obtained under a light intensity of 100 µmol photons m^{-2} s^{-1}without optimisation of cultivation parameters. Data in this study were generated by a doubled light intensity of 200 µmol photons m^{-2} s^{-1}. The resulting higher biomass productivity led to higher lipid productivity.

The productivity of lipids is the product of biomass productivity and content of lipids. To determine the impact of both values on resulting lipid productivity, correlation studies were undertaken. Biomass productivity influenced the lipid productivity (r^2=0.683; Fig. 6a) more than the lipid content (r^2=0.563), but the distinction was low.

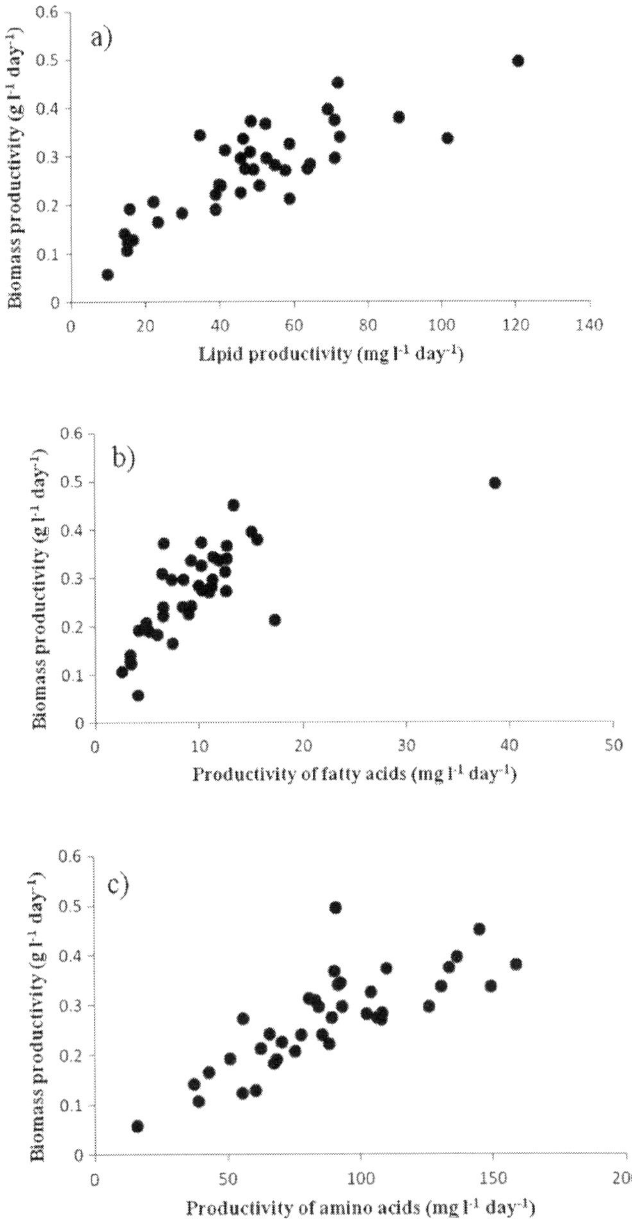

Figure 6: Correlation of biomass productivity with lipid productivity (a; $r^2=0.683$), with fatty acid productivity (b; $r^2=0.510$) and with amino acid productivity (c; $r^2=0.638$) in 39 microalgae strains including the reference strain

Most promising strains in terms of productivity of fatty acids

In tested strains, the content of fatty acids ranged from 1.8% dw (*Chlorella* sp. 391) to 8.1% dw (*C. saccharophila* 477). Renaud et al. (1994) obtained comparable results for total fatty acid content of microalgae strains ranging from 1.8% to 10.5% dw. For resulting productivity of fatty acids, values between 2.6 mg L^{-1} day^{-1} (*Chlorella* sp. 572) and 38.7 mg L^{-1} day^{-1} (*Chlorella* sp. 800) were calculated. Comparable fatty acid productivity was reported in literature (Lee et al. 2010; Meireles et al. 2008). Please note that these results were achieved under defined non-limiting cultivation conditions. In response to nitrogen deficiency, an accumulation of triacylglycerols containing fatty acids was observed in numerous strains (Basova 2005; Cobelas and Lechado 1989; Merzlyak et al. 2007). It is also worth pointing out that phosphorous limitation or sulphur limitation can lead to increasing amount of fatty acids in algae biomass (Khozin-Goldberg and Cohen 2006; Sato et al. 2000). To estimate the influence of nutrient limitation on screening strains of the current study, further experiments will be conducted.

The correlation investigations gave results as follows: Similar to lipid productivity, the productivity of fatty acids was more influenced by biomass productivity (r^2=0.51; Fig. 6b) than by content of fatty acids (r^2=0.417). The results indicate that biomass productivity is a good indicator of suitability for biodiesel production, but the content of fatty acids is also a factor. For direct transesterification reaction, the amount of fatty acids is crucial. To determine the indication function of lipid content, the correlation between content of fatty acids and content of lipids was compared. A coefficient of determination of r^2=0.29 was estimated, with both values only moderately linked with each other. A good accordance was given for *C. saccharophila* 477 (27.6% dw lipids and 8.1% dw fatty acids). In contrast, *Chlorella* sp. 589 reached, with 30.2% dw, the highest lipid content of all screened strains. However, with 3.6% dw, a low content of fatty acids was also recorded. In all tested strains, the fatty acid content was, on average, 20.9±6.4% of lipid content. However, reasons for discrepancy between lipid content and fatty acid content are complex. Based on extraction method, a wide range of lipophilic substances (soluble in chloroform/methanol mixture) were detectable: neutral or polar lipids, wax esters, sterols, hydrocarbons and prenyl derivatives (Hu et al. 2008). In the lipophilic fraction of *Selenastrum gracile*, chlorophyll *a* and *b*, β-carotene, lutein, violaxanthin and neoxanthin were observed (Czygan 1970). Which substances were responsible for high lipid content of *Chlorella* sp. 589 will be determined in further studies.

All identified fatty acids had predominantly even-numbered carbon fatty acid chain lengths, existing as saturated, mono-, di- and tri-unsaturated ones. In accordance for microalgae, most commonly synthesised fatty acids with chain length between C16 and C18 were reported (Hu et al. 2008; Ohlrogge and Browse 1995). In Chlorophyceae, Borowitzka (1997) identified palmitic acid and oleic acid as the main components of fatty acid fraction. Under defined screening conditions, algae mainly synthesise membrane lipids such as glycosylglycerides and phosphoglycerides (Hu et al. 2008; Piorreck and Pohl 1984). Palmitic and oleic fatty acids function as precursors for major constituents of membrane glycerolipids, provided by aerobic desaturation and chain elongation (Erwin 1973). Compared to analysis of *Neochloris oleoabundans* (Tomabene et al. 1983), in nearly all screened strains in the present study, substantial quantities of fatty acids with odd-numbered chain lengths of pentadecanoic acid (C15:0) and of margaric acid (C17:0) were detected. A significant amount of elaidic acid (C18:1 t), a trans-configured fatty acid, was identified in *Chlorella fusca*, *Chlorella* sp. 17-1, *Chlorella*sp. 313, *Chlorella* sp. 418, *C. vulgaris* C1 and *Scenedesmus rubescens*.

With different temperatures of cultivation, different numbers of double bonds in fatty acids were found (Fig. 4). For *Chlorella* sp. 800, the number of double bonds per fatty acid molecule decreased with increasing cultivation temperature. This effect is in accordance to literature data. Sato et al. (1979) described in *Anacystis nidulans* similar changes in the desaturation grade of fatty acids. This is a common strategy in many organisms in an attempt to tolerate various temperatures. By increasing the number of double bounds, the temperature for transition of fluid-to-solid phase of membranes was lowered. In this manner, organisms can prevent adverse solid status. In existing literature, it has been described that the average chain length of fatty acids decreased with a decreasing of temperature. However, in this study and with strain *Chlorella* sp. 800 the opposite was found to be true. The content of C16 fatty acids compared to C18 fatty acids increased with increasing cultivation temperatures (Fig. 4).

The composition of fatty acids strongly influences the quality of biodiesel (Yoo et al. 2010). The most important properties are cetane number, cold-flow properties, oxidative stability and iodine value, derived from the structure of fatty acid methyl esters (Damiani et al. 2010). Despite the high content of polyunsaturated fatty acids in most algal oils (Damiani et al. 2010), fatty acid spectra found in present study are very suitable for use as biodiesel. High amounts of C16 and C18 fatty acids and a high saturation grade were found in the strain with an outstanding productivity for fatty acids *Chlorella* sp. 800 as well as other top producers. Increasing the cultivation temperature indicated an additional quality-increasing effect in the phase of increased lipid synthesis, for instance during nitrogen starvation. A positive influence of increasing light intensity on fatty acid productivity without a quality-downgrading effect on the fatty acid spectrum was furthermore postulated.

Role of proteins and amino acid spectrum

For economic biodiesel production, residual algae biomass plays an important role. Lipidless biomass contains proteins, which can enhance the amount of nutritions of conventional food preparations (Spolaore et al. 2006). These were analysed in this study.

Contents of amino acids in screened strains ranged from between 18% dw (*Chlorella* sp. 800) and 47% dw (*S. platensis*). Quantification of amino acids described in 'Materials and methods' excluded analysis of any tryptophane amount in probes. Despite this fact, contents of summarised amino acids are in accordance with protein contents in literature. Boyd (1973) reported protein contents between 11 and 46% dw in species of fresh water algae. In another publication, the biomass of microalgae contained proteins between 20 and 67% dw (Renaud et al. 1994). These data tend to higher values compared to values reported in this study. A similar lowering of protein contents has been observed by Brown (1991): Continuous illumination conditions decreased protein content. Higher protein contents were obtained under a 12:12 h light–dark cycle. However, proteins are major organic constituents in algae biomass and are the main reason for suitability of food preparations (Cornet 1998; Soletto et al. 2005).

In the screened microalgae strains, all detectable amino acids were quantified. *C. vulgaris* 132, which reached the highest productivity of amino acids, also reached the highest content of all amino acids, ranging from 0.22% dw (cysteine) to 4.5% dw (arginine). For tested *Chlorella* and *Nannochloropsis*, similar amino acid spectra were reported in literature (Sato et al. 1979).

Biomass productivity was clearly identified as the main influencing factor for productivity of amino acids, resulting in high correlation ($r^2 = 0.638$; Fig. 6c). For absolute content of amino acids, only a weak correlation of $r^2 = 0.137$ was analysed. For increasing productivity of amino acids, the cultivation temperature was varied for *C. vulgaris* 132. Based on high biomass productivity at an optimal temperature of 25°C, the productivity reached 163 mg L^{-1} day^{-1} at its highest value. Productivity decreased at 35°C to 147 mg L^{-1} day^{-1} and at 15°C to 62 mg L^{-1} day^{-1}. An influence of cultivation temperature on amino acid spectra was not detected. The relative amount of all amino acids was unchanged during cultivation between 15 and 35°C.

NOTES

Acknowledgements

This work has been supported by the European Union within the European Fund for Regional Development EFRE (Project—'Fixation of carbon dioxide from exhaust gases by algae'), co-financed by the Land Brandenburg. We would like to thank Janine Kleye and Michael Henschel for excellent technical assistance and helpful discussions.

REFERENCES

1. Basova MM (2005) Fatty acid composition of lipids in microalgae. Int J Algae 7:33–57
2. Bhatnagar A, Bhatnagar M, Chinnasamy S, Das K (2010) Chlorella minutissima—a promising fuel alga for cultivation in municipal wastewaters. Appl Biochem Biotechnol 161:523–536
3. Borowitzka MA (1997) Microalgae for aquaculture: opportunities and constraints. J Appl Phycol 9:393–401
4. Boyd CE (1973) Amino acid composition of freshwater algae.

Arch Hydrobiol 72:1–9

5. Brown MR (1991) The amino acid and sugar composition of sixteen species of microalgae used in mariculture. J Exp Mar Biol Ecol 145:79–99
6. Brown LM, Zeiler BG (1993) Aquatic biomass and carbon dioxide trapping. Energy Convers Manag 34:1005–1013
7. Chinnasamy S, Bhatnagar A, Hunt RW, Das KC (2010) Microalgae cultivation in a wastewater dominated by carpet mill effluents for biofuel applications. Bioresour Technol 101:3097–3105
8. Chisti Y (2007) Biodiesel from microalgae. Biotechnol Adv 25:294–306
9. Cobelas MA, Lechado JZ (1989) Lipids in microalgae. A review. I. Biochemistry. Grasas y Aceites 40:118–145
10. Cornet JF (1998) Le technoscope: les photobioréacteurs. Biofutur 176:1–10
11. Crutzen PJ, Moiser AR, Smith KA, Winiwarter W (2007) N2O release from agro-biofuel production negates global warming reduction by replacing fossil fuels. Atmos Chem Phys Discuss 7:11191–11205
12. Czygan FC (1970) Untersuchungen über die Bedeutung der Biosynthese von Sekundär-Carotinoiden als Artmerkmal bei Grünalgen. Arch Mikrobiol 74:77–81
13. Damiani MC, Popovich CA, Constenla D, Leonardi PI (2010) Lipid analysis in Haematococcus pluvialis to assess its potential use as a biodiesel feedstock. Bioresour Technol 101:3801–3807
14. Doucha J, Straka F, Lívanský K (2005) Utilization of flue gas for cultivation of microalgae (Chlorella sp.) in an outdoor open thin-layer photobioreactor. J Appl Phycol 17:403–412
15. Erwin JA (1973) Comparative biochemistry of fatty acids in eukaryotic microorganisms. In: Erwin JA (ed) Lipids and biomembranes of eukaryotic microorganisms. Academic, New York, pp 141–143
16. Griffiths MJ, Harrison STL (2009) Lipid productivity as a key characteristic for choosing algal species for biodiesel production. J Appl Phycol 21:493–507

17. Hammouda O, Gaber A, Abdel-Raouf N (1995) Microalgae and wastewater treatment. Ecotoxicol Environ Saf 31:205–210
18. Hankamer B, Lehr F, Rupprecht J, Mussgnug JH, Posten C, Kruse O (2007) Photosynthetic biomass and H2 production: from bioengineering to bioreactor scale up. Physiol Plantarum 131:10–21
19. Hoffart MI, Caldeira K, Benford G, Criswell DR, Green C, Herzog H, Jain AK, Kheshgi HS, Lackner KS, Lewis JS, Lightfoot HD, Manheimer W, Mankins JC, Mauel ME, Perkins LJ, Schlesinger ME, Volk T, Wigley TML (2002) Advanced technology paths to global climate stability: energy for a greenhouse planet. Science 298:981–987
20. Hu Q, Sommerfeld M, Jarvis E, Ghirardi M, Posewitz M, Seibert M, Darzins A (2008) Microalgal triacylglycerols as feedstocks for biofuel production: perspectives and advances. Plant J 54:621–639
21. Huerlimann R, de Nys R, Heimann K (2010) Growth, lipid content, productivity, and fatty acid composition of tropical microalgae for scale-up production. Biotechnol Bioeng 107:245–257
22. Khozin-Goldberg I, Cohen Z (2006) The effect of phosphate starvation on the lipid and fatty acid composition of the fresh water eustigmatophyte Monodus subterraneus. Phytochemistry 67:696–701
23. Koonin SE (2006) Getting serious about biofuels. Science 311:435Kruse O, Rupprecht J, Mussgnug JR, Dismukes GC, Hankamer B (2005) Photosynthesis: a blueprint for solar energy capture and biohydrogen production technologies. Photochem Photobiol Sci 4:957–970
24. Lee JY, Yoo C, Jun SY, Ahn CY, Oh HM (2010) Comparison of several methods for effective lipid extraction from microalgae. Bioresour Technol 101:75–77
25. Mallick N (2002) Biotechnological potential of immobilized algae for wastewater N, P and metal removal: a review. Biometals 15:377–390
26. Meireles LA, Guedes AC, Barbosa CR, Azevedo JL, Cunha JP, Malcata FX (2008) On-line control of light intensity in a microalgal bioreactor using a novel automatic system. Enz Microb Technol

42:554–559

27. Merzlyak MN, Chivkunova OB, Gorelova OA, Reshetnikova IV, Solovchenko AE, Khozin-Goldberg I, Cohen Z (2007) Effect of nitrogen starvation on optical properties, pigments, and arachidonic acid content of the unicellular green alga Parietochloris incisa (Trebouxiophyceae, Chlorophyta). J Phycol 43:833–843
28. Muñoz R, Guieysse B (2006) Algal–bacterial processes for the treatment of hazardous contaminants: a review. Water Res 40:2799–2815
29. Ohlrogge J, Browse J (1995) Lipid biosynthesis. Plant Cell 7:957–970
30. Pacala S, Socolow R (2004) Stabilization wedges: solving the climate problem for the next 50 years with current technologies. Science 305:968–972
31. Patil V (2007) The relevance of biofuels. Curr Sci 92:707
32. Patil V, Tran KQ, Giselrød HR (2008) Towards sustainable production of biofuels from microalgae. Int J Mol Sci 9:1188–1195
33. Piorreck M, Pohl P (1984) Formation of biomass, total protein, chlorophylls, lipids and fatty acids in green and blue-green algae during one growth phase. Phytochem 23:217–223
34. Pirt SJ (1986) The thermodynamic efficiency (quantum demand) and dynamics of photosynthetic growth. New Phytol 102:3–37
35. Pulz O (2001) Photobioreactors: production systems for phototrophic microorganisms. Appl Microbiol Biotechnol 57:287–293
36. Pulz O, Gross W (2004) Valuable products from biotechnology of microalgae. Appl Microbiol Biotechnol 65:635–648
37. Rasoul-Amini S, Montazeri-Najafabady N, Mobasher MA, Hoseini-Alhashemi S, Ghasemi Y (2011) Chlorella sp.: a new strain with highly saturated fatty acids for biodiesel production in bubble-column photobioreactor. Appl Energy. doi:10.1016/j.apenergy.2010.12.040
38. Renaud SM, Parry DL, Thinh LV (1994) Microalgae for use in tropical aquaculture I: gross chemical and fatty acid composition of twelve species of microalgae from the Northern Territory,

Australia. J Appl Phycol 6:337–345

39. Richmond A (2000) Microalgal biotechnology at the turn of the millennium: a personal view. J Appl Phycol 12:441–451
40. Righelato R, Spracklen DV (2007) Carbon mitigation by biofuels or by saving and restoring forests? Science 327:902
41. Rodolfi L, Zittelli GC, Bassi N, Padovani G, Biondi N, Bonini G, Tredici MR (2008) Microalgae for oil: strain selection, induction of lipid synthesis and outdoor mass cultivation in a low-cost photobioreactor. Biotechnol Bioeng 102:100–112
42. Sato N, Murta N, Miura Y, Ueta N (1979) Effect of growth temperature on lipid and fatty acid compositions in the blue-green algae Anabaena variabilis and Anacystis nidulans. Biochim Biophys Acta 572:19–28
43. Sato N, Hagio M, Wada H, Tsuzuki M (2000) Environmental effects on acidic lipids of thylakoid membranes. In: Harwood JL and Quinn, PJ (eds) Recent advances in the biochemistry of plant lipids. Portland, London, pp 912–914
44. Schaub G, Kolb T, Steiger W (2007) Biofuels. Second-generation biofuels. Chemie Ingenieurtechnik 79:1325–1326
45. Schenk PM, Thomas-Hall SR, Stephens E, Marx UC, Mussgnug JH, Posten C, Kruse O, Hankamer B (2008) Second generation biofuels: high-efficiency microalgae for biodiesel production. Bioenerg Res 1:20–43
46. Shay EG (1993) Diesel fuel from vegetable oils: status and opportunities. Biomass Bioenergy 4:227–242
47. Soletto D, Binaghi L, Lodi A, Carvalho JCM, Converti A (2005) Batch and fed-batch cultivations of Spirulina platensis using ammonium sulphate and urea as nitrogen sources. Aquaculture 243:217–224
48. Spolaore P, Joannis-Cassan C, Duran E, Isambert A (2006) Commercial applications of microalgae. J Biosci Bioeng 101:87–96
49. Tang H, Chen M, Garcia MED, Abunasser N, Ng KYS, Salley SO (2011) Culture of microalgae Chlorella minutissima for biodiesel feedstock production. Biotechnol Bioeng 108:2280–2287

50. Tomabene TG, Holzer G, Lien S, Burris N (1983) Lipid composition of the nitrogen starved green alga Neochloris oleoabundans. Enzyme Microb Technol 5:435–440

51. Ugwu CU, Aoyagi H, Uchiyama H (2008) Photobioreactors for mass cultivation of algae. Bioresour Technol 99:4021–4028

52. Van Gerpen J (2009) Biofuel. Methods in molecular biology. In: Mielenze JR (ed) Biodiesel: small scale production and quality requirements. Humana, New Jersey, pp 281–290

53. Xu L, Weathers PJ, Xiong XR, Liu CZ (2009) Microalgal bioreactors: challenges and opportunities. Eng Life Sci 9:178–189

54. Yoo C, Jun SY, Lee JY, Ahn CY, Oh HM (2010) Selection of microalgae for lipid production under high levels carbon dioxide. Bioresour Technol 101:71–74

CHAPTER

14

POTENTIAL AND PROPERTIES OF MARINE MICROALGAE NANNOCHLOROPSIS OCULATA AS BIOMASS FUEL FEEDSTOCK

Sukarni[1,2], Sudjito[3], Nurkholis Hamidi[3], Uun Yanuhar[4], and I. N. G. Wardana[3]

[1] Doctoral Program of Mechanical Engineering, Faculty of Engineering, University of Brawijaya, Malang, Indonesia

[2] Department of Mechanical Engineering, Faculty of Engineering, State University of Malang, Malang, Indonesia

[3] Department of Mechanical Engineering, Faculty of Engineering, University of Brawijaya, Malang, Indonesia

[4] Biotechnology Laboratory, Department of Water Resources Management, Faculty of Fisheries and Marine Sciences, University of Brawijaya, Malang, Indonesia

ABSTRACT

Microalgal biomass is the most promising and attractive alternative to replace the terrestrial crop utilization for renewable biomass fuel feedstock. The potential for biomass fuel is due to its fast growth rate and high ability for CO_2 fixation as compared to terrestrial vegetation. There are many species in the globe, growing both in marine and freshwater. In this work, the marine microalgae *Nannochloropsis oculata* (*N. oculata*) had been investigated in terms of potential abundance and physicochemical properties, which determine its feasibility as biomass fuel feedstock. The chemical composition was evaluated by energy-dispersive X-ray spectrometry, and the proximate analysis was done by performing experiments in the thermal gravimetric analyzer. During 7 days of cultivation, the average rate of increase in algal biomass was about 1.5×10^6 cells/ml/day. The proximate analysis of *N. oculata* indicated that it had compositions of low moisture content and fixed carbon, whereas high volatile matter and ash content, i.e., 3.99, 8.08, 67.45, and 24.47 %, respectively. The energy content, which was calculated through the proximate analysis results, was 16.80 MJ/kg. The algal biomass and its residue after 1,200 °C were characterized by Fourier transform infrared spectroscopy to investigate their chemical macromolecular compounds. This present study concludes that *N. oculata* is feasible as biomass fuel feedstock, either to direct or co-combustion mode by giving special attention to high ash

Keywords: Renewable Biofuel Microalgae *Nannochloropsis oculata* Biomass fuel feedstock

INTRODUCTION

Since fossil fuel sources across the globe are fast depleting, renewable and non-conventional energy sources are urgent to explore to ensure the world is not threatened by a vacuum of energy. Biomass is generally defined as any hydrocarbon material, mainly consists of carbon, hydrogen, oxygen, and nitrogen [1]. It has been recognized as a major world renewable energy source to supplement declining fossil fuel resources [2]. However, the use of first-generation biofuels, which have been mostly taken from food and oil crops, such as rapeseed oil, sugar cane, sugar beet, maize, vegetable oils, and animal fats [3], has generated a lot of controversies, primarily because of their impact on global food markets and food security [4]. Hence, it drives the need to base non-food biomass resource biofuels. As a result, interests in aquatic biomass as the feasible fuel source have recently gained prominence. One of the potential aquatic biomass for fuel is microalgae.

Microalgae will be potential as an important energy source in the future due to their potentially abundant, renewable, and CO_2 neutral fuel [5]. Compared with wood and land crops, microalgae are able to reproduce their biomass very rapidly and are very rich of oil content. Biomass productivity of microalgae can be 50 times more than that of the fastest growing terrestrial plant, namely switch grass [6]. Microalgae multiply their biomass about 1–3 per day [7] and theoretically can produce oil per hectare 100 times higher than that of terrestrial crops [8]. In the appropriate culture media, some algae can achieve very high lipid content, such as *Chlorella emersonii* [9], *Neochloris oleabundans* [10], and *Scenedesmus abundans* [11], which their lipid content up to 63, 56, and 67 %, by dry weight, respectively. Microalgae consume 2 g of CO_2 to produce 1 g of biomass when considering the glucose conversion into starch or lipids. Thus, by assuming a growth rate of 50 g/m^2/day, it is possible for one hectare pond algae to absorb up to one ton of CO_2 per day [12]. Thereby, microalgae cultures will neutralize CO_2 from their combustion.

Beyond the potential of microalgae for fuel in the future mentioned above, harnessing energy from microalgae must be considered in full transparence of physical and chemical properties. These properties are essential parameters for biomass fuel feedstock because they will affect both their combustion characteristics and proper handling modes in the furnace. The moisture content is very important property affecting the burning characteristics of biomass [13], mainly owing to its necessary of energy for releasing before a combustion process takes place. Hence, it will reduce the temperature in the combustion chamber. To ensure the sustained combustion process, this parameter must be exactly well known. Volatile matter content has also been indicated to affect the thermal behavior of solid fuels. The rate of volatile releases and its quantity determine the flame ignition, stability, and temperature profile in the radiation part of the furnace. Chemical properties of biomass are also a critical parameter as they determine the amount of energy contained in biomass. Likewise, the presence of the inorganic component in the biomass influence in the formation of ash, slag deposits, corrosion of boiler components, and aerosol [14]. Generally the inorganic constituents of biomass are Si, Ca, K, Na, Mg, S, P, Fe, Mn, and Al [15, 16], and the concentration of each element in biomass varies in accordance with the kind of species and growing environment. During combustion processes, part of the ash-forming compounds in the biomass mainly alkali metals and chlorine are volatilized into a gas phase, such as HCl(g), KCl(g), $(KCl)_2$(g), K_2SO_4(g), KOH(g), K(g), Na_2SO_4(g), and NaCl(g) [17, 18]. As cooling proceeds in the heat exchanger section of boilers, subsequent nucleation or condensation lead to the gases, together with solid phase submicron particles released from the fuel bed, form a large part of the fine ash particles. Otherwise, the compounds composed of Mg, Ca, Fe, Al, and non-metallic elements such as S, P, and Si are relatively more stable and less volatilized [19]. The fractions of non-volatilized compounds are remaining in the charcoal, and through a process of coalescence or melting form ash particles with a variety of shapes, sizes, and compositions, parallel to the characteristics of the parent material [20]. Therefore, to make biomass thermal conversion processes effective, these parameters must be clearly studied.

This work aimed to provide fundamental information for biomass fuel feedstock of *N. oculata*by investigating the biomass growth in the culture medium and its properties which affect the performance of biomass fuel including the calorific value, moisture content and physicochemical properties. The Fourier transform infrared (FTIR) spectroscopy analysis of both biomass and its residue were also conducted to identify the cellular macromolecular content.

The description of species

N.oculata is a species of a genus *Nannochloropsis*. It is a unicellular small green alga, found in both marine and freshwater [21, 22]. This alga is characterized by spherical or slightly ovoid cells [23] with a diameter of 2–5 μm [24] and absence of chlorophyll *b* and cellular xanthophyll pigment [25, 26, 27]. There is one chloroplast in each cell, without pyrenoid. Chloroplast endoplasmic reticulum is continuous with the nuclear envelope. Inside the chloroplast are lamellae, consisting of three thylakoids for each. Only one chlorophyll is present, namely chlorophyll *a* and the main pigment is violaxanthin. Its cell walls are composed of two components, the fibrillar component and the amorphous component. Fibrillar part forms the skeleton of the wall, which is embedded in the matrix of amorphous component. The most common fibrillar component is cellulose, a polymer of 1,4 linked β-d-glucose. An amorphous mucilaginous material is arranged from proteins, lipids, and polysaccharides. Occasionally, silica, calcium carbonate, or sporopollenin is also present as encrusting substances [28].

Nannochloropsis is widely distributed in oceans worldwide. It plays significant roles in the global carbon and mineral cycles [29]. This microalgae contains rich proteins, pigments, and polyunsaturated fatty acids [26, 30, 31], and it is commonly used in aquaculture as feed [32, 33]. In recent years, *Nannochloropsis* has been proposed as an excellent candidate for biofuel production [34, 35].

METHODS

Microorganism and culture conditions

The *N. oculata* strain was obtained from the Institute of Brackish Water Aquaculture (BBAP) Situbondo, East Java, Indonesia. This species was cultured in an open pond with seawater medium in which its salinity and pH were 34 ‰ and 8.6, respectively. The culture medium was enriched with fertilizers whose composition was presented in Table 1. The use of fertilizers was 1 ml/l of the cultivation medium. The cultivation had been performed for 7 days.

Table 1: The composition of fertilizers used for culturing *Nannochloropsis oculata*

Substances	Composition (g/l of water)
KNO_3	100
NaH_2PO_4	10
Na_2EDTA	10
$FeCl_3$	1.3

Growth evaluation

The biomass productivity of *N. oculata* was evaluated daily by means of counting the number of cells in the culture medium using hemocytometer and optical microscope (CH2 Olympus optical Co. Ltd, Tokyo, Japan). Measurement was carried out every day at 11:00 AM during the cultivation period. Each measurement was repeated 3 times and the results were averaged.

The *N. oculata* specific growth rate, which was the change of the natural log of the cell number density (Cn) with time (t), was calculated from the slope of the linear regression of time (days) and cell density (cells/ml) [36]:

$$\mu = d \ln C_n / dt = (\ln C_t - \ln C_o)/(t - t_o) \quad (1)$$

where Ct is cell density at time (t), and Co is cell density at the start of the exponential phase (to).

The time required to double the cell of *N. oculata*, termed the doubling time (day), could be estimated by equation:

$$t_d = \ln 2/\mu \quad (2)$$

Harvesting and drying

The biomass was harvested by precipitating it with caustic soda (NaOH); then it was filtered and washed twice with distilled water. Subsequently, the biomass sediment was dried in an oven at 80 °C for 24 h. The dried microalgae chunks were pulverized by the mortar to be fine particles and then finally stored in desiccators.

Chemical composition

Energy-dispersive X-ray (EDX) spectrometry was used to determine the chemical composition of algal biomass. Scanning electron microscope images were taken by FEI Inspect S50 equipped with X-ray microanalysis capabilities (AMETEK EDAX^{TSL}). Samples were gold coated to minimize the effect of electron charging of the surface, which could distort the images.

Lipid extraction

The total lipid of *N. oculata* was extracted using a Soxhlet extractor system. Approximately 3 g of biomass powder was weighed into a cellulose thimble inside the extraction chamber. A total of 150 ml pure *n*-hexane was used to extract the lipid for 6 h at the rate of 10 refluxes per hour to achieve maximum extraction efficiency. The extracted lipid was measured after removing the solvent using vacuum rotary evaporator to evaporate the *n*-hexane at 35 °C for 60 min, and then lipid content was calculated. The adopted Soxhlet unit description could be found in different literature [37].

Proximate analysis

The moisture, volatile matter, fixed carbon, and ash content of the dry algal biomass were determined by thermal gravimetric analyzer (STA PT1600 by Linseis). The adopted method is basically described by Beamish [38] and Mayoral et al. [39]. Approximately 20 mg samples were loaded into an Al_2O_3 ceramic crucible. Initially, the sample was kept isothermally at 25 °C for 8 min. The heating ramp was set at 50 °C/min, from 25 °C up to 125 °C then held for 3 min. A new 80 °C/min heating ramp was programmed up to 950 °C and maintained at that temperature for a total duration of 40 min. The whole process was done using Argon as purge gas to guarantee a non-oxidative environment. However, this gas was then switched to the air atmosphere when it was achieved isothermally at 40 min. Once a constant mass loss was reached, the isothermally final temperature was maintained for a total duration of 65 min in the air atmosphere to complete the combustion process. The moisture, volatile matter, fixed carbon, and ash were calculated by the difference from slope to slope of thermogravimetric (TG) curve.

Determination of higher heating value (HHV)

The heating value of microalgal biomass was evaluated by an equation proposed by Nhuchhen and Abdul Salam [40]. This calculation was based on values resulted from the proximate analysis.

$$HHV=19.2880-0.2135 \cdot VM/FC+0.0234 \cdot FC/A-1.9584 \cdot A/VM \quad (3)$$

where, VM, FC, and *A* were volatile matter, fixed carbon, and ash content of biomass, respectively, by dry weight basis.

FTIR analysis

The infrared spectrum of dried algal biomass and its residue were recorded by Shimadzu FTIR spectroscopy. Both kinds of samples were mixed with potassium bromide (KBr) powder, and then pressed into tablets before measurement. Samples were scanned from 400 to 4,000/cm.

RESULTS AND DISCUSSION

One of microalgae potential's benchmarks as biomass fuel feedstock is determined by the abundance of the biomass so the availability will be assured. Therefore, the biomass productivity of algae in terms of growth rate is one of the most significant parameters that must be examined.

The biomass yields of *N. oculata* during 7-day cultivation period were depicted in Fig. 1. These results showed that *N. oculata* multiplied its biomass rapidly, starting at around 2 × 10^6 cells/ml on day 1 to approximately 12 × 10^6 cells/ml on day 7. It meant the average increase in biomass was approximately 1.5 × 10^6 cells/ml/day, higher than that of *Selenastrum*sp., which increase in cell number per day was about 1.6 × 10^4 up to 2.2 × 10^4 cells/ml/day [41]. If the productivity of *N. oculata* was expressed in growth kinetics, then it had specific growth rate of 0.27/day and doubling time of 2.59 days as shown in Table 2. Its growth kinetics revealed that *N. oculata* biomass productivity was higher than *Dunaliella salina* [42], *S. obliquus* [43], and *C. vulgaris* [44], which their growth rates were 0.18, 0.22, and 0.14/day, and their doubling time were 3.85, 3.15, and 4.95 days, respectively. This result was important to note by considering the *N. oculata* had been cultivated in the traditionally natural open pond as illustrated in Fig. 2a. The products of biomass sediment, the dried biomass chunk and the biomass powder of *N. oculata*, were shown at Fig. 2b–d, respectively.

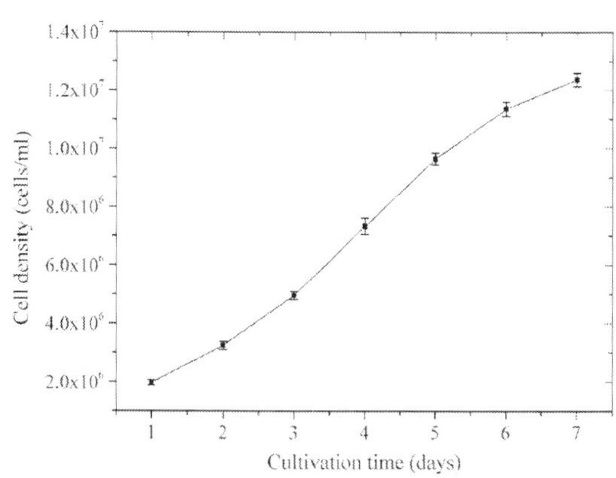

Figure1: The cell density of *Nannochloropsis oculata* during 7-day cultivation period. Relative standard error of means were below 5 % for all situations

By fitting the curve of Fig. 1, the growth pattern of *N. oculata* could be estimated as an order 3 of a polynomial equation as $Y = 2{,}018{,}570 - 736{,}666.67X + 783{,}095.24X^2 - 66{,}666.67X^3$ with the fitting degree (R^2) of 0.999, where Y and X were cell density and cultivation time, respectively.

Table 2: The growth kinetics parameters and properties of *Nannochloropsis oculata*

Parameters/properties	**Values**
Specific growth rate (/day)	0.27
Doubling time (day)	2.59
Lipid content (%)	11.44
EDX analysis (wt%)	
C	28.32
O	43.80
Na	2.04
Mg	13.16
Al	0.92
Si	1.60
Cl	1.97
Ca	8.20
Proximate analysis (wt%)	
M (adb)	3.99
VM (db)	67.45
FC (db)	8.08
A (db)	24.47
HHV analysis (MJ/kg)	16.80

adb air-dried basis, *db* dry basis

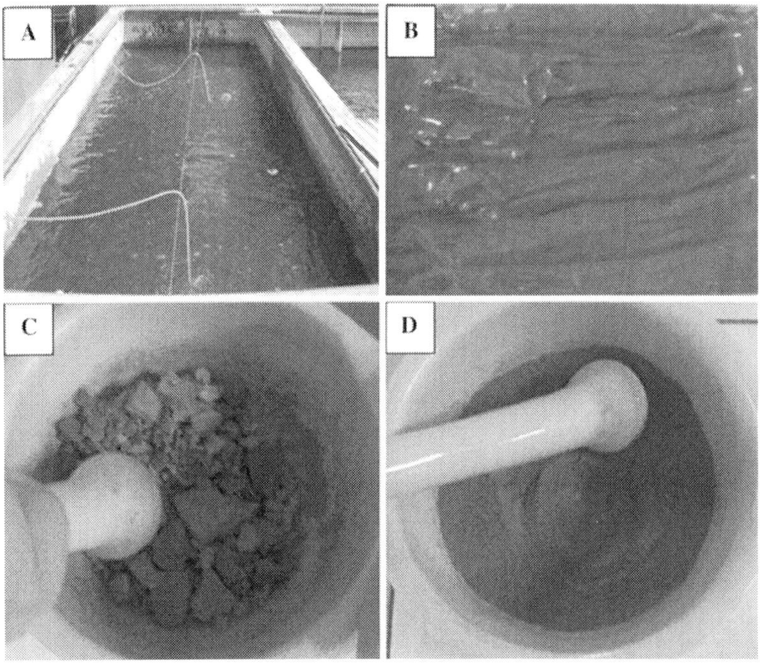

Figure 2: The cultivation pond (**a**), the biomass sediment (**b**), the dried biomass chunk (**c**), and biomass powder (**d**) of *Nannochloropsis oculata*

The properties of algal biomass as a biofuel source were taken into consideration in this study. These properties varied widely, depended on the kind of algae, environmental culturing, cultivation periods, and harvesting conditions. Hence, investigating these properties was essential to determine the proper combustion technology.

Figure 3 showed the EDX spectrograms of *N. oculata* biomass for determination of its compositions as compiled in Table 2. Generally, the investigated *N. oculata* biomass has low carbon, high oxygen, and high inorganic elements compared to the result of Patil et al. [45]. These compositions had an impact on biomass combustion and the residual characteristics. They were also needed for estimating the air required for the combustion process and for predicting the released gases. Carbon was oxidized during combustion by exothermic reactions and formed CO_2. The organically bound of oxygen is released as the result of thermal processes, and it supplies a part of the overall oxygen needed for the combustion reactions [14].

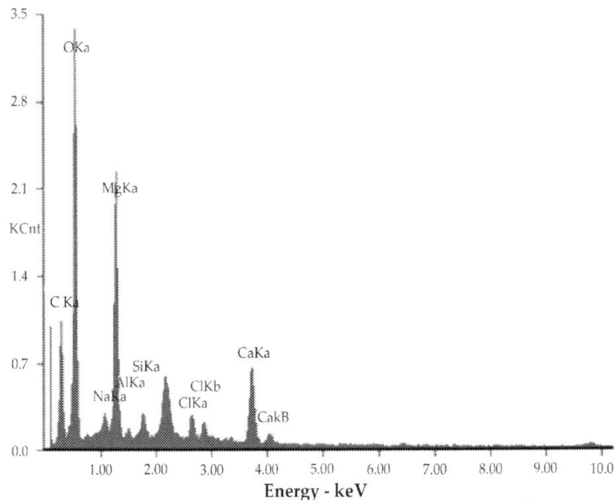

Figure 3: EDX spectrograms of *Nannochloropsis oculata* biomass

The high amount presence of inorganic materials in this biomass, especially magnesium, calcium, sodium, silicon, and chlorine needs to be considered because they frequently results in slagging and fouling problems in furnaces. The reaction of alkali metals with silica presented in the ash can lead to the formation of low-melting silicates in fly ash particles yielding a sticky, mobile liquid phase, which causes blockages of airways in the furnace and boiler plant [46]. Cl vaporized during combustion, forming HCl, Cl_2, and alkali chlorides. Alkali and alkaline earth chlorides are condensed in the boiler section as fly- ash particles or on the heat exchanger surfaces when flue gas temperature decreases [47]. Parts of the Cl were bounded in the fly ash while the rests were emitted as HCl in the flue gas. HCl emissions could play in the corrosion of metal component plant such as boilers and air pollution control devices.

The TG curve for determining the proximate value was shown in Fig. 4. The values of moisture (*M*), volatile matter (VM), fixed carbon (FC), and ash (*A*) were directly determined from the TG curve by the difference while higher heating value (HHV) of the biomass was estimated by Eq. (3). All of these results were tabulated in Table 2.

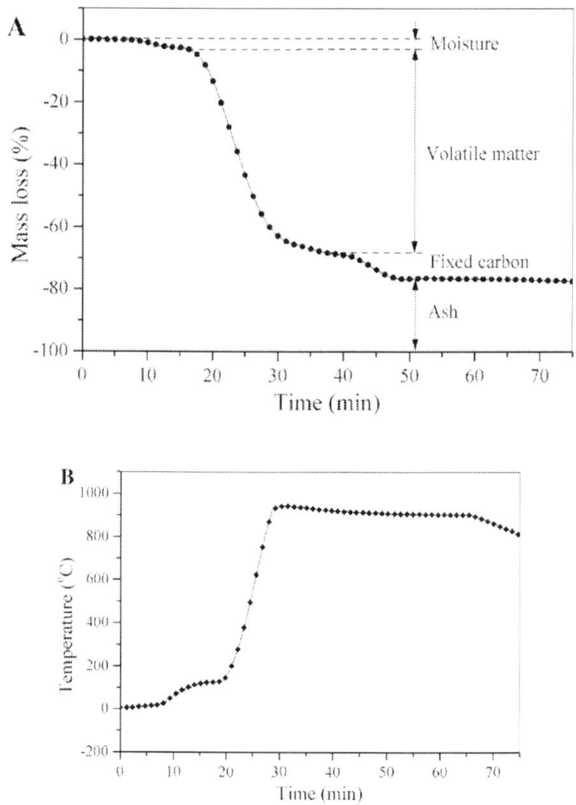

Figure 4: TG curve for determining the proximate value. **a** Mass loss (%) vs. time (min) and **b**temperature setting (°C) vs. time (min)

As shown in Fig. 4, the first mass loss, which was indicated by the reaching of the first constant mass line, corresponded to moisture loss. The *N. oculata* moisture content was 3.99 %, which seemed to be a potential candidate for direct combustion. The moisture content influenced the combustion behavior and the amount of energy to evaporate water. The water content in the fuel had to be released before the first step combustion starting; hence high moisture content of biomass meant more energy for evaporation, and subsequently for heating the water vapor. Removal of water in the biomass would decrease the maximum possible combustion temperature and overall system efficiency. Aside from moisture evaporation that consumed a great deal of heat, it also caused the evaporation time to be extended during the combustion process. Hence, it resulted in the decay of ignition and affected the necessary residence time of biomass fuel in the combustion chamber before gasification and combustion taking place. Consequently, the high moisture fuel contents would require the larger combustion chambers and would result in an expansive boiler. Moreover, the amount of moisture also affected the burning reaction rate; thus, it influenced the generated gas emission as NO and CO, as well. In the devolatilization process, due to evaporated moisture moves away from the devolatilization front, the amount of volatile determined the reaction zone thickness [48], and it influenced the heat transfer to the devolatilization zone which affected the devolatilization rate. Thus, the moisture content was very important fuel parameter, and investigating this parameter was required to adjust the temperature control system of the furnace properly.

In line with the rising temperature, a major part of biomass organic component experienced thermal cracking then decomposed and volatilized to be the volatile matter. In inert gas, the volatile is released at temperature up to 900 °C [39]. For microalgal biomass, it was especially resulted from the thermal cleavage of proteins, carbohydrates, and lipids. Table 2 showed that *N. oculata* volatile matter content was 67.45 %, lower than *Chlorella* sp. MP-1 that was 77.45 % [49] and higher than *C. vulgaris* that was 55.37 % [50]. The amount of biomass volatile matter strongly influenced the thermal decomposition and the combustion behavior. As it had been previously stated that the biomass, which had a high amount of volatiles, would be degraded as a result of the heating process. A significant part of the biomass was vaporized earlier before homogeneous gas phase combustion reactions took place and subsequently the remaining char undergone heterogeneous combustion reactions. Because of volatile matter was the reactive substance, hence, more volatile meant easier biomass to be burned. The aforementioned phenomenon would determine both of the ignition and burnout time of volatiles in the first stage of combustion processes, as well as the char combustion in the second stage. Moreover, in the context of co-combustion with other solid fuel, higher volatile could create larger fuel-rich regions in the near-burner region. The fuel-rich regions were very critical area. They were used for flame stabilization but, the formation of undesired prompt NO_x, H_2S, and soot could occur in this zone, as well [51, 52, 53]. Hence, finding out its amount is necessary to adjust the sufficiency of air supplied in the zone where volatile is released to improve combustion efficiency. Furthermore, the appropriate composition of supplied air in the released volatile zone will affect the formation of soot and NO emissions as discussed earlier by Liu et al. [54, 55].

The fixed carbon content, which was the mass left over after the releases of volatiles excluding the ash and moisture contents, produced a char and was burned as a solid material in the combustion system. The *N. oculata* fixed carbon content was 8.08 %. This content was lower than both of *C. vulgaris* [50] and *Chlorella* sp. MP-1 [49], which were 34.35 and 16.95 %, respectively. The volatile matter and fixed carbon contents significance were that they provided a measure of the ease which the biomass could be gasified and oxidized in the combustion process. The *N. oculata* biomass, which had high volatile matter and low fixed carbon contents, was highly reactive fuel and would give a faster combustion rate during the devolatilization phase. This amount of fixed carbon was also needed to estimate the heating value of biomass, and it was acted as the main heat generator during the burning process.

The final residual material after the combustion process of biomass was ash, and it formed a standard measurement parameter for solid fuels. Table 2 indicated that *N. oculata* ash content was 24.47 %, which was very high, four times higher than *Chlorella* sp. MP-1 that was 6.36 % [49]. It was in accordance with the EDX analysis results, which showed the inorganic components content of *N. oculata* was high, as well. Inorganic components in biomass had a direct influence in the ash formation. Ash-forming elements exist in biomass as salts [14], bounded in the carbon structure, or they present as mineral particles from cultivation environment and introduce into the biomass fuel during harvest. In the combustion process, the part of the ash-forming compounds in biomass was vaporized and released to the gas phase while non-vaporized ash compounds that were left over in the char might melt and coalesce inside and on the surface of the char. The vaporized compounds would either be condensed or reacted on the surface of pre-existing ash particles in the flue gas when its temperature decreased. All of these could contribute to bed agglomeration, heat transfer surface fouling, and system corrosion. Furthermore, in the combustion process, the ash resulted from char burning would be formed as the layer surrounding their surface; hence, it inhibited the oxygen diffusion during char combustion. It affects both burning rate and mass loss rate, as well [56]. To address the problem of

N. oculata's high ash content during thermal conversion processes, it might suitably blend this biomass in terms of co-combustion with other feedstocks to obtain the overall low ash content. One of the low ash biomass was rice straw, which its ash content was around 0.1–0.7 % by dry weight [57]. It might appropriate be blended with *N. oculata* to be mixed biomass fuel feedstock. However, their characteristics have not been studied yet.

Higher heating value was a total of heat yielded by the complete combustion of a unit quantity of fuels, including the latent heat contained in the water vapor. Therefore, it represented the maximum amount of energy which was potentially recoverable from a given biomass source. Based on the proximate result, HHV of *N. oculata* was 16.80 MJ/kg. This value was lower than both of *C. vulgaris* and *C. minutissima* that were 18 and 21 MJ/kg, respectively [9]. According to Illman et al. [9] the heating value of algae is in correlation with the lipid content of their biomass rather than with other components such as carbohydrates and proteins. As a comparison, the lipid content of *N. oculata* in this work was about 11 %, whereas *C. vulgaris*and *C. minutissima* lipid content were 18 and 31 %, respectively [9].

The FTIR spectrum of *N. oculata* biomass and its residue were shown in Fig. 5. For the biomass, it could be observed fifteen distinct transmission bands over the wave number range of 4,000–400/cm. These bands were assigned to specific molecular groups and were tentatively identified on the basis of biochemical reference standards and published FTIR spectra as quoted in Table 3. The cellular macromolecules (proteins, lipids, and carbohydrates) could be identified by their distinct transmittance in different frequency regions. The "carbohydrate band spectra" was characterized by weak and medium features at around 1,101–839/cm due to C–O and C–O–C stretching, Si–O stretching of siloxane silicate frustules, Si–OH bond stretching and bending of silanol. The "protein band spectra" was presented as less pronounced band at about 1,242/cm owing to asymmetric stretching of phosphodiester P=O. The spectrum of 1,435/cm related to CH_2 and CH_3 bending of methyl and C–O stretching of carboxylic group. The band at around 1,504/cm was amide II band spectra mainly ν(C–H) and δ(N–H). The medium amide I band at around 1,811–1,599/cm was primarily for C=O stretching of esters. The less pronounced band at 2,523/cm and around 3,659–3,003/cm because of amine salts $-NH_3^+$ stretching and N–H stretching of amide A, respectively. The bands associated with "lipid band spectra" were indicated by two weak peaks at around 2,940–2,835/cm related to CH_2 stretching of methylene and a strong peak at 1,788/cm as a result of C=O stretching of esters. The aforementioned bands spectra have specified the presence of the principal components in the *N. oculata*, namely carbohydrate, protein, and lipid. These components derived from various microalgae cell structure, such as a cell wall, plasma membrane, chloroplasts, mitochondrion, etc. Each cell structure substance would be progressively destroyed during the thermal processes when the temperature reached its decomposition temperature, and part of it released and formed volatile matter, whereas the remaining substance formed char as a solid material. Both of volatile and char would be burned through the combustion processes. The last rests after their complete burning processes were ash as residual materials.

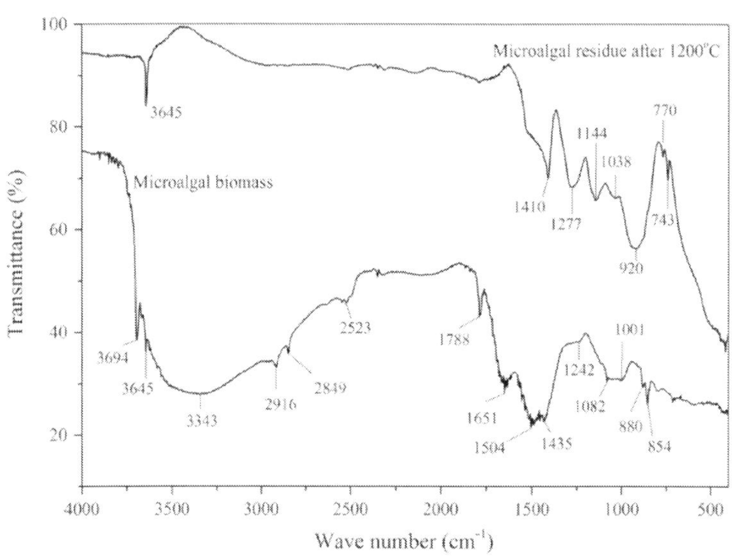

Figure 5: FTIR spectra of *Nannochloropsis oculata* biomass and its residue

Table 3: Tentative assignments of bands found in the FTIR spectra of *Nannochloropsis oculata*biomass

Band	Main peak (/cm)	Wave number range (/cm)	Typical band assignment	References
1	3,694	3,759–3,676	Water	[62, 63]
			v(O–H) stretching of silanol and adsorbed water	
2	3,645	3,659–3,636	Water	[62, 64, 65]
			v(O–H) stretching of silanol and adsorbed water	
			Protein	
			v(N–H) stretching (amide A)	
3	3,343	3,512–3,003	Water	[64, 65, 66, 67, 68]
			v(O–H) stretching	
			Protein	
			v(N–H) stretching (amide A)	
4	2,916	2,940–2,897	Lipid	[49, 64, 69]
			vas(CH_2) stretching of methylene	
5	2,849	2,860–2,835	Lipid	[49, 64, 68, 69, 70]
			vs(CH_2) stretching of methylene	
6	2,523	2,571–2,509	Protein	[63]
			Superimposed (O–H) and $–NH_3^+$ stretching	
7	1,788	1,811–1,765	Lipid (fatty acids)	[49, 66, 68, 69, 71, 72, 73, 74]
			Protein	
			v(C=O) stretching of esters	
8	1,651	1,692–1,599	Protein	[49, 64, 66, 67, 68, 69, 71, 74, 75, 76, 77, 78, 79, 80]
			Amide I band mainly v(C=O) stretching	

9	1,504	1,545–1,460	Protein Amide II band mainly ν(C–H) and δ(N–H) associate with proteins DNA (cytosine) In-plane double bond base vibrations, which include C=C, C=N and C=O stretching	[64, 65, 67, 68, 69, 71, 74, 75, 76, 80, 81, 82, 83]
10	1,435	1,445–1,416	DNA (Adenine) Protein δas(CH_2) and δas(CH_3) bending of methyl Carboxylic group ν(C–O) stretching	[64, 66, 67, 69, 74, 75, 80, 82]
11	1,242	1,312–1,204	DNA-phospholipids Phosphodiester of nucleic acids and phospholipid Antisymmetric PO−2 stretching (vas P=O)	[64, 65, 66, 67, 68, 73, 75, 79, 80, 82, 83]
12	1,082	1,101–1,070	Phospholipids, DNA and RNA νs(P=O) Carbohydrates ν(C–O) stretching ν(C–O–C) stretching of polysaccharides Siloxane, silicate frustules ν(Si–O)	[62, 64, 65, 66, 67, 68, 69, 71, 73, 74, 75, 80, 81, 82, 83, 84, 85, 86]

13	1,001	1,020–957	Siloxanes	[62, 67, 68, 69, 74, 75, 79, 80, 81, 82, 84, 85, 86]
			ν(Si–O)	
			Silanes	
			Si–OH bond stretching	
			Carbohydrates	
			ν(C–O) stretching	
			ν(C–O–C) stretching of polysaccharides	
			RNA–DNA	
			Ribose	
			νas(P=O)	
14	880	893–870	Polysaccharides of the cell wall	[62, 69, 70, 78, 83, 84, 87, 88, 89, 90, 91]
			β-d-glucans deformation	
			δ(C–H)	
			Silanol	
			δ(Si–OH) bending	
15	854	866–839	Polysaccharides of the cell wall	[62, 70, 88, 92, 93, 94, 95, 96]
			α-d-glucans deformation	
			δ(C–H) aromatic deformation	
			Silanol	
			δ(Si–OH) bending	

ν stretching, δ deformation, *subscripts* symmetric, *subscript as* asymmetric

Microalgal residue bands spectra were characterized by a strong feature at 3,645/cm related to O–H stretching of adsorbed water by silica presented in the ash, and strong, medium, and weak features at the fingerprint area associated with the rest of the inorganic component which was left in the ash. The band spectra between 1,490 and 1,410/cm correlated with calcium carbonate, mainly due to antisymmetric CO_3– stretching [58]. The features at around 1,200–900/cm region associated with the silica structure, especially due to Si–O stretching modes [59]. The peaks below 800/cm were vibrations of Ca–O and Mg–O [60]. The strong band at around 400–500/cm is caused by Si–O or Al–O bending vibrations [61]. The elements of ash content were critical parameter in that case of byproduct handling and utilization, as well as they have an impact on the overall processing cost of biomass conversion. However, the ash characterization, other than FTIR that have been previously stated, was not performed in this research, and it will be extensively investigated in the future studies.

Figure 5 revealed the progressive degradation intensity in the spectra of biomass residue after 1,200 °C, mainly at around 3,590–1,630/cm. The declined intensity was caused by the loss of protein and lipid components of biomass during thermal processes. The feature changes at fingerprint area associated with protein degradation of biomass mainly amide II. It also correlated with degradation of cell wall structures including their fibrillar component that composed of polymer of β-d-glucose and their amorphous mucilaginous material which constituted polysaccharides, lipids, and proteins.

CONCLUSIONS

The potential and properties of *N. oculata* biomass for futuristic renewable fuel have been studied. With traditionally natural and low-cost nutrient cultivation, this algae species grows rapidly. The physicochemical properties of this biomass also indicated that it is feasible as biofuel feedstock, except for its high content of inorganic components, which need special attention. The aforementioned attributes allow *N. oculata* to be a strong alternative renewable fuel, either by direct combustion under the proper handling for its residue or by co-combustion with other biomass feedstocks, which have low ash content such as rice straw, for reducing ash product. To optimize the usefulness of this biomass related to its productivity and appropriateness of combustion technology, it requires further in-depth research.

NOTES

Acknowledgments

This work was supported by Directorate General of Higher Education, the Ministry of Education and Culture, Republic of Indonesia; the Fundamental Research Funds, in accordance with the Agreement on Implementation and Assignment Research Program Number 023.04.1.673,453/2012, December 5, 2012, the second revision in May 1,

2013. The authors would like to thank Mrs. Wiwie, Mr. Praptono, and Mr. Sugeng Joko Purnomo from the Institute of Brackish Water Aquaculture (BBAP) Situbondo for facilitating the microalgae cultivation.

Conflict of interest

The authors declare that they have no competing interests.

REFERENCES

1. Yaman, S.: Pyrolysis of biomass to produce fuels and chemical feedstocks. Energy Convers. Manag. 45(5), 651–671 (2004)
2. Özçimen, D., Karaosmanoğlu, F.: Production and characterization of bio-oil and biochar from rapeseed cake. Renew. Energy 29(5), 779–787 (2004)
3. Brennan, L., Owende, P.: Biofuels from microalgae—a review of technologies for production, processing, and extractions of biofuels and co-products. Renew. Sustain. Energy Rev. 14(2), 557–577 (2010)
4. Tabatabaei, M., Tohidfar, M., Jouzani, G.S., Safarnejad, M., Pazouki, M.: Biodiesel production from genetically engineered microalgae: future of bioenergy in Iran. Renew. Sustain. Energy Rev. 15(4), 1918–1927 (2011)
5. Gonçalves, A., Pires, J., Simões, M.: Lipid production of Chlorella vulgaris and Pseudokirchneriella subcapitata. Int. J. Energy Environ. Eng. 4(1), 1–6 (2013)
6. Li, Y., Horsman, M., Wu, N., Lan, C.Q., Dubois-Calero, N.: Biofuels from microalgae. Biotechnol. Prog. 24(4), 815–820 (2008)
7. Hu, Q., Sommerfeld, M., Jarvis, E., Ghirardi, M., Posewitz, M., Seibert, M., et al.: Microalgal triacylglycerols as feedstocks for biofuel production: perspectives and advances. Plant J. 54(4), 621–639 (2008)
8. Chisti, Y.: Biodiesel from microalgae. Biotechnol. Adv. 25(3), 294–306 (2007)
9. Illman, A., Scragg, A., Shales, S.: Increase in chlorella strains calorific values when grown in low nitrogen medium. Enzyme Microb. Technol. 27(8), 631–635 (2000)

10. Gouveia, L., Marques, A.E., da Silva, T.L., Reis, A.: Neochloris oleabundans utex #1185: a suitable renewable lipid source for biofuel production. J. Ind. Microbiol. Biotechnol. 36(6), 821–826 (2009)
11. Mandotra, S.K., Kumar, P., Suseela, M.R., Ramteke, P.W.: Fresh water green microalga Scenedesmus abundans: a potential feedstock for high quality biodiesel production. Bioresour. Technol. 156, 42–47 (2014)
12. Schenk, P.M., Thomas-Hall, S.R., Stephens, E., Marx, U.C., Mussgnug, J.H., Posten, C., et al.: Second generation biofuels: high-efficiency microalgae for biodiesel production. Bioenergy Res. 1(1), 20–43 (2008)
13. Akowuah, J.O., Kemausuor, F., Mitchual, S.J.: Physico-chemical characteristics and market potential of sawdust charcoal briquette. Int. J. Energy Environ. Eng. 3(1), 1–6 (2012)
14. Van Loo, S., Koppejan, J.: The Handbook of Biomass Combustion and Co-firing. Earthscan, London (2008)
15. Raveendran, K., Ganesh, A., Khilar, K.: Influence of mineral matter on biomass pyrolysis characteristics. Fuel 74(12), 1812–1822 (1995)
16. Xiao, R., Chen, X., Wang, F., Yu, G.: The physicochemical properties of different biomass ashes at different ashing temperature. Renew. Energy 36(1), 244–249 (2011)
17. Doshi, V., Vuthaluru, H.B., Korbee, R., Kiel, J.H.A.: Development of a modeling approach to predict ash formation during co-firing of coal and biomass. Fuel Process. Technol. 90(9), 1148–1156 (2009)
18. Wei, X., Schnell, U., Hein, K.: Behaviour of gaseous chlorine and alkali metals during biomass thermal utilisation. Fuel 84(7–8), 841–848 (2005)
19. Du, S., Yang, H., Qian, K., Wang, X., Chen, H.: Fusion and transformation properties of the inorganic components in biomass ash. Fuel 117, 1281–1287 (2014)
20. Obernberger, I., Brunner, T., Barnthaler, G.: Chemical properties of solid biofuels—significance and impact. Biomass Bioenergy 30(11), 973–982 (2006)
21. Hibberd, D.J.: Notes on the taxonomy and nomenclature of the

algal classes eustigmatophyceae and tribophyceae (synonym xanthophyceae). Bot. J. Linn. Soc. 82(2), 93–119 (1981)
22. Karlson, B., Potter, D., Kuylenstierna, M., Andersen, R.A.: Ultrastructure, pigment composition, and 18 s rrna gene sequence for Nannochloropsis granulata sp. nov. (monodopsidaceae, eustigmatophyceae), a marine ultraplankter isolated from the Skagerrak, northeast atlantic ocean. Phycologia 35(3), 253–260 (1996)
23. Gwo, J.-C., Chiu, J.-Y., Chou, C–.C., Cheng, H.-Y.: Cryopreservation of a marine microalga, Nannochloropsis oculata (eustigmatophyceae). Cryobiology 50(3), 338–343 (2005)
24. Hu, H., Gao, K.: Optimization of growth and fatty acid composition of a unicellular marine picoplankton, Nannochloropsis sp., with enriched carbon sources. Biotechnol. Lett. 25(5), 421–425 (2003)
25. Whittle, S.J., Casselton, P.J.: The chloroplast pigments of the algal classes eustigmatophyceae and xanthophyceae. I. Eustigmatophyceae. Br. Phycol. J. 10(2), 179–191 (1975)
26. Volkman, J.K., Brown, M.R., Dunstan, G.A., Jeffrey, S.: The biochemical composition of marine microalgae from the class eustigmatophyceae. J. Phycol. 29(1), 69–78 (1993)
27. Adl, S.M., Simpson, A.G.B., Farmer, M.A., Andersen, R.A., Anderson, O.R., et al.: The new higher level classification of eukaryotes with emphasis on the taxonomy of protists. J. Eukaryot. Microbiol. 52(5), 399–451 (2005)
28. Barsanti, L., Gualtieri, P.: Algae: Anatomy, Biochemistry, and Biotechnology. CRC Press, Boca Raton (2006)
29. Fogg, G.E.: Some comments on picoplankton and its importance in the pelagic ecosystem. Aquat. Microb. Ecol. 9, 33–39 (1995)
30. Lubián, L.M., Montero, O., Moreno-Garrido, I., Huertas, I.E., Sobrino, C., González-del Valle, M., et al.: Nannochloropsis (eustigmatophyceae) as source of commercially valuable pigments. J. Appl. Phycol. 12(3), 249–255 (2000)
31. Lee, M.-Y., Min, B.-S., Chang, C.-S., Jin, E.: Isolation and characterization of a xanthophyll aberrant mutant of the green alga Nannochloropsis oculata. Mar. Biotechnol. 8(3), 238–245 (2006)

32. Osinga, R., Kleijn, R., Groenendijk, E., Niesink, P., Tramper, J., Wijffels, R.H.: Development of in vivo sponge cultures: particle feeding by the tropical sponge Pseudosuberites aff. andrewsi. Mar. Biotechnol. New York NY 3(6), 544–554 (2001)
33. Ferreira, M., Coutinho, P., Seixas, P., Fábregas, J., Otero, A.: Enriching rotifers with "premium" microalgae. Nannochloropsis gaditana. Mar. Biotechnol. 11(5), 585–595 (2009)
34. Rodolfi, L., Chini Zittelli, G., Bassi, N., Padovani, G., Biondi, N., Bonini, G., et al.: Microalgae for oil: strain selection, induction of lipid synthesis and outdoor mass cultivation in a low-cost photobioreactor. Biotechnol. Bioeng. 102(1), 100–112 (2009)
35. Griffiths, M.J., Harrison, S.T.L.: Lipid productivity as a key characteristic for choosing algal species for biodiesel production. J. Appl. Phycol. 21(5), 493–507 (2009)
36. Andersen, R.A. (ed.): Algal Culturing Techniques. Elsevier Academic Press, London (2005)
37. Wang, L., Weller, C.L.: Recent advances in extraction of nutraceuticals from plants. Trends Food Sci. Technol. 17(6), 300–312 (2006)
38. Beamish, B.B.: Proximate analysis of New Zealand and Australian coals by thermogravimetry. New Zeal. J. Geol. Geophys. 37(4), 387–392 (1994)
39. Mayoral, M.C., Izquierdo, M.T., Andres, J.M., Rubio, B.: Different approaches to proximate analysis by thermogravimetry analysis. Thermochim. Acta 370, 91–97 (2001)
40. Nhuchhen, D.R., Abdul Salam, P.: Estimation of higher heating value of biomass from proximate analysis: a new approach. Fuel 99, 55–63 (2012)
41. Goswami, R., Kalita, N., Kalita, M.C.: A study on growth and carbon dioxide mitigation by microalgae Selenastrum sp.: its growth behavior under different nutrient environments and lipid production. Ann. Biol. Res. 3(1), 499–510 (2012)
42. García-González, M., Moreno, J., Manzano, J.C., Florencio, F.J., Guerrero, M.G.: Production of Dunaliella salina biomass rich in 9-cis-beta-carotene and lutein in a closed tubular photobioreactor. J. Biotechnol. 115(1), 81–90 (2005)

43. De Morais, M.G., Costa, J.A.V.: Biofixation of carbon dioxide by Spirulina sp. and Scenedesmus obliquus cultivated in a three-stage serial tubular photobioreactor. J. Biotechnol. 129(3), 439–445 (2007)

44. Converti, A., Casazza, A.A., Ortiz, E.Y., Perego, P., Del Borghi, M.: Effect of temperature and nitrogen concentration on the growth and lipid content of Nannochloropsis oculataand Chlorella vulgaris for biodiesel production. Chem. Eng. Process. Process Intensif. 48(6), 1146–1151 (2009)

45. Patil, P.D., Gude, V.G., Mannarswamy, A., Cooke, P., Munson-McGee, S., Nirmalakhandan, N., et al.: Optimization of microwave-assisted transesterification of dry algal biomass using response surface methodology. Bioresour. Technol. 102(2), 1399–1405 (2011)

46. McKendry, P.: Energy production from biomass (part 1): overview of biomass. Bioresour. Technol. 83(1), 37–46 (2002)

47. Biedermann, F., Obernberger, I.: Ash-related problems during biomass combustion and possibilities for a sustainable ash utilisation. In: Proceedings of Int. Conf. 'World Renew. Energy Congr. Elsevier Ltd, Aberdeen (2005)

48. Yang, Y.B., Sharifi, V.N., Swithenbank, J.: Effect of air flow rate and fuel moisture on the burning behaviours of biomass and simulated municipal solid wastes in packed beds. Fuel 83(11–12), 1553–1562 (2004)

49. Phukan, M.M., Chutia, R.S., Konwar, B.K., Kataki, R.: Microalgae chlorella as a potential bio-energy feedstock. Appl. Energy 88, 3307–3312 (2011)

50. Chen, C., Ma, X., Liu, K.: Thermogravimetric analysis of microalgae combustion under different oxygen supply concentrations. Appl. Energy 88(9), 3189–3196 (2011)

51. Wiinikka, H.: High Temperature Aerosol Formation and Emission Minimisation during Combustion of Wood Pellets. Luleå University of Technology, Luleå (2005)

52. Hill, S., Douglas Smoot, L.: Modeling of nitrogen oxides formation and destruction in combustion systems. Prog. Energy Combust. Sci. 26(4–6), 417–458 (2000)

53. Toftegaard, M.B., Brix, J., Jensen, P.A., Glarborg, P., Jensen, A.D.: Oxy-fuel combustion of solid fuels. Prog. Energy Combust. Sci. 36(5), 581–625 (2010)
54. Liu, F., Guo, H., Smallwood, G.J., Gülder, Ö.L.: The chemical effects of carbon dioxide as an additive in an ethylene diffusion flame: implications for soot and NOx formation. Combust. Flame 125(1–2), 778–787 (2001)
55. Liu, J., Gao, S., Jiang, X., Shen, J., Zhang, H.: NO emission characteristics of superfine pulverized coal combustion in the O2/CO2 atmosphere. Energy Convers. Manag. 77, 349–355 (2014)
56. Sukarni, Sudjito, Hamidi, N., Yanuhar, U., Wardana, I.N.G.: Thermogravimetric kinetic analysis of Nannochloropsis oculata combustion in air atmosphere. Front. Energy (2014) (accepted manuscript)
57. Liu, Z.: Energy from combustion of rice straw: status and challenges to china. Energy Power Eng. 03(03), 325–331 (2011)
58. Bellamy, L.: The infra-red spectra of complex molecules. Chapman and Hall, London (1975)
59. Dodson, J.: Wheat straw ash and its use as a silica source. Dissertation, University of York (2011)
60. Abraham, R., George, J., Thomas, J., Yusuff, K.K.M.: Physicochemical characterization and possible applications of the waste biomass ash from oleoresin industries of India. Fuel 109, 366–372 (2013)
61. Bai, J., Li, W., Li, B.: Characterization of low-temperature coal ash behaviors at high temperatures under reducing atmosphere. Fuel 87(4–5), 583–591 (2008)
62. Parida, S.K., Dash, S., Patel, S., Mishra, B.K.: Adsorption of organic molecules on silica surface. Adv. Colloid Interface Sci. 121(1–3), 77–110 (2006)
63. Silverstein, R.M., Webster, F.X., Kiemle, D.: Spectrometric Identification of Organic Compounds, 7th edn. Wiley, New Jersey (2005)
64. Sigee, D.C.D., Dean, A., Levado, E., Tobin, M.J.: Fourier-transform infrared spectroscopy of Pediastrum duplex: characterization of a

micro-population isolated from a eutrophic lake. Eur. J. Phycol. 37(1), 19–26 (2002)

65. Duygu, D.Y., Udoh, A.U., Ozer, T.B., Akbulut, A., Acikgoz, I., Yildiz, K., et al.: Fourier transform infrared (FTIR) spectroscopy for identification of Chlorella vulgaris Beijerinck 1890 and Scenedesmus obliquus (turpin) kützing 1833. Afr. J. Biotechnol. 11(16), 3817–3824 (2012)
66. Murdock, J.N., Wetzel, D.L.: FT-IR microspectroscopy enhances biological and ecological analysis of algae. Appl. Spectrosc. Rev. 44(4), 335–361 (2009)
67. Benning, L.G., Phoenix, V., Yee, N., Konhauser, K.: The dynamics of cyanobacterial silicification: an infrared micro-spectroscopic investigation. Geochim. Cosmochim. Acta 68(4), 743–757 (2004)
68. Mayers, J.J., Flynn, K.J., Shields, R.J.: Rapid determination of bulk microalgal biochemical composition by Fourier-transform infrared spectroscopy. Bioresour. Technol. 148, 215–220 (2013)
69. Jiang, Y., Yoshida, T., Quigg, A.: Photosynthetic performance, lipid production and biomass composition in response to nitrogen limitation in marine microalgae. Plant Physiol. Biochem. 54, 70–77 (2012)
70. Marshall, C., Javaux, E., Knoll, A., Walter, M.: Combined micro-Fourier transform infrared (FTIR) spectroscopy and micro-Raman spectroscopy of proterozoic acritarchs: a new approach to palaeobiology. Precambrian Res. 138(3–4), 208–224 (2005)
71. Gao, Y., Yang, M., Wang, C.: Nutrient deprivation enhances lipid content in marine microalgae. Bioresour. Technol. 147, 484–491 (2013)
72. Tan, S.-T., Balasubramanian, R.K., Das, P., Obbard, J.P., Chew, W.: Application of mid-infrared chemical imaging and multivariate chemometrics analyses to characterise a population of microalgae cells. Bioresour. Technol. 134, 316–323 (2013)
73. Mecozzi, M., Pietroletti, M., Di Mento, R.: Application of FTIR spectroscopy in ecotoxicological studies supported by multivariate analysis and 2d correlation spectroscopy. Vib. Spectrosc. 44(2), 228–235 (2007)
74. Stehfest, K., Toepel, J., Wilhelm, C.: The application of micro-FTIR

spectroscopy to analyze nutrient stress-related changes in biomass composition of phytoplankton algae. Plant Physiol. Biochem. 43(7), 717–726 (2005)

75. Benning, L.G., Phoenix, V., Yee, N., Tobin, M.: Molecular characterization of cyanobacterial silicification using synchrotron infrared micro-spectroscopy. Geochim. Cosmochim. Acta 68(4), 729–741 (2004)
76. Ragusa, S., Cambria, M.T., Pierfederici, F., Scirè, A., Bertoli, E., Tanfani, F., et al.: Structure–activity relationship on fungal laccase from Rigidoporus lignosus: a Fourier-transform infrared spectroscopic study. Biochim. Biophys. Acta 1601(2), 155–162 (2002)
77. Wysokowski, M., Behm, T., Born, R., Bazhenov, V.V., Meissner, H., Richter, G., et al.: Preparation of chitin-silica composites by in vitro silicification of two-dimensional Ianthella basta demosponge chitinous scaffolds under modified Stöber conditions. Mater. Sci. Eng. C. Mater. Biol. Appl. 33(7), 3935–3941 (2013)
78. Kamnev, A., Ristić, M., Antonyuka, L.P., Chernyshev, A.V., Ignatov, V.V.: Fourier transform infrared spectroscopic study of intact cells of the nitrogen-fixing bacterium Azospirillum brasilense. J. Mol. Struct. 409, 201–205 (1997)
79. Dean, A.P., Sigee, D.C., Estrada, B., Pittman, J.K.: Using FTIR spectroscopy for rapid determination of lipid accumulation in response to nitrogen limitation in freshwater microalgae. Bioresour. Technol. 101(12), 4499–4507 (2010)
80. Dean, A.P., Nicholson, J.M., Sigee, D.C.: Impact of phosphorus quota and growth phase on carbon allocation in Chlamydomonas reinhardtii: an FTIR microspectroscopy study. Eur. J. Phycol. 43(4), 345–354 (2008)
81. Meng, Y., Yao, C., Xue, S., Yang, H.: Application of Fourier transform infrared (FT-IR) spectroscopy in determination of microalgal compositions. Bioresour. Technol. 151, 347–354 (2014)
82. Banyay, M., Sarkar, M., Gräslund, A.: A library of IR bands of nucleic acids in solution. Biophys. Chem. 104, 477–488 (2003)
83. Jangir, D.K., Charak, S., Mehrotra, R., Kundu, S.: FTIR and circular dichroism spectroscopic study of interaction of 5-fluorouracil with

DNA. J. Photochem. Photobiol. B 105(2), 143–148 (2011)

84. Goo, B.G., Baek, G., Choi, D.J., Synytsya, A., Park, Y., Bleha, R., et al.: Characterization of a renewable extracellular polysaccharide from defatted microalgae Dunaliella tertiolecta. Bioresour. Technol. 129, 343–350 (2013)
85. Guo, H., Daroch, M., Liu, L., Qiu, G., Geng, S., Wang, G.: Biochemical features and bioethanol production of microalgae from coastal waters of pearl river delta. Bioresour. Technol. 127, 422–428 (2013)
86. El-Toni, A.M., Khan, A., Ibrahim, M.A., Labis, J.P., Badr, G., Al-Hoshan, M., et al.: Synthesis of double mesoporous core-shell silica spheres with tunable core porosity and their drug release and cancer cell apoptosis properties. J. Colloid Interface Sci. 378(1), 83–92 (2012)
87. Lim, J.M., Joo, J.H., Kim, H.O., Kim, H.M., Kim, S.W., Hwang, H.J., et al.: Structural analysis and molecular characterization of exopolysaccharides produced by submerged mycelial culture of Collybia maculata tg-1. Carbohydr. Polym. 61(3), 296–303 (2005)
88. Lecellier, A., Mounier, J., Gaydou, V., Castrec, L., Barbier, G., Ablain, W., et al.: Differentiation and identification of filamentous fungi by high-throughput FTIR spectroscopic analysis of mycelia. Int. J. Food Microbiol. 168–169, 32–41 (2014)
89. Huang, G.L.: Extraction of two active polysaccharides from the yeast cell wall. Z. Naturforsch. C. 63(11–12), 919–921 (2008)
90. Han, M., Han, J., Hyun, S., Shin, H.: Solubilization of water-insoluble beta-glucan isolated from Ganoderma lucidum. J. Environ. Biol. 29(March), 237–242 (2008)
91. Jung, H.-K., Hong, J.-H., Park, S.-C., Park, B.-K., Nam, D.-H., Kim, S.-D.: Production and physicochemical characterization of β-glucan produced by Paenibacillus polymyxa jb115. Biotechnol. Bioprocess Eng. 12, 713–719 (2007)
92. Peng, Y., Zhang, L., Zeng, F., Kennedy, J.F.: Structure and antitumor activities of the water-soluble polysaccharides from Ganoderma tsugae mycelium. Carbohydr. Polym. 59(3), 385–392 (2005)
93. Huang, Q., Jin, Y., Zhang, L., Cheung, P.C.K., Kennedy, J.F.: Structure, molecular size and antitumor activities of polysaccharides

from poria cocos mycelia produced in fermenter. Carbohydr. Polym. 70(3), 324–333 (2007)

94. De Lourdes Corradida Silva, M., Fukuda, E.K., Vasconcelos, A.F.D., Dekker, R.F.H., Matias, A.C., Monteiro, N.K., et al.: Structural characterization of the cell wall d-glucans isolated from the mycelium of Botryosphaeria rhodina mamb-05. Carbohydr. Res. 343(4), 793–798 (2008)

95. Tianqi, W., Hanxiang, L.I., Manyi, W., Tianwei, T.A.N.: Integrative extraction of ergosterol, (1 → 3)—α-d-glucan and chitosan from penicillium chrysogenum mycelia*. Chin. J. Chem. Eng. 15(5), 725–729 (2007)

96. Peng, Y., Zhang, L., Zeng, F., Xu, Y.: Structure and antitumor activity of extracellular polysaccharides from mycelium. Carbohydr. Polym. 54(3), 297–303 (2003)

CHAPTER

15

EFFECTIVE CULTIVATION OF MICROALGAE FOR BIOFUEL PRODUCTION: A PILOT-SCALE EVALUATION OF A NOVEL OLEAGINOUS MICROALGA GRAESIELLA SP. WBG-1

Xiaobin Wen[1], Kui Du[1,2], Zhongjie Wang[1], Xinan Peng[1,2], Liming Luo[1,3], Huanping Tao[1], Yan Xu[1,2], Dan Zhang[1], Yahong Geng[1] and Yeguang Li[1]

[1]. Key Laboratory of Pant Germplasm Enhancement and Specialty Agriculture, Wuhan Botanical Garden, Chinese Academy of Sciences, Wuhan, China

[2]. University of Chinese Academy of Sciences, Beijing, China

[3]. Department of Pathology and Immunology, Baylor College of Medicine, Houston, USA

ABSTRACT

Background

Commercial production of microalgal biodiesel is not yet economically viable, largely because of low storage lipid yield in microalgae mass cultivation. Selection of lipid-rich microalgae, thus, becomes one of the key research topics for microalgal biodiesel production. However, the laboratory screening protocols alone cannot predict the ability of the strains to dominate and perform in outdoor ponds. Comprehensive assessment of microalgae species should be performed not only under the laboratory conditions, but also in the fields.

Results

Laboratory investigations using a bubbled column photobioreactor indicated the microalga *Graesiella* sp. WBG-1 to be the most productive species among the 63 Chlorophyta strains. In a 10 L reactor, mimicking the industrial circular pond, *Graesiella* sp. WBG-1 produced 12.03 g biomass m^{-2} day^{-1} and 5.44 g lipids (45.23 % DW) m^{-2} day^{-1} under 15 mol m^{-2} day^{-1}artificial light irradiations. The lipid content decreased to ~34 % DW when the microalga was cultured in 30 L tank PBR under natural solar irradiations, but the decline of lipid content with scaling up was the minimum among the tested strains. Based on these results, the microalga was further tested for its lipid production and culture competitiveness using a pilot-scale raceway pond (200 m^2 illuminated area, culture volume 40,000 L). Consequently, *Graesiella*sp. WBG-1 maintained a high lipid content (33.4 % DW), of which ~90 % was storage TAGs. Results from the outdoor experiments indicated the nice adaptability of the *Graesiella* sp. WBG-1 to strong and fluctuating natural solar irradiance and temperature, and also demonstrated several other features, such as large cell size (easy for harvest and resistant to swallow by protozoa) and tolerance to high culture pH (helpful to CO_2 fixation).

Conclusions

Graesiella sp. WBG-1 was a promising strain capable of accumulating large amount of storage lipid under nature solar irradiance and temperature. The high lipid content of 33.4 % DW was achieved for the first time in pilot-scale raceway pond. The results also provide evidence for the feasibility of using low-cost raceway pond for autotrophic cultivation of microalgae for biodiesel production.

Keywords: *Graesiella* Microalgae selection Lipid Mass culture Open raceway pond

BACKGROUND

In recent years, mass production of microalgae has been extensively exploited for applications in the industry of food, aquaculture, and bioenergy, and especially for biodiesel production [1, 2]. However, there is still a large margin to improve economic yield for large-scale biofuel production [3, 4], since fossil fuel is still much cheaper and serves as the dominant energy source nowadays.

One of the obstacles making microalgal biodiesel commercially unfeasible is the low yield of storage lipid in outdoor microalgae mass cultivation [3, 5]. So far, the algal biomass productivities achieved in large-scale raceway pond do not exceed 20–40 g dry weight m^{-2} day^{-1}, and a maximum solar-to-biomass conversion efficiency of 3 % was reported [6, 7]. Furthermore, high lipid accumulation is more difficult to achieve. Although *Chlorella*, *Scenedesmus*, *Neochloris*, and *Nanochloropsis* strains are repeatedly reported to give an average lipid content of 40–60 % in dry cell mass in laboratory [8, 9], outdoor cultivation of these strains has only been accomplished on very small scale and a total lipid content of 30 % has rarely been achieved [10, 11, 12]. The Aquatic Species Program (ASP) spent considerable effort in isolation, screening, genetic improvement, and outdoor cultivation of microalgae strains. At its peak, the collection of ASP contained over 3000 strains of lipid-rich microalgae [13]. However, there is still lack of robust strains capable of high lipid productivity in outdoor large-scale cultivation [6, 13]. In its close-out report, ASP pointed out that the laboratory-level screening protocols have relatively little predictive power for the ability of the strains to dominate and perform in outdoor ponds [13].

To fill the gap between laboratory experiment and field test, future research should cover the whole chain of process development in an integrated and iterative way [6]. In this context, selection and assessment of robust microalgal strains for large-scale cultivation have become one of the key research topics for biodiesel production. Small-scale open systems should be used as selection devices for microalgae strains suitable for outdoor mass culture [13]. Full-scale assessment of lipid-rich microalgae should be performed not only under the laboratory conditions, but also in the fields, to test their adaptability to changes in temperature, strong light irradiance, and other chemical and biological environment conditions.

Graesiella sp. WBG-1 is a unicellular green microalga with broadly ellipsoidal or globose cells. This strain was originally isolated from Chenghai Lake, Yunnan province, China, by the researchers in our lab. Molecular analysis contributed the WBG-1 strain mainly to genus *Graesiella* and showed 99.8 % similarity with two *Graesiella* species: *Graesiellaemersonii* and *Graesiellavacuolata* based on the 18S rDNA/ITS sequence. Preliminary investigations using a bubbled column photobioreactor showed *Graesiella* sp. WBG-1 to be one of the most productive microalgae among the 63 studied Chlorophyta strains. Furthermore, this strain also possesses some other desirable features, such as large cell size (easy to harvest) and high adaptive capacity to a wide range of culture pH. All the above advantages together encouraged us to perform a comprehensive selection and assessment using this promising strain *Graesiella* sp. WBG-1.

Given the above considerations, we carried out both indoor and outdoor experiments to study the effects of several fundamental factors on the growth and lipid accumulation of the microalga *Graesiella* sp. WBG-1, and to test its robustness under the outdoor environmental conditions in a 200 m^2 raceway pond. CO_2 utilization and other features of *Graesiella* sp. WBG-1 were also discussed.

METHODS

Growth medium and preparation

The basal growth medium for microalgae culture was a modified BG-11 medium which had the following components (per liter): $NaNO_3$ (100 mg), $K_2HPO_4 \cdot 3H_2O$ (40 mg), $MgSO_4 \cdot 7H_2O$ (75 mg), $CaCl_2 \cdot 2H_2O$ (36 mg), citric acid (6 mg), Fe-ammonium citrate (6 mg), EDTA·Na_2(1 mg), $NaHCO_3$ (20 mg), H_3BO_3 (2.86 mg), $MnCl_2 \cdot 4H_2O$ (1.8 mg), $ZnSO_4 \cdot 7H_2O$ (0.22 mg), $CuSO_4 \cdot 5H_2O$(0.08 mg), $Na_2MoO_4 \cdot 2H_2O$ (0.391 mg), and $Co(NO_3)_2 \cdot 6H_2O$ (0.0494 mg). Sodium nitrate and sodium bicarbonate concentrations were modified as indicated in the text.

The BG-11 medium was prepared with deionized water for laboratory experiments, and with water from Chenghai Lake (containing ~1 g L^{-1} $NaHCO_3$) for outdoor experiments. The deionized water was sterilized by autoclave, and the water from Chenghai Lake was purified by filtration, and then, the sterile nutrient stock solutions were added to either the deionized water or the water from Chenghai Lake to make final medium. The culture pond was exposed to direct solar irradiation for at least 12 h before medium preparation.

Bubbled column PBR cultivation

Each column of the bubbled column PBR used in this study has an inner diameter of 3 cm and a working volume of 200 mL. The columns were illuminated with white fluorescent tubes for 14 h every day, and the light intensity at the reactor surface was 300 μmol m^{-2} s^{-1}(15 mol photons m^{-2} day^{-1}). CO_2-enriched (1 %, v/v) air was injected at the bottom of the column through a glass tube, that is used for CO_2 supplement, culture mixing, and O_2exchange. The air flow rate was maintained at 250 mL min^{-1}. A thermostatic water circulator was used as water bath to keep the temperature of the culture columns at 30 °C.

The algal cells were harvested from seed culture and resuspended into sterilized BG-11 medium for inoculation. The cells were grown in batch with an initial cell density of 0.5 ± 0.05 (optical density at 540 nm). To investigate the effects of nitrate and bicarbonate concentration on cell growth and lipid accumulation, different doses of sodium nitrate or sodium bicarbonate were added into each column before inoculation. All of the experiments were carried out in triplicate.

Scale-up cultivation in a 10 L circular pond

A mini circular pond cultivation system was used to simulate the outdoor open pond cultivation under artificial illumination. The circular pond was equipped with four units to control mixing, irradiance, temperature, and pH, as described in previous study [14].

The cells were grown in batch with an initial cell density of 0.1 ± 0.05 (optical density at 540 nm). Total culture volume was about 10 L giving a culture depth (light path) of 10 cm, and the effective illuminated area was 0.1 m^2. Banks of white fluorescent tubes were placed above the pond, and provided 300 μmol photons m^{-2} s^{-1} with 14 h:10 h light–dark cycle (15 mol photons m^{-2} day^{-1}). The cultures were kept at 30 °C and mixed at 50 r min^{-1}continuously. Pure CO_2 was dispersed into the culture suspension through a gas diffuser under the programming control of an online pH sensor, to maintain the culture pH within a desired range. In another study, we found that *Graesiella* sp. WBG-1 was well adapted to culture pH range of 7–10 (unpublished data). Therefore, the culture pH of 9.0 ± 0.5 was maintained in the simulate experiments to facilitate external CO_2 uptake. The experiments were carried out in duplicate.

Scale-up cultivation in a 30 L tank PBR

To test the adaptability of the strain to variable environmental conditions, outdoor scale-up cultivations of the *Graesiella* sp. WBG-1 were carried out in 30 L tank PBR under natural light illumination.

The PBR used in this experiment consisted of four open-top polyethylene tanks placed side-by-side in a single row on a concrete platform. The working volume is 30 L for each culture tank, giving an illuminated area of 0.19 m^2. An electromagnetic air compressor continuously blew sterilized air into the culture suspension during cultivation. The air flow was maintained at 3 L min^{-1}. The air was enriched with CO_2 (1 %, v/v) during light period to support cell growth and maintain pH within the desired range (9.0 ± 0.5). Natural solar irradiance and environmental temperature were monitored and logged on site with an automatic weather station.

Outdoor cultivation in 200 m^2 raceway pond

Pilot-scale evaluation of *Graesiella* sp. WBG-1 was carried out at Chenghai Lake, Yunnan province, China, (N26°29′29.64″ E100°40′56.12″) using a traditional raceway pond.

The raceway pond was an open system constructed of concrete blocks. The pond was 20 m long, 12 m wide, giving an effective culture area of 200 m^2. Two large paddle wheels were installed apart within the raceway for mixing, and were driven by speed-adjustable motors and able to provide a flow velocity of 10–60 cm s^{-1}. The liquid flow velocity was set at 45 cm s^{-1} in this study, and the paddle wheels were turned on in daytime and turned off in night. Pure CO_2 was automatically injected into the culture via a 3 m-long microporous polymer tube (gas diffuser) that was placed at the bottom of the culture pond. The automatic injection of CO_2, which was controlled by an online pH sensor, maintained the culture pH within the desired range (pH 9.0 ± 0.5). The working culture depth (light path) was 20 cm, corresponding to 40,000 L culture volume.

Natural solar irradiance, environmental temperature, and suspension temperature were monitored and logged on site with an automatic weather station.

Two similar but smaller raceway ponds (20 m^2) covered by greenhouse were used for high-quality seed culture preparation. Independent batch cultures in the 200 m^2 pond were carried out three times in June 2013, July 2013, and May 2014.

Analytical procedure

Culture growth was estimated by measuring the dry biomass concentration of the culture broth. About 10 mL culture broth was filtered through a pre-dried GF/C glass microfiber filter paper (pore size 0.45 μm), and dried at 105 °C for 4 h, and then weighed to calculate dry biomass concentration (DW, in grams of biomass per liter of culture broth). Daily biomass productivity was calculated by dividing the difference between the DWs at the start time and the end time by its duration (days). The microalga was stained with Nile Red according to Li's method [10] and observed using a Nikon Eclipse 80i microscope.

A spectrophotometric method, described by Collos et al. [15], was used to monitor residual nitrate concentration. Briefly, the absorbance of culture filtrate (0.22 μm filter) at 220 nm and a pre-constructed standard curve were used to determine residual nitrate.

Cells were collected by centrifugation (5000 rpm for 5 min) and lyophilized (−56 °C cryotrapping, 10–14 Pa vacuum) for biochemical analysis. For lipid quantification, 50 mg of dry algal biomass was fully grounded, transferred to a covered centrifuge tube, and then extracted with a mixture of *n*-hexane and ethyl acetate (1:1, v:v) for 20 min. The extraction was repeated three times, and all extracts were combined into a pre-weighed glass tube, and then dried under nitrogen protection. The lipids were determined gravimetrically.

Neutral lipid (Triacylglycerides, TAGs) was fractionated from the lipid extracts by column chromatography using a 2 cm × 20 cm column packed with 4 g silica gel 60 [16]. The lipid extracts (~100 mg) were dissolved in 2 mL chloroform and loaded onto the column. TAGs were eluted from the column by 20 mL chloroform. The eluted TAGs fraction were confirmed by TLC, and then dried and quantified gravimetrically.

To analyze fatty acids, about 20 mg of the lipid extract was dissolved in 3 mL *n*-hexane and then transmethylated by adding 3 mL methanol-KOH (0.5 % KOH) and heating at 50 °C for 60 min. After cooling to room temperature, the hexane layer was separated and dried with anhydrous sodium sulfate. Fatty acid methyl esters were analyzed by gas chromatography (Agilent 7890A) using an HP-5 Phenyl Methyl Siloxan column (30 m × 0.32 mm × 0.25 μm) and a flame ionization detector. 1 μL fatty acid methyl esters solution was injected to the sampler with a splitting ratio of 5:1. The heating program was 150 °C held for 2 min, then increased to 250 °C at a rate of 10 °C per min, and held for 8 min. A standard FAME Mix (Sigma-Aldrich) was used for fatty acid identification.

About 10 mg of lyophilized algal powder was used for total carbohydrate quantification. The algal powder was transferred to a covered tube, fully mixed with 1 mL hydrochloric acid (6 M), to digest at 105 °C for 1 h. After cooling to RT, about 1 mL NaOH (6 M) was added into the solution to neutralize acid, followed by centrifugation at 3500*g* for 5 min. The supernatant was collected into a new tube and its volume was brought to 2 mL with deionized H_2O. 100 μL of the diluted supernatant was diluted again with deionized H_2O to 2 mL, mixed with 1 mL phenol (6 %), and then 5 mL sulfuric acid was trickled into the sample for color development. Finally, the optical density at 490 nm was measured on a spectrophotometer. To quantify total carbohydrate content, glucose was used to establish the standard curve.

Protein content was determined as described by Slocombe et al. [17]. Briefly, 10 mg freeze-dried algal powder was suspended in 500 μL of 24 % (w/v) trichloroacetic acid (TCA) and then incubated at 95 °C for 15 min. The lysate containing 24 % (w/v) TCA were cooled to RT and diluted to 6 % (w/v) with 1.5 mL deionized water. The homogenate was centrifuged at 15,000*g*for 20 min and the supernatant was discarded. The pellets were resuspended in 1 mL NaOH (1 M) by repeated pipetting and then incubated at 40 °C for 2 h. The protein concentration was then spectrophotometrically measured according to standard Bradford assay.

Calculations

Areal biomass concentration ($C_{biomass}$, g m^{-2}) was calculated by the following:

$$C_{\text{biomass}} = \frac{\text{DW} \times V}{S} \quad (1)$$

with DW as measured biomass dry weight (g L^{-1}), V as culture volume (L), and S as illuminated area (m^2).

Biomass productivity (P_{biomass}, g m^{-2} day^{-1}) was calculated according to the following:

$$P_{\text{biomass}} = \frac{C_{\text{biomass},t2} - C_{\text{biomass},t1}}{t2 - t1} \quad (2)$$

with $C_{\text{biomass},t2}$ and $C_{\text{biomass},t1}$ as biomass concentrations at culture time $t2$ (day) and $t1$ (day), respectively.

Lipid productivity (P_{lipid}, g m^{-2} day^{-1}) was calculated according to the following:

$$P_{\text{lipid}} = \frac{C_{\text{biomass},t2} \times C_{\text{lipid},t2} - C_{\text{biomass},t1} \times C_{\text{lipid},t1}}{t2 - t1} \quad (3)$$

with $C_{\text{lipid},t2}$ and $C_{\text{lipid},t1}$ as lipid content at culture time $t2$ (day) and $t1$ (day), respectively.

Average daily light intensity (I_{av}, mol m^{-2} day^{-1}) was calculated according to the following:

$$I_{\text{av}} = \frac{\sum_0^t I_{\text{inc}}}{t} \quad (4)$$

with I_{inc} as daily incident light intensity (mol m^{-2} day^{-1}) measured by the automatic weather station.

Biomass-specific light availability (I_{biomass}, mol g^{-1} day^{-1}) was calculated according to Ref. [18] by the following:

$$I_{\text{biomass}} = \frac{I_{\text{av}}}{C_{\text{biomass}}} \quad (5)$$

CO_2 utilization rate (R_c, %) was calculated by the following:

$$R_c = \frac{W_b \times 1000 \times 0.5 \div 12}{C_1 - C_2 + C_3} \times 100\,\% \qquad (6)$$

with W_b as net increase of biomass (kg) during cultivation; C_1 as dissolved inorganic carbon ($[H_2CO_3] + [CO_3^{2-}] + [HCO_3^-]$, mol) in the medium at the begin of culture; C_2 as dissolved inorganic carbon (mol) in the medium at the end of culture; and C_3 as the total CO_2 (mol) used during culture. The bio-fixated carbon (50 % of the biomass is *C*) divided by consumed carbon was defined as CO_2 utilization rate in this study.

RESULTS

Effects of nitrate and bicarbonate on cell growth of *Graesiella* sp. WBG-1 under the laboratory conditions

The microalga *Graesiella* sp. WBG-1 was cultivated in a 200 mL column photobioreactor under different bicarbonate and nitrate concentrations, to study the effects of carbon and nitrogen on the cell growth and lipid accumulation, which were essential for mass production optimization.

As shown in Fig. 1a, the biomass concentration ($C_{biomass}$) of *Graesiella* sp. WBG-1 increased almost linearly with increasing concentrations of bicarbonate from 0 to 11.9 mM, and the increase was no more significant when the concentration of bicarbonate was above 11.9 mM. When the concentration of bicarbonate was increased to 47.62 mM, the $C_{biomass}$ of *Graesiella*sp. WBG-1 dropped significantly due to excessive salinity. In contrast, the lipid content (C_{lipid}) of *Graesiella* sp. WBG-1 did not change under all tested bicarbonate concentrations (Fig. 1a).

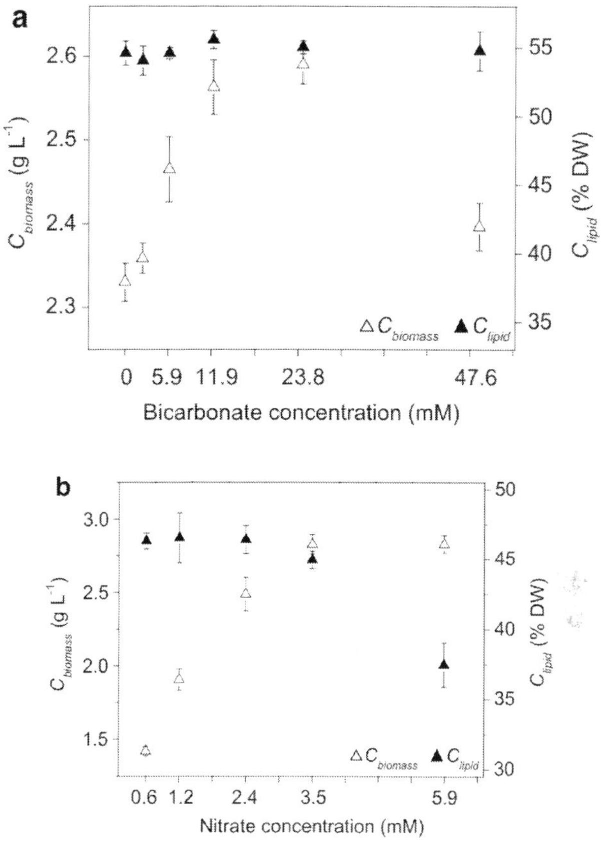

Figure 1: Effects of bicarbonate (**a**) and nitrate (**b**) on cell growth (*open triangle*) and lipid accumulation (*filled triangle*)

The effects of nitrate on cell growth and lipid accumulation were shown in Fig. 1b. The C_{biomass} of *Graesiella* sp. WBG-1 increased with increasing nitrate concentrations, while the lipid content (C_{lipid}) decreased gradually (Fig. 1b). The maximum C_{biomass} of ~2.8 g L^{-1} was achieved at 5.88 mM nitrate, while the C_{lipid} further decreased to 37.46 % DW at this concentration.

Lipid production of *Graesiella* sp. WBG-1 in small-scale reactor

As shown in Fig. 2a, in a 10 L circular pond, *Graesiella* sp. WBG-1 showed fast and continued growth (C_{biomass}) during 8 days of cultivation under 15 mol m^{-2} day^{-1} artificial light irradiation. Although the nitrate in the algal suspension got depleted from day 2, the cells still kept fast growth ($C_{biomass}$) and the growth slightly slowed down from day 5. The overall biomass productivity ($P_{biomass}$) was 12.03 g m^{-2} day^{-1}. From day 2 to day 5, the biomass productivities maintained at ~14 g m^{-2} day^{-1}. The maximum biomass productivity (15.65 g m^{-2} day^{-1}) was achieved on day 2.

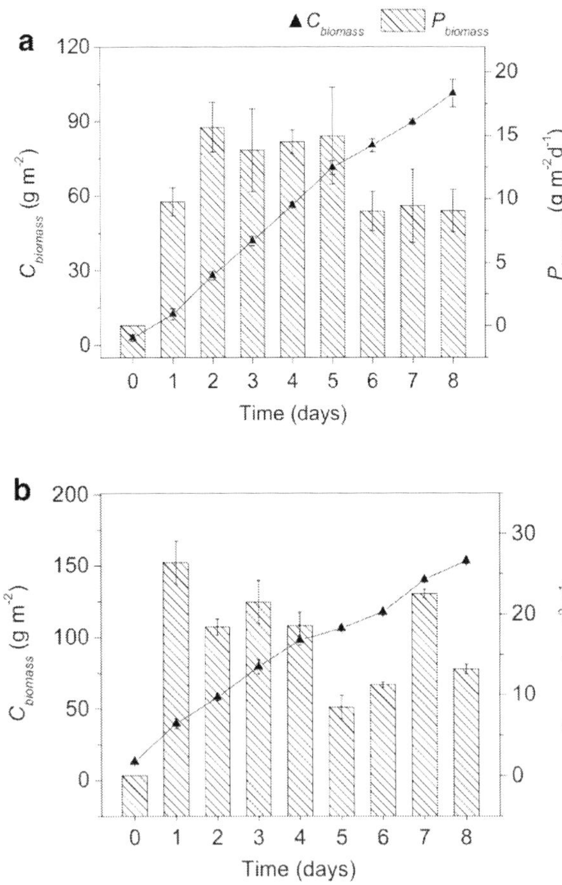

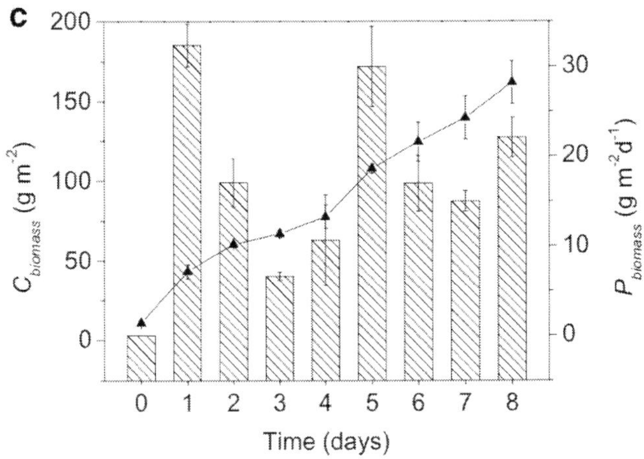

Figure 2: Microalgal growth and biomass productivity in small-scale reactor. **a** Cultivation in 10 L laboratory circular pond; **b** cultivation in 30 L outdoor tank PBR in June 2012; and **c**cultivation in 30 L outdoor tank PBR in July 2012. Time evolution of biomass concentrations and biomass productivities are indicated by *line* and *column,*respectively. The areal biomass concentration ($C_{biomass}$, g m^{-2}) was used in all outdoor experiments

Cultivation of *Graesiella* sp. WBG-1 in 30 L tank PBR was carried out twice under direct sun light irradiation. The first cultivation (Fig. 2b) was carried out in June 2012 with an average light intensity of 28 mol photon m^{-2} day^{-1} and air temperature of 23.2 °C. The second time (Fig. 2c) was carried out in July 2012 with an average light intensity of 31 mol photon m^{-2} day^{-1}and air temperature of 24.8 °C.

For both cultivations, no obvious adaptation phase was observed in the 30 L tank PBR, and the $C_{biomass}$ increased continuously with slight fluctuation in growth rate, perhaps, due to fluctuating natural irradiance. The biomass productivity fluctuated even more through time with fluctuating light. At the end of cultivation, very similar $C_{biomass}$(~155 g m^{-2}) was achieved in both experiments. The average $P_{biomass}$ was 17.53 g m^{-2} day^{-1} for the first cultivation (June 2012) and 18.85 g m^{-2} day^{-1} for the second cultivation (July 2012), respectively.

The seed *Graesiella* sp. WBG-1 cultured under nitrate sufficient condition only contained a small proportion of lipids (18.01 % DW), of which 47.22 % were TAGs. After 8 days of cultivation in the 10 L circular pond, the C_{lipid} increased to 45.23 % DW and over 86 % of the lipids were TAGs (Table 1). Compared to the small-scale cultivation, lower C_{lipid} was achieved in 30 L outdoor cultivations, partly because of lower availability of biomass-specific light (Discuss in detail later). However, the lipid productivity (P_{lipid}) achieved in outdoor 30 L cultivations was higher than those of 10 L cultivations, because more biomass (though containing lower lipid) was produced per illuminated area.

Table 1: Lipid production in the two small-scale reactors and corresponding light conditions

Trials	$I_{biomass}$ **(mol g^{-1} DW day^{-1})**	C_{lipid} **(% DW)**	**TAGs (% DW)**	P_{lipid} **(g m^{-2} day^{-1})**
10 L	0.2–5.3	45.2 ± 1.2	39.3 ± 0.5	5.7 ± 0.8
30 L (June 2012)	0.2–2.1	35.6 ± 1.0	30.8 ± 1.0	6.5 ± 0.8
30 L (July 2012)	0.2–2.8	32.9 ± 1.3	30.0 ± 0.4	6.4 ± 1.2

Cultivation of *Graesiella* sp. WBG-1 in outdoor 200 m^2 raceway pond

Three batch cultures of the *Graesiella* sp. WBG-1 in a 200 m^2 raceway pond were carried out three times in June 2013, July 2013 and May 2014. Figure 3 shows the changes of the culture color as well as the morphologic appearance of the cells. At the early stage of the batch culture, the microalgae suspension looked dark green and a pyrenoid could be observed in the cell. With the aging of culture, the pyrenoid was disappeared and the culture color was yellowish. After 15 days of cultivation, yellow-green or orange color could be seen in the raceway pond.

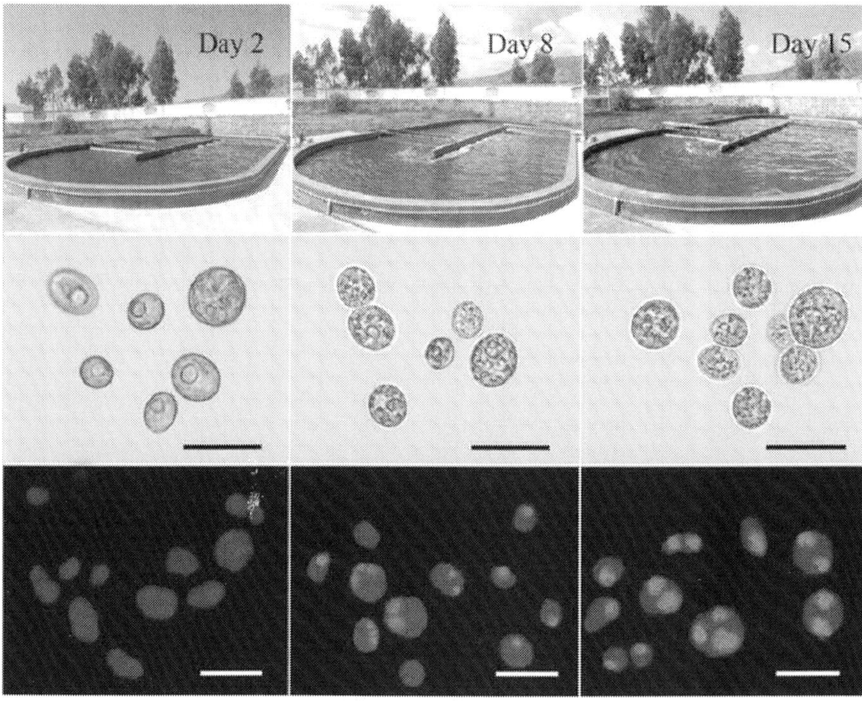

Figure 3: Cultivation of *Graesiella* sp. WBG-1 in 200 m^2 raceway pond. Changes in culture color and morphology of the cells during cultivation were shown. Intracellular lipid bodies were stained using Nile Red (*yellow*) and auto-fluorescence of chloroplasts is seen in *red*. *Scale bar* 10 µm

The time evolution of biomass concentration, lipid productivity, as well as natural light intensity, during the first cultivation in June 2013, were shown in Fig. 4 as a typical example, and a summary ($C_{biomass}$, C_{lipid}, $P_{biomass}$, P_{lipid}, et al.) of the three cultivations was illustrated in Table 2.

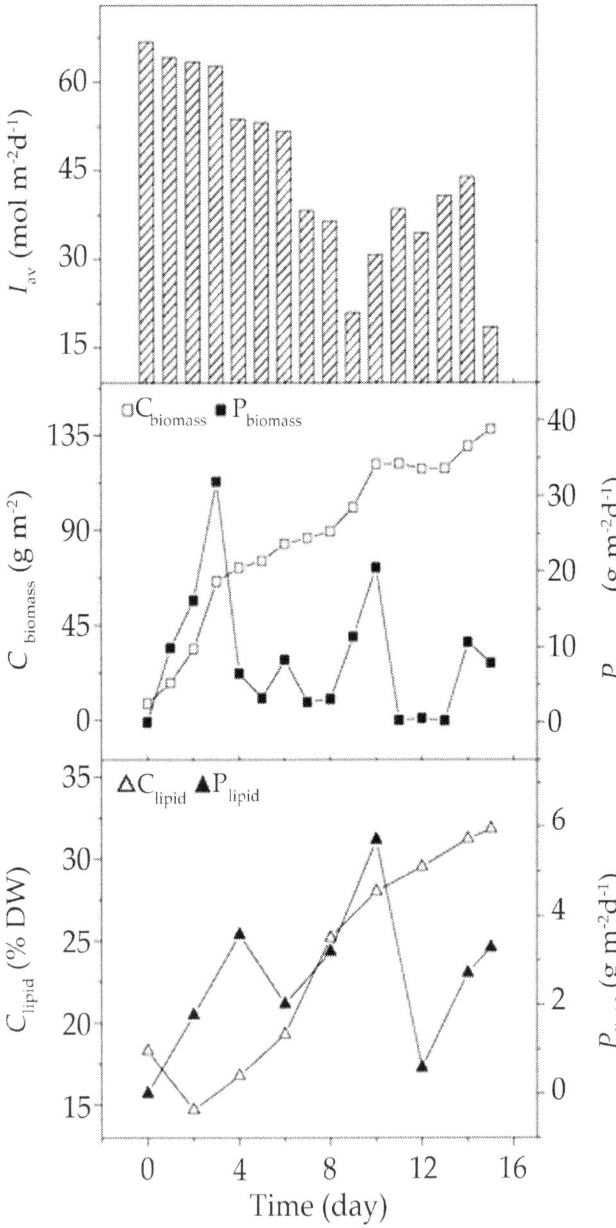

Figure 4: Time evolution of biomass concentration ($C_{biomass}$, *open square*), biomass productivity ($P_{biomass}$, *filled square*), lipid content (C_{lipid}, *open triangle*), lipid productivity (P_{lipid}, *filled triangle*), and natural light intensity (I_{av}) during the outdoor raceway cultivation (June 2013)

Table 2: Summary of the cultivations at different scale

Cultures	Duration (days)	I_{av} (mol m^{-2} day^{-1})	$I_{biomass}$ (mol g^{-1} DW day^{-1})	$C_{biomass}$ (g m^{-2})	C_{lipid} (% DW)	$P_{biomass}$ (g m^{-2} day^{-1})	P_{lipid} (g m^{-2} day^{-1})
10 L circular pond	8	15	0.2–5.38	101.2	45.2	12.0	5.7
30 L tank (June 2012)	8	33	0.2–2.1	153.3	35.6	17.5	6.5
30 L tank (July 2012)	8	29	0.2–2.8	161.8	32.9	18.9	6.4
Raceway (June 2013)	15	45	03–5.7	137.8	31.8	8.7	2.9
Raceway (July 2013)	13	36	0.3–4.4	102.3	33.4	7.2	2.4
Raceway (May 2014)	14	39	0.4–3.4	97.7	29.4	6.2	2.0

As shown in the middle panel of Fig. 4, the microalga grew rapidly in the first 4 days of cultivation in the raceway pond, and the $C_{biomass}$ increased from 7.86 to 65.67 g m^{-2}. After day 4, the cell growth slowed down, because the nitrate in the algal suspension was depleted and the natural light intensity was reduced (upper panel of Fig. 4).

As shown in the lower panel of Fig. 4, the C_{lipid} of *Graesiella* sp. WBG-1 decreased in the first 2 days of cultivation and then increased afterwards, from 14.70 to 31.82 % DW. The P_{lipid}showed positive correlation with the $P_{biomass}$, suggesting the essential role of biomass productivity in lipid production. Fluorescence microscopic observation (Fig. 3) also demonstrated the increase in lipid content in *Graesiella* sp. WBG-1 cells.

The biochemical compositions and fatty acid profiles of *Graesiella* sp. WBG-1 in 200 m^{2}raceway cultivation were also examined on the day 4 and day 15 of cultivation, when the cells were under the nitrate starvation condition. The carbohydrate content as a percentage of dry biomass did not show a significant change in all three experiments, while the protein content decreased with aging of the culture (Fig. 5a). A significant increase was observed in total lipids, especially the storage TAGs (Fig. 5a; Table 2), which made up 28.76 % of the dry biomass on the day 15. In contrast, the storage TAGs on the day 4 were only made up 16.67 % of the dry biomass. The fatty acid profile was analyzed using total lipids instead of TAGs, because DAGs and other free fatty acids in the total lipids can also be used to form fatty acid methyl esters. As shown in Fig. 5b, the most abundant fatty acids in *Graesiella* sp. WBG-1 cultivated in the 200 m^2 raceway pond were C16–C18 (especially C16:0 and C18:1), and the amounts of C5–C15 and C1–9C24 fatty acids were relatively low.

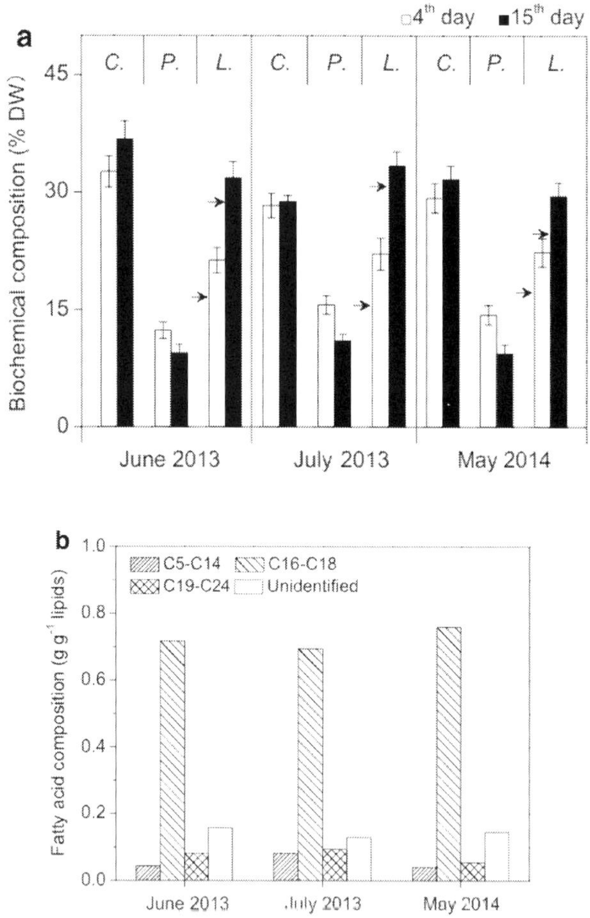

Figure 5: Biochemical compositions (**a**) and fatty acid profiles (**b**) of *Graesiella* sp. WBG-1 in 200 m^2 raceway cultivation. Carbohydrate, protein, and lipid at the 4th day (*open column*) and 15th day (*filled column*) are denoted with *C.*, *P.*, and *L.*, respectively. *Arrows* indicate TAGs content on dry biomass bases (% DW)

DISCUSSION

Optimization of nitrate and bicarbonate concentration

Nitrogen deficiency is the primary stress to alter algal metabolism to TAG synthesis, while limit cell growth at the same time [8, 19]. These general trends were also observed in *Graesiella* sp. WBG-1, as shown in Fig. 1b. Adams et al. [20] pointed out that culturing either many cells with low lipid content or few cells with high lipid content will not result in an economically viable biodiesel feedstock, although both cases give high lipid productivity. In this sense, the initial nitrate concentrations of 0.59 and 5.88 mM, in this study (Fig. 1b), will not suitable for microalgal lipid production. The optimal nitrate concentration range for lipid production should be 1.18–3.53 mM in this study (Fig. 1b) to balance the trade-off between the growth and lipid accumulation of *Graesiella* sp. WBG-1.

On the other side, initial nitrate concentration of a simple bath culture also put strong influence on biomass density. To some extent, the higher the initial nitrate concentration, the higher the biomass density could be achieved (Fig. 1b). However, when biomass density is low, light has good penetration and individual cells are exposed to a large quantity of light energy [18], resulting in more metabolic carbon flux to be channeled to lipid accumulation [21, 22]. Our observation was consistent with these conclusions, showing that low nitrate concentration not only resulted in high C_{lipid}, but also low biomass density (Fig. 1b), which helped to form high light availability ($I_{biomass}$) in the culture (Fig. 6). $I_{biomass}$ was in turn favorable for high lipid content (C_{lipid}) [18]. Therefore, the optimal nitrate concentration should be able to give sufficient light availability per gram of biomass.

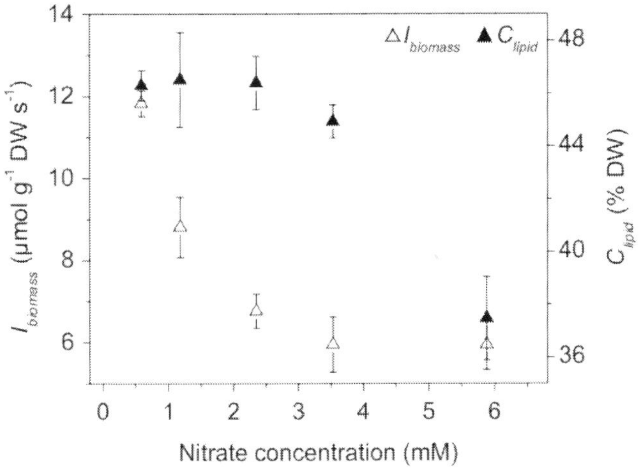

Figure 6: Lipid content (C_{lipid}, *filled triangle*) and biomass-specific light availability ($I_{biomass}$, *open triangle*) as functions of nitrate concentration

$I_{biomass}$ varies depending on reactor setup as well as light intensity and initial nutrition. A rough simulation revealed a huge difference in $I_{biomass}$ among four reactors used in this study. Given the same volumetric biomass concentration, $I_{biomass}$ in 200 mL column was four times higher than that in the outdoor 30 L tank PBR, and $I_{biomass}$ in 30 L tank PBR and 200 m^2 raceway pond were very similar. Thus, compared to the cultivation in 200 mL column, a lower biomass concentration ($C_{biomass}$) is needed for 30 L tank PBR and 200 m^{2}raceway pond, to give a relatively higher $I_{biomass}$ to facilitate lipid accumulation. Therefore, among the optimal nitrate concentration range of 1.18–3.53 mM for lipid production, 1.18 mM nitrate is the best concentration for following outdoor cultivation. By sacrificing some of the algal growth using reduced initial nitrate concentration, fairly high light availability ($I_{biomass}$) could be achieved.

Several studies have demonstrated the importance of photosynthesis efficiency on oil formation [18, 21, 23]. In other words, normal photosynthesis is needed to produce enough carbon flux and drive the biosynthesis of TAGs under nitrogen stress. Actually, carbon in cultures with air-blowing or paddle wheel stirring is far from sufficient [24, 25]. Thus, addition of carbon is another possible way to improve microalgal lipid production. White et al. [26] reported that addition of bicarbonate could significantly increase cell abundance in culture of *Tetraselmis suecica* and *Nannochloropsis salina*. We got a similar result using *Graesiella* sp. WBG-1 (Fig. 1a), although the $C_{biomass}$ were statistically insignificant between 11.90 and 23.81 mM bicarbonate. Lipid accumulation, however, was unaffected by different bicarbonate concentrations in this study. The optimal bicarbonate concentration for *Graesiella* sp. WBG-1 cultivation, therefore, is 11.90 mM considering the culture cost.

Performance of *Graesiella* sp. WBG-1 in scale-up cultivation

The 10 L photobioreactor used in this study is a mini-type industrial circular pond, equipped with light, temperature, and pH control systems. In this reactor, the strain *Graesiella* sp. WBG-1 showed good growth (12.03 g m^{-2} $day^{-1}$$P_{biomass}$) and high C_{lipid} (45.23 % DW) (Table 2), consistent with the previous results observed in our laboratory. Compared to the 10 L photobioreactor, higher $P_{biomass}$ (17.53–18.85 g m^{-2} day^{-1}) was achieved in outdoor cultivation in 30 L tank PBR. However, the C_{lipid} (32.89–35.63 % DW) decreased in 30 L tank PBR. The decrease in lipid content could be partly interpreted as a result of lower availability of biomass-specific light ($I_{biomass}$, the average light irradiance received by each gram of biomass) [18]. Light is an important stress factor for microalgal lipid accumulation [8]. It is also reported that the de novo synthesis of intracellular TAGs is highly correlated with photosynthetic activity as well as light energy available for microalgal cells [21, 23]. In this sense, sufficient light supply (higher $I_{biomass}$) is helpful to get a high lipid content.

In fact, the 30 L tank PBR has a longer light path (15 cm) than the 10 L circular pond (10 cm), giving a much lower biomass-specific light availability ($I_{biomass}$). Although lower C_{lipid} was achieved in 30 L tank PBR, $P_{biomass}$ was significantly higher than achieved in 10 L circular pond, because there were more growing cells per illuminated area in the 30 L tank PBR. As a result, higher lipid productivity was achieved in the 30 L tank PBR. On the other hand, outdoor light irradiance was fluctuated in a wide range of 16–44 mol m^{-2} day^{-1} during the experiments, and the highest instant light irradiance reached 2500 μmol m^{-2} s^{-1}, but no lag phase was observed in both laboratory and outdoor conditions. These results indicated the nice adaptability of *Graesiella* sp. WBG-1 to strong and fluctuating natural solar irradiance. Cultivation of *Graesiella* sp. WBG-1 in 200 m^2 raceway pond was successfully carried out three times in the summer at Chenghai town, where most days are sunny throughout the spring and early summer. The average daily light intensity during these three cultivations was 45, 36, and 39 mol m^{-2} day^{-1}, respectively, which was two-to-three times as high as the artificial light intensity in the 10 L circular pond. Under such light irradiance, the daily average $P_{biomass}$ achieved in 200 m^2 raceway pond was 6.17–8.66 g m^{-2} day^{-1}, substantially lower than the productivity previously reported in studies of microalgal mass production in food/feed industry [27]. Nevertheless, a maximum $P_{biomass}$ of 31.86 g m^{-2} day^{-1} in 200 m^2 raceway pond (Fig. 4) was reached in the first 3 days when nitrate was sufficient. The marked difference between average $P_{biomass}$ and maximum $P_{biomass}$ suggested that the cell growth was severely limited by nitrate depletion in the later 12 days of cultivation. Actually, the initial nitrate concentration was only 1.18 mM, which was insufficient for the alga to grow 15 days in 200 m^2 raceway pond. On the other hand, lower initial nitrate concentration only supported lower $C_{biomass}$, and thus resulted in higher $I_{biomass}$, which was found to promote lipid accumulation [18]. As a result, relatively high C_{lipid} (29.41–33.42 % DW) was achieved in the 200 m^2 raceway pond. If the initial nitrate concentration was increased from 1.18 to 1.76 mM, no increase in the C_{lipid} of *Graesiella* sp. WBG-1 could be observed after 15 days of cultivation in raceway pond (unpublished data). Therefore, the biological processes of cell growth and lipid accumulation should be further studied to optimize momentum, energy, and mass transfer in raceway pond.

Although the lipid content of 29.41–33.42 % DW is much lower than achieved in the small-scale cultivation (nitrate-deplete) in laboratory, it is of great importance for two reasons. First, Chlorophyta strains are reported to have a C_{lipid} of 10–25 % DW under nitrate-replete conditions [28]. For the microalga *Graesiella* sp. WBG-1, the C_{lipid} is about 14–17 % DW under nitrate-replete conditions (Fig. 4). The final C_{lipid} of 29.41–33.42 % DW obtained in the 200 m^2raceway pond under nitrate-depleted conditions is nearly two times the initial value. Furthermore, TAGs contributed a large proportion (84.38–91.88 %) of the total lipids in the harvested cell mass, and C16–C18 were the major fatty acids, comprising more than 70 % of total fatty acids. These data clearly indicated that lipid was markedly accumulated in the cells of *Graesiella* sp. WBG-1 cultured in the 200 m^2 raceway pond.

Second, sharp declines in C_{lipid} with increasing culture volumes have been observed in many studies. For example, Li et al. [10] reported that *Parachlorella kessleri* CCALA 255 had a lipid content of 55 % DW in laboratory, but only 25 % DW when cultured in a Large-Scale Industrial Thin-Layer Bioreactor. In a 2000 L raceway pond, the green microalga *Botryococcus bruanii*AP103 reached a total lipid content of only 30.8 % DW, after 15 days of outdoor cultivation [11]. More recently, more than 50 % decrease in lipid content (from 40 to ~19.8 % DW) was observed in *Chlorella* strain CH2, when its cultivation was scaled up from 0.25 GWP to 1 ha GWP [12]. In other studies in our lab, although some strains of *Chlorella* and *Scenedesmus* can accumulate large amounts of lipids (over 50 % DW) in laboratory, their lipid content are similar to the initial values (~20 % DW), and there is almost no net increase in TAGs when cultivated in the raceway pond with the same technology described in this study. However, the decline in C_{lipid} of the microalga *Graesiella* sp. WBG-1 cultured in the 200 m^2raceway pond is less than the other tested microalgae. Moreover, more than one-fold increase in TAGs was observed. These phenomena suggest that lipid accumulation does not always occur for any oleaginous microalgae in mass open culture. One possible explanation is the loss of culture competitiveness [13]. This is why the field test is necessary for oleaginous microalgae selection. As discussed above, *Graesiella* sp. WBG-1 is a potential oil producer with wide tolerance to natural environmental conditions.

Other features of *Graesiella* sp. WBG-1 for large-scale cultivation in open ponds

As indicated in many studies, lipid productivity is not the sole criterion for the selection of lipid-rich microalgae. One of the desirable characteristics for mass culture is large cell size [28], which is assumed to be easier to sediment in liquid culture [29] and reduce harvesting and down-stream processing costs. *Graesiella* sp. WBG-1 cells are broadly ellipsoidal or globose with a diameter of 8–12 μm, which is significantly larger than other reported oleaginous microalgae, such as *Chlorella* (diameter 3–8 μm). It was observed in this study that *Graesiella*sp. WBG-1 cells settle down easily in medium in a short time, and this feature becomes more obvious with the aging of culture. The high culture pH (9.0 ± 0.5 in this study) may also play a role in the auto-participation [30]. Over 98 % of the cells settled down to the bottom of the raceway pond when the algal suspension was kept without stirring over night. By removing the upper layer medium from the pond, the algal biomass was concentrated more than 50 times (Fig. 7). This feature potentially reduce liquid volume for harvesting if a standing procedure is employed before centrifugation or filtration, and the harvesting cost is reduced then.

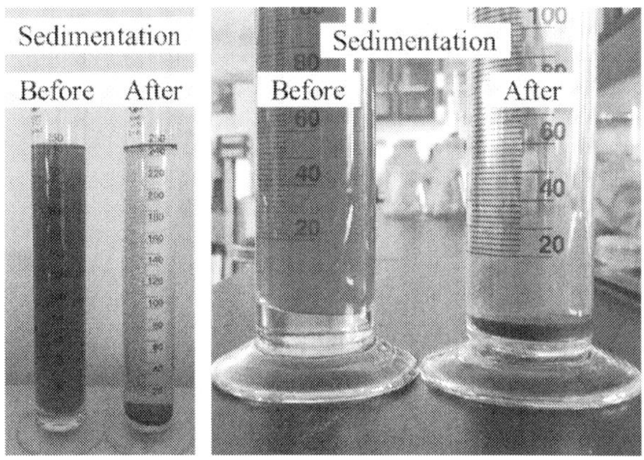

Figure 7: Sedimentation of *Graesiella* sp. WBG-1 cells over night. The homogeneous cell suspension (*left*) was kept still in a glass cylinder and over 98 % of biomass sedimented to the bottom after about 8 h (*right*)

Biological contamination is a serious problem for microalgae cultivation in open pond [31]. Protozoa may be predators on algae, bacteria and fungi, may be parasites, and other non-target microalgae may compete with the cultured algae for light and nutrients [32]. Theoretically, the ability to resist contamination is different among different algal strains, because there are many factors that can potentially inhibit the growth of contaminated organism, such as microalgal cell size and their tolerance to extreme environmental conditions, including temperature, pH, and salinity. Recently polysaccharides from algae are reported to have antibacterial activity [33], suggesting allelopathy is a prospective way for contamination control in microalgae mass cultivation [34]. In our experience, *Graesiella* sp. WBG-1 is more resistant to contamination compared to other tested algae. Although invading organisms, such as bacteria and zooplankton, could be identified under microscope in *Graesiella* sp. WBG-1 culture, but the number of these organisms was limited and the negative effects on productivity was negligible. In contrast, zooplankton contamination often led to culture collapse in the open cultivation of *Chlorella*, *Scenedesmus*, *Desmodesmus*, and *Chlorococcum*. It is not clear if *Graesiella* sp. WBG-1 has some specific chemical defense system or not, but *Graesiella* sp. WBG-1 is an indigenous strain isolated from Chenghai Lake and well adapted to the environment. Maybe, its large cell size can cause swallowing difficulty for protozoa.

Coupling bio-fixation of carbon dioxide with microalgae triacylglycerides production is a hot research topic worldwide [35, 36]. It is known that sufficient gas–liquid contact is important for efficient absorption of CO_2 [36], and high pH can prevent the escape of CO_2 from culture [37]. In this study, a microporous tubular gas diffuser (average pore size <100 μm) was used to maximize CO_2-water contact, and a high culture pH (9.0 ± 0.5), which had no negative effects on *Graesiella* sp. WBG-1 growth and lipid accumulation (unpublished data), was maintained in 200 m^2 raceway to prevent the escape of CO_2. Under such condition, a CO_2 utilization rate (bio-carbon divide by consumed carbon) of 65.7 % was achieved. This utilization rate is even higher than that those achieved in the mass culture of *Spirulina* (60 %) [38, 39], *Chlorella* (46 %) [40], and *Phaeodactylum* (63 %) [41]. These results indicate that *Graesiella* sp. WBG-1 can readily be used for microalgal mass production coupled with CO_2 fixation.

CONCLUSIONS

This study identified a lipid-rich microalgal strain *Graesiella* sp. WBG-1. We assessed its growth and lipid production under laboratory and outdoor conditions, and demonstrated that *Graesiella* sp. WBG-1 is an industrial strain capable of high lipid productivity in outdoor mass cultivation. The results can greatly enrich our knowledge of the behavior of oleaginous microalga in outdoor open raceway pond. To the best of our knowledge, this is one of the few reports on integrated indoor and outdoor cultivation for oleaginous microalgae selection. A high lipid content of 33.4 % DW was achieved for the first time in a 200 m^2 raceway pond. The results also provide evidence for the feasibility of using low-cost raceway pond for autotrophic cultivation of microalgae for lipid production.

NOTES

Authors' contributions

XW conceived, designed the experiments, performed the laboratory work, analyzed, and interpreted the data and drafted the paper. KD and ZW performed the outdoor cultivations and revised the manuscript. XP, HT, and YX analyzed the biomass lipid, protein, and carbohydrate, and performed the statistical analysis. LL designed the measurement of CO_2 utilization rate, interpreted the data. and critically revised the manuscript. DZ and YG maintained the seed culture and contributed to the scale-up cultivation. YL contributed to experimental design and data interpretation and critically revised the manuscript. All authors read and approved the final manuscript.

Acknowledgements

The authors acknowledge the financial support for this work from the Ministry of Science and Technology of China (Nos. 2012FY112900, 2013AA065805) and China Petrochemical Corporation (Nos. 210080, 215104).

REFERENCES

1. Hu Q. Industrial production of microalgal cell mass and secondary products-major industrial species. In: Richmond A, editor. Handbook of microalgal culture: biotechnology and applied phycology. Oxford: Wiley; 2004. p. 264–72.
2. Pulz O, Gross W. Valuable products from biotechnology of microalgae. Appl Microbiol Biotechnol. 2004;65:635–48.
3. Singh B, Guldhe A, Rawat I, Bux F. Towards a sustainable approach for development of biodiesel from plant and microalgae. Renew Sustain Energy Rev. 2014;29:216–45.
4. Wobbe L, Remacle C. Improving the sunlight-to-biomass conversion efficiency in microalgal biofactories. J Biotechnol. 2014;201:28–42.
5. Klok AJ, Martens DE, Wijffels RH, Lamers PP. Simultaneous growth and neutral lipid accumulation in microalgae. Bioresour Technol. 2013;134:233–43.

6. Wijffels RH, Barbosa MJ. An outlook on microalgal biofuels. Science. 2010;329:796–9.
7. Stephens E, Ross IL, Mussgnug JH, Wagner LD, Borowitzka MA, Posten C, Kruse O, Hankamer B. Future prospects of microalgal biofuel production systems. Trends Plant Sci. 2010;15:554–64.
8. Hu Q, Sommerfeld M, Jarvis E, Ghirardi M, Posewitz M, Seibert M, Darzins A. Microalgal triacylglycerols as feedstocks for biofuel production: perspectives and advances. Plant J. 2008;54:621–39.
9. Gouveia L, Oliveira AC. Microalgae as a raw material for biofuels production. J Ind Microbiol Biotechnol. 2009;36:269–74.
10. Li XL, Pribyl P, Bisova K, Kawano S, Cepak V, Zachleder V, Cizkova M, Branyikova I, Vitova M. The microalga Parachlorella kessleri—a novel highly efficient lipid producer. Biotechnol Bioeng. 2013;110:97–107.
11. Ashokkumar V, Rengasamy R. Mass culture of Botryococcus braunii Kutz. under open raceway pond for biofuel production. Bioresour Technol. 2012;104:394–9.
12. Guccione A, Biondi N, Sampietro G, Rodolfi L, Bassi N, Tredici M. Chlorella for protein and biofuels: from strain selection to outdoor cultivation in a Green Wall Panel photobioreactor. Biotechnol Biofuels. 2014;7:84.
13. Sheehan J, Dunnahay T, Benemann J, Roessler P. A look back at the US Department of Energy's Aquatic Species Program: biodiesel from algae. Golden: National Renewable Energy Laboratory. 1998; Report No. NREL/TP-580-24190.
14. Wen XB, Liang F, Geng YH, Li YG. Two-stage characteristics of lipid production in batch culture of two green microalgae. Fresen Environ Bull. 2014;23:2253–8.
15. Collos Y, Mornet F, Sciandra A, Waser N, Larson A, Harrison PJ. An optical method for the rapid measurement of micromolar concentrations of nitrate in marine phytoplankton cultures. J Appl Phycol. 1999;11:179–84.
16. Wen XB, Geng YH, Li YG. Enhanced lipid production in Chlorella pyrenoidosa by continuous culture. Bioresour Technol. 2014;161:297–303.

17. Slocombe SP, Ross M, Thomas N, McNeill S, Stanley MS. A rapid and general method for measurement of protein in microalgal biomass. Bioresour Technol. 2013;129:51–7.

18. Benvenuti G, Bosma R, Klok AJ, Ji F, Lamers PP, Barbosa MJ, Wijffels RH. Microalgal triacylglycerides production in outdoor batch-operated tubular PBRs. Biotechnol Biofuels. 2015;8:100.

19. Takagi M, Watanabe K, Yamaberi K, Yoshida T. Limited feeding of potassium nitrate for intracellular lipid and triglyceride accumulation of Nannochloris sp. UTEX LB1999. Appl Microbiol Biot. 2000;54:112–7.

20. Adams C, Godfrey V, Wahlen B, Seefeldt L, Bugbee B. Understanding precision nitrogen stress to optimize the growth and lipid content tradeoff in oleaginous green microalgae. Bioresour Technol. 2013;131:188–94.

21. Dillschneider R, Steinweg C, Rosello-Sastre R, Posten C. Biofuels from microalgae: photoconversion efficiency during lipid accumulation. Bioresour Technol. 2013;142:647–54.

22. Courchesne N, Parisien A, Wang B, Lan C. Enhancement of lipid production using biochemical, genetic and transcription factor engineering approaches. J Biotechnol. 2009;141:31–41.

23. Pan YY, Wang ST, Chuang LT, Chang YW, Chen CNN. Isolation of thermo-tolerant and high lipid content green microalgae: oil accumulation is predominantly controlled by photosystem efficiency during stress treatments in Desmodesmus. Bioresour Technol. 2011;102:10510–7.

24. Su Z, Kang R, Shi S, Cong W, Cai Z. An economical device for carbon supplement in large scale microalgae production. Bioprocess Biosyst Eng. 2008;31:641–5.

25. Berger R, Libby WF. Equilibration of atmospheric carbon dioxide with sea water: possible enzymatic control of the rate. Science. 1969;164:1395–7.

26. White DA, Pagarette A, Rooks P, Ali ST. The effect of sodium bicarbonate supplementation on growth and biochemical composition of marine microalgae cultures. J Appl Phycol. 2012;25:153–65.

27. Tredici MR. Mass production of microalgae: photobioreactors. In: Richmond A, editor. Handbook of microalgal culture: biotechnology

and applied phycology. Oxford: Wiley; 2004. p. 178–214.

28. Griffiths MJ, Harrison STL. Lipid productivity as a key characteristic for choosing algal species for biodiesel production. J Appl Phycol. 2009;21:493–507.
29. Chatsungnoen T, Chisti Y. Harvesting microalgae by flocculation–sedimentation. Algal Res. 2016;13:271–83.
30. Wu ZC, Zhu Y, Huang WY, Zhang CW, Li T, Zhang YM, Li AF. Evaluation of flocculation induced by pH increase for harvesting microalgae and reuse of flocculated medium. Bioresour Technol. 2012;110:496–502.
31. McBride RC, Lopez S, Meenach C, Burnett M, Lee PA, Nohilly F, Behnke C. Contamination management in low cost open algae ponds for biofuels production. Ind Biotechnol. 2014; 10:221–7.
32. Wang H, Zhang W, Chen L, Wang J, Liu T. The contamination and control of biological pollutants in mass cultivation of microalgae. Bioresour Technol. 2013;128:745–50.
33. de Jesus Raposo MF, de Morais AMB, de Morais RMSC. Marine polysaccharides from algae with potential biomedical applications. Mar Drugs. 2015;13:2967–3028.
34. Bacellar Mendes LB, Vermelho AB. Allelopathy as a potential strategy to improve microalgae cultivation. Biotechnol Biofuels. 2013;6:1–14.
35. Gutierrez-Arriaga CG, Serna-Gonzalez M, Ponce-Ortega JM, El-Halwagi MM. Sustainable integration of algal biodiesel production with steam electric power plants for greenhouse gas mitigation. ACS Sustain Chem Eng. 2014;2:1388–403.
36. Kumar A, Ergas S, Yuan X, Sahu A, Zhang QO, Dewulf J, Malcata FX, van Langenhove H. Enhanced CO2 fixation and biofuel production via microalgae: recent developments and future directions. Trends Biotechnol. 2010;28:371–80.
37. Zhang D, Zhu XY, Wen XB, Geng YH, Li YG. Studies on equilibrium pH values of microalgal medium. Acta Hydrobiol Sin. 2014;3:401–6 (in Chinese with english abstract).
38. Li YG, Hu HJ, Gong XM. Studies on the mechanism of pH value change and carbon conversion ratio of Spirulina media. Chin J

Biotechnol. 1996;12:242–8 (in Chinese with english abstract).

39. Belay A. Mass culture of Spirulina outdoors—the earthrise farms experience. In: Vonshak A, editor. Spirulina platensis (Arthrospira): physiology, cell biology and biotechnology. London: Taylor and Francis; 1997. p. 131–58.
40. Ramanan R, Kannan K, Deshkar A, Yadav R, Chakrabarti T. Enhanced algal CO2sequestration through calcite deposition by Chlorella sp. and Spirulina platensis in a mini-raceway pond. Bioresour Technol. 2009;101:2616–22.
41. Sobczuk TM, Camacho FG, Rubio FC, Fernandez FGA, Grima EM. Carbon dioxide uptake efficiency by outdoor microalgal cultures in tubular airlift photobioreactors. Biotechnol Bioeng. 2000;67:465–75.

SECTION IV

EFFLUENT TREATMENT AND FISHMEAL REPLACEMENT: A SUSTAINABLE APPROACH

CHAPTER

16

MICROALGAL CULTIVATION IN SECONDARY EFFLUENT: RECENT DEVELOPMENTS AND FUTURE WORK

Junping Lv, Jia Feng, Qi Liu and Shulian Xie

School of Life Science, Shanxi University, Taiyuan 030006, China

ABSTRACT

Eutrophication of water catchments and the greenhouse effect are major challenges in developing the global economy in the near future. Secondary effluents, containing high amounts of nitrogen and phosphorus, need further treatment before being discharged into receiving water bodies. At the same time, new environmentally friendly energy sources need to be developed. Integrating microalgal cultivation for the production of biodiesel feedstock with the treatment of secondary effluent is one way of addressing both issues. This article provides a comprehensive review of the latest progress in microalgal cultivation in secondary effluent to remove pollutants and accumulate lipids. Researchers have discovered that microalgae remove nitrogen and phosphorus effectively from secondary effluent, accumulating biomass and lipids in the process. Immobilization of appropriate microalgae, and establishing a consortium of microalgae and/or bacteria, were both found to be feasible ways to enhance pollutant removal and lipid production. Demonstrations of pilot-scale microalgal cultures in secondary effluent have also taken place. However there is still much work to be done in improving pollutants removal, biomass production, and lipid accumulation in secondary effluent. This includes screening microalgae, constructing the consortium, making use of flue gas and nitrogen, developing technologies related to microalgal harvesting, and using lipid-extracted algal residues (LEA).

Keywords: Microalgae; the treatment of secondary effluent; pollutants removal; biomass production; lipid accumulation

INTRODUCTION

Wastewater treatment plants (WWTPs) have played a significant role in improving water environments and the efficient use of water resources. Secondary effluent from the anaerobic–anoxic–oxic (A^2/O) process (the most commonly used wastewater treatment process in China), or processes derived from this, is considered to be low in organics, nitrogen, and phosphorus [1]. Generally the biological removal of nitrogen via an A^2/O process is mainly dependent on nitrifying bacteria and denitrifying bacteria [2]. However some features of the process, such as slow growth rate, sensitivity to toxic shocks, pH, and temperature change, are not conducive to the stable removal of nitrogen from wastewaters [3]. In China in particular, many WWTPs are facing pressures to improve effluent quality to meet the Chinese National First A-level Sewage Discharge Standard. This means that secondary effluent should not exceed the threshold of chemical oxygen demand (COD) 50 mg/L, ammonium 5 mg/L, total nitrogen (TN) 15 mg/L, and total phosphorous (TP) 0.5 mg/L. More importantly, some reports show that the threshold of nitrogen and phosphorus causing eutrophication in streams is 0.21–1.2 and 0.01–0.1 mg/L, respectively [4]. Tertiary treatment is, therefore, necessary to reduce the risk of eutrophication. Currently tertiary treatment technologies are developing rapidly, focusing on coagulation–sedimentation, adsorption, ion exchange, membrane technology, biological filtering, ozonation, and biological nutrient removal (BNR) process, etc. [5,6,7,8,9,10]. Specifically facilities have been upgrading their BNR process to meet the new standards for nitrogen and phosphorous discharges. At present, the most effective BNR processes are the Modified Bardenpho and the University of Cape Town (UCT) treatment process, which exhibit excellent nitrogen and phosphorus removal efficiency, as reviewed by Arita et al. [11]. Nevertheless these techniques have many disadvantages; for example, the instability of the treatment effect, high investment, high treatment cost, and difficulty in being popularized in large scale. Thus efficient, stable, low cost tertiary wastewater treatment systems are essential in reducing nitrogen and phosphorus concentrations in secondary effluent.

Fossil fuels are, currently, the main source of energy for human production activities. Excessive consumption of them accelerates the emission and accumulation of CO_2, which results in the greenhouse effect and aggravates global climate change [12]. The combustion of fossil fuels also contributes to gaseous pollutants, such as SO_2, NO_x, CO, ozone, and volatile organic compounds and may have adverse effects on human health and the environment [13]. Biodiesel appears to be an attractive partial alternative to fossil fuels and a way to reduce carbon emissions and reduce the risk of environment pollution from nitrogen oxides and sulfur oxides. At present, biodiesel derived from oil crops, animal fats and waste oil plays an effective part in addressing the problems caused by the use of fossil fuels [14,15,16,17]. These bioresources do, however, have some disadvantages. For example, the production of oil crops requires huge tracts of arable land and fresh water [15], and the production of animal fats may not be sufficient to replace fossil fuels [17]. It is essential that we seek new bioresources for biodiesel production that need only limited or no additional arable land; that need minimal clean water or can use wastewater; and that, simultaneously, have high biomass and lipid productivity.

Microalgae consist of a wide range of autotrophic organisms. They have comparable photosynthetic efficiency to higher plants, rapid growth rate, and notable adaptability. Carbon, nitrogen, and phosphorus are essential elements for microalgal growth and can be effectively used via different metabolic pathways [18]. They therefore have potential advantages for the removal of pollutants from wastewater.

Some microalgae have been shown to be capable of doubling their biomass several times per day. For example, *Ochromonas danica*, grown in a system containing 40 g/L waste cooking oil with acid values of 10.7 mg KOH/g, has a short doubling time of 12.1 h [19]. This short harvesting life will allow multiple and continuous harvesting of biomass throughout the year. Compared to many types of plants, microalgae are easy to cultivate and can produce a high yield of lipid for biodiesel production. As indicated by Nascimento et al. [20], microalgae with a lipid content varying from 13% to 49% have biodiesel productivity ranging from 3.4 to 23.0 $m^3 \cdot ha^{-1} \cdot year^{-1}$ in open raceway pond systems. By contrast, the biodiesel productivity of soybeans is only 0.446 $m^3 \cdot ha^{-1} \cdot year^{-1}$ [21]. Also, as reviewed by Bohutskyi and Bouwer [22], Harun et al. [23], and Sialve et al. [24], the microalgal biomass remaining after lipid extraction can be used to generate methane. Coupling biodiesel and methane production provides significant energy advantages, along with the sustainability and economic benefits from nutrient recycling. Microalgae can also fix CO_2 from flue gas to enhance biomass and lipid production [25,26]. This mitigates the greenhouse effect. Microalgae can produce neutral lipids, which are particularly suitable as a potential alternative to fossil fuel. Accordingly, compared to existing bioresources, microalgae are regarded as an excellent alternative for biodiesel production.

ADVANCES IN MICROALGAL CULTIVATION IN SECONDARY EFFLUENT

The Feasibility of Microalgal Cultivation in Secondary Effluent

Cai et al. [18], Chen et al. [27], and Zeng et al. [28] reviewed many investigations into lipid accumulation and the biological removal of nitrogen and phosphorus via microalgae in diverse wastewaters. There are some challenges in cultivating microalgae in wastewater, including unbalanced N/P ratio, the presence of bacterial contamination and competitors, low biomass density and lipid content, and incomplete removal of nutrients [29,30,31]. Nevertheless these studies effectively demonstrate that wastewaters can be an excellent medium for biomass production and lipid accumulation and that nitrogen and phosphorus are also removed from wastewaters. The physical and chemical characteristics of secondary effluent are completely different from those of municipal wastewater, anaerobic digestion effluent, industrial wastewater, and agricultural wastewater. Nearly all pollutants with high biodegradability in wastewaters are degraded in WWTPs, and the biodegradability of the remaining organic compounds in secondary effluent is poor. It has been reported that the ratio of BDOC/DOC (BDOC: biodegradable dissolved organic carbon; DOC: dissolved organic carbon) in secondary effluent is around 0.25 [32]. The lack of a carbon source is, therefore, a serious challenge for microalgal cultivation with secondary effluent. Li et al. [33] isolated a freshwater microalga, *Scenedesmus* sp. (LX1), and cultivated it with secondary effluent to investigate its ability to remove pollutants and accumulate lipids. The results indicated that *Scenedesmus* sp. (LX1) was well adapted to secondary effluent. The microalgal biomass and lipid content was 0.11 g/L and 31%–33%, respectively, after 15 days batch cultivation. At the same time, concentrations of nitrogen and phosphorus in the secondary effluent were decreased by 98% and 98.5%, respectively. Several other microalgae, such as *Botryococcus braunii* and *Chlorella ellipsoidea* (YJ1), have also been grown in secondary effluent for biomass production and pollutant removal, although the microalgal biomass was below 0.5 g/L [34,35]. Importantly the lipid content of *C. ellipsoidea* (YJ1) was up to 43% [26]. In other studies, microalgal lipid accumulation has been strengthened by nutrient starvation [35]. Consequently, although biomass production may be low, high lipid content can be achieved with low nutrient concentrations in secondary effluent.

Some filamentous, mat-forming cyanobacteria affiliated to *Phormidium* have been isolated from Arctic, subarctic, and Antarctic environments to investigate their potential for tertiary wastewater treatment in cool climates. These cyanobacteria exhibited superior growth and high phosphate uptake rates under cold temperatures (10 and 5 °C, respectively) and formed aggregates that could readily be harvested by sedimentation [36,37]. These interesting studies provide avenues for further research and lay the foundations for microalgal cultivation and tertiary wastewater treatment in regions of high altitudes and latitudes with cold climates.

Microalgal Lipid Production in Secondary Effluent

Although low nutrient concentrations in secondary effluent are beneficial to microalgal lipid accumulation, microalgal biomass production is inhibited (as seen in the studies described in Section 2.1), and this shortcoming needs to be addressed. Some studies have found that relatively high biomass was obtained when microalgae were cultivated in secondary effluent. For example, biomass production of *S. obliquus* and *B. braunii*, cultivated in secondary effluent, was up to 1.684 and 1.88 g/L, respectively [38,39]. The possible reason for this was supplementation with CO_2 as an inorganic carbon source, promoting microalgal growth. Park et al. [40] found that three strains, *Chlamydomonas debaryana* AMB1, *C. sorokiniana* RBD8 and *Micractinium* sp. RB1b, showed large increases in biomass productivity when cultivated mixotrophically in secondary municipal wastewater supplemented with glycerol. This indicates that the exogenous supplement of an organic carbon source helped to strengthen the biomass production of microalgae. The adaption to secondary effluent was species-dependent. Compared to the high biomass production of *Muriellopsis* sp. and *S. subpicatus* in secondary effluent, *C. vulgaris*, *C. fusca*, *Chlorella* sp., and *Pseudokirchneriella subcapitata* had low biomass production in the same conditions [41]. Similar results were also found in other investigations [39]. Selection of the most appropriate microalgae is clearly critical to producing high biomass in secondary effluent.

Microalgal Immobilization

Immobilization technology is a good choice for harvesting or separating microalgal biomass from reactors. However the main focus of microalgal immobilization for secondary effluent treatment has been to enhance removal of pollutants. At present, the most widely used material for microalgal immobilization is alginate. A major advantage of alginate beads containing immobilized microalgae is that immobilized microalgae do not suffer extreme physical and/or chemical changes in the process of immobilization. In addition, the excellent permeability, low or null toxicity, and high transparency of alginate matrix provide a suitable environment for immobilized microalgae. As reported, *Chlorella* sp. and *Scenedesmus* sp., immobilized by alginate matrix, showed high removal efficiency of nitrogen and phosphorus (more than 90%), when some other factors (cell density and starvation) were simultaneously optimized [42,43,44]. Chitosan- and carrageenan- immobilized microalgae (*Phormidium*, *S. bicellularis*, *S. quadricauda*) were also efficient in removing nitrogen and phosphorus from secondary effluent [45,46,47]. Although the removal of pollutants is mainly dependent on microalgal assimilation, immobilized matrixes also promote the removal of pollutants. As discussed by Tam and Wong [48], the presence of calcium ions in alginate matrix, together with elevated wastewater pH, favored the precipitation of phosphate as calcium phosphate. Similarly the calcium dissolved as a result of abrasion of the chitosan particles in wastewater seemed to be conducive in reducing orthophosphate levels in the effluent [46]. Of course, the possible linkages or ionic exchanges between the orthophosphate and chitosan amide groups may provide exchange or fixation sites for orthophosphate [46]. Covarrubias et al. [49] found that the populations of *C. sorokiniana* and *Azospirillum brasilense* in non-sterile secondary effluent were significantly lower than in sterile wastewater when they were cultivated as free suspensions, and the population of wastewater bacteria and natural microfauna increased. However immobilization of *C. sorokiniana* and *A. brasilense* in polymer Ca-alginate beads significantly enhanced their populations in non-sterile secondary effluent. Alginate beads clearly provided a beneficial physical barrier against native microorganisms in secondary effluent.

Attaching to surfaces or matrices in the natural environment is normal for many microorganisms [50], and microalgae can attach to different kinds of materials. *C. vulgaris* and *Scenedesmus* sp. have been found to attach to carriers made from bundles of polypropylene fibers, and the microalgal biofilm photobioreactor thus created significantly reduced more than 90% of ammonium and total phosphorus in secondary effluent [51,52]. Concrete slabs have also been used as carriers for microalgal growth when pollutants from secondary effluent were removed effectively [53]. A rotating algal biofilm reactor (RABR) for wastewater treatment with in situ biomass harvest has been reported in some studies [54,55]. The RABR consists of a cylinder provided with a growth surface partially submerged in wastewater. The cylinder is rotated to alternately expose the growth surface to the wastewater and to air. Of course, effective growth substrata are crucial to microalgal attachment and biofilm formation in RABR [54,55]. Additionally the Algal turf scrubber (ATS), using a natural mixed assemblage of attached periphyton, microalgae, and bacteria, has been developed [56,57]. The ATS is a long inclined flow-way that supports a biofilm of microalgae and bacteria. When wastewater washes over the flow-way in a series of pulses, pollutants are removed effectively. To date, ATSs have been used to provide tertiary treatment of municipal wastewaters and have performed very well [58,59].

Although all the above examples demonstrate that immobilized microalgae were beneficial in improving pollutants removal, partial biological degradation of alginate beads was observed during tertiary wastewater treatment [60]. Immobilization also had some negative effects on microalgal growth rate and biomass productivity [61]. Other unfavorable factors, such as low light penetration and high cost, have limited the commercial application of microalgal immobilization in tertiary wastewater treatment [62].

Construction of a Microalgal Consortium

Microalgae are generally grown as monocultures for harvesting high-value products. However to take advantage of the synergetic growth of microalgae and bacteria, a consortium of the two has been developed to treat wastewater with high levels of organic pollutants [63,64,65]. The consortium is also robust in the face of environmental fluctuations, resists the invasion of other species, and enhances the stability of reaction systems [66]. De-Bashan et al. [67] found that a mixed culture of microalgae (*C. vulgaris* or *C. sorokiniana*) and a bacterium (*A. brasilense* strain Cd) immobilized into small alginate beads improved the efficiency of nitrogen and phosphorus removal from secondary effluent, compared to microalgae alone. The mixed culture removed 100% ammonium, 15% nitrate, and 36% phosphorus, a result that was superior to microalgae alone (75% ammonium, 6% nitrate, and 19% phosphorus). To identify the appropriate combination of microalgae to enhance biomass productivity in secondary effluent, *Scenedesmus* sp. (LX1), *C. ellipsoidea*, and *Haematococcus pluvislis*, were chosen for a study of the growth characteristic of different microalgal combinations in secondary effluent [68]. The intrinsic growth rate of the three microalgae in the mixed culture was higher than when they were grown as monocultures. Biomass productivity in the mixed cultures of *Scenedesmus* sp. (LX1) and *H. pluvialis* increased by 64% and 42%, respectively, compared to monocultures of *Scenedesmus* sp. (LX1) and *H. pluvialis*. These cases suggest that the efficiency of pollutants removal and biomass productivity are strengthened by constructing the consortium.

Pilot-Scale Culture of Microalgae in Secondary Effluent

Pilot-scale culture of microalgae in secondary effluent has been developed gradually. Van Coillie et al. [69] tested the feasibility of a tertiary treatment using *Scenedesmus* sp. at pilot-scale level. An outdoor tank with a capacity of 15,000 L was used for batch culture. Results showed that nutrient removal was 95% for total inorganic nitrogen and 60% for PO_4^{3-} and that the biomass productivity was 0.39 $mg \cdot L^{-1} \cdot h^{-1}$. McGinn et al. [70] operated a 300 L proprietary Brite-Box photobioreactor for microalgal cultivation in secondary effluent. They showed that biomass productivity averaged 130 $mg \cdot L^{-1} \cdot day^{-1}$, and nitrogen and phosphorus removal could be up to 90%, when average hydraulic retention times ranged from 6.55 to 6.65 days and from 6.50 to 6.56 days, respectively. Arbib et al. [71] operated a 530 L high rate alga pond (HRAP) and a 380 L airlift tubular photobioreactor (TPBR) to remove nitrogen and phosphorous from WWTP effluent. Maximum areal productivity was 8.26 and 21.76 g suspension solid (SS)$\cdot m^{-2} \cdot day^{-1}$ for HRAP and TPBR, respectively; total nitrogen removal averaged 89.68% and 65.12% for TPBR and HRAP, respectively; and, for total phosphorus removal, TPBR and HRAP averaged 86.71% and 58.78%, respectively. There was no significant lipid content difference between the reactors, which was about 20.8%. Based on these studies, it appears that tertiary wastewater treatment and biomass production by microalgal cultivation is feasible at a pilot-scale level, and this provides a basis for large-scale cultivation of microalgae in secondary effluent.

FUTURE RESEARCH INTO MICROALGAL CULTIVATION IN SECONDARY EFFLUENT

The progress in research examined here clearly indicates that microalgae have some potential to biomass production, lipid accumulation and pollutant removal in secondary effluent. However considerable work remains to be done to enhance lipid productivity, wastewater treatment efficiency, and microalgal harvesting in the future.

Screening Microalgae and Constructing the Microalgal Poly-Culture and Microalgal–Bacterial Co-Culture Consortium

Many microalgae, such as *Chlorella*, *Scenedesmus*, and *Botryococcus*, have been screened for biomass production [72,73,74]. Nevertheless, compared to those microalgae growing well in municipal wastewater, the diversity of microalgae growing in secondary effluent is low. The main groups focused on *Chlorella* and *Scenedesmus* (Table 1). In addition, the removal efficiency of nitrogen and phosphorous and lipid productivity during long-term operation was not good (Table 2). It was found to be essential to screen more microalgae to determine their suitability for growth in secondary effluent and to evaluate their lipid accumulation and biomass production potential. Whether oil-rich microalgae growing well in municipal wastewater could adapt to the new environmental conditions of secondary effluent remains to be explored.

Table 1: Biomass production, lipid productivity, and pollutant removal of microalgae in secondary effluent under batch culture.

Microalgae Species	Free Cell (F) or Immobilization (I)	Preliminary Treatment of Wastewater	The Volume of the Cultivation (L)	Cultivation Time (h)	CO_2 (%)	Biomass Energy				Nutrients Removal Efficiency (%)					References
						Biomass Production ($g \cdot L^{-1}$)	Biomass Productivity ($mg \cdot L^{-1} \cdot day^{-1}$)	Lipid Productivity ($mg \cdot L^{-1} \cdot day^{-1}$)	Lipid Content (%)	COD	TN	NH_4^+	NO_3^-	TP	
a natural algal bloom	F	no treatment	2	240	5	1.884	200.4	-	26.82	-	79	-	-	>98	[38]
Botryococcus braunii	F	filtration (0.2 μm) and autoclaving	3	240	1	0.35	-	-	-	-	-	-	>99	>99	[34]
Botryococcus braunii	F	filtration	0.5	1000	-	-	288–345.6	-	17.85	-	-	-	-	-	[75]
Botryococcus braunii	F	no treatment	9	336	5	1.88	-	-	36.14	-	-	-	79.63	100	[39]
*Chlorella ellipsoidea*YJ1	F	filtration (0.45 μm) and autoclaving	0.3	528	-	0.425	-	12.7	43	-	>99	-	-	>90	[35]
Chlorella kessleri	F	no treatment	2	240	5	1.172	132.3	-	20.55	-	>90	-	-	>98	[38]
Chlorella sorokiniana	F	autoclaving	1	96	12	0.25	62.5	8	32	-	-	-	100	0	[76]
Chlorella sorokiniana	F	no treatment	0.45	240	-	0.1	-	-	-	-	80	-	-	40	[77]
Chlorella sp. 227	F	filtration (0.45 μm) or UV-radiation	0.5	216	-	0.41–0.67	-	6.9–22.9	15–31	13.8–24.8	75–92	-	-	84–86	[78]
Chlorella vulgaris	F	no treatment	2	120	-	0.76–0.82	73.88–79.82	-	-	-	-	-	-	92	[79]
Chlorella vulgaris	F	filtration (0.2 μm)	0.2	168	15	0.29	-	-	30	-	>99	-	-	>99	[80]

Chlorella vulgaris	F	no treatment	2	240	5	1.303	116	-	22.02	-	>90	-	-	>98	[38]
Chlorella vulgaris	F	filtration	2	168	air [a]	1.03	171.33	43.52	27.6	-	-	-	94	-	[81]
Desmodesmus communis	F	filtration	1	360	2	0.79	23	-	9.3				100		[82]
Neochloris oleoabundans	F	filtration (1.2 μm) and autoclaving	0.4	240	5	2.1	233.3	-	-	-	-	>90	78–99	100	[83]
Ourococcus multisporus	F	filtration (0.2 μm)	0.2	168	15	0.31	-	-	31	-	>99	-	-	>99	[80]
Scenedesmus obliquus	F	filtration and autoclaving	1	192	-	-	-	-	31.4	-	-	>90	-	>90	[84]
Scenedesmus obliquus	F	filtration (0.2 μm)	0.2	168	15	0.31	-	-	27	-	>99	-	-	>99	[80]
Scenedesmus obliquus	F	no treatment	2	240	5	1.684	201.4	-	19.38	-	>90	-	-	>98	[38]
Scenedesmussp. AMDD	F	filtration (0.2 μm)	0.15	288	unknown concentration	0.13	127.22–132.73	-	11.72–12.08	-	-	>90	-	>90	[40]
Scenedesmussp. LX1	F	filtration (0.45 μm) and autoclaving	0.2	360	-	0.11	-	35	31–33	-	98.5	-	-	98	[33]
Scenedesmussp. LX1	F	autoclaving	0.2	336	5	0.77	-	-	35	-	-	-	-	-	[85]
Chlorella sp.	I	filtration and autoclaving	0.35	8	air	-	-	-	-	-	-	100	-	100	[44]
Phormidiumsp.	I	no treatment	0.5	24	-	-	-	-	-	-	-	>90	>90	>90	[46]
Scenedesmus bicellularis	I	unknown	unknown	2	-	1.57–1.86	-	-	-	-	-	100	-	88–100	[42]
Scenedesmus bicellularis	I	autoclaving	2.5	2	750–1500 [b]	-	-	-	-	-	-	42.1–100	-	19.1–99.1	[86]

Seenedesmus quadricauda	I	roughly screened	1	3	-	-	-	-	-	-	-	85–100	-	-	[45]
Scenedes-mussp.	I	filtration and auto-claving	0.35	4	air	-	-	-	-	-	-	100	-	100	[43]
Scenedes-mussp.	I	no treat-ment	96	72	-	-	-	-	-	0	47.86	96	-	>90	[52]

[a] Except air, the system was supplied by $NaHCO_3$; [b] ppm.

Table 2: Biomass production, lipid productivity, and pollutant removal of microalgae in secondary effluent under continuous culture.

Microalgae Species	Free cell (F) or Im-mobiliza-tion (I)	Culti-vation Time (d)	Hy-draulic Reten-tion Time (d)	Biomass Energy				Nutrients Removal Efficiency (%)					Refer-ences
				Biomass Produc-tion ($g \cdot L^{-1}$)	Biomass Productivity ($mg \cdot L^{-1} \cdot day^{-1}$)	Lipid Productivity ($mg \cdot L^{-1} \cdot day^{-1}$)	Lipid Content (%)	COD	TN	NH_4^+	NO_3^-	TP	
Chlorella vulgaris	F	240	0.04–2	0.69–1.289 [a]	47.5–131.7	-	-	-	54–95.3	-	-	84.4–94.9	[87]
Scenedes-mussp. AMDD	F	24	1.48	0.312–0.356	234–267	11.91–15.19	5.14–5.70	-	-	100	-	100	[70]
Scenedesmus obliquus	F	112	5	-	21.76 [b]	-	20.8	-	89.68	-	-	86.71	[71]
Scenedesmus obliquus	F	104	10	-	8.26 [b]	-	20.8	-	58.78	-	-	58.78	[71]
Scenedes-mussp.	F	-	5.2	0.5	20 [b]	-	-	-	-	-	-	-	[88]
Scenedes-mussp.	I	91	2	0.1–0.3	-	-	-	21–48.36	36	24–55	-	40–80	[52]

[a] $mg \cdot COD/L$; [b] $g \cdot m^{-2} \cdot day^{-1}$.

The screening of native microalgae has been reported to be a suitable technique for integrated wastewater treatment and biomass accumulation [89]. Bohutskyi et al. [77] reported that the microalgae performing best in unsterilized wastewater were found to be clustered on the same branch of the phylogenetic tree. Combining microalgal screening and molecular phylogeny will be important in identifying the phenotypic traits and genes responsible for superior growth in wastewater. Following the fast development of omic technologies (e.g., genomics, transcriptomics, and proteomics), a large number of genomic, proteomic, and transcriptional data have become available for elucidating gene properties relevant to oil accumulation [90,91,92]. These will be valuable in the genetic engineering of microalgae for lipid production. Nevertheless there have been no reports of combining microalgal screening and omic technologies for microalgal growth and pollutant removal. Revealing the genetic traits of microalgal growth and nutrients removal via omic technologies and applying genetic engineering to improve productivity and nutrients removal abilities will be the focus of future research.

Even though the appropriately screened microalgae have shown excellent growth performance, their scaled cultivation with real secondary effluents in open systems is subject to strong competition (from local microalgal and microbial communities, when single microalgal bioprocesses are considered) and predation [93]. Single microalgal cultivation is also particularly sensitive to sudden changes in environmental conditions (such as light, temperature and nutrient availability) [93]. As reviewed by Fouilland [93], microalgal communities with high species richness and specific metabolic capacities were not only able to capture a high proportion of available resources for growth but also had their resilience enhanced. Microalgal–bacterial co-cultures also helped to ensure successful, intensive, stable microalgal production [93]. At present, preliminary attempts to establish a consortium for biomass production and pollutant removal in secondary effluent are being carried out, as seen in Section 2.4. Both microalgal poly-cultures and microalgal–bacterial co-cultures exhibited better performance than microalgal mono-cultures. However this research was in the early stages of tertiary wastewater treatment; importantly the consortium only contained two microalgae or an artificial combination of microalgae and bacteria. Constructing the consortium by natural selection, or by using an artificial assemblage of robust microalgal poly-cultures and microalgal–bacterial co-cultures with higher species richness and specific metabolic capacities, is of important practical value for tertiary wastewater treatment and lipid production.

Carbon Supplementation by CO_2 *Sequestration in Secondary Effluent*

For microalgal cultivation, aeration provided by CO_2 is not only beneficial to carbon accumulation via photosynthesis, but also contributes to mixing the culture, preventing settlement, and maintaining homogeneous conditions. Currently supplying pure CO_2 or flue gas for microalgal cultivation prevails, and biomass and lipid content have thereby been increased in many microalgae [94]. Considering the serious lack of a carbon sources in secondary effluent, it has been necessary to investigate the relationship between CO_2 supplements and biomass and lipid accumulation. Compared to microalgal cultivation without CO_2 supplement, the biomass production and lipid content were enhanced in most cases when the concentration of pure CO_2 supplied in secondary effluent ranged from 1% to 5% (Table 1). Flue gas produced by human activities is huge and can be regarded as a rich source of CO_2. Apart from CO_2, flue gas contains N_2, NO_x, SO_x, C_xH_y, CO, particulate matter, halogen acids, and heavy metals, which are toxic and likely to inhibit the microalgal growth if added to secondary effluent [95]. CO_2 concentrations in flue gas were also found to be up to 6%–15%, which can inhibit the growth of some microalgae [96]. This phenomenon was also found when microalgae were cultivated in secondary effluent with 15% CO_2 (Table 1). It is, therefore, critical to optimize the use of CO_2 from flue gas for biomass production and lipid accumulation of microalgae in secondary effluent.

Improving the Use of Nitrogen in Secondary Effluent

Ammonium has been the preferred form of nitrogen because a redox reaction is not involved in its assimilation by microalgae [18]. Nitrate is used by microalgae only when the ammonium is low or depleted in the wastewater [18]. For WWTPs, the difference in processes or operation parameters resulted in differences in the abundance and richness of nitrifying and denitrifying bacteria [97,98]. These led to different nitrifying and denitrifying activities and then influenced the concentration of ammonium and nitrate in secondary effluent. Seasonal variation was also found likely to influence nitrifying and denitrifying activity via the temperature fluctuation [99]. Thus, to some extent, the ratio of ammonium/nitrate in secondary effluent was not constant and showed dynamic changes (Table 3). Enhancing the efficiency with which different forms of nitrogen (especially nitrate) are removed, in the light of possible nitrogen fluctuations in secondary effluent, will be a major concern.

Table 3: Characteristics of secondary effluent from different sources.

The Number of Secondary Effluent	COD (mg/L)	TOC (mg/L)	BOD (mg/L)	TN (mg/L)	NH_4^+-N (mg/L)	NO_3^--N (mg/L)	TP (mg/L)	NO_3^--N/ NH_4^+-N	References
1	-	7.4	-	6.3	<0.01	4.48	0.39	>448	[34]
2	-	5.5	-	8.9	0.17	7.67	0.04	45.11	[34]
3	45–60	-	-	12.5–23.8	3.8–7.6	-	0.82–1.67	-	[52]
4	24	-	-	15.5	2.5	-	0.5	-	[33]
5	-	-	-	-	0.24	4.94	<0.01	20.58	[82]
6	56	-	-	22.13	4.10	15.12	-	3.69	[100]
7	-	-	-	20.0	7.6	10.3	1.95	1.36	[101]
8	24	-	-	7.0	0.50	-	0.46	-	[35]
9	22.1	-	-	15.5	2.5	-	0.05	-	[35]
10	24.5	-	-	16.7	3.7	-	0.08	-	[35]
11	49.7	-	-	11.9	15.0	0.9	11.5	0.06	[75]
12	-	-	10–19	-	21.62–28.85	-	2.22–3.51	-	[70]
13	-	8.1	-	8.7	9.4	8.5	1.71	0.90	[80]
14	100	-	-	-	21	1.6	5.6	0.08	[38]

TOC: Total organic carbon; BOD: Biochemical oxygen demand.

Growth of Microalgae in Unsterilized Secondary Effluent

It has been difficult to sterilize or filter large amounts of secondary effluent to eliminate the potential effects of native organisms on microalgae in large-scale microalgal cultivation. Secondary effluent commonly contains some microorganisms, viruses, and predatory zooplankton, although their diversity and abundance are lower than in untreated wastewater. Microalgal growth is, therefore, likely to be affected when microalgae are cultivated in unsterilized secondary effluent. Lee et al. [102] found that *Chlorogonium* sp. outgrew other species in non-sterile secondary effluent. Zhang et al. [103] showed that *Scenedesmus* sp. ZTY1 exhibited good adaptability to secondary effluent. Although the growth rates of *Chlorella* sp. under conditions of non-sterilization were lower than under sterilization, non-sterile circumstances were beneficial for accumulating lipids and removing nutrients from secondary effluent [104]. In contrast, Bohutskyi et al. [77] found that most of the miroalgal species they tested were unable to grow efficiently in unsterilized secondary wastewater effluents. Yu et al. [105] showed that the growth of *C. ellipsoidea* was inhibited in unsterilized, domestic, secondary effluent. The potential mechanism was that soluble algal products accumulated in microalgal culture were used as a carbon source by bacteria and promoted bacterial growth. The overgrowth of bacteria then significantly inhibited the activity of microalgae. Based on the above results, there are some challenges for microalgal cultivation in unsterilized secondary effluent.

There has been considerable effort to promote the growth and lipid accumulation of microalgae in unsterilized secondary effluent. Supplementation with centrate from an anaerobic digester was found to be helpful in promoting biomass production in secondary effluent [106,107]. Appropriate doses of microalgal inoculum and correct light intensity also increased final biomass density and productivity in secondary wastewater [108]. The isolation of specific microalgae with excellent resistance to biotic pollution will be an important future goal. A successful example of screening microalgae with good biomass and lipid production in unsterilized wastewater has been reported [109]. Apart from improving the lipid content of microalgae via genetic engineering [110], it may be possible to enhance microalgal resistance to biotic pollution by genetic manipulation. Ultraviolet pretreatment of wastewater was found to increase microalgal growth rate and significantly reduce native bacterial densities [111]. However there has been almost no use of the above control technologies to enhance the growth of microalgae in unsterilized secondary effluent. Considerable work remains to be done.

Development of Microalgal Harvesting Technologies

Harvesting of the microalgal biomass has been a major bottleneck for biodiesel production, largely because of the small size of some cells (typically in the range of 3–30 μm); the relatively low cell density, especially in the raceways (<0.5 kg/m^{3}of dry biomass); and the large volume of water being harvested [112]. The recovery of microalgae generally requires one or more solid–liquid separation steps. Currently microalgal harvesting technologies include chemical flocculation, centrifugation, gravity sedimentation, filtration and screening, flotation, electrophoresis, and immobilization [113,114]. The immobilization technique, in particular, has been used effectively to separate microalgal cells from secondary effluent. However the huge consumption of chemicals and energy were major challenges. These drawbacks limited the use of these technologies for large-scale microalgal harvesting as they accounted for 20%–40% of the total costs of lipid production [112]. Some chemicals also showed a certain degree of biomass toxicity [115], and the high density of the immobilization matrix was likely to reduce the light penetrating through the reactor, thus affecting metabolic activity [62]. Theoretically self-settlement induced by exogenous selection pressures was beneficial to biomass–water separation. A typical case was the development of aerobic granular sludge by regulating some parameters such as settling time and the mixed liquor volume exchange ratio [116,117]. Valigore et al. [118] regulated hydraulic retention times and solids retention times of laboratory sequencing batch reactors to cultivate settleable microbial (microalgal–bacterial) biomass grown on primary treated wastewater as a biodiesel feedstock. They showed that biomass settleability was typically 70%–95%, and the microbes, aggregated into compact flocs as cultures, aged up to 4 months. Related investigations, however, were only just beginning. In the future, it will be interesting to consider strengthening microalgal–water separation in secondary effluent, screening microalgae of high settleability via exogenous selection pressures.

Use of Lipid-Extracted Algal Residues

A scaled biodiesel production process will generate enormous amounts of lipid-extracted algal residues (LEA), containing major parts of the energy and all nutrients captured by microalgae. Rational use of LEA will help to reduce costs and develop a sustainable microalgal cultivation and biodiesel production process. Anaerobic digestion (AD) is a good choice, as this can convert LEA into a biogas, mainly consisting of CH_4 and CO_2, with traces of other gases such as H_2S [119]. Passing the biogas into the microalgae culture will not only benefit microalgal growth and nutrient removal, but also the purification of the biogas [120,121]. The biogas may be burned to produce electricity and to generate an onsite source for CO_2 to supplement microalgae [122,123]. AD effluent, which contains many nutrients, can also be partially used as a chemical fertilizer during microalgal growth [106,107]. As elucidated by Bohutskyi et al. [107], the methane production from LEA increased the energy yield from microalgal biomass by more than 30%. Additionally the supplementation of AD effluent during microalgae culture can reduce fertilizer costs by 45%. Coupling biodiesel production and LEA use provides significant energy advantages as well as sustainability and economic benefits from nutrient recycling. However some factors will have a significant impact on the methane yield and productivity, notably (1) the difficulty in biodegrading microalgal cell walls; (2) the high protein content of the microalgal biomass, which results in high ammonium release; and (3) the high sodium concentration in microalgal cells, which may alter the anaerobic process. It will, therefore, be important to optimize LEA anaerobic digestion to ensure the absence of solvent residues inhibiting methanogens and to enhance LEA biodegradability for maximum methane yield, since some LEA fraction may be recalcitrant, as reviewed by some literatures [24,119].

CONCLUSIONS

Sustainability is a key principle in microalgal-based wastewater treatment and biodiesel production. It is important to consider how to minimize the environmental impact and strengthen the economic and social benefits of the process. This review underlines the viability of using secondary effluent as a potential medium for simultaneous microalgal growth and pollutant removal. As elucidated in the review, pollutant from secondary effluent can be effectively removed, and the lipid, as a renewable energy resource, is also accumulated by microalgae via fixing CO_2. Immobilization of microalgae has significant advantages in promoting pollutants removal and also provides a beneficial physical barrier against native microorganisms in secondary effluent. A consortium of microalgae and/or bacteria results in greater biomass and lipid productivity and more effective removal of pollutants than is found in a mono-culture. Obviously it is feasible to integrate microalgal cultivation, as biodiesel production feedstock, with tertiary wastewater treatment. It is beneficial to reduce the risk of the eutrophication and carbon emissions. In addition, compared to oil crops for biodiesel production, it doesn't involve the conversion of agricultural land for biodiesel production or affect food security. Therefore it exhibits huge environmental and social benefits. However the high economic cost and the low efficiency (mainly in microalgal growth, lipid accumulation, and pollutant removal) have become the bottleneck of microalgal cultivation in secondary effluent. As potential solutions to these problems, considerable effort will be needed in the future to screen microalgae, construct the consortium, improve the use of flue gas and nitrogen, develop technologies for harvesting microalgae, and use LEA. Coupling biodiesel production and LEA use especially provides significant energy advantages, along with sustainability and economic benefits from nutrient recycling. In general, all these investigations represent potential approaches to integrating tertiary wastewater treatment and microalgal lipid accumulation for biodiesel production.

Acknowledgments

This research project was financed by the Key Scientific Development Project of Shanxi Province, China (No. FT-2014-01-15), the Natural Science Foundation of Shanxi Province, China (No. 2015021159), and the Social Development Foundation of Shanxi, China (No. 201603D321008).

REFERENCES

1. Jin, L.; Zhang, G.; Tian, H. Current state of sewage treatment in China. Water Res. 2014, 66, 85–98.

2. Wagner, M.; Loy, A. Bacterial community composition and function in sewage treatment systems. Curr. Opin. Biotechnol. 2002, 13, 218–227.
3. Wagner, M.; Rath, G.; Amann, R.; Koops, H.P.; Schleifer, K.H. In situ identification of ammonia-oxidizing bacteria. Syst. Appl. Microbiol. 1995, 18, 251–264.
4. Chambers, P.A.; McGoldrick, D.J.; Brua, R.B.; Vis, C.; Culp, J.M.; Benoy, G.A. Development of environmental thresholds for nitrogen and phosphorus in streams. J. Environ. Qual. 2012, 41, 7–20.
5. Acero, J.L.; Benitez, F.J.; Real, F.J.; Teva, F. Micropollutants removal from retentates generated in ultrafiltration and nanofiltration treatments of municipal secondary effluents by means of coagulation, oxidation, and adsorption processes. Chem. Eng. J. 2016, 289, 48–58.
6. Daigger, G.T.; Waltrip, G.D.; Romm, E.D.; Morales, L.M. Enhanced secondary treatment incorporating biological nutrient removal. J. Water Pollut. Control Fed. 1988, 60, 1833–1842.
7. Kalkan, C.; Yapsakli, K.; Mertoglu, B.; Tufan, D.; Saatci, A. Evaluation of biological activated carbon (BAC) process in wastewater treatment secondary effluent for reclamation purposes. Desalination 2011, 265, 266–273.
8. Sun, J.; Li, X.; Quan, Y.; Yin, Y.; Zheng, S. Effect of long-term organic removal on ion exchange properties and performance during sewage tertiary treatment by conventional anion exchange resins. Chemosphere 2015, 136, 181–189.
9. Zhang, J.; Yu, H.; Quan, X.; Chen, S.; Zhang, Y. Ceramic membrane separation coupled with catalytic ozonation for tertiary treatment of dyestuff wastewater in a pilot-scale study. Chem. Eng. J. 2016, 301, 19–26.
10. Zheng, S.K.; Chen, J.J.; Jiang, X.M.; Li, X.F. A comprehensive assessment on commercially available standard anion resins for tertiary treatment of municipal wastewater. Chem. Eng. J. 2011, 169, 194–199.
11. Arita, C.E.Q.; Peebles, C.; Bradley, T.H. Scalability of combining microalgae-based biofuels with wastewater facilities: A review.

Algal Res. 2015, 9, 160–169.

12. Höök, M.; Tang, X. Depletion of fossil fuels and anthropogenic climate change—A review. Energy Policy 2013, 52, 797–809.
13. Kampa, M.; Castanas, E. Human health effects of air pollution. Environ. Pollut. 2008, 151, 362–367.
14. Adewale, P.; Dumont, M.J.; Ngadi, M. Recent trends of biodiesel production from animal fat wastes and associated production techniques. Rene. Sustain Energy. Rev. 2015, 45, 574–588.
15. Beringer, T.I.M.; Lucht, W.; Schaphoff, S. Bioenergy production potential of global biomass plantations under environmental and agricultural constraints. GCB Bioenergy 2011, 3, 299–312.
16. Petrou, E.C.; Pappis, C.P. Biofuels: A survey on pros and cons. Energy Fuel 2009, 23, 1055–1066.
17. Ahmad, A.L.; Yasin, N.H.; Derek, C.J.C.; Lim, J.K. Microalgae as a sustainable energy source for biodiesel production: A review. Renew. Sustain Energy Rev. 2011, 15, 584–593.
18. Cai, T.; Park, S.Y.; Li, Y. Nutrient recovery from wastewater streams by microalgae: Status and prospects. Renew. Sustain Energy. Rev. 2013, 19, 360–369.
19. Hosseini, M.; Ju, L.K. Use of phagotrophic microalga Ochromonas danica to pretreat waste cooking oil for biodiesel production. J. Am. Oil Chem. Soc. 2015, 92, 29–35.
20. Nascimento, I.A.; Marques, S.S.I.; Cabanelas, I.T.D.; de Carvalho, G.C.; Nascimento, M.A.; de Souza, C.O.; Druzian, J.I.; Hussain, J.; Liao, W. Microalgae versus land crops as feedstock for biodiesel: Productivity, quality, and standard compliance. Bioenergy Res. 2014, 7, 1002–1013.
21. Demirbas, A.; Demirbas, M.F. Importance of algae oil as a source of biodiesel. Energy Convers. Manag. 2011, 52, 163–170.
22. Bohutskyi, P.; Bouwer, E. Biogas production from algae and cyanobacteria through anaerobic digestion: A review, analysis, and research needs. In Advanced Biofuels and Bioproducts; Lee, J.W., Ed.; Springer: New York, NY, USA, 2012; pp. 873–975.
23. Harun, R.; Davidson, M.; Doyle, M.; Gopiraj, R.; Danquah, M.; Forde, G. Technoeconomic analysis of an integrated microalgae

photobioreactor, biodiesel and biogas production facility. Biomass Bioenergy 2011, 35, 741–747.

24. Sialve, B.; Bernet, N.; Bernard, O. Anaerobic digestion of microalgae as a necessary step to make microalgal biodiesel sustainable. Biotechnol. Adv. 2009, 27, 409–416.
25. Jeon, H.; Lee, Y.; Chang, K.S.; Lee, C.G.; Jin, E. Enhanced production of biomass and lipids by supplying CO2 in marine microalga Dunaliella sp. J. Microbiol. 2013, 51, 773–776.
26. Kao, C.Y.; Chen, T.Y.; Chang, Y.B.; Chiu, T.W.; Lin, H.Y.; Chen, C.D.; Chang, J.H.; Lin, C.S. Utilization of carbon dioxide in industrial flue gases for the cultivation of microalga Chlorella sp. Bioresour. Technol. 2014, 166, 485–493.
27. Chen, G.; Zhao, L.; Qi, Y. Enhancing the productivity of microalgae cultivated in wastewater toward biofuel production: A critical review. Appl. Energy 2015, 137, 282–291.
28. Zeng, X.; Guo, X.; Su, G.; Danquah, M.K.; Zhang, S.; Lu, Y.; Sun, Y.; Lin, L. Bioprocess considerations for microalgal-based wastewater treatment and biomass production. Renew. Sustain Energy. Rev. 2015, 42, 1385–1392.
29. Li, X.; Hu, H.Y.; Gan, K.; Sun, Y.X. Effects of different nitrogen and phosphorus concentrations on the growth, nutrient uptake, and lipid accumulation of a freshwater microalga Scenedesmus sp. Bioresour. Technol. 2010, 101, 5494–5500.
30. Min, M.; Wang, L.; Li, Y.; Mohr, M.J.; Hu, B.; Zhou, W.; Chen, P.; Ruan, R. Cultivating Chlorella sp. in a pilot-scale photobioreactor using centrate wastewater for microalgae biomass production and wastewater nutrient removal. Appl. Biochem. Biotechnol. 2011, 165, 123–137.
31. Zhang, Y.; Su, H.; Zhong, Y.; Zhang, C.; Shen, Z.; Sang, W.; Yan, G.; Zhou, X. The effect of bacterial contamination on the heterotrophic cultivation of Chlorella pyrenoidosa in wastewater from the production of soybean products. Water Res. 2012, 46, 5509–5516.
32. Li, L.; Zhu, W.; Zhang, P.; Zhang, Z.; Wu, H.; Han, W. Comparison of AC/O3–BAC and O3–BAC processes for removing organic pollutants in secondary effluent. Chemosphere 2006, 62, 1514–

1522.

33. Li, X.; Hu, H.Y.; Yang, J. Lipid accumulation and nutrient removal properties of a newly isolated freshwater microalga, Scenedesmus sp. LX1, growing in secondary effluent. New Biotechnol. 2010, 27, 59–63.
34. Sawayama, S.; Minowa, T.; Dote, Y.; Yokoyama, S. Growth of the hydrocarbon-rich microalga Botryococcus braunii in secondarily treated sewage. Appl. Microbiol. Biotechnol. 1992, 38, 135–138.
35. Yang, J.; Li, X.; Hu, H.Y.; Zhang, X.; Yu, Y.; Chen, Y. Growth and lipid accumulation properties of a freshwater microalga, Chlorella ellipsoidea YJ1, in domestic secondary effluents. Appl. Energy 2011, 88, 3295–3299.
36. Chevalier, P.; Proulx, D.; Lessard, P.; Vincent, W.F.; de la Noüe, J. Nitrogen and phosphorus removal by high latitude mat-forming cyanobacteria for potential use in tertiary wastewater treatment. J. Appl. Phycol. 2000, 12, 105–112.
37. Tang, E.P.; Vincent, W.F.; Proulx, D.; Lessard, P.; de la Noüe, J. Polar cyanobacteria versus green algae for tertiary waste-water treatment in cool climates. J. Appl. Phycol. 1997, 9, 371–381.
38. Arbib, Z.; Ruiz, J.; Álvarez-Díaz, P.; Garrido-Pérez, C.; Perales, J.A. Capability of different microalgae species for phytoremediation processes: Wastewater tertiary treatment, CO2 bio-fixation and low cost biofuels production. Water Res. 2014, 49, 465–474.
39. Sydney, E.B.; Da Silva, T.E.; Tokarski, A.; Novak, A.C.; de Carvalho, J.C.; Woiciecohwski, A.L.; Larroche, C.; Soccol, C.R. Screening of microalgae with potential for biodiesel production and nutrient removal from treated domestic sewage. Appl. Energy 2011, 88, 3291–3294.
40. Park, K.C.; Whitney, C.; McNichol, J.C.; Dickinson, K.E.; MacQuarrie, S.; Skrupski, B.P.; Zou, J.T.; Wilson, K.E.; O'Leary, S.J.B.; McGinn, P.J. Mixotrophic and photoautotrophic cultivation of 14 microalgae isolates from Saskatchewan, Canada: Potential applications for wastewater remediation for biofuel production. J. Appl. Phycol.2012, 24, 339–348.
41. Gómez-Serrano, C.; Morales-Amaral, M.M.; Acién, F.G.; Escudero, R.; Fernández-Sevilla, J.M.; Molina-Grima, E. Utilization of

secondary-treated wastewater for the production of freshwater microalgae. Appl. Microbiol. Biotechnol.2015, 99, 6931–6944.

42. Kaya, V.M.; de la Noüe, J.; Picard, G. A comparative study of four systems for tertiary wastewater treatment by Scenedesmus bicellularis: New technology for immobilization. J. Appl. Phycol. 1995, 7, 85–95.

43. Zhang, E.; Wang, B.; Wang, Q.; Zhang, S.; Zhao, B. Ammonia–nitrogen and orthophosphate removal by immobilized Scenedesmus sp. isolated from municipal wastewater for potential use in tertiary treatment. Bioresour. Technol. 2008, 99, 3787–3793.

44. Zhang, E.; Wang, B.; Ning, S.; Sun, H.; Yang, B.; Jin, M.; Hou, L. Ammonia-nitrogen and orthophosphate removal by immobilized Chlorella sp. isolated from municipal wastewater for potential use in tertiary treatment. Afr. J. Biotechnol.2012, 11, 6529–6534.

45. Chevalier, P.; de la Noüe, J. Efficiency of immobilized hyperconcentrated algae for ammonium and orthophosphate removal from wastewaters. Biotechnol. Lett. 1985, 7, 395–400.

46. De la Noüe, J.; Proulx, D. Biological tertiary treatment of urban wastewaters with chitosan-immobilized Phormidium. Appl. Microbiol. Biotechnol. 1988, 29, 292–297.

47. Kaya, V.M.; Picard, G. Stability of chitosan gel as entrapment matrix of viable Scenedesmus bicellularis cells immobilized on screens for tertiary treatment of wastewater. Bioresour. Technol. 1996, 56, 147–155.

48. Tam, N.F.Y.; Wong, Y.S. Effect of immobilized microalgal bead concentrations on wastewater nutrient removal. Environ. Pollut. 2000, 107, 145–151.

49. Covarrubias, S.A.; de-Bashan, L.E.; Moreno, M.; Bashan, Y. Alginate beads provide a beneficial physical barrier against native microorganisms in wastewater treated with immobilized bacteria and microalgae. Appl. Microbiol. Biotechnol. 2012, 93, 2669–2680.

50. Palmer, J.; Flint, S.; Brooks, J. Bacterial cell attachment, the beginning of a biofilm. J. Ind. Microbiol. Biotechnol. 2007, 34, 577–588.

51. Gao, F.; Yang, Z.H.; Li, C.; Zeng, G.M.; Ma, D.H.; Zhou, L. A novel

algal biofilm membrane photobioreactor for attached microalgae growth and nutrients removal from secondary effluent. Bioresour. Technol. 2015, 179, 8–12.

52. He, S.; Xue, G. Algal-based immobilization process to treat the effluent from a secondary wastewater treatment plant (WWTP). J. Hazard. Mater. 2010, 178, 895–899.

53. Sukačová, K.; Trtílek, M.; Rataj, T. Phosphorus removal using a microalgal biofilm in a new biofilm photobioreactor for tertiary wastewater treatment. Water Res. 2015, 71, 55–63.

54. Christenson, L.B.; Sims, R.C. Rotating algal biofilm reactor and spool harvester for wastewater treatment with biofuels by-products. Biotechnol. Bioeng. 2012, 109, 1674–1684.

55. Gross, M.; Henry, W.; Michael, C.; Wen, Z. Development of a rotating algal biofilm growth system for attached microalgae growth with in situ biomass harvest. Bioresour. Technol. 2013, 150, 195–201.

56. Adey, W.H.; Kangas, P.C.; Mulbry, W. Algal turf scrubbing: Cleaning surface waters with solar energy while producing a biofuel. Bioscience 2011, 61, 434–441.

57. Bohutskyi, P.; Chow, S.; Ketter, B.; Shek, C.F.; Yacar, D.; Tang, Y.; Bouwer, E.J. Phytoremediation of agriculture runoff by filamentous algae poly-culture for biomethane production, and nutrient recovery for secondary cultivation of lipid generating microalgae. Bioresour. Technol. 2016, 222, 294–308.

58. Craggs, R.J.; Adey, W.H.; Jenson, K.R.; John, M.S.S.; Green, F.B.; Oswald, W.J. Phosphorus removal from wastewater using an algal turf scrubber. Water Sci. Technol. 1996, 33, 191–198.

59. Craggs, R.J.; Adey, W.H.; Jessup, B.K.; Oswald, W.J. A controlled stream mesocosm for tertiary treatment of sewage. Ecol. Eng. 1996, 6, 149–169.

60. Cruz, I.; Bashan, Y.; Hernàndez-Carmona, G.; de-Bashan, L.E. Biological deterioration of alginate beads containing immobilized microalgae and bacteria during tertiary wastewater treatment. Appl. Microbiol. Biotechnol. 2013, 97, 9847–9858.

61. Hoffmann, J.P. Wastewater treatment with suspended and

nonsuspended algae. J. Phycol. 1998, 34, 757–763.

62. Ruiz-Marin, A.; Mendoza-Espinosa, L.G.; Stephenson, T. Growth and nutrient removal in free and immobilized green algae in batch and semi-continuous cultures treating real wastewater. Bioresour. Technol. 2010, 101, 58–64.

63. Mahdavi, H.; Prasad, V.; Liu, Y.; Ulrich, A.C. In situ biodegradation of naphthenic acids in oil sands tailings pond water using indigenous algae–bacteria consortium. Bioresour. Technol. 2015, 187, 97–105.

64. Maza-Márquez, P.; Martinez-Toledo, M.V.; Fenice, M.; Andrade, L.; Lasserrot, A.; Gonzalez-Lopez, J. Biotreatment of olive washing wastewater by a selected microalgal-bacterial consortium. Int. Biodeterior. Biodegrad. 2014, 88, 69–76.

65. Zhao, X.; Zhou, Y.; Huang, S.; Qiu, D.; Schideman, L.; Chai, X.; Zhao, Y. Characterization of microalgae-bacteria consortium cultured in landfill leachate for carbon fixation and lipid production. Bioresour. Technol. 2014, 156, 322–328.

66. Subashchandrabose, S.R.; Ramakrishnan, B.; Megharaj, M.; Venkateswarlu, K.; Naidu, R. Consortia of cyanobacteria/microalgae and bacteria: Biotechnological potential. Biotechnol. Adv. 2011, 29, 896–907.

67. De-Bashan, L.E.; Hernandez, J.P.; Morey, T.; Bashan, Y. Microalgae growth-promoting bacteria as “helpers” for microalgae: A novel approach for removing ammonium and phosphorus from municipal wastewater. Water Res.2004, 38, 466–474.

68. Zhu, S.F.; Yu, Y.; Hu, H.Y. Effects on the growth characteristic by mixed culture of three energy microalgae species using domestic secondary effluent. Ecol. Environ. Sci. 2014, 23, 642–648. (In Chinese)

69. Van Coillie, R.; del la Noue, J.; Thellen, C.; Pouliot, Y. Tertiary domestic waste water treatment by Scenedesmus sp. pilot-scale culture. Revue des Sciences de l’Eau 1990, 3, 441–455. (In French)

70. McGinn, P.J.; Dickinson, K.E.; Park, K.C.; Whitney, C.G.; MacQuarrie, S.P.; Black, F.J.; Frigon, J.C.; Guiot, S.R.; O’Leary, S.J.B. Assessment of the bioenergy and bioremediation potentials of the microalga Scenedesmus sp. AMDD cultivated in municipal wastewater effluent in batch and continuous mode. Algal Res.

2012, 1, 155–165.

71. Arbib, Z.; Ruiz, J.; Álvarez-Díaz, P.; Garrido-Pérez, C.; Barragan, J.; Perales, J.A. Long term outdoor operation of a tubular airlift pilot photobioreactor and a high rate algal pond as tertiary treatment of urban wastewater. Ecol. Eng.2013, 52, 143–153.
72. Wu, Y.H.; Hu, H.Y.; Yu, Y.; Zhang, T.Y.; Zhu, S.F.; Zhuang, L.L.; Zhang, X.; Lu, Y. Microalgal species for sustainable biomass/lipid production using wastewater as resource: A review. Renew. Sustain Energy Rev. 2014, 33, 675–688.
73. Chiu, S.Y.; Kao, C.Y.; Chen, T.Y.; Chang, Y.B.; Kuo, C.M.; Lin, C.S. Cultivation of microalgal Chlorella for biomass and lipid production using wastewater as nutrient resource. Bioresour. Technol. 2015, 184, 179–189.
74. Maity, J.P.; Bundschuh, J.; Chen, C.Y.; Bhattacharya, P. Microalgae for third generation biofuel production, mitigation of greenhouse gas emissions and wastewater treatment: Present and future perspectives–A mini review. Energy 2014, 78, 104–113.
75. Órpez, R.; Martínez, M.E.; Hodaifa, G.; El Yousfi, F.; Jbari, N.; Sánchez, S. Growth of the microalga Botryococcus braunii in secondarily treated sewage. Desalination 2009, 246, 625–630.
76. Lizzul, A.M.; Hellier, P.; Purton, S.; Baganz, F.; Ladommatos, N.; Campos, L. Combined remediation and lipid production using Chlorella sorokiniana grown on wastewater and exhaust gases. Bioresour. Technol. 2014, 151, 12–18.
77. Bohutskyi, P.; Liu, K.; Nasr, L.K.; Byers, N.; Rosenberg, J.N.; Oyler, G.A.; Betenbaugh, M.J.; Bouwer, E.J. Bioprospecting of microalgae for integrated biomass production and phytoremediation of unsterilized wastewater and anaerobic digestion centrate. Appl. Microbiol. Biotechnol. 2015, 99, 6139–6154.
78. Cho, S.; Luong, T.T.; Lee, D.; Oh, Y.K.; Lee, T. Reuse of effluent water from a municipal wastewater treatment plant in microalgae cultivation for biofuel production. Bioresour. Technol. 2011, 102, 8639–8645.
79. Cabanelas, I.T.D.; Ruiz, J.; Arbib, Z.; Chinalia, F.A.; Garrido-Pérez, C.; Rogalla, F.; Nascimento, I.A.; Perale, J.A. Comparing the use of different domestic wastewaters for coupling microalgal

production and nutrient removal. Bioresour. Technol. 2013, 131, 429–436.

80. Ji, M.K.; Abou-Shanab, R.A.; Kim, S.H.; Salama, E.S.; Lee, S.H.; Kabra, A.N.; Lee, Y.S.; Hong, S.; Jeon, B.H. Cultivation of microalgae species in tertiary municipal wastewater supplemented with CO2 for nutrient removal and biomass production. Ecol. Eng. 2013, 58, 142–148.

81. Ebrahimian, A.; Kariminia, H.R.; Vosoughi, M. Lipid production in mixotrophic cultivation of Chlorella vulgaris in a mixture of primary and secondary municipal wastewater. Renew. Energ. 2014, 71, 502–508.

82. Samorì, G.; Samorì, C.; Guerrini, F.; Pistocchi, R. Growth and nitrogen removal capacity of Desmodesmus communisand of a natural microalgae consortium in a batch culture system in view of urban wastewater treatment: Part I. Water Res. 2013, 47, 791–801.

83. Wang, B.; Lan, C.Q. Biomass production and nitrogen and phosphorus removal by the green alga Neochloris oleoabundans in simulated wastewater and secondary municipal wastewater effluent. Bioresour. Technol. 2011, 102, 5639–5644.

84. Martınez, M.E.; Sánchez, S.; Jimenez, J.M.; El Yousfi, F.; Munoz, L. Nitrogen and phosphorus removal from urban wastewater by the microalga Scenedesmus obliquus. Bioresour. Technol. 2000, 73, 263–272.

85. Li, X.; Hu, H.Y.; Zhang, Y.P. Effect of inorganic carbon supplement on the growth and lipid accumulation properties of Scenedesmus sp. LX1. Environ. Sci. 2011, 32, 2260–2266. (In Chinese)

86. Kaya, V.M.; Goulet, J.; de la Noüe, J.; Picard, G. Effect of intermittent CO2 enrichment during nutrient starvation on tertiary treatment of wastewater by alginate-immobilized Scenedesmus bicellularis. Enzym. Microb. Technol. 1996, 18, 550–554.

87. Xu, M.; Li, P.; Tang, T.; Hu, Z. Roles of SRT and HRT of an algal membrane bioreactor system with a tanks-in-series configuration for secondary wastewater effluent polishing. Ecol. Eng. 2015, 85, 257–264.

88. Wu, Y.H.; Li, X.; Yu, Y.; Hu, H.Y.; Zhang, T.Y.; Li, F.M. An integrated microalgal growth model and its application to optimize

the biomass production of Scenedesmus sp. LX1 in open pond under the nutrient level of domestic secondary effluent. Bioresour. Technol. 2013, 144, 445–451.

89. Lynch, F.; Santana-Sánchez, A.; Jämsä, M.; Sivonen, K.; Aro, E.M.; Allahverdiyeva, Y. Screening native isolates of cyanobacteria and a green alga for integrated wastewater treatment, biomass accumulation and neutral lipid production. Algal Res. 2015, 11, 411–420.
90. Choi, Y.E.; Hwang, H.; Kim, H.S.; Ahn, J.W.; Jeong, W.J.; Yang, J.W. Comparative proteomics using lipid over-producing or less-producing mutants unravels lipid metabolisms in Chlamydomonas reinhardtii. Bioresour. Technol.2013, 145, 108–115.
91. De Lomana, A.L.G.; Schäuble, S.; Valenzuela, J.; Imam, S.; Carter, W.; Bilgin, D.D.; Price, N.D. Transcriptional program for nitrogen starvation-induced lipid accumulation in Chlamydomonas reinhardtii. Biotechnol. Biofuels 2015, 8, 207.
92. Gao, C.; Wang, Y.; Shen, Y.; Yan, D.; He, X.; Dai, J.; Wu, Q. Oil accumulation mechanisms of the oleaginous microalga Chlorella protothecoides revealed through its genome, transcriptomes, and proteomes. BMC Genom. 2014, 15, 582.
93. Fouilland, E. Biodiversity as a tool for waste phycoremediation and biomass production. Rev. Environ. Sci. Biol. 2012, 11, 1–4.
94. Cheah, W.Y.; Show, P.L.; Chang, J.S.; Ling, T.C.; Juan, J.C. Biosequestration of atmospheric CO2 and flue gas-containing CO2 by microalgae. Bioresour. Technol. 2015, 84, 190–201.
95. Van Den Hende, S.; Vervaeren, H.; Boon, N. Flue gas compounds and microalgae: (Bio-)chemical interactions leading to biotechnological opportunities. Biotechnol. Adv. 2012, 30, 1405–1424.
96. Zhao, B.; Su, Y. Process effect of microalgal-carbon dioxide fixation and biomass production: A review. Renew. Sustain Energy. Rev. 2014, 31, 121–132.
97. Krasner, S.W.; Westerhoff, P.; Chen, B.; Rittmann, B.E.; Nam, S.N.; Amy, G. Impact of wastewater treatment processes on organic carbon, organic nitrogen, and DBP precursors in effluent organic matter. Environ. Sci. Technol. 2009, 43, 2911–2918.

98. Pochana, K.; Keller, J. Study of factors affecting simultaneous nitrification and denitrification (SND). Water Sci. Technol. 1999, 39, 61–68.

99. Zhang, B.; Yamamoto, K. Seasonal change of microbial population and activities in a building wastewater reuse system using a membrane separation activated sludge process. Water Sci. Technol. 1996, 34, 295–302.

100. Xiong, J.; Guo, G.; Mahmood, Q.; Yue, M. Nitrogen removal from secondary effluent by using integrated constructed wetland system. Ecol. Eng. 2011, 37, 659–662.

101. Abe, K.; Komada, M.; Ookuma, A.; Itahashi, S.; Banzai, K. Purification performance of a shallow free-water-surface constructed wetland receiving secondary effluent for about 5 years. Ecol. Eng. 2014, 69, 126–133.

102. Lee, K.Y.; Ng, T.W.; Li, G.; An, T.; Kwan, K.K.; Chan, K.M.; Huang, G.; Yip, H.Y.; Wong, P.K. Simultaneous nutrient removal, optimised CO2 mitigation and biofuel feedstock production by Chlorogonium sp. grown in secondary treated non-sterile saline sewage effluent. J. Hazard. Mater. 2015, 297, 241–250.

103. Zhang, T.Y.; Wu, Y.H.; Hu, H.Y. Domestic wastewater treatment and biofuel production by using microalga Scenedesmus sp. ZTY1. Water Sci. Technol. 2014, 69, 2492–2496.

104. Zhang, Q.; Hong, Y. Comparison in growth, lipid accumulation, and nutrient removal capacities of Chlorella sp. in secondary effluents under sterile and non-sterile conditions. Water Sci. Technol. 2012, 69, 573–579.

105. Yu, Y.; Wu, Y.H.; Zhu, S.F.; Hu, H.Y. The bioavailability of the soluble algal products of different microalgal strains and its influence on microalgal growth in unsterilized domestic secondary effluent. Bioresour. Technol. 2015, 180, 352–355.

106. Bohutskyi, P.; Chow, S.; Ketter, B.; Betenbaugh, M.J.; Bouwer, E.J. Prospects for methane production and nutrient recycling from lipid extracted residues and whole Nannochloropsis salina using anaerobic digestion. Appl. Energ. 2015, 154, 718–731.

107. Bohutskyi, P.; Ketter, B.; Chow, S.; Adams, K.J.; Betenbaugh, M.J.; Allnutt, F.T.; Bouwer, E.J. Anaerobic digestion of lipid-extracted

Auxenochlorella protothecoides biomass for methane generation and nutrient recovery. Bioresour. Technol.2015, 183, 229–239.

108. Bohutskyi, P.; Kligerman, D.C.; Byers, N.; Nasr, L.K.; Cua, C.; Chow, S.; Su, C.; Tang, Y.; Betenbaugh, M.J.; Bouwer, E.J. Effects of inoculum size, light intensity, and dose of anaerobic digestion centrate on growth and productivity of Chlorella and Scenedesmus microalgae and their poly-culture in primary and secondary wastewater. Algal Res. 2016, 19, 278–290.
109. Chinnasamy, S.; Bhatnagar, A.; Hunt, R.W.; Das, K.C. Microalgae cultivation in a wastewater dominated by carpet mill effluents for biofuel applications. Bioresour. Technol. 2010, 101, 3097–3105.
110. Ho, S.H.; Ye, X.; Hasunuma, T.; Chang, J.S.; Kondo, A. Perspectives on engineering strategies for improving biofuel production from microalgae-A critical review. Biotechnol. Adv. 2014, 32, 1448–1459.
111. Passero, M.L.; Cragin, B.; Hall, A.R.; Staley, N.; Coats, E.R.; McDonald, A.G.; Feris, K. Ultraviolet radiation pre-treatment modifies dairy wastewater, improving its utility as a medium for algal cultivation. Algal Res. 2014, 6, 98–110.
112. Olguín, E.J. Dual purpose microalgae-bacteria-based systems that treat wastewater and produce biodiesel and chemical products within a biorefinery. Biotechnol. Adv. 2012, 30, 1031–1046.
113. Amaro, H.M.; Guedes, A.; Malcata, F.X. Advances and perspectives in using microalgae to produce biodiesel. Appl. Energy 2011, 88, 3402–3410.
114. Rawat, I.; Ranjith Kumar, R.; Mutanda, T.; Bux, F. Dual role of microalgae: Phycoremediation of domestic wastewater and biomass production for sustainable biofuels production. Appl. Energy 2011, 88, 3411–3424.
115. Rawat, I.; Kumar, R.R.; Mutanda, T.; Bux, F. Biodiesel from microalgae: A critical evaluation from laboratory to large scale production. Appl. Energy 2013, 103, 444–467.
116. Adav, S.S.; Lee, D.J.; Show, K.Y.; Tay, J.H. Aerobic granular sludge: Recent advances. Biotechnol. Adv. 2008, 26, 411–423.
117. Wang, Z.W.; Liu, Y.; Tay, J.H. The role of SBR mixed liquor volume exchange ratio in aerobic granulation. Chemosphere2006,

62, 767–771.

118. Valigore, J.M.; Gostomski, P.A.; Wareham, D.G.; O'Sullivan, A.D. Effects of hydraulic and solids retention times on productivity and settleability of microbial (microalgal–bacterial) biomass grown on primary treated wastewater as a biofuel feedstock. Water Res. 2012, 46, 2957–2964.

119. Tan, C.H.; Show, P.L.; Chang, J.S.; Ling, T.C.; Lan, J.C.W. Novel approaches of producing bioenergies from microalgae: A recent review. Biotechnol. Adv. 2015, 33, 1219–1227.

120. Conde, J.L.; Moro, L.E.; Travieso, L.; Sanchez, E.P.; Leiva, A.; Dupeiron, R.; Escobedo, R. Biogas purification process using intensive microalgae cultures. Biotechnol. Lett. 1993, 15, 317–320.

121. Heubeck, S.; Craggs, R.J.; Shilton, A. Influence of CO2 scrubbing from biogas on the treatment performance of a high rate algal pond. Water Sci. Technol. 2007, 55, 193–200.

122. Doušková, I.; Kaštánek, F.; Maléterová, Y.; Kaštánek, P.; Doucha, J.; Zachleder, V. Utilization of distillery stillage for energy generation and concurrent production of valuable microalgal biomass in the sequence: Biogas-cogeneration-microalgae-products. Energy Convers. Manag. 2010, 51, 606–611.

123. Holm-Nielsen, J.B.; Al Seadi, T.; Oleskowicz-Popiel, P. The future of anaerobic digestion and biogas utilization. Bioresour. Technol. 2009, 100, 5478–5484.

CHAPTER

17

REMOVAL OF NITRATE AND PHOSPHATE FROM AQUEOUS SOLUTIONS BY MICROALGAE: AN EXPERIMENTAL STUDY

M.H. Sayadi [1], N. Ahmadpour [1], M. Fallahi Capoorchali [2], M.R. Rezaei [1]

[1]Department of Environmental Science, Faculty of Natural Resources and Environment, University of Birjand, Birjand, Iran

[2]Research Center, National Inland Water Aquaculture Institute, Anzali Port, Iran

ABSTRACT

The aim of this study was to evaluate the ability of microalgae *Spirulina platensis* and *Chlorella vulgaris* to remove nitrate and phosphate in aqueous solutions. *Spirulina platensis* and *Chlorella vulgar is* microalgae was collected in 1000 ml of municipal water and KNO_3, K_2HPO_4 was added as sources of nitrate and phosphate in three different concentrations (0.25, 0.35 and 0.45g/L). During the growth period, the concentration of nitrate and phosphate was recorded at 1, 4, 6 and 8 days. The highest nitrate removal on the 8 day for *Chlorella vulgaris* was 89.80% at the treatment of 0.25g/L and for *Spirulina platensis* was 81.49% at the treatment of 0.25g/L. The highest phosphate removal for *Spirulina platensis* was 81.49% at the treatment of 0.45g/L and for *Chlorella vulgaris* was 88% at the treatment of 0.45g/L. The statistical results showed that the amount of phosphate and nitrate removal during different time periods by *Chlorella vulgaris* depicted a significant difference at $P<0.01$, while *Spirulina platensis* demonstrated a significant difference at $P<0.05$.Thus, *Spirulina platensis* and *Chlorella vulgaris* can be effectively used to remove nitrate and phosphate from effluent and waste water treatments, although it demands more research in different climatic conditions.

KEYWORDS: *Aqueous solutions; Chlorella vulgaris; Microalgae; Spirulina platensis; Wastewater*

INTRODUCTION

One of the essential needs of human life is clean water to create healthy conditions for human life. The rapid development of human activity is the result of increased nitrogen and phosphorus entering the groundwater and surface water resources (Sayadi *et al.*, 2012; Sayadi*et al.*, 2014; Abhijna, 2016). The access of this resource to water bodies on one hand aggravates the phenomena such as algal blooms and on the other hand affects the ecological balance of water bodies and subsequently declines the water quality (Rezaei *et al.*, 2015). Various physical and chemical methods are used such as adsorption, chemical deposition, solvent extraction and ion exchange to separate these nutrients. Biological treatment of nitrogen and phosphorus wastewater has been widely studied, and organisms such as bacteria, fungi, protozoa and microalgae have been used for this purpose (Shahriari Moghadam*et al.*, 2016) . Microalgae have the potential to remove mineral nutrients such as nitrogen and phosphorus from wastewater and therefore received considerable attention in the recent years (Khan and Yoshida, 2008) .Wastewater treatment by algal cultures besides controlling additional pollution does not produce effective results in openwheel food, and is a cheap and efficient tool for the rcmoval of nutrients which cause ecological safety in aquatic ecosystems (Aslan and Kapdan, 2006). Many studies related to the viability and usefulness of microalgae in sewage has been reported (Caswell and

Zilberman, 2002).Yang *et al.,* in 2011, studied biomass production and nitrogen and phosphorus removal from municipal wastewater by green algae *Oleoabundans Neochloris*. The results indicated that the algae can completely remove the high nitrogen and phosphorus concentrations with subsequent increase of biomass production. *Chlorella vulgaris* and *Scenedesmus dimorphus are* unicellular microalgae that depicted an ability to eliminate 55% of phosphate from wastewater, dairy and pig farms in Colombia. Other studies related to evaluate the capability of nitrogen and phosphorus wastewater removal by two microalgae*Dunaliella viridis* and *Tetraselmis chuii* reported that the two species of algae are able to traceconcentrations of dissolved salts in the 90%amount or less than that at the beginning of the decrease of biological tests (Koening, and Demacedo, 2004). Accumulation of nutrients cause algae growth and at the same time separation phenomenon occurs in other nutrients such as ammonium evaporation and phosphate deposition produced from algae as a result of high pH (Ahmadpour *et al.,* 2014).

The study was carried out in 2015 at Birjand city which is located in south of Khorasan Province and attempts to evaluate the rate of nitrate and phosphate removal from aquatic solutions using microalgae *Spirulina platensis* and *Chlorella vulgaris.*

MATERIALS AND METHODS

Medium of microalgae

Spirulina platensis is a cyanobacterium (bluegreen algae) that can be growth in brackish water and consumed by humans and other animals. *Chlorella vulgaris* is a genus of single cell green algae belonging to the phylum Chlorophyta which can serve as a potential source of food and energy. Micro-algae used in this study *viz,* pure blue-green algae *Spirulina platensis* and *Chlorella vulgaris* (50ml) where was prepared from the Aquatic Research Institute of Bandar Anzali. Microalgae *Spirulina platensis* and *Chlorella vulgaris* were cultured to 1000ml in Zarrok's Medium and on the Bold's Basal Medium (BBM) respectively (De Morais and Costa, 2007). Agitation and aeration were accomplished using air from a compressor (RESUN AC-9603-0.12MPa) and two fluorescent lamps light from the intensity of 3,500 lux in periods of 12 hours of darkness / brightness Erlenmeyer flask exposure levels was used to determine (Shabani *et al.*, 2016).

Experiments

350 ml *Spirulina platensis* and *Chlorella vulgaris* microalgae were collected in 1000 ml of distilled water and due to the concentration of nitrate and phosphate in municipal waste water treatment plant, KNO_3, K_2HPO_4 was added as sources of nitrate and phosphate in three different concentrations 0.25, 0.35 and 0.45 g/L, conversely in higher concentration of the nutrient algae growth is limited (Aslan and Kapdan, 2006) . The experiment carried out for 8 days because the maximum growth rate of these types of algae occurs in 8 days (Shabani *et al.*, 2016) . Additionally the experiment was conducted in triplicate.

Chemical analysis methods

Primarily, in period of 1-4, 4-6 and 6-8 days, 50ml of the solution was separated *via* filter paper (0.45 micrometres) and later subjected to nitrate and phosphate analysis. All tests were performed at a constant temperature of 25°C and the pH was recorded daily. For determination of residual nitrate and phosphate, the microalgae was separated by centrifugation for 15 minutes at 5000 RPM, and the resultant was passed through a membrane filter with a pore size 0.45 mμ and a diameter of 1 inch (Shabani *et al.*, 2016) . The concentration of nitrate and phosphate was measured by spectrophotometer Bio spec-1601 with wavelength 690nm for phosphate and 220-225nm for nitrate according standard methods (APHA, 1998).

Calculation of the specific growth rate

Specific growth rate was obtained, based on the dry weight using Eq. 1 (Tang *et al.*, 2011):

$$\mu \text{ (1/day)} = ln\,(x_t/x_0)/(t_t-t_0) \quad \text{..............................(1)}$$

Where parameters μ represents the special rate of growth (1/day), X_0 primary biomass (g/L) and biomass amount of time t_t is the X_t. For data analysis, the data obtained by Excel and SPSS were analyzed. Statistical analyzes based on three replicates per test was carried out *via* one way ANOVA and Tukey plan.

RESULTS AND DISCUSSION

Specific growth rate

According to the results of Tables (1, 2) which calculated by Eq.1, the growth rate increased with increasing concentrations of nitrate and phosphate in the treatments.

Table 1: Mean specific growth rate trend observed in the treatments of Spirulina platensis

Time (Day)	1-4	4-6	6-8
PO_4(0.25g/L)	0.029±0.123	0.072±0.288	0.046±0.233
PO_4(0.35g/L)	0.024±0.125	0.082±0.299	0.058±0.268
PO_4(0.45g/L)	0.033±0.167	0.038±0.169	0.017±0.571
NO_3(0.25g/L)	0.015±0.765	0.024±0.123	0.016±0.83
NO_3(0.35g/L)	0.024±0.816	0.026±0.134	0.019±0. 097
NO_3(0.45g/L)	0.16±0.819	0.04±0.148	0.04±0.225

Table 2: Mean specific growth rate trend in the treatments of Chlorella vulgaris

Time (Day)	1-4	4-6	6-8
PO_4(0.25g/L)	0.029±0.127	0.072±0.291	0.046±0.221
PO_4(0.35g/L)	0.024±0.139	0.082±0.302	0.058±0.298
PO_4(0.45g/L)	0.033±0.172	0.038±0.238	0.017±0.602
NO_3(0.25g/L)	0.015±0.767	0.024±0.248	0.016±0.924
NO_3(0.35g/L)	0.024±0.924	0.026±0.365	0.019±0.137
NO_3(0.45g/L)	0. 16±0.992	0.04±0.216	0. 04±0.334

As demonstrated, the highest growth rate of *Spirulina platensis* and *Chlorella vulgaris* was in the treatments; 0.45 g/Lnitrate*viz.*0.503 (1/day) and 0.58 (1/day) and 0.45 g/L phosphate *viz.*0.244 (1/day) and 0.354 (1/day), respectively. In a study conducted by Afshari *et al* (2011), on the ability of algae *Tetraselmis suecica* in urban secondary wastewater treatment to reduce nitrate and phosphate the similar trend in the different days was achieved. 78.1% and 70.3% nitrate in the aquatic solutions were removed by *Chlorella vulgaris* and *Scenedesmus* respectively (Kshyrsagr, 2013). Similarly, the results of the current study showed that the highest growth rate was in the concentration 0.45 g/L, with a higher nitrate density. In a study conducted by Mousavi *et al* (2009) on the growth of *Chlorella vulgaris*in different cell densities it was stated that less phytoplankton density resulted in low nitrogen removal efficiency which is in concurrence with the present study where is very high densities in the treatments reduce light penetration and increases shade effects which in turn limits the growth and metabolic activities in the phytoplankton cells. In this study, the growth rate of microalgae was effective for nitrate and phosphate removal of aqueous solutions of different treatments where by the highest growth rate and higher removal rate was obtained on the eight day. The average growth rate comparison showed no significant difference between the two microalgae ($P<0.05$). The average growth rate of *Chlorella vulgaris* in both phosphate and nitrate treatments was more in relation to *Spirulina platensis*. However, with reference to time period, a significant difference was observed on the eighth day ($P<0.05$) in different nitrate and phosphate concentrations.

pH changes pH is an important phenomenon especially in the removal of nutrients that were assessed this study. Increased alkalinity and pH indicator phosphate and nitrate; in other terms it tends to reduce nutrients (Ji *et al*, 2014) . However, the pH value in microalgae growth, showed an upward trend. Table 3 and 4 exhibit the pH changes in microalgae *Spirulina platensis* and

Chlorella vulgaris, with increased cell growth, besides with an increase in nitrate and phosphate, the pH value increases(Mousavi *et al*, 2009). However, the highest pH value was indicated in the highest growth, and the highest cell density and growth was observed in the final days i.e. on the 8 day. In a study carried out by Zamani *et al.* (2010) on efficiency removal of nitrate, nitrogen and orthophosphate from wastewater by using several species of microalgae to check the pH changes during the 12 days culturing period, the results showed that for most of the studied species, this amount gradually increased from the initiation to the termination of the experimental period which was in concurrence with the results of this study. Increasing pH leads to phosphate precipitation in the medium. If the pH is reduced, it reduces the phosphate dissolved in the medium. Nitrogen absorption by microalgae influences the pH value. Absorption of nitrate ions lead to increased pH, but if ammonia is used as the nitrogen source then it reduces to very acidic pH of 3, which is good for growth. (Caswell and Zilberman, 2002) . One of the negative effects of high pH in microalgae medium is that it reduces the removal of nutrients and therefore growth should be increased to avoid turbulence for gas exchange between water and air and pH should be adjusted to a certain extent (Suva, 1999). Thus, in the present study, it can be suggested that the increase in pH is due to the photosynthetic uptake of carbon dioxide. In case of lack of carbon dioxide supply, the pH value of 10 is not uncommon and if carbon dioxide is limited and bicarbonate is used as a carbon source, then pH can be 11 or even higher (Oswald, 1988).

Nitrate removal

Removal of nutrients by microalgae cells is in accordance with their cell density and metabolic activity. In this study, the ability of microalgae to grow and remove nitrate at different concentrations of KNO_3, K_2HPO_4 were tested in the course of 8 days at three different periods of 4, 6 and 8 days respectively. The results of this study showed that microalgae *Chlorella vulgaris* and *Spirulina platensis* grown in nitrogenrich treatments, and as depicted in the Figs. 1 and 2, an optimal decrease in nitrate concentrations was shown in all the treatments on the 8 day. The highest rate pertained to 0.45 g/L where in a nitrate removal percentage of 89.97% and 89.90% were reported by microalgae Spirulina platensis and *Chlorella vulgaris* respectively. The removal process in both treatments increased with time. The lowest removal rate was observed in the treatment 0.25 g/L with 1-4 days viz. 28.71% and 20.08% for microalgae *Spirulina platensis* and *Chlorella vulgaris* respectively. Based on previous studies removed during the eight days of constant and numerous changes do not occur with the

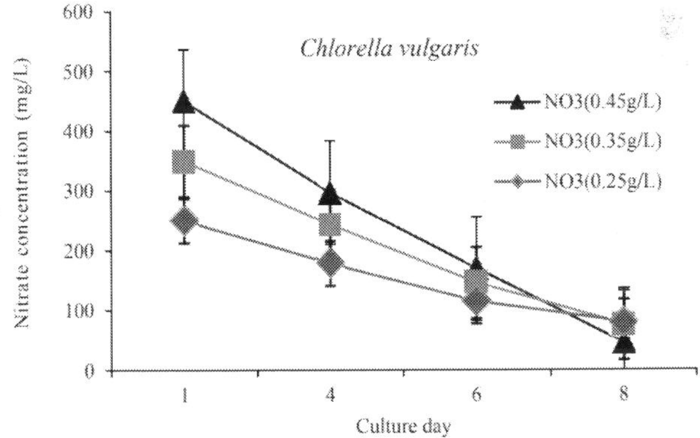

Figure 1: Comparing the mean and standard deviation of remaining nitrate treatments

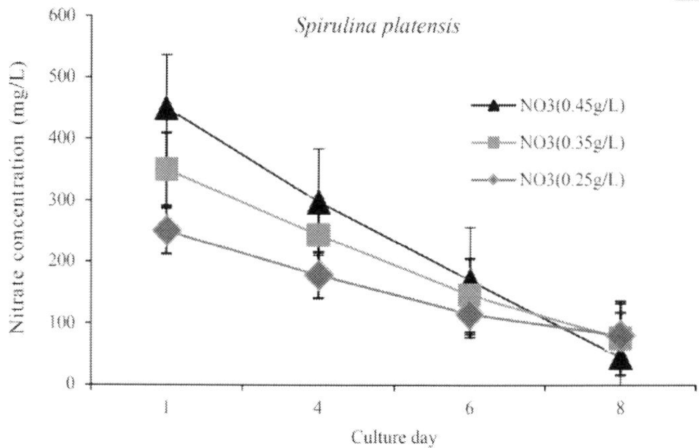

Figure 2: Comparing the mean and standard deviation of remaining nitrate treatments

passage of time (Tam and Wong, 1994). Rasolamini *et al.* (2011) in a study conducted on the nitrate and phosphate removal from wastewater by microalgae reported that the percentage of nitrogen removal by *Chlorella*was84.11% within 14 days and phosphorus removal rate by algae *Chlamydomonas* was 100%, which is a higher percentage in comparison to the results of the present study. Studies on the nitrogen and phosphorus removal from wastewater by microalgae *Scenedes musobliquus* under different temperature and mixing conditions showed that the highest nitrogen removal was at a temperature of 25 ° C and vigorous mixing at a rate of 100% and the maximum phosphorus removal (98%) was obtained in the same conditions, relating a higher ability of microalgae *Spirulina* to remove nitrate and phosphate. In a similar study conducted by Dickinson *et al.* (2013) the ratio of nitrogen /phosphorus was reported 5: 1 to 12: 1 i.e. nitrogen and phosphorus removal rates were between 99-83% and 99% of phosphorus respectively. Tang *et al.* (2011) studied the reduction of phosphorus and nitrogen in wastewater using *Spirulina* algae and stated that *Spirulina* needs nitrate and phosphorus in the culture and reported that culturing of *Spirulina* in wastewater could reduce nitrate content from 16 ± 4.30 to 5.5 ± 1.50 ppm. In the present study, the highest nitrate and phosphate removal by the microalgae *Spirulina platensis* was on the eighth day, with a nitrate (115mg/L) and phosphate (54mg/L) reduction in the concentration of 0.45 g/L. Cheunbarn and Peerapornpisal, (2010) successfully cultured *Spirulina* in the wastewater, and after 12 days of retention, measured the maximum efficiency, which was 67, 23, 45, 49, 92 percent for COD, BOD, NO^-_3 and NH^+_4respectively.In this study, the maximum efficiency of NO_3 and PO_4within 8 days of microalgae culture was 65/78 and 49/81 respectively. Chngany *et al.* (2012) used micro-algae *Spirulina platensis* for wastewater treatment and showed that this microalga grows in nutrient-rich wastewater and can act as a substitute for appropriate secondary treatment. Nevertheless, in the present study, the maximum NO_3 and PO_4removal were, 65.78% and 49.81 %, respectively.

Phosphate removal

Phosphate ions, including chemical contaminants enter surface water and groundwater resources through municipal, industrial and agricultural wastewater. In this study, the ability of microalgae to grow and remove phosphate at different concentrations of K_2HPO_4 during the 8 days experimental course was tested at different time periods *viz.* 4, 6 and 8 days. The results showed that the highest percentage of phosphate removal was on the eighth day. The highest phosphate removal in the three different concentrations of PO_4 *viz.* 0.25, 0.35 and 0.45g/L on the 8 days for *Chlorella* was (29.12%), (40.88%) and (82.02%) and for *Spirulina* was

(32.89%) , (40.40%) and (88%) for days 1 to 4, 4 to 6, and 6 to 8, respectively. The lowest removal rates were observed in the treatment 0.25 g/L with 1-4 days viz. 24% and 20.06% for microalgae *Spirulina platensis* and *Chlorella vulgaris* respectively.

Figs. 3 and 4 represent the phosphate removal at various time intervals in *Chlorella vulgaris* and *Spirulina platensis.*

As the Figs. 3 and 4 indicate, with the passage of time, the phosphate concentration in the aqueous solution reduced and the lowest phosphate concentration was reported at (0.45 g/L) concentration of PO_4 i.e. (80.89 mg/L) in 8 days, thereby indicating 82.02% removal by microalgae *Chlorella vulgaris*. Dickinson *et al.* (2013) reported an optimal phosphate absorption rate for the microalgae *Scenedesmusviz.*PO_4-P 0.104 mg/L/h.

However, in this study, the absorption rate of phosphate by the microalgae *Chlorella vulgaris* and *Spirulina platensis* within 8 days was 80.89 mg/L and 54 mg/L respectively. Moreover, the largest decline in eight days was reported in the (0.45 g/L) concentration of PO_4. Wang and Lan in 2011 studied the biomass production besides nitrogen and phosphorus removal from municipal wastewater by green algae *Neochloris oleoabundans*. The results indicated that the algae had a potential to completely remove the high concentrations of nitrogen and phosphorus with an increase in biomass production, so it can be safely considered as a viable option for the treatment of municipal wastewater. In this study, the micro-algae *Chlorella vulgaris* and *Spirulina platensis* could successfully remove nitrate and phosphate during the experiment. Likewise, Aravantinou *et al.* (2013) reported phosphate absorption rate of 0.475 mg/L for microalgae *Chlorococcum*. Statistical results based on the Tukey test showed that the amount of phosphate and nitrate removal during different time periods by microalgae *Chlorella vulgaris* exhibited a significant difference in $P<0.01$, while microalgae *Spirulina platensis* showed significant differences at $P<0.05$. Tukey test also showed that the removal efficiency of microalgae *Chlorella vulgaris* and *Spirulina platensis* are significantly different ($P<0.01$) (Patel *et al.*, 2012). However, higher ability to remove phosphate and nitrate was shown by micro-algae *Spirulina platensis* in relation to *Chlorella vulgaris*.

CONCLUSION

Adsorptions of metal by bio-absorbents are considered as an effective process to remove metal ions from aqueous solution. In the present study, the ability of *Chlorella vulgaris* and *Spirulina platensis* absorbent to remove nitrate and phosphate ions influenced by different nitrate and phosphate concentrations was investigated. The results showed that micro-algae *Spirulina platensis* have a higher ability to remove nitrate and phosphate. Thus, it can be safely concluded thatthoughsome algae can remove phosphate and nitrate from the aqueous solution, but *Chlorella vulgaris* and *Spirulina platensis* have a high ability to reduce nitrate and phosphate from aqueous solutions, and thus can effectively remove phosphate and nitrate from the effluent at different climatic conditions, however, this finding demands more research.

ACKNOWLEDGEMENT

The valuable collaboration of faculty authorities of Natural Resources and Environmental Sciences, University of Birjand to provide necessary facilities for conduction of this study is highly appreciated.

CONFLICT OF INTEREST

The authors declare that there are no conflicts of interest regarding the publication of this manuscript.

REFERENCES

1. Abhijna, U.G., (2016). Monitoring and assessment of a eutrophicated coastal lake using multivariate approaches. Global J. Environ. Sci. Manage., 2 (3): 275-288 (**14 pages**). Afshari, A.; Yahyavi, M.; Shybk, H.; Abdolallyan, A., (2011). Check availabilityofTetraselmis suecica microalgae in municipal secondary wastewater treatment. Aquat. Fish., 2 (8): 1-8 (**8 papges**).
2. Ahmadpour, N.; Sayadi, M.H.; Fallahi, M.; Rezaei, M. R., (2014) . Removal of phosphate by microalgae from municipal wastewater effluents: Lab Experiment. Biotechnology Tarbiat Modares University, 6 (2):40-50 (In Persian) (**11 pages**).
3. APHA, (1998). Standard methods for examination of water and wastewater. American Public Health Association Publication, Washington DC.
4. Aravantinou, A.F.; Theodorakopoulos, M. A.; Manariotis, I.D., (2013). Selection of microalgae for wastewater treatment and potential lipids production. Biores. Tech.,147: 130 – 134 (**5 pages**).
5. Aslan, S.; Kapdan, I.K., (2006). Batch kinetics of nitrogen and phosphorus removal from synthetic wastewater by algae. Ecol. Eng., 28:64–70 (**7 pages**).
6. Caswell, M.; Zilberman, D., (2002). Algol – culture. University of California at Berkeley., 6:1-12 (**13 pages**).
7. Cheunbarn, S.; Peerapornpisal. Y., (2010). Cultivation of Spirulina platensis using anaerobically swine wastewater treatment effluent. Int. J. Agricul. Bio.,12: 586–590 (**5 pages**).
8. Chngany, Z.; Modares, A.; Afsharzadeh, S., (2012).Wastewater treatment using micro-algae Spirulina platensis culture. First National Conference on Phytoremediation. Kerman, Iran 16 February.

9. De Morais, M. G.; Costa, J.A., (2007). Biofixation of carbon dioxide by Spirulina sp. and Scenedesmus obliquus cultivated in a three-stage serial tubular photobioreactor. J. Biotech., 129:439–445 (**7 pages**).
10. Dickinson, K.E.; Whitney, C. G.; McGinn, P.J., (2013). Nutrient remediation rates in municipal wastewater and their effect on biochemical composition of the microalga Scenedesmus sp. AMDD. Algal Res.,2: 127–134 (**8 pages**).
11. Khan, M.; Yoshida, N., (2008). Effect of L-glutamic acid on the growth and ammonium removal from ammonium solution and natural wastewater by Chlorella vulgaris NTM06. Biores. Tech.,99: 575–582 (**8 pages**).
12. Koening, L. M.; Demacedo, J.S., (2004). Urban secondary sewage: an alternative medium for the culture of Tetraselmis chuii(Prasinophyceae) and Dunaliella viridis (Chlorophyceae). Braz. Arch. Biol. Tech., 3:451-459 (**9 pages**).
13. Kshirsagar, A.D., (2013). Bioremediation of wastewater by using microalgae, an experimental study. Int. J. Life Sci. Biotech. Pharma. Res., 2: 339–346 (**7 pages**).
14. Mousavi, A.; Yahyavi, M.; Taherizadeh, M., (2009). Evaluation of Chlorella vulgarisphytoplankton growth and its effects on urban wastewater nutrients. Aquat. Fish., 1: 64-72 (**8 pages**).
15. Oswald, W.J., (1988). Micro-algae and wastewater treatment. Micro-algal Biotech. Cambridge University Press, 22:305– 328 (**24 pages**).
16. Patel, A.; Barrington, S.; Lefsrud, M., (2012). Microalgae for phosphorus removal and biomass production: a six species screen for dual purpose organisms. GCB Bioenergy, 4(5):485495(**11pages**).
17. Rasoulamini, S.; Montazeri-Najafabady, N.; Mobasher, M.A.; Hoseini-Alhashemi, S.; Ghasemi, Y., (2011). Chlorella sp., a new strain with highly saturated fatty acids for biodiesel production inbubble-column photobioreactor. Appl. Energy, 88 (10): 3354–3356(**3 pages**).
18. Rezaei, A.; Sayadi, M.H., (2015). Long-term evolution of the composition of surface water from the River Gharasoo, Iran: a case study using multivariate statistical techniques. Environ. Geochem.

Health., 37(2): 251-261(**11 pages**).

19. Sayadi, M. H.; Kargar, R.; Doosti, M.R.; Salehi, H., (2012). Hybrid constructed wetlands for wastewater treatment: A worldwide review. Proc. Int. Acad. Eco. Environ. Sci.,2(4): 204-222 (**18 pages**).
20. Sayadi, M.H.; Rezaei, A.; Rezaei, M.R.; Nourozi, K., (2014). Multivariate statistical analysis of surface water chemistry: A case study of Gharasoo River, Iran. Proc. Int. Acad. Eco. Environ. Sci., 4(3):114-122 (**8 pages**).
21. Shabani, M.; Sayadi M.H.; Rezaei M.R., (2016) CO_2 biosequestration by Chlorella vulgaris and Spirulina platensis in response to different levels of salinity and CO_2. Proc. Int. Acad. Eco. Environ. Sci., 6(2): 53-61(**8 pages**).
22. Shahriari Moghadam, M., Safaei, N., Ebrahimipour, G.H., (2016) . Optimization of phenol biodegradation by efficient bacteria isolated from petrochemical effluents. Global J. Environ. Sci. Manage., 2 (3): 249-256 (**7 pages**).
23. Suva, F. (1999). Technical guidance on pearl hatchery development in the Kingdom of Tonga. FAO. GCP/RAS/ 116 /JPN.
24. Tam, N.F. Y.; Wong, Y.S., (1994). Effect of immobilized microalgal bead concentrations on wastewater nutrient removal. Environ. Poll.,107: 145-151(**7 pages**).
25. Tang, D.; Han, W.; Li, P.; Miao, X.; Zhong, J., (2011). CO_2 biofixation and fatty acid composition of Scenedesmus obliquus and Chlorella pyrenoidosa in response to different CO_2 levels. Biores. Tech.,102: 3071–3076 (**7 pages**).
26. Wang, B.; Lan, C. Q., (2011). Biomass production and nitrogen and phosphorus removal by the green alga *Neochloris oleoabundans* in simulated wastewater and secondary municipal wastewater effluent. Biores. Tech.,102(10), 5639-5644 (**5 pages**).
27. Yang, J.; Xu, M.; Zhang, X.Z.; Hu, M., (2011). Sommerfeld and Y.S. Chen: Life-cycle analysis on biodiesel production from microalgae: Water footprint and nutrients balance. Biores. Tech., 102, 159-165 (**6 pages**).
28. Zamani, N.; Nowshadi, M.; Amin, S.; Ghasemi, Y.; Niyazi, A., (2010). Removal of nitrogen-nitrate and ortho phosphate from wastewater using microalgae biotechnology. The second

International Symposium on Environmental Engineering. Tehran, Iran 18-20 February.

CHAPTER

18

DEFATTED BIOMASS OF THE MICROALGA, DESMODESMUS SP., CAN REPLACE FISHMEAL IN THE FEEDS FOR ATLANTIC SALMON

Viswanath Kiron[1], Mette Sørensen[1], Mark Huntley[2,3], Ghana K. Vasanth[1], Yangyang Gong[1,4], Dalia Dahle[1] and Anjana M. Palihawadana[1]

[1]Faculty of Biosciences and Aquaculture, Nord University, Bodø, Norway

[2]Department of Biological and Environmental Engineering, Cornell University, Ithaca, NY, USA

[3]Marine Laboratory, Nicholas School of the Environment, Duke University, Beaufort, NC, USA

[4]Key Laboratory of East China Sea and Oceanic Fishery Resources Exploitation, East China Sea Fisheries Research Institute, Shanghai, China

Keywords: Atlantic salmon, microalga, Desmodesmus, feed ingredient, fishmeal replacement, growth, feed performance, intestinal health

Microalgal biomass is a potential feed ingredient that can replace fishmeal and ensure sustainability standards in aquaculture. To understand the efficacy of the defatted biomass from the marine microalga,

Desmodesmus sp. a 70-day feeding study was performed with Atlantic salmon (*Salmo salar*) smolts. Three groups of fish (av. wt. 167 g) were offered either a control feed (without the microalga) or the microalga-containing (10/20%) feeds. At the end of the feeding period, the growth indices (condition factor, specific growth rate) and survival of the microalga-fed fish were not significantly different from the respective values of the control fish, but the feed conversion ratios were inferior. The proximate composition of the whole body of salmon from the three groups did not vary significantly. Compared to the control fish, the 10% alga-fed fish had lower lipid content in their filet. The protein and lipid digestibility in the three feeds did not differ significantly, but the digestibility of energy in the 10% alga-feed was significantly lower than that of the control feed. Furthermore, comparison of the distal intestinal proteome of Atlantic salmon revealed that the expressions of Alpha-2-HS-glycoprotein-like (Ahsg), Myosin-11 isoform X1 (My11) and Dihydrolipoyl dehydrogenase, mitochondrial-like (Dld) were altered by the microalgal feeding. Examination of the physiological status of the fish based on the serum antioxidant capacities did not reveal any alga-feed-related differences. Moreover, the expression of the selected immune and inflammatory marker genes and the micromorphological observations did not indicate any aberration in the intestinal health of the microalga-fed fish. It is possible to include 20% of defatted *Desmodesmus* sp. in the feeds of Atlantic salmon.

INTRODUCTION

Production of farmed aquatic species has increased significantly and therefore the aquafeed industry is actively engaged in finding alternatives for the highly used fishmeal. Such a strategy is imperative for the sustainable growth of the aquaculture industry. The aquafeed sector has been using terrestrial plant-derived ingredients as substitutes for fishmeal and fish oil in the feeds of various farmed fish including those for Atlantic salmon, Salmo salar. Terrestrial plant ingredients employed in the feeds of Atlantic salmon include soyprotein concentrate, sunflower expeller, wheat gluten, fava beans, pea protein, maize gluten, horse beans, rapeseed oil, wheat starch, pea starch and tapioca starch (Ytrestøyl et al., 2015). However, some of these ingredients can compromise the intestinal health of carnivorous species like Atlantic salmon (Baeverfjord and Krogdahl,

1996; Knudsen et al., 2007; Bakke, 2011). Hence, it would be ideal to use ingredients from microalgae, which are present in the aquatic food web, in the feeds of Atlantic salmon.

Microalgae are rich in essential nutrients and bioactive molecules, and therefore, are valuable nutrient sources for humans and animals (Becker, 2004). They contain high-quality protein, 50–70% of which is likely to be bioavailable to animals (Becker, 2007). The bioactive components include polyunsaturated fatty acids, sugars, polysaccharides, vitamins, minerals, and pigments (Becker, 2004). Thus, microalgae can be high quality feed components.

In Europe and the United States, efforts are underway to commercially exploit the full potential of microalgae (Enzing et al., 2014; Energy.Gov: Office of Energy Efficiency Renewable Energy, 2016). The co-products from the cultivated microalgae such as Staurosira sp. and Desmodesmus sp. include "biocrude" oil and residual protein-rich fractions (Huntley et al., 2015). The defatted protein-rich biomass obtained after the extraction of the biofuel could serve as an ingredient in animal feeds. Even though some studies have proved the potential of using microalgae in fish feeds, the protein quality of an alga can be determined only through elaborate feeding studies on animals (Becker, 2004). The effect of current plant-derived fishmeal replacers on farmed animals varies widely. Therefore, feeding studies should evaluate the potential of the microalgae products as fishmeal substitutes by investigating not only the growth of the animals but also their health status.

In the present study, the ability of the defatted biomass of Desmodesmus sp. to replace fishmeal in the feeds of Atlantic salmon was evaluated by assessing the growth, feed performance, antioxidant status, and intestinal health of the fish. The hypothesis is that the Desmodesmus biomass in the feeds of Atlantic salmon could impart good growth and health to the fish.

MATERIALS AND METHODS

Fish and Rearing

The study was approved by the National Animal Research Authority (FDU: Forsøksdyrutvalget ID - 5887) in Norway. The fish rearing and handling procedures were according to the approved protocols of FDU.

Atlantic salmon, *Salmo salar* smolts obtained from Cermaq, Hopen, Bodø, Norway were maintained on commercial feeds at the Research Station of the Nord University, Bodø, Norway for 2 months. Before the start of the feeding experiment, the fish were transferred to the indoor rearing tanks (800 L), which were part of a flow-through seawater system. The tanks had feed traps (Aquatic Eco-Trap, Pentair Aquatic Eco-Systems®, Florida, and USA) to collect the uneaten feed. Each of the 18 tanks used for the study was stocked with 25 fish of average weight 167.6 g. The temperature (7.6 ± 0.9°C) and dissolved oxygen saturation (av. 90%) of the rearing water were recorded daily. A 24-h light regime was employed throughout the experimental period.

Experimental Feeds, Design and Feeding

The three types of experimental feeds (one control feed and two microalgal feeds, Table 1) were made to order (Center for Feed Technology, Norwegian University of Life Sciences, Ås, Norway). In the algal feeds, the algal biomass of *Desmodesmus* sp., a co-product from biofuel production (Cellana, San Diego, USA), replaced a portion of the fishmeal. The extruded feed pellets (3 mm size) were stored at 4°C until they were used for feeding the fish.

Table 1: Ingredients and proximate composition of the experimental feeds.

Feed ingredients (g/100 g)	Experimental feeds		
	4C	4L	4H
Fish meal	69	60	51
Algal meal	0	10	20
Wheat	12	12	12
Wheat gluten	5	5	5
Fish oil	13.5	12.5	11.5
Microingredients*	0.5	0.5	0.5
Marker (Yttrium)	0.01	0.01	0.01
Proximate composition (g/100 g dry basis)			
Moisture	3.5	7.6	7.1
Crude protein	53.8	50.4	47.8
Crude lipid	20.1	19.8	19.5
Ash	11.1	11.5	12.2
Energy (MJ/100g)	2.3	2.3	2.3

*Vitamin and mineral mix is a proprietary formulation of Polarfeed, Leknes, Norway.

The three fish groups in this study were: a control group that received alga-devoid feed (4C), and 2 alga-fed groups that were offered feeds containing the defatted *Desmodesmus* biomass at inclusion levels of 10% (4L) and 20% (4H). Fish in six tanks were randomly allotted to one feed group and they were offered the respective feeds daily for a period of 10 weeks. Automatic feeder (Arvo-Tec, Huutokoski, Finland) of a particular tank was programmed to deliver daily doses of feed (until the end of the study) corresponding to 1.2% of the body mass of fish in the tank. The daily ration was delivered intermittently, between 08:00 and 09:00, and between 14:00 and 15:00. The left-over feeds were collected from the trap, 1 h after each meal. They were then dried and weighed to determine the actual feed intake. The fish in the tanks were visually inspected in the afternoon, after the feeding session, to ascertain their well-being.

Sampling

At the start of the experiment, individual weights and lengths of the fish were recorded. The fish were starved for 24 h and anesthetized with MS-222 (Tricaine methane sulphonate; Argent Chemical Laboratories, Redmond, USA; 80 mg/l), prior to taking the measurements. Similarly, the final weights and lengths were recorded at day 70. Representative samples were collected at day 70 from fish in all tanks—blood (serum), distal intestine, filet and whole body—to assess the different parameters of interest. In addition, the weight of the visceral organs and liver of the sampled fish were recorded. After the termination of the feeding study, the feces for the digestibility determination was collected by stripping the fish, as described by Austreng (1978). Feces from individual fish in one tank of a particular group were pooled during collection. Prior to chemical analysis, freeze-dried feces from two tanks had to be pooled to get enough material for analyses. The sample size is indicated in the figure legends and table footnotes.

Study Parameters

Growth and Feed Performance

Initial and final weights, and feed consumption of the fish were used to calculate the growth and feed performance indices. The formulae for calculating the condition factor, specific growth rate (SGR), feed conversion ratio (FCR), protein efficiency ratio (PER) and survival are provided in the footnote of Table 3.

Hepatosomatic index [HSI = (liver weight/final weight) × 100] and viscerosomatic index [VSI = (visceral organ weight/final weight) × 100] of the fish from different groups were calculated.

To determine the proximate composition of the whole body and filet, the samples were thawed and homogenized in a grinder (Grindomix GM 200, Retsch, Haan, Germany). Moisture (gravimetry), ash (incineration at 550°C), crude protein (N*6.25, Kjeldahl Autoanalyser, Tecator, Sweden) and total lipid (in fish -ethyl acetate extraction method, Norges Standardiseringsforbund, 1994) contents of the fish and the feeds were determined using standard protocols (Kiron et al., 2012). The energy content in the feeds and feces were determined by bomb calorimetry (C200 calorimeter, IKA, Staufen, Germany). The total lipid and yttrium content in the feeds and feces were determined by Eurofins AS (Moss, Norway) and they were used to calculate the digestibility coefficients. The total lipid content was determined using the Soxhlet method with acid hydrolysis. Yttrium content was analyzed after combustion of the samples at 550°C and boiling with hydrochloric acid and nitric acid. Thereafter the samples were cooled and dissolved in 5% nitric acid. Yttrium was then detected with a spectrometer (Perkin Elmer, Massachusetts, USA).

Histological Studies

The processing and sectioning techniques of the distal intestinal samples (n = 6 fish) are described in Vasanth et al. (2015). Approximately 5 μm sections were fixed for the histological studies. The sections were stained with Alcian Blue-Periodic Acid Schiff›s reagent (AB-PAS, pH 1.0) (Bancroft and Gamble, 2007). Photomicrographs were captured employing an Olympus BX61/Camera Color View *III*u (Olympus Europa GmbH, Hamburg, Germany) and processed using the imaging system, photo program Cell P (Soft Imaging System GmbH, Munster, Germany).

Antioxidant Indices

The frozen (−80°C) serum aliquots were thawed prior to conducting the different assays, which were performed using commercially available kits employing a microtiter plate (Nuclon sterile 96 flat, Fisher Scientific,

Massachusetts, USA). The respective absorbances were measured at the suggested wavelengths employing a microplate reader (Fluostar Optima; BMG Labtech, Ortenberg, Germany).

Total antioxidant capacity (TAC): Aliquots (20 μL) of the serum samples were used for determining the TAC, employing a kit from Cell Biolabs (STA-360, Cell Biolabs Inc., San Diego, CA, USA). The initial and final absorbances (490 nm) of the samples and standards were measured, and based on the net absorbance values and the uric acid standard curve, the uric acid equivalence (UAE) of the samples was obtained. The μM copper reducing equivalents (CRE), which is proportional to TAC, were calculated by multiplying the UAE concentrations by a factor of 2189 μM Cu^{++}/μM uric acid.

Catalase (CAT) activity: The Cayman Chemicals assay kit 707002 (Cayman Chemicals, Ann Arbor, Michigan, USA) was employed to determine the CAT activity of serum aliquots (20 μL), following the manufacturer›s protocol. The enzyme activity was calculated based on the absorbance at 540 nm and the formaldehyde standard curve. The activity (nmol/min/ml) is expressed as the amount of enzyme that will cause the formation of 1.0 nmol of formaldehyde per minute at 25°C.

Superoxide dismutase (SOD) activity: A commercially available kit (STA-340, Cell Biolabs Inc.) was employed to determine the SOD activity in 20 μL serum samples. The assay was performed following the manufacturer›s instructions. The SOD activity (inhibition %) of the sample is inversely proportional to the colorimetric signal and is calculated using the absorbance values of the sample and blank at 490 nm. The activity is expressed as SOD units U/μL.

Gene Expression Study

The mRNA levels of selected genes - antioxidant-related (superoxide dismutase1 - sod1; nuclear factor erythroid 2-related factor - nrf2), mucosa-related (immunoglobulin T - igt), inflammation-related (interleukin 1b - il1b; interleukin 10 - il10; interleukin 17a and 17d - il17a, il17d; transforming growth factor beta - tgfb; tumor necrosis factor a - tnfa), antimicrobial (cathelicidin 1 and 2 - cathl1, cathl2; defensin beta 2 - defb2) - were assessed in this study.

RNA extraction and cDNA synthesis

The total RNA from the distal intestine was extracted using E-Z 96 Total RNA Kit (OMEGA Bio-Tek Inc., Norcross, GA, USA), following the instructions of the manufacturer. About 100 mg of the tissue was homogenized with Zirconium oxide beads (1.4 mm; Percellys, Tarnos, France) in TRK lysis buffer in a precooled free standing tube (VWR International, Oslo, Norway) at 6000 rpm. The resulting crude mixture was centrifuged (18,000 × g, 20°C) and approximately 300 μl of the supernatant was collected and mixed with equal volume of 70% ethanol. This mixture was added to the assigned wells of the E-Z 96 RNA plate mounted on an E-Z 96® Vacuum Manifold (VAC-03, Omega Bio-tek Inc.). A pressure of 1100 kPa was applied to draw the sample through the well, which had an RNA HiBind® matrix. This step was followed by buffer washes; first with RWF Wash Buffer (500 μl) and then twice with 600 μL of RNA Wash Buffer II. Pressure was applied each time the wells were treated with the wash buffers. Finally, the purified RNA was recovered from the HiBind® matrix by adding 50 μl of RNase-free water (5 Prime GmbH, Hilden, Germany) to each well and applying a final round of pressure.

The cDNA was synthesized as described by Vasanth et al. (2015). The obtained cDNA was 30-fold diluted and used in qPCR.

Real-time PCR (qPCR) and the quantification of gene expression

The primers for the reference and target genes are given in Table 2. The PCR reactions were carried out using LightCycler® 96 Real-Time PCR System (Roche Holding AG, Basel, Switzerland). Each reaction from one fish sample was done in duplicate, and to quantify the mRNA level of a particular gene, reactions were done on samples from 12 fish (2 from each tank). All the reactions were performed using a total volume of 10 μl, containing 5 μl of FastStart Universal SYBR Green Master (ROX) mixed with 1 μl primer mix (200 nM), and 4 μl cDNA (5 ng/μl). Each reaction was done under the following thermal cycling conditions: preincubation at 95°C for 600 s followed by 40 cycles of two step amplification at 95°C (10 s), and 60°C (30 s) and melting at 95°C (10 s), 65°C (60 s) and 97°C (1 s). Specificity of the amplified products was checked by performing melt curve analyses for each sample and the specificity of the primers was confirmed by sequencing the amplified products. For relative quantification, the total RNA from all samples was pooled, and normalized to 1000 ng, from which a 6-point 3-fold serial dilution was prepared (1:1 to 1:243) to make a standard curve. The efficiency of the primers was calculated using the equation $E = (10^{-1/m} - 1)\ 100$. A normalization factor was computed for each of the samples based on the relative quantities of the two most stable genes, rpl13 and rps29 from among the set of four reference genes—elongation factor 1ab (ef1ab), ribosomal protein L13 (rpl13), ribosomal protein S29 (rps29) and ubiquitin (ubi), employing geNorm (Vandesompele et al., 2002). The expression levels of all the target genes were then calculated relative to the normalization factor.

Table 2: List of primers for the differentially expressed genes and reference genes.

Gene name	Sequence (5′-3′)	Amplicon size (bp)	PCR efficiency (%)	Reference
sod1	CCACGTCCATGCCTTTGG **F** TCAGCTGCTGCAGTCACGTT **R**	141	95.3	GenBank : AY736282.1
nrf2	TCAACAGGACATCGACCTAAT **F** GGCAGTAGTCAAACACCTCT **R**	83	91.2	GenBank : BT059007.1
cathl1	CTAGCAACAACCTGAACACTG **F** CTTCTTGTCCGAATCTTCTGCATA **R**	88	96.4	GenBank : AY728057.1
cathl2	TGCTTCAAATCAGCTGCTTCC **F** CTGTATTCAAAGTCTCGACTTCCA **R**	83	104.4	GeneBank : AY542961.1
il17d	CTTGTCTCCCTGGGCATACAG **F** CAATATGCCTCGGGTATGAACCT **R**	201	112.7	GenBank : EU689087.1
tgfb	GGTACTGGCTCTGTATAAGCAC **F** CTGCTCCACCTTGTGTTGTC **R**	120	100.9	GenBank : BT059581
igt	CAACACTGACTGGAACAACAAGGT **F** CGTCAGCGGTTCTGTTTTGGA **R**	97	107.7	GenBank : GQ307004
REFERENCE GENES				
ef1ab	TGCCCCTCCAGGATGTCTAC **F** CACGGCCCACAGGTACTG **R**	59	96	GenBank : BG933853
rpl13	CGCTCCAAGCTCATCCTCTTCCC **F** CCATCTTGAGTTCCTCCTCAGTGC **R**	79	96.4	GenBank : BT048949.1
rps29	GGGTCATCAGCAGCTCTATTGG **F** AGTCCAGCTTAACAAAGCCGATG **R**	167	94.5	GenBank : BT043522.1
ubi	AGCTGGCCCAGAAGTACAACTGTG **F** CCACAAAAAGCACCAAGCCAAC **R**	162	92.7	GenBank : AB036060.1

Protein Expression Study

Two-dimensional electrophoresis (2-DE) was performed on the distal intestinal samples (n = 6 fish from 6 tanks) from each of the three groups. The various steps such as protein extraction, two-dimensional electrophoresis and capturing of the gel images are described previously (Vasanth et al., 2015).

The gel images were analyzed using Same Spots (Totallab, Newcastle, UK). The spots were automatically aligned and those with significantly different volumes (with a minimum fold-change of 1.5 and statistical significance at $p < 0.05$) were selected as the spots of interest. To identify the proteins corresponding to the selected spots, 300 µg of protein was loaded onto a preparative Sypro® Ruby stained gel. The selected spots were excised, and sent to the Tromsø University Proteomics Platform (TUPP), Norway—for performing liquid chromatography and tandem mass spectrometry (LC-MS/MS). The peak list files obtained from TUPP were used to determine the protein identity. Mascot search engine (version 2.5.00) was used to search the NCBInr database or the whole genome sequences fasta files from NCBI. The search gave the matched peptides of Atlantic salmon. At least two unique peptides were considered for the protein inference and contaminants were eliminated from the search list.

Statistical Analysis

GraphPad Prism V6.03 was used to carry out the statistical analyses. The tanks were considered as the experimental units. Normality of the data was tested by KS normality test, and the assumption of equal variance was checked by the Brown-Forsythe test. One-way ANOVA followed by Tukey's multiple comparisons test was employed to detect the significant differences between the means of interest. For non-parametric data, the Kruskal-Wallis test followed by Dunn's multiple comparisons test was employed to understand the difference between the study groups.

RESULTS

Growth and Feed Performance

The experimental feeds (Table 1) were formulated to provide the nutrient and energy needs of Atlantic salmon. At the end of the feeding period, the fish of 4C, 4L, and 4H attained mean weights of 359, 344, 360 g, respectively, compared to the initial weight of 167 g (Table 3). The condition factors of the 4C, 4L, and 4H groups were 1.06, 1.05, and 1.05, respectively. The viscero- and hepato-somatic indices of the alga-fed groups were not significantly different from the values of the control group (Table 3). The SGR of the alga-fed groups were similar to that of the control group. The FCR was significantly lower for the control group compared to the algae-fed groups. PER also did not vary significantly among the groups.

Table 3: Growth, feed performance, organosomatic indices and survival rate of Atlantic salmon from the different groups.

	4C	4L	4H
Initial weight (g)	166.61 ± 2.10	165.90 ± 1.95	170.30 ± 2.00
Final weight (g)	358.60 ± 3.16^{a}	343.90 ± 4.61^{b}	359.80 ± 4.57^{a}
Condition factor	1.06 ± 0.01	1.05 ± 0.01	1.05 ± 0.02
SGR (% day-1)	1.10 ± 0.01	1.04 ± 0.01	1.07 ± 0.01
FCR (g feed/g gain)	0.81 ± 0.01^{a}	0.86 ± 0.01^{b}	0.90 ± 0.01^{b}
PER	2.29 ± 0.03	2.32 ± 0.03	2.36 ± 0.03
HSI	1.30 ± 0.08	1.29 ± 0.08	1.30 ± 0.05
VSI	8.11 ± 0.26	8.06 ± 0.36	8.29 ± 0.24
Survival (%)	100	99.33 ± 0.67	100

n = 6 tanks, values from 25 fish/tank, mean ± s.e.m. Different superscripts in a row indicate statistically significant differences.
Condition factor, CF = (Final weight/Total length^3) × 100
Specific growth rate, SGR = 100 × ((ln Final mean fish weight in a tank - ln Initial mean fish weight in a tank)/Number of feeding days)
Feed conversion ratio, FCR = (Apparent feed intake in dry basis/(Final biomass-Initial biomass + Biomass of dead fish))
Protein efficiency ratio, PER = ((Final biomass-Initial biomass)/Protein intake)
Survival = 100 × (Final fish number/Initial fish number).

The protein and lipid digestibility of the experimental feeds did not differ significantly. The digestibility of energy in the 10% alga-containing feed was lower compared to that of the control feed (Table 4).

Table 4: Digestibility of the different feeds.

	4C	4L	4H
Protein	84.57 ± 0.33	83.08 ± 0.47	83.61 ± 0.32
Lipid	89.04 ± 1.74	88.30 ± 1.05	93.55 ± 0.16
Energy	81.02 ± 0.75^{a}	78.28 ± 0.63^{b}	79.70 ± 0.07^{ab}

All values are calculated using the dry basis values.
Faeces from 23 fish/tank were pooled and freeze dried. Faeces from 2 tanks of a particular group were pooled prior to analysis; n = 3, mean ± s.e.m. Different superscripts in a row indicate statistically significant differences.

The proximate composition of the whole body of fish from the alga-fed groups was similar to that of the control group (Table 5). In the case of the filet from fish fed 10% of the alga the lipid content was lower compared to the control group. Furthermore, the ash content of the filet of fish fed 20% of the alga was relatively lower compared to those of the control group and the fish fed 10% of the alga. The protein content in the whole body and filet of the alga-fed groups were higher and lower, respectively, compared to the control group, although not statistically significant.

Table 5: Proximate composition of Atlantic salmon (g/100 g dry basis) from the different groups.

Group	Moisture	Protein	Lipid	Ash
WHOLE BODY				
Initial	74.75 ± 0.51	65.89 ± 1.17	27.66 ± 1.30	8.94 ± 0.31
Final				
4C	69.13 ± 0.30	56.44 ± 0.73	32.68 ± 0.91	6.21 ± 0.21
4L	68.77 ± 0.49	57.81 ± 1.01	31.76 ± 0.80	6.62 ± 0.23
4H	68.64 ± 0.32	58.02 ± 0.43	32.46 ± 0.82	6.57 ± 0.21
FILLET				
Initial	73.17 ± 0.67	78.94 ± 2.07	7.48 ± 0.30	6.07 ± 0.37
Final				
4C	73.84 ± 0.22	79.71 ± 1.30	16.73 ± 0.56^{a}	6.76 ± 0.24^{a}
4L	73.08 ± 0.39	75.00 ± 1.97	12.75 ± 0.39^{bc}	6.53 ± 0.08^{a}
4H	73.47 ± 0.25	75.19 ± 1.20	14.84 ± 0.75^{ac}	5.57 ± 0.18^{b}

n = 6 tanks, values from 6 fish/tank; mean ± s.e.m. Different superscripts in a column indicate statistically significant differences.

Antioxidant Status

The total antioxidant capacity as well as the catalase activity of the alga-fed fish was similar to that of the control group (Figure 1). The superoxide dismutase activity of the alga-fed groups was higher than that in the control group, although a statistical significance was not detected (Figure 1). In addition, the mRNA levels of *sod1* and *nrf2* in the different experimental groups were similar (Figure 2).

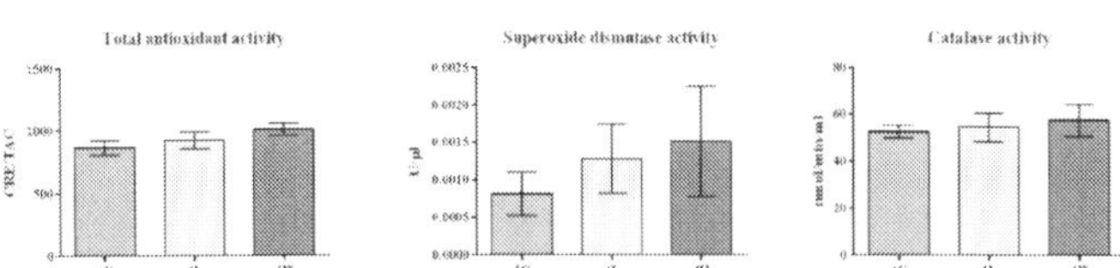

Figure 1: Serum antioxidant capacity of Atlantic salmon from the different groups. Data is generated from 12 fish of each group (n = 6 tanks, 2 fish/tank) and the values are expressed as mean ± s.e.m.

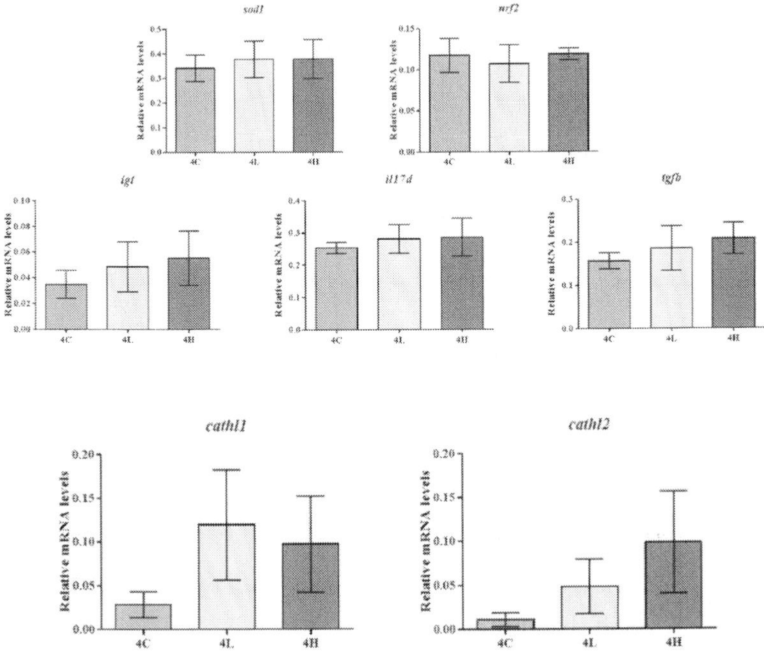

Figure 2: Relative mRNA levels of *sod, nrf2, igt, il17d, tgfb, cathl1, cathl2* in the distal intestine of Atlantic salmon from the different groups. The values are presented as mean ± s.e.m.; *n* = 6 tanks, 2 fish/tank.

Intestinal Health

Concerning the expression of the selected genes in the distal intestine of Atlantic salmon, the mRNA levels of the gut mucosa-linked immunoglobulin *igt* in the alga-fed groups were similar to that of the control group (Figure 2). The expression of inflammation-related genes, *il1b, il17a, il10, tnfa* were below the detection limit. The mRNA levels of *il17d* in the alga-fed groups were similar to the value in the control group (Figure 2). The expression of *tgfb* was similar in all the groups (Figure 2). Among the antimicrobial genes examined, the expression of *defb2* was not detectable. The expression of *cathl1* and *cathl2* were higher in the alga-fed groups compared to the values of the control group (Figure 3). However, a statistical significance was not detected.

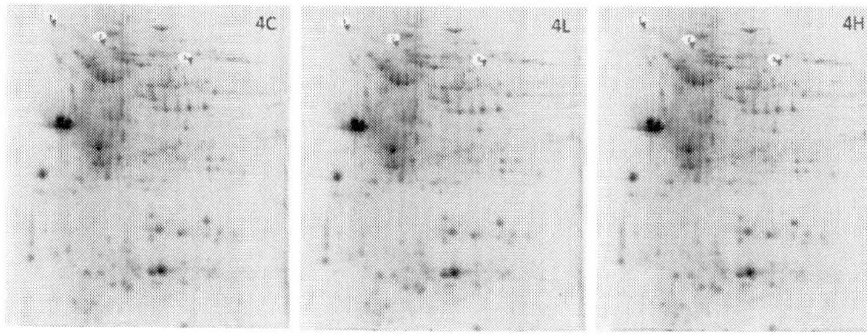

Figure 3: Representative 2-DE gels showing the spots of proteins from the distal intestine of Atlantic salmon. For generating the gels, the intestinal proteins from the distal intestine of Atlantic salmon from the 4C, 4L, and 4H groups were first isoelectrically focused on 17 cm IPG strips (pI 3-10) and then subjected to 12.5% SDS-PAGE. The Sypro®Ruby protein gel stain was used to stain the gels. The spots 1, 2, and 3 corresponds to Ahsg, My11 and Dld, respectively (see Table 6). $n = 6$ tanks, 1 fish/tank.

Analysis of the intestinal proteome of Atlantic salmon revealed that the expressions of Alpha-2-HS-glycoprotein-like (Ahsg), Myosin-11 isoform X1 (My11), Dihydrolipoyl dehydrogenase, mitochondrial-like (Dld) were altered by algal feeding (Figure 3, Table 6; Supplementary Figure 2). Dld was underexpressed in the alga-fed groups (10% – 0.45-fold, $p < 0.05$; 20% – 0.49-fold, $p > 0.05$) compared to the expression in the control group. Ahsg was under expressed in the alga-fed groups (0.55 and 0.50-fold, $p < 0.05$) compared to the protein expression in the control group. My11 was under expressed in the 20% alga-fed group (0.6-fold, $p < 0.05$) compared to the expression in the control group.

Table 6: Information of the proteins that were differentially expressed in the distal intestine of Atlantic salmon from different groups.

Spot no.	Protein name	Apparent pI/MW (kDa)	Peptides[a]
1↓	Alpha-2-HS-glycoprotein-like (Ahsg)	3.2/143.5	**YALNQIDDIK** **VVTAVEGDCDVVLR** **KVGGVMSVTAFK** **VGGVMSVTAFK** **ESLFAIMEVGR**
2↓	Myosin-11 isoform X1 (My11)	5.1/99.9	**DVSTLSAQVQDTQELLAEETR** **NSLQEQLEEEVEAK** **KLEEMAGNNELLEESK** **DLEASNTQFEEK** KFDQMLAEEK **ALEEAQDSREELER** **TEMEDLVSSK** **GLEAQVEEMK** **LEVNMQALK** **ELETELEDER** **DLEGQIEMSNK** **ISQLEEELEEESSNMEILNDR** **KSTQQVDQLNNELQTER** **STQQVDQLNNELQTER**
3↓	Dihydrolipoyl dehydrogenase, mitochondrial-precursor (Dld)	7.6/88.6	**GIEISGITLNLEK** **VTAVEFLGHVGGLGIDMEISK** **RPDGQIDVAVEAAAGGK** **NLGLDTVGLELDNR** **VPSIYAIGDVIAGPMLAHK** **EANLAASFGK**

[a]Unique peptides in bold; ↓ indicates underexpression in the algal groups compared to the control group.

Algal feeding did not affect the distal intestinal morphology of Atlantic salmon. The normal structure of lamina propria and the appearance of goblet cells are evident from the photomicrographs (Figure 4; Supplementary Figure 1).

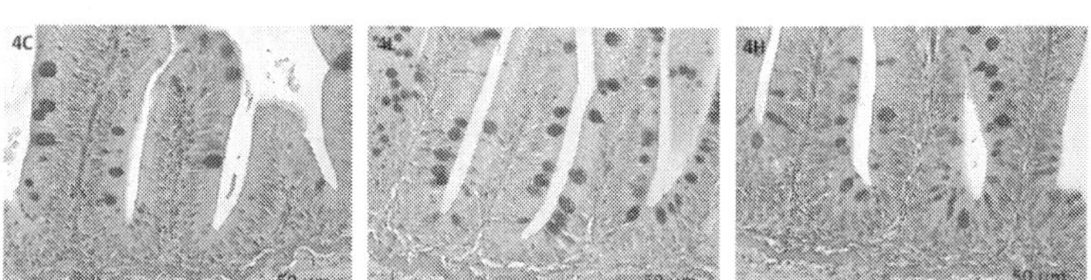

Figure 4: Photomicrographs of the distal intestine of Atlantic salmon from the different groups. Representative images show the normal architecture of the intestine. Scale bar: 50 µm. $n = 6$ tanks, 1 fish/tank.

DISCUSSION

The present study reports the effects of replacement (10 or 20%) of fishmeal with the microalgal biomass of Desmodesmus sp. in the feeds of Atlantic salmon. Desmodesmus-containing feeds did not significantly alter the specific growth rate and health indices of Atlantic salmon.

Atlantic salmon is a carnivorous fish and requires quality protein for their normal growth and development. As per the National Research Council report, Atlantic salmon should be provided with 36% digestible protein via feeds (National Research Council of the National Academies, 2011). The experimental feeds used in the present study had protein content (dry basis) in the range 478–538 g/kg.

Growth and Feed Conversion Efficiency upon Feeding *Desmodesmus* to Atlantic salmon

Overall, the fish performed well, and the weights of the fish after 70 days were twice that of the initial values. The results indicate that defatted *Desmodesmus* sp. can be included (up to 20%) in the feeds for Atlantic salmon. The final weights of the 4C and 4H groups were similar while that of the 4L group was significantly lower than that of the other groups. Higher FCR of the alga-fed groups indicates that nutrient availability in the alga-containing feeds was poorer than that in the 4C feed, although this was reflected only in the energy digestibility of the 4L group. Different studies have reported the effects of feeding algae to omnivorous fish. Nile tilapia (*Oreochromis niloticus*) larvae/juveniles, when fed with 44% *Spirulina* for 11 weeks, had higher SGR and PER, and good FCR values compared to the corn gluten-based control feed (Hussein et al., 2013). The fry of the same fish fed on 10 g of live *Spirulina* (*Arthrospira platensis*)/kg of feed had optimum growth and feed efficiency (Abdel-Tawwab and Ahmad, 2009). On the contrary, feeding 25% of *Chlorella* spp. or 23% of *Scenedesmus* spp. for 90 days led to poor growth, FCR and PER of Nile tilapia fingerlings (Badwy et al., 2008).

Studies have indicated the ability of carnivorous fish to make use of proteins from microalgae. Rainbow trout (*Oncorhynchus mykiss*) can tolerate up to 68% *Spirulina* (*A. maxima*, 35% algal protein) (Matty and Smith, 1978). On the other hand, an eight-week feeding study on rainbow trout fry indicated a significant reduction in the growth of the fish that received 25 and 50% algae (a consortium of algal biomass consisting of Chlorophyceae and Cyanobacteria) in their diets (Dallaire et al., 2007). The authors concluded that the maximum incorporation level should be 12.5% to avoid negative effects on growth, body lipid and energy content of the fish. The growth and feed performance of longfin yellowtail (*Seriola rivoliana*) was affected upon feeding with the microalga *Schizochytrium limacinum* (Kissinger et al., 2016). Feeding European sea bass (*Dicentrarchus labrax*) with freeze-dried *Isochrysis* sp. did not alter the SGR, FCR, and PER of the fish (Tibaldi et al., 2015). These indices in gilthead sea bream (*Sparus aurata*) juveniles were also not affected by feeding with *Scenedesmus almeriensis* (38%) (Vizcaíno et al., 2014). Similarly, feeding juvenile Atlantic salmon with land animal- or plant-protein based feeds that contained *Spirulina* meal (11%) did not alter the growth, feed performance, and survival of the fish (Burr et al., 2012). On the contrary, *Entomoneis* spp. (5%) in the diets of Atlantic salmon did not alter the growth or PER, but lowered the FCR (Norambuena et al., 2015). In the present feeding study, we did not observe any difference in either the specific growth rate (1.04–1.07) or protein efficiency ratio (2.32–2.36) of salmon that consumed the defatted meal of *Desmodesmus* sp.

Organosomatic indices are indicators of the general well-being of an organism (Dethloff and Schmitt, 2000). The HSI and VSI of sturgeon (*Acipenser baeri*) fed on 60% of an Indian strain of *Spirulina* were 3.07 and 9.49, respectively; the VSI values were significantly higher than that of the control group (Palmegiano et al., 2005). Fifteen percent replacement of fishmeal with *Spirulina pacifica* in the diets of parrot fish (*Oplegnathus fasciatus*) also did not alter the HSI but increased the VSI compared to the control group (Sung-Sam et al., 2013). Furthermore, the HSI value of longfin yellowtail was not altered by feeding with *S. limacinum* (Kissinger et al., 2016). In the present study, the VSI (8.1–8.3) and HSI (1.3) of the algal groups were not significantly different from the values of the control group.

The apparent digestibility coefficients of diets containing microalgae have been evaluated for farmed fishes. A study on European sea bass reported that the ADC values of protein, lipid and energy in the diets containing freeze-dried *Isochrysis* sp. (14%) were 92.6, 87.6, and 85.0, respectively (Tibaldi et al., 2015). The digestibility of protein in fingerling mirror carp and rainbow trout feeds that contained *A. maxima* (68%) were 87.1 and 83.1, respectively (Atack et al., 1979). In Atlantic salmon, the digestibility of protein and lipid of feeds containing *Entomoneis* spp. (5%) were 83 and 87, respectively (Norambuena et al., 2015). The protein, lipid and energy digestibility values of an algal product made from *Schizochytrium* sp. (20%) in the feeds of rainbow trout were 87.3, 53, and 60, respectively (Zhang, 2013). In the present study, the protein, lipid and energy digestibility values in the 20% alga-containing feeds were 84, 94, and 80, respectively. The digestibility of energy of the feeds containing 10% alga was significantly different from that of the control feed. The lower digestibility of energy in the 4L feed coincided with the lower final weights and the filet lipid content of the group. However, the higher inclusion of the alga did not cause a significant reduction in energy digestibility.

The protein, lipid and ash content of the whole body and filet of Atlantic salmon were analyzed. Algal feeding did not significantly alter the proximate composition of the whole body of Atlantic salmon, but the protein content in the whole body and fillet of the alga-fed groups was relatively higher and lower, respectivley. Feeding red sea bream (*Pagrus major*) for 95 days on a diet supplemented with 2% *Spirulina* sp. elevated the protein synthesis of the fish, and the stromal fraction of the muscle protein was significantly higher in the *Spirulina*-fed group (Mustafa et al., 1994). A study on Nile tilapia revealed that feeding 25% of *Chlorella* spp. or 23% of *Scenedesmus* spp. for 90 days lowered carcass protein content, but increased both fat and ash contents (Badwy et al., 2008). On the other hand, the protein and lipid content of the whole body of Nile tilapia fry fed on 10 g live *A. platensis*/kg of feed were significantly higher than the values of the control group (Abdel-Tawwab and Ahmad, 2009). The body protein content was higher, but the lipid content was lower in the *Spirulina*-fed Nile tilapia larvae/juveniles than the values in the control fish that were fed on corn gluten meal-based feed (Hussein et al., 2013). In the current study, the lipid and ash content were lower in the filet of Atlantic salmon fed 10 and 20% of *Desmodesmus* sp., respectively.

Health Status upon Feeding *Desmodesmus* to Atlantic Salmon

Plant ingredients are associated with anti-nutritional factors, which can reduce the bioavailability of some nutrients and negatively affect the growth and health of fish. For instance, there are many anti-nutritional substances in soybean, and these are suspected to induce allergy and alter the growth of fish. It is reported that feeding rainbow trout with high levels (60%) of soya or offering Atlantic salmon with 3.5% soybean agglutinin led to binding of the agglutinin to the intestinal enterocytes, which in turn caused aberrations of the intestinal structure of the carnivorous fishes (Buttle et al., 2001). One of the antinutritional factors, β-conglycinin was found to decrease the antioxidant capacity indices such as SOD, CAT, glutathione-S-transferase, glutathione peroxidase and glutathione reductase activities and glutathione content in the intestine and enterocytes of juvenile Jian carp (*Cyprinus carpio*) (Zhang et al., 2013). Furthermore, feeding Atlantic salmon with genetically modified maize is reported to elevate the SOD and CAT activity in the distal intestine of the fish (Sagstad et al., 2007). However, based on the results of the present study, *Desmodesmus*-feeding did not significantly alter the serum antioxidant indices (TAC, SOD and CAT) and the associated genes (*sod1, nrf2*) in the intestine of alga-fed groups. *sod1* instructs the synthesis of SOD enzyme and nuclear factor erythroid 2–related factor 2 (Nrf2) controls the expression of antioxidant response element–dependent genes (Ma, 2013). GSTA2 (glutathione S-transferase A2) and NQO1 (NADPH: quinone oxidoreductase 1) are two Nrf2-regulated genes (Ma, 2013), and *gstp1* and *nqo1* were altered in zebrafish exposed to toxins (Lam et al., 2008).

Cytokines, namely TGFβ and IL10 are mediators of anti-inflammatory responses, and tumor necrosis factor (TNF) and IL-17A are responsible for pro-inflammatory reactions. IL-17D is important in local immune responses and repair of tissues after structural damage (Starnes et al., 2002). *il17a* and *tfgb* were upregulated in the intestine of Atlantic salmon during the early and/or late inflammatory stages (Marjara et al., 2012), and *il1b* was reduced after recovery from inflammation (Vasanth et al., 2015). Oral intubation of 10 mg of *A. platensis* suspension per carp (*C. carpio*) increased the phagocytic activity, superoxide anion production, *il1b* and *tnfa*, but decreased *il10* in head kidney leukocytes (Watanuki et al., 2006). In the present study, the expression of *il1b, tnfa, il17a* and *il10* were below the detection limit. Marjara et al. (2012) reported the low expression of certain genes including *tnfa* in the distal intestine of Atlantic salmon. In the present study, the mRNA levels of *tgfb* and *il17d* of the algal groups were similar to those of the control fish. The transcripts of *il1b* and *tnfa* were not altered in the intestine of Atlantic salmon fed olive oil, rapeseed oil or soybean oil (Moldal et al., 2014). However, soybean oil feeding was found to lower the *igm* and *igt* levels in the distal intestine of the fish (Moldal et al., 2014). The feeding of the algal biomass in this study did not alter the level of *igt* in the distal intestine of Atlantic salmon. Cathelicidins, a group of antimicrobial peptides, have been implicated in intestinal inflammatory conditions in humans (Schauber et al., 2006). In the present study, the mRNA levels of *cathl1* and *cathl2* were higher than that in the control group, but not significantly different.

We did not observe any inflammatory signals in the micromorphology of the distal intestine. Similarly, feeding longfin yellowtail with *S. limacinum* (Kissinger et al., 2016) or gilthead sea bream with *S. almeriensis* (Vizcaíno et al., 2014) did not alter the intestinal morphology of the fishes.

There are not many studies that have examined the diet-related changes in the intestinal proteins of fish. Distal intestinal proteins (Ahsg, My11, Dld) associated with immune defense, motility of actin filaments and energy metabolism were underexpressed in Atlantic salmon that consumed *Desmodesmus* sp. The functions of AHSG in humans include regulation of protease activity, pathogenesis of colorectal cancer, monocyte recruitment to bone, and bone metabolism (Crawford, 1984; Malone and Richards, 1987; Katsuhito et al., 2011; Fan et al., 2014). Upregulation of AHSG is considered as an indication of squamous cell carcinoma (Tian et al., 2015). Reduction in levels of this glycoprotein in the serum of protein-energy-malnourished children was associated with stunted growth, and compromised defense ability (Abiodun et al., 1985). Although Ahsg was significantly lower in the 4L and 4H groups, the growth of the fish and the expression of immune defense genes in the groups were not significantly different from those of the control group. A myosin type was also altered in the algal groups—myosins are motor proteins that bind to actin, and they convert chemical energy into mechanical work, which enables them to move along actin filaments (Heissler and Sellers, 2016). My11 heavy chain is expressed in the smooth muscle cells of the gastrointestinal tract of mammals, and the expression of the protein was unchanged by hypertrophy (Park et al., 2015), which is associated with inflammation (Blennerhassett et al., 1992). In the present study, My11 was underexpressed in the distal intestine of salmon fed 20% of *Desmodesmus* sp. and we did not observe any inflammation in the distal intestine. DLD is associated with energy metabolism, and this TCA cycle-associated protein was overexpressed in the gut of black tiger shrimp, *Penaeus monodon*, that were orally intubated with white spot syndrome viral antigens (Kulkarni et al., 2014). On the other hand, Dld was underexpressed in the alga-fed groups, significantly in the case of fish that received 10% of the algal biomass. Interestingly, the energy digestibility of feeds that contained the algal biomass was lower compared to that of the control feed, significantly in the case of feeds with 10% of the algal biomass. A decrease in Dld was suggested to help probiotic bacteria such as *Lactobacillus fermentum* to survive in the small intestine (Yang et al., 2007). The physiological relevance of the underexpression of the three proteins in the distal intestine needs to be verified through further studies.

CONCLUSION

Atlantic salmon is able to tolerate the defatted biomass from *Desmodesmus* sp. The results indicate that 20% inclusion level in the feeds of Atlantic salmon does not have any adverse effects on the specific growth rate, condition factor, protein efficiency ratio and whole body proximate composition of fish. Furthermore, the algal inclusion did not impair the health of Atlantic salmon, as evident from the physiological indices, molecular markers of inflammation and intestinal micromorphology. Future studies should consider evaluating the effects of algal inclusion above 20% in long-term feeding studies on salmon, after optimizing the ingredient and feed processing to increase the digestibility of algal nutrients and reduce the feed conversion ratio. Thus, the *Desmodesmus* biomass can effectively replace fishmeal in Atlantic salmon feeds. The defatted microalgal biomass can also be a potential ingredient in the feeds of other aquatic animals.

Conflict of Interest Statement

The authors declare that the research was conducted in the absence of any commercial or financial relationships that could be construed as a potential conflict of interest.

ACKNOWLEDGMENTS

We thank the Department of Energy (DoE), USA for funding the project "Large-scale production of fuels and feed from marine microalgae" (DE-EE0003371) that was awarded to the Cornell Marine Algae Biofuels Consortium. We thank all the members of the Cornell Marine Algae Biofuels Consortium, especially Charles Green, who led the team. Cellana, San Diego, USA is thanked for providing the algal biomass produced at their facilities in Kailua-Kona, Hawaii. The authors would like to acknowledge the support of the staff at the Research Station of Nord University.

REFERENCES

1. Abdel-Tawwab, M., and Ahmad, M. H. (2009). Live Spirulina (Arthrospira platensis) as a growth and immunity promoter for Nile tilapia, Oreochromis niloticus (L.), challenged with pathogenic Aeromonas hydrophila. Aquacult. Res. 40, 1037–1046. doi: 10.1111/j.1365-2109.2009.02195.x
2. Abiodun, P. O., Ihongbe, J. C., and Dati, F. (1985). Decreased levels of alpha2 HS-glycoprotein in children with protein-energy-malnutrition. Eur. J. Pediatr. 144, 368–369. doi: 10.1007/BF00441779
3. Atack, T. H., Jauncey, K., and Matty, A. J. (1979). The utilization of some single cell proteins by fingerling mirror carp (Cyprinus carpio). Aquaculture 18, 337–348. doi: 10.1016/0044-8486(79)90037-1
4. Austreng, E. (1978). Digestibility determination in fish using chromic oxide marking and analysis of contents from different segments of the gastrointestinal tract. Aquaculture 13, 265–272. doi: 10.1016/0044-8486(78)90008-X
5. Badwy, T. M., Ibrahim, E. M., and Zeinhom, M. M. (2008). "Partial replacement of fish meal with dried microalga (Chlorella spp. and Scenedesmus spp.) in Nile tilapia (Oreochromis niloticus) diets," in 8th International Symposium on Tilapia in Aquaculture (Cairo), 801–811.
6. Baeverfjord, G., and Krogdahl, A. (1996). Development and regression of soybean meal induced enteritis in Atlantic salmon, Salmo salarL., distal intestine: a comparison with the intestines of

fasted fish. J. Fish Dis. 19, 375–387. doi: 10.1111/j.1365-2761.1996.tb00376.x

7. Bakke, A. (2011). "Pathophysiological and immunological characteristics of soybean meal-induced enteropathy in salmon: contribution of recent molecular investigations," in Avances en Nutrición Acuícola XI - Memorias del Décimo Primer Simposio Internacional de Nutrición Acuícola, eds L. E. Cruz-Suárez, D. Ricque-Marie, M. Tapia-Salazar, M. G. Nieto-López, D. A. Villarreal-Cavazos, J. Gamboa-Delgado, and L. Hernández-Hernández (Monterrey: Universidad Autónoma de Nuevo León), 345–372.
8. Bancroft, J. D., and Gamble, M. (2007). Theory and Practice of Histological Techniques. London, UK: Churchill Livingstone.
9. Becker, E. W. (2007). Micro-algae as a source of protein. Biotechnol. Adv. 25, 207–210. doi: 10.1016/j.biotechadv.2006.11.002
10. Becker, W. (2004). "Microalgae in human and animal nutrition," in Microalgal Culture, ed A. Richmond (Oxford: Blackwell Science Ltd.), 312–351.
11. Blennerhassett, M. G., Vignjevic, P., Vermillion, D. L., and Collins, S. M. (1992). Inflammation causes hyperplasia and hypertrophy in smooth muscle of rat small intestine. Am. J. Physiol. 262, G1041–G1046.
12. Burr, G. S., Wolters, W. R., Barrows, F. T., and Hardy, R. W. (2012). Replacing fishmeal with blends of alternative proteins on growth performance of rainbow trout (Oncorhynchus mykiss), and early or late stage juvenile Atlantic salmon (Salmo salar). Aquaculture 334–337, 110–116. doi: 10.1016/j.aquaculture.2011.12.044
13. Buttle, L. G., Burrells, A. C., Good, J. E., Williams, P. D., Southgate, P. J., and Burrells, C. (2001). The binding of soybean agglutinin (SBA) to the intestinal epithelium of Atlantic salmon, Salmo salar and rainbow trout, Oncorhynchus mykiss, fed high levels of soybean meal. Vet. Immunol. Immunopathol. 80, 237–244. doi: 10.1016/S0165-2427(01)00269-0
14. Crawford, S. M. (1984). Alpha-2 HS glycoprotein in the hypercalcaemia of multiple myeloma. Br. J. Cancer 49, 813–815. doi: 10.1038/bjc.1984.127

15. Dallaire, V., Lessard, P., Vandenberg, G., and de la Noüe, J. (2007). Effect of algal incorporation on growth, survival and carcass composition of rainbow trout (Oncorhynchus mykiss) fry. Bioresour. Technol. 98, 1433–1439. doi: 10.1016/j.biortech.2006.05.043
16. Dethloff, G. M., and Schmitt, C. J. (eds.). (2000). "Condition factor and organo-somatic indices," in Information and Technology Report USGS/BRD-2000–0005: Biomonitoring of Environmental Status and Trends (BEST) Program: Selected Methods for Monitoring Chemical Contaminants and Their Effects in Aquatic Ecosystems (Columbia, IN: U.S. Geological Survey, Biological Resources Division), 13–18.
17. Energy.Gov: Office of Energy Efficiency Renewable Energy (2016). Biomass Feedstocks [Online]. U.S. Department of Energy. Available online at: http://energy.gov/eere/bioenergy/biomass-feedstocks (Accessed March 24, 2016).
18. Enzing, C., Ploeg, M., Barbosa, M., and Sijtsma, L. (2014). "Microalgae-based products for the food and feed sector: an outlook for Europe," eds M. Vigani, C. Parisi, and E. R. Cerezo (Luxembourg: European Commission, Joint Research Centre - Institute for Prospective Technological Studies), 78. Available online at: http://ftp.jrc.es/EURdoc/JRC85709.pdf
19. Fan, N. J., Kang, R., Ge, X. Y., Li, M., Liu, Y., Chen, H. M., et al. (2014). Identification alpha-2-HS-glycoprotein precursor and tubulin beta chain as serology diagnosis biomarker of colorectal cancer. Diagn. Pathol. 9, 53. doi: 10.1186/1746-1596-9-53
20. Heissler, S. M., and Sellers, J. R. (2016). "Myosins," in Encyclopedia of Cell Biology, eds R. A. Bradshaw and P. D. Stahl (Waltham, MA: Academic Press), 597–607. doi: 10.1016/B978-0-12-394447-4.20059-X
21. Huntley, M. E., Johnson, Z. I., Brown, S. L., Sills, D. L., Gerber, L., Archibald, I., et al. (2015). Demonstrated large-scale production of marine microalgae for fuels and feed. Algal Res. 10, 249–265. doi: 10.1016/j.algal.2015.04.016
22. Hussein, E. E.-S., Dabrowski, K., El-Saidy, D. M. S. D., and Lee, B.-J. (2013). Enhancing the growth of Nile tilapia larvae/juveniles by replacing plant (gluten) protein with algae protein. Aquacult.

Res. 44, 937–949. doi: 10.1111/j.1365-2109.2012.03100.x

23. Katsuhito, M., Masanori, E., and Masaaki, I. (2011). Fetuin-A: a multifunctional protein. Recent Pat. Endocr. Metab. Immune Drug Discov. 5, 124–146. doi: 10.2174/187221411799015372
24. Kiron, V., Phromkunthong, W., Huntley, M., Archibald, I., and De Scheemaker, G. (2012). Marine microalgae from biorefinery as a potential feed protein source for Atlantic salmon, common carp and whiteleg shrimp. Aquacult. Nutr. 18, 521–531. doi: 10.1111/j.1365-2095.2011.00923.x
25. Kissinger, K. R., García-Ortega, A., and Trushenski, J. T. (2016). Partial fish meal replacement by soy protein concentrate, squid and algal meals in low fish-oil diets containing Schizochytrium limacinum for longfin yellowtail Seriola rivoliana. Aquaculture 452, 37–44. doi: 10.1016/j.aquaculture.2015.10.022
26. Knudsen, D., Urán, P., Arnous, A., Koppe, W., and Frøkiaer, H. (2007). Saponin-containing subfractions of soybean molasses induce enteritis in the distal intestine of Atlantic salmon. J. Agric. Food Chem. 55, 2261–2267. doi: 10.1021/jf0626967
27. Kulkarni, A. D., Kiron, V., Rombout, J. H. W. M., Brinchmann, M. F., Fernandes, J. M. O., Sudheer, N. S., et al. (2014). Protein profiling in the gut of Penaeus monodon gavaged with oral WSSV-vaccines and live white spot syndrome virus. Proteomics 14, 1660–1673. doi: 10.1002/pmic.201300405
28. Lam, S. H., Mathavan, S., Tong, Y., Li, H., Karuturi, R. K. M., Wu, Y., et al. (2008). Zebrafish whole-adult-organism chemogenomics for large-scale predictive and discovery chemical biology. PLoS Genet. 4:e1000121. doi: 10.1371/journal.pgen.1000121
29. Ma, Q. (2013). Role of Nrf2 in oxidative stress and toxicity. Annu. Rev. Pharmacol. Toxicol. 53, 401–426. doi: 10.1146/annurev-pharmtox-011112-140320
30. Malone, J. D., and Richards, M. (1987). α2HS glycoprotein is chemotactic for mononuclear phagocytes. J. Cell. Physiol. 132, 118–124. doi: 10.1002/jcp.1041320116
31. Marjara, I. S., Chikwati, E. M., Valen, E. C., Krogdahl, Å., and Bakke, A. M. (2012). Transcriptional regulation of IL-17A and other inflammatory markers during the development of soybean meal-

induced enteropathy in the distal intestine of Atlantic salmon (Salmo salar L.). Cytokine 60, 186–196. doi: 10.1016/j.cyto.2012.05.027

32. Matty, A. J., and Smith, P. (1978). Evaluation of a yeast, a bacterium and an alga as a protein source for rainbow trout: I. Effect of protein level on growth, gross conversion efficiency and protein conversion efficiency. Aquaculture 14, 235–246. doi: 10.1016/0044-8486(78)90097-2

33. Moldal, T., Løkka, G., Wiik-Nielsen, J., Austbø, L., Torstensen, B. E., Rosenlund, G., et al. (2014). Substitution of dietary fish oil with plant oils is associated with shortened mid intestinal folds in Atlantic salmon (Salmo salar). BMC Vet. Res. 10:60. doi: 10.1186/1746-6148-10-60

34. Mustafa, M. G., Umino, T., and Nakagawa, H. (1994). The effect of Spirulina feeding on muscle protein deposition in red sea bream, Pagrus major. J. Appl. Ichthyol. 10, 141–145. doi: 10.1111/j.1439-0426.1994.tb00153.x

35. National Research Council of the National Academies (2011). Nutrient Requirements of Fish and Shrimp. Washington, DC: The National Academies Press.

36. Norambuena, F., Hermon, K., Skrzypczyk, V., Emery, J. A., Sharon, Y., Beard, A., et al. (2015). Algae in fish feed: Performances and fatty acid metabolism in juvenile Atlantic salmon. PLoS ONE 10:e0124042. doi: 10.1371/journal.pone.0124042

37. Norges Standardiseringsforbund (1994). "Norsk Standard 9402E," in Atlantic Salmon Colour and Fat Measurement (Oslo).

38. Palmegiano, G. B., Agradi, E., Forneris, G., Gai, F., Gasco, L., Rigamonti, E., et al. (2005). Spirulina as a nutrient source in diets for growing sturgeon (Acipenser baeri). Aquacult. Res. 36, 188–195. doi: 10.1111/j.1365-2109.2005.01209.x

39. Park, C., Lee, M. Y., Park, P. J., Ha, S. E., Berent, R. M., Fuchs, R., et al. (2015). Serum response factor is essential for prenatal gastrointestinal smooth muscle development and maintenance of differentiated phenotype. J. Neurogastroenterol. Motil. 21, 589–602. doi: 10.5056/jnm15063

40. Sagstad, A., Sanden, M., Haugland, Ø., Hansen, A. C., Olsvik, P. A., and Hemre, G. I. (2007). Evaluation of stress- and immune-response

biomarkers in Atlantic salmon, Salmo salar L., fed different levels of genetically modified maize (Bt maize), compared with its near-isogenic parental line and a commercial suprex maize. J. Fish Dis. 30, 201–212. doi: 10.1111/j.1365-2761.2007.00808.x

41. Schauber, J., Rieger, D., Weiler, F., Wehkamp, J., Eck, M., Fellermann, K., et al. (2006). Heterogeneous expression of human cathelicidin hCAP18/LL-37 in inflammatory bowel diseases. Eur. J. Gastroenterol. Hepatol. 18, 615–621. doi: 10.1097/00042737-200606000-00007
42. Starnes, T., Broxmeyer, H. E., Robertson, M. J., and Hromas, R. (2002). Cutting edge: IL-17D, a novel member of the IL-17 family, stimulates cytokine production and inhibits hemopoiesis. J. Immunol. 169, 642–646. doi: 10.4049/jimmunol.169.2.642
43. Sung-Sam, K., Rahimnejad, S., Kang-Woong, K., and Kyeong-Jun, L. (2013). Partial replacement of fish meal with Spirulina pacifica in diets for parrot fish (Oplegnathus fasciatus). Turkish J. Fish. Aquat. Sci. 13, 197–204. doi: 10.4194/1303-2712-v13_2_01
44. Tian, W.-D., Li, J.-Z., Hu, S.-W., Peng, X.-W., Li, G., Liu, X., et al. (2015). Proteomic identification of alpha-2-HS-glycoprotein as a plasma biomarker of hypopharyngeal squamous cell carcinoma. Int. J. Clin. Exp. Pathol. 8, 9021–9031.
45. Tibaldi, E., Chini Zittelli, G., Parisi, G., Bruno, M., Giorgi, G., Tulli, F., et al. (2015). Growth performance and quality traits of European sea bass (D. labrax) fed diets including increasing levels of freeze-dried Isochrysis sp. (T-ISO) biomass as a source of protein and n-3 long chain PUFA in partial substitution of fish derivatives. Aquaculture 440, 60–68. doi: 10.1016/j.aquaculture.2015.02.002
46. Vandesompele, J., De Preter, K., Pattyn, F., Poppe, B., Van Roy, N., De Paepe, A., et al. (2002). Accurate normalization of real-time quantitative RT-PCR data by geometric averaging of multiple internal control genes. Genome Biol. 3, research0034.0031–research0034.0011. doi: 10.1186/gb-2002-3-7-research0034
47. Vasanth, G., Kiron, V., Kulkarni, A., Dahle, D., Lokesh, J., and Kitani, Y. (2015). A microbial feed additive abates intestinal inflammation in Atlantic salmon. Front. Immunol. 6:409. doi: 10.3389/fimmu.2015.00409

48. Vizcaíno, A. J., López, G., Sáez, M. I., Jiménez, J. A., Barros, A., Hidalgo, L., et al. (2014). Effects of the microalga Scenedesmus almeriensis as fishmeal alternative in diets for gilthead sea bream, Sparus aurata, juveniles. Aquaculture 431, 34–43. doi: 10.1016/j.aquaculture.2014.05.010
49. Watanuki, H., Ota, K., Tassakka, A. C. M. A. R., Kato, T., and Sakai, M. (2006). Immunostimulant effects of dietary Spirulina platensison carp, Cyprinus carpio. Aquaculture 258, 157–163. doi: 10.1016/j.aquaculture.2006.05.003
50. Yang, F., Wang, J., Li, X., Ying, T., Qiao, S., Li, D., et al. (2007). 2-DE and MS analysis of interactions between Lactobacillus fermentumI5007 and intestinal epithelial cells. Electrophoresis 28, 4330–4339. doi: 10.1002/elps.200700166
51. Ytrestøyl, T., Aas, T. S., and Åsgård, T. (2015). Utilisation of feed resources in production of Atlantic salmon (Salmo salar) in Norway. Aquaculture 448, 365–374. doi: 10.1016/j.aquaculture.2015.06.023
52. Zhang, C. (2013). Determination of the Digestibility of a Whole-Cell DHA-Rich Algal Product and Its Effect on the Lipid Composition of Rainbow Trout and Atlantic salmon. Saskatoon: University of Saskatchewan.
53. Zhang, J.-X., Guo, L.-Y., Feng, L., Jiang, W.-D., Kuang, S.-Y., Liu, Y., et al. (2013). Soybean β-conglycinin induces inflammation and oxidation and causes dysfunction of intestinal digestion and absorption in fish. PLoS ONE 8:e58115. doi: 10.1371/journal.pone.0058115

CHAPTER

19

TOWARDS SUSTAINABLE AQUAFEEDS: COMPLETE SUBSTITUTION OF FISH OIL WITH MARINE MICROALGA SCHIZOCHYTRIUM SP. IMPROVES GROWTH AND FATTY ACID DEPOSITION IN JUVENILE NILE TILAPIA (OREOCHROMIS NILOTICUS)

Pallab K. Sarker, Anne R. Kapuscinski, Alison J. Lanois, Erin D. Livesey, Katie P. Bernhard, Mariah L. Coley

Environmental Studies Program, Dartmouth College, Hanover, NH 03755, United States of America

ABSTRACT

We conducted a 84-day nutritional feeding experiment with dried whole cells of DHA-rich marine microalga *Schizochytrium* sp. (Sc) to determine the optimum level of fish-oil substitution (partial or complete) for maximum growth of Nile tilapia. When we fully replaced fish oil with *Schizochytrium* (Sc100 diet), we found significantly higher weight gain and protein efficiency ratio (PER), and lower (improved) feed conversion ratio (FCR) and feed intake compared to a control diet containing fish oil (Sc0); and no significant change in SGR and survival rate among all diets. The Sc100 diet had the highest contents of 22:6n3 DHA, led to the highest DHA content in fillets, and consequently led to the highest DHA: EPA ratios in tilapia fillets. *Schizochytrium* sp. is a high quality candidate for complete substitution of fish oil in juvenile Nile tilapia feeds, providing an innovative means to formulate and optimize the composition of tilapia juvenile feed while simultaneously raising feed efficiency of tilapia aquaculture and to further develop environmentally and socially sustainable aquafeeds. Results show that replacing fish oil with DHA-rich marine Sc improves the deposition of n3 LC PUFA levels in tilapia fillet. These results support further studies to lower *Schizochytrium* production costs and to combine different marine microalgae to replace fish oil and fishmeal into aquafeeds.

INTRODUCTION

Aquaculture has been rapidly expanding all over the world in the last few decades, at an average rate of 8–10% per year [1]. Between 1980 and 2010, aquaculture contribution to global fishery output for human consumption rose from nine percent to 47 percent and its use of artificial feeds rose from 50 to 66 percent of production [2,3]. Analysts expect aquaculture to provide an additional 23 million tonnes of aquatic foods by 2030 to maintain the current per-capita level of consumption of aquatic foods for a growing global population [3]. Responsible expansion of aquafeeds, *inter alia*, requires finding sustainable alternatives to fishmeal and fish oil ingredients, of which aquaculture is the largest user. Aquaculture feeds currently use over 80% of the world's fishmeal and fish oil [3], which are extracted from small ocean-caught fish. To date, fish oils (FO) are still regarded as good lipid sources of aquafeed formulations [4]. However, due to the short-fall from wild-harvested marine fish stocks for the production of fish oil [3], alternative lipid sources for the production of fish feed are essential to meet the demand of the growing aquaculture industry [5].

This research project focuses on reformulation of diets for Nile tilapia (*Oreochromis niloticus*), the second most farmed fish worldwide [6] and the most widely farmed in the United States [7]. This species is pre-adapted to thrive on fish-ingredient-free aquafeed because of its low trophic status. Commercial diets for tilapia have already greatly reduced inclusion levels of fish oil by replacing them with terrestrial vegetable oils, which have relatively low prices and large production volumes. Scientists have reported considerable success in partially or totally replacing fish oil with vegetable oil in many fish species [8, 9]. Several studies have shown that replacement of high-quality fish oil with vegetable oil sources negatively affects growth performance, fatty acid profile or health parameters of fish, such as gilthead seabream (*Sparus aurata*) [10] and European sea bass (*Dicentrarchus labrax* L.) [11], the magnitude of which depends on the source and level of replacement. Moreover, unbalanced essential amino acids, low levels of n3 PUFAs, lack of DHA and EPA, and high levels of anti-nutritional factors [12] limit the inclusion rates of terrestrial plant ingredients, even in diets for omnivorous species like tilapia [13,14].

Recent commercial-scale production of microalgae for biofuels and human nutritional supplements has also stimulated interest in microalgae for animal feeds [15]-[16]. Increasing attention has focused on marine microalgae for aquaculture feeds because of their elevated fatty acid profiles. Additionally, microalgae are at the base of the aquatic food chain that produce the food resources that tilapia are adapted to consume. In contrast to terrestrial plant sources, microalgae are relatively high in essential long chain n-3 polyunsaturated fatty acids (n3 LC PUFA) such as DHA (C22:6n3) and EPA (C20:5n3), which are important both for maintaining fish health and imparting neurological, cardiovascular and anticancer benefits to humans [17,18]. Our recent evaluation of a marine microalga, *Schizochytrium* sp. (Sc), in tilapia found significantly higher digestibility of lipid and all unsaturated fatty acid fractions from Sc compared to freshwater microalgae [19]. This was the first study to report that dried whole cells of Sc are a highly digestible lipid source for Nile tilapia. However, the response of fish growth performance, feed utilization efficiency, and fatty acid deposition in muscle to dietary inclusion of Sc is unknown. Building on our previous findings on digestibility of Sc in tilapia [19], the study reported below determined whether whole-cell Sc is a quality fish oil substitute in Nile tilapia feed. We conducted a nutritional feeding trial with diets containing dried whole cells of this microalga designed to identify the optimum level of fish oil substitution (partial or complete) to support high growth and feed efficiency, and to measure whether inclusion of dried whole-cell Sc can raise n3 LC PUFAs deposition in juvenile tilapia fillets.

MATERIALS AND METHODS

Experimental design, fish rearing and feeding

We obtained Nile tilapia (*O. niloticus*) juveniles from Americulture Inc. (Animas, New Mexico, 88020, USA). We conducted this experiment in a temperature controlled wet lab at the Life Sciences Center, Dartmouth College (Hanover, NH, USA) using fifteen indoor, static-water 114-L cylindro-conical tanks. We filled each tank with charcoal filtered de-chlorinated tap water and provided aeration through an air stone diffuser via a low-pressure electrical blower. Each tank contained bio-ball and sponge biological filters. Prior to the start of the experiment, we randomly assigned 40 tilapia to each tank, with an initial mean weight of 1.52 ± 0.2 g/fish and accustomed the fish to a photoperiod cycle of 10 h light and 14 h dark. We acclimated fish to the experimental conditions for two weeks before starting the experiment, during which we fed them the control diet (see section, feed formulation and pellet preparation). We randomly allocated the five experimental diets to 15 tanks and fed each diet to three replicate tanks ($n = 3$) in a completely randomized design. We hand-fed fish three times daily at 0930, 1300 and 1700h for 84 days. At the start of the trial, we administered feed at a rate of 10% of body weight. We recorded the daily satiation ration and used this to guide a gradual reduction of the feeding rate to 4% of body weight at the end of the trial, following the approach of Hussein et al. 2012 and Karapanagiotidis et al. 2007 for tilapia [20,21], and NRC 2011 [4]. The experimental design and fish use protocol were approved by the Institutional Animal Care and Use Committee (IACUC) of Dartmouth College.

All water quality parameters, monitored during the course of the study, confirmed that we maintained excellent conditions for tilapia. We exchanged 10–15% of the tank water each week. We kept water temperature throughout the experiment within the range 26.4–28.2°C with a thermostat-regulated, immersion heater in each tank (Hagen Marina Submersible Pre-Set Aquarium Heater, 150W, model UPC 015561112345). The range of values for other variables were pH 7.17 to

7.60, dissolved oxygen 6.18 to 7.13 mg L^{-1}, nitrite 0.10 to 0.20 mg L^{-1}, and total ammonia nitrogen 0.23 to 0.53 mg L^{-1}.

Feed formulation and pellet preparation

We formulated five iso-nitrogenous (38% crude protein), iso-energetic (14 kJ/g) and iso-lipidic (10% lipid) experimental diets, following the requirements for optimum growth of juvenile Nile tilapia [4]. We prepared a control diet for juvenile tilapia, adapted from a proven high quality formulation [22]. The diets differed from each other in their relative amounts of fish oil (menhaden-derived) and dried whole cells of *Schizochytrium* sp. (Sc). We used wheat flour (only 11% protein) as filler [20] and all experimental diets were iso-nitrogenous, iso-energetic, iso-lipidic composition, and also balance protein/lipid ratio differences between fish oil and its replacer (Sc) in the diets. We designated control feed containing fish oil (Sc0), and substituted 25% fish oil (Sc25), 50% fish oil (Sc50), 75% fish oil (Sc75), and 100% fish oil (Sc100) with dried whole cells of *Schizochytrium* sp. (Table 1). We obtained dried *Schizochytrium* sp. from Algamac™, Aquafauna Bio-marine, Inc., Hawthorne, CA, USA. We obtained menhaden fish oil from Double Liquid Feed Service, Inc., Danville, IL, USA. We produced the diets in accordance with our previous work [19], by weighing and mixing oil and dry ingredients in a stand mixer (Hobart Corporation, Tory, OH, USA) for 15 min; then blending water (330 ml kg^{-1} diet) into the mixture to attain a texture appropriate for pelleting; and running each diet through a meat grinder (Panasonic, MK-G20NR) to create 2 mm-diameter pellets. After pelleting, we dried the diets to a moisture content of 80–100 g kg^{-1} under a chemical fume hood at room temperature for 12 h and then stored the finished diets at -20°C. Table 2 reports the proximate composition, gross energy, and amino acid profiles of dried Sc and Table 3 reports the fatty acid profile of dried Sc and menhaden fish oil. Table 4 reports the fatty acid profiles of the five experimental diets.

Table 1: Formulation (g/100g diet) and proximate composition (%) and essential amino acids (% in the weight of diet as is) of five experimental diets for juvenile tilapia. https://doi.org/10.1371/journal.pone.0156684.t001

	Diet				
Ingredient	**Control (Sc0)**	**Sc25**	**Sc50**	**Sc75**	**Sc100**
Fish meal*	20	20	20	20	20
Corn gluten meal	20	20	20	20	20
Soybean meal	20	20	20	20	20
Wheat flour	26.25	24.5	22.75	20.5	19.15
CaH_2PO_4	0.75	0.75	0.75	0.75	0.75
Vitamin mix[1]	1	1	1	1	1
Mineral mix[2]	1	1	1	1	1
Fish oil	9	6.75	4.5	2.25	0
Schizochytrium sp.	0	4	8	12.5	16.1
Choline chloride	2	2	2	2	2
Proximate composition (%)					
Dry matter	88.2	89	90.2	89.4	91.2
Crude protein	38.2	38.6	39.3	39.1	39.3
Lipid	11.1	11.1	10.6	10.7	10.2
Ash	7	7.4	7.2	8	8.6
NFE	31	31.8	32.2	31.4	32
Gross energy (kJ g-1)	14	14.1	14.2	14	14.1
Amino acids (% in the weight of diet as is)					
Arginine	2	2	2.1	2	2.2
Lysine	2	2	2.1	1.8	2.1
Isoleucine	1.4	1.4	1.6	1.4	1.5
Leucine	3.7	3.9	4.1	3.7	4.2
Histidine	0.8	0.8	0.8	0.8	0.8
Methionine	0.8	0.8	0.9	0.9	0.8
Cystine	0.5	0.5	0.5	0.5	0.5
Phenyle alanine	1.8	1.9	2	1.8	2.1
Threonine	1.3	1.3	1.5	1.2	1.5
Tryptophan	0.2	0.2	0.2	0.1	0.1
Valine	1.6	1.6	1.8	1.6	1.8

[1]Vitamin premix (mg kg-1 dry diet unless otherwise stated):vitamin A (as acetate), 7500 IU kg-1 dry diet; vitamin D3 (as cholecalcipherol), 6000 IU kg-1 dry diet; vitamin E (as DL-a-tocopherylacetate), 150 IU kg-1 dry diet; vitamin K (as menadione Na-bisulphate), 3; vitamin B12 (as cyanocobalamin), 0.06; ascorbic acid (as ascorbyl polyphosphate), 150; D-biotin, 42; choline (as chloride), 3000; folic acid, 3; niacin (as nicotinic acid), 30; pantothenic acid, 60; pyridoxine, 15; riboflavin, 18; thiamin, 3

[2]Mineral premix (mg kg-1 dry diet unless otherwise stated):ferrous sulphate, 0.13; NaCl, 6.15; copper sulphate, 0.06; manganese sulphate, 0.18; potassium iodide, 0.02; zinc sulphate, 0.3; carrier (wheat middling or starch).

*Omega Protein, Inc. Houston, Texas 77042, as manufacturer specification, the guaranteed gross composition analysis: crude protein, 60%; crude fat, 6%; fiber, 2%.

doi:10.1371/journal.pone.0156684.t001

Table 2: Proximate chemical composition, gross energy, essential, and non-essential amino acids of whole dried cell of *Schizochytrium* sp. https://doi.org/10.1371/journal.pone.0156684.t002

	Ingredient
Composition	***Schizochytrium* sp.**
Dry matter	96.5
Crude protein	11.9
Lipid	54.1
Ash	8.7
Crude fibre	2.4
Energy	17.7
Essential amino acids (% in the weight of ingredient as is)	
Arginine	0.8
Lysine	0.5
Isoleucine	0.4
Leucine	0.7
Histidine	0.3
Methionine	1.2
Phenylealanine	0.4
Threonine	0.4
Tryptophan	0.2
Valine	0.6
Non-essential amino acids ((% in the weight of ingredient as is)	
Aspartic acid	1.2
Serine	0.4
Glutamic acid	1.9
Glycine	0.5
Tyrosine	0.3
Alanine	0.8
Hydroxyproline	0.0
Proline	0.5

doi:10.1371/journal.pone.0156684.t002

Table 3: Fatty acid content (% of total fatty acids) of lipid sources (menhaden fish oil and whole cell dried *Schizochytrium* sp) used in the experimental diets. https://doi.org/10.1371/journal.pone.0156684.t003

	Lipid source	
Fatty acids (% of TFA)	**Fish oil**	***Schizochytrium* sp**
14:00	8.1	9.3
15:00	0.6	0.5
16:00	17.9	24.4
17:00	0.6	ND
18:00	3.1	0.5
20:00	0.2	0.1
22:00	0.1	0.1
24:00	ND	ND
Total SFA	6.6	9.2
16:1n9	0.2	ND
16:1n7	13.9	0.2
18:1n9	5.2	0.1
18:1n7	3.3	ND
20:1n9	0.6	ND
20:1n7	0.2	ND
22:1n11	ND	ND
22:1n9	0.1	ND
24:1n9	0.4	1.4
Total MUFA	23.9	1.7
18:2n6	1.5	ND
18:3n6	0.3	0.2
20:2n6	0.2	ND
20:3n6	0.2	0.3
20:4n6 ARA	1.3	1.4
22:4n6	0.3	0.1
22:5n6	0.5	15.8
Total n6 PUFA	4.3	17.8
18:3n3 ALA	1.5	ND
18:4n3	2.7	0.6
20:3n3	0.2	0.1
20:4n3	1.4	0.8
20:5n3 EPA	14.9	0.8
22:5n3	2.6	0.4
22:6n3 DHA	13	43.2
Total n3 PUFA	36.3	45.9
Total PUFA	40.6	63.7
Total n6 LCPUFA	2.5	17.6
Total n3 LCPUFA	32.1	45.3
n3:n6 PUFA ratio	8.4	2.6

SFA, saturated fatty acids (sum of all fatty acids without double bonds); MUFA, monounsaturated fatty acids (sum of all fatty acids with a single bond); PUFA, polyunsaturated fatty acids (sum of all fatty acids with ≥2 double bonds); n6 PUFA, omega 6 polyunsaturated fatty acids (18:2, 18:3, 20:2, 20:3, 20:4, 22:4, 22:5); n6 LCPUFA, omega 6 long chain polyunsaturated fatty acids (20:2, 20:3, 20:4, 22:4, 22:5), n3 PUFA, omega 3 polyunsaturated fatty acids (18:3, 18:4, 20:3, 20:4, 20:5, 22:5, 22:6); n3 LCPUFA, omega 3 long chain polyunsaturated fatty acids (20:3, 20:4, 20:5, 22:5, 22:6); EPA, eicosapentaenoic acid; DHA, docosahexaenoic acid; ND, not detectable (<0.1% of total fatty acids).

doi:10.1371/journal.pone.0156684.t003

Table 4: Fatty acid content (% of total fatty acids) of experimental diets. https://doi.org/10.1371/journal.pone.0156684.t004

			Diet		
Fatty acids (% of TFA)	Control (Sc0)	Sc25	Sc50	Sc75	Sc100
14:00	7.6	8	8.2	8	8
15:00	0.5	0.6	0.5	0.5	0.5
16:00	18.4	19.6	20.7	21.7	22.8
17:00	0.6	0.5	0.4	0.3	0.1
18:00	3.2	3.0	2.4	1.8	1.3
20:00	0.2	0.2	0.2	0.1	0.2
22:00	0.2	0.1	0.1	0.1	0.2
24:00	0.1	0.1	0.1	0.1	0.2
Total SFA	30.8	32.1	32.6	32.6	33.3
16:1n9	0.2	0.2	0.2	ND	ND
16:1n7	11	9.1	6.8	4.1	2.0
18:1n9	7.8	6.7	5.6	5.0	4.2
18:1n7	2.9	2.4	1.8	1.2	0.7
20:1n9	0.5	0.4	0.3	0.2	0.1
20:1n7	0.2	0.1	0.1	0.3	ND
22:1n11	ND	ND	ND	ND	ND
22:1n9	ND	ND	ND	ND	ND
24:1n9	0.3	0.5	0.6	0.9	1.0
Total MUFA	23	19.4	15.4	11.7	8.0
18:2n6	10.1	9.4	9.1	9.2	9.6
18:3n6	0.3	0.2	0.2	0.2	0.1
20:2n6	0.2	0.1	0.1	ND	ND
20:3n6	0.2	0.2	0.2	ND	0.3
20:4n6 ARA	1.2	1.3	1.2	1.2	1.3
22:4n6	0.2	0.2	0.1	0.1	0.1
22:5n6	0.8	3	5.7	8.5	10.8
Total n6 PUFA	13	14.4	16.6	19.2	22.2
18:3n3 ALA	1.6	1.4	1.1	0.9	0.7
18:4n3	1.9	1.6	1.2	0.8	0.4
20:3n3	0.2	0.1	0.1	0.1	ND
20:4n3	1.0	1.0	0.8	0.8	0.6
20:5n3 EPA	11.1	9.22	7.0	4.6	2.5
22:5n3 DPA	2.1	1.8	1.3	1.0	0.6
22:6n3 DHA	10.4	15	20.7	26.2	30.8
Total n3 PUFA	28.3	30.12	32.2	34.4	35.6
Total PUFA	41.3	44.52	48.8	53.6	57.8
Total n6 LCPUFA	2.6	4.8	7.3	9.8	12.5
Total n3 LCPUFA	24.8	27.12	29.9	32.7	34.5
n3:n6 PUFA ratio	2.2	2.1	1.9	1.8	1.6

SFA, saturated fatty acids (sum of all fatty acids without double bonds); MUFA, monounsaturated fatty acids (sum of all fatty acids with a single bond); PUFA, polyunsaturated fatty acids (sum of all fatty acids with ≥2 double bonds); n6 PUFA, omega 6 polyunsaturated fatty acids (18:2, 18:3, 20:2, 20:3, 20:4, 22:4, 22:5); n6 LCPUFA, omega 6 long chain polyunsaturated fatty acids (20:2, 20:3, 20:4, 22:4, 22:5), n3 PUFA, omega 3 polyunsaturated fatty acids (18:3, 18:4, 20:3, 20:4, 20:5, 22:5, 22:6); n3 LCPUFA, omega 3 long chain polyunsaturated fatty acids (20:3, 20:4, 20:5, 22:5, 22:6); EPA, eicosapentaenoic acid; DHA, docosahexaenoic acid; ND, not detectable (<1% of total fatty acids).

doi:10.1371/journal.pone.0156684.t004

Biological sampling procedures, fillet preparations and growth measurements

We bulk weighed the fish at the beginning of the experiment, and then every 3 weeks until the end of the experiment (84 days). We stopped feeding the fish for 24 h prior to each bulk weight-sampling event. We sampled 5 fish per tank at day 42 (middle) and day 84 (final) for fillet fatty acid composition analysis. We immediately euthanized the fish by single cranial pithing, filleted the fish from a standardized dorso-anterior landmark, packaged in sterile polythene bags (Whirl-pak, Naso, Fort Atkinson, Wisconsin) and stored frozen (-20°C) until fatty acid analysis. At the end of the feeding experiment, we sampled 10 fish per tank, pooled, ground into a homogeneous slurry, freeze-dried, reground and stored at—20°C until analyzed for the whole-body proximate analysis.

We determined the dietary effects on growth by evaluating final weight, weight gain, feed conversion ratio (FCR), specific growth rate (SGR), protein efficiency ratio (PER) and percent survival. We calculated the indices as follows: Weight gain (g) = final weight–initial weight; FCR, feed conversion ratio = feed intake (as fed basis)/ weight gain; SGR, specific growth rate (%/day) = 100 x (ln final wet weight (g)–ln initial wet weight (g)) / Time (days); PER, protein efficiency ratio = weight gain (g)/protein fed (g); and % survival rate = (Final number of fish / Initial number of fish) x 100.

Biochemical analysis

We sent the five types of samples (pure microalgae, fish oil, diets, wholebody and fillet) to New Jersey Feed Laboratory, Inc. (Ewing, NJ, USA) for the following types of analysis [23]: moisture (AOAC, 930.15), crude protein (AOAC 990.03), lipid (AOAC 920.39), ash (AOAC 942.05), crude fiber (AOAC 1978.10), The nitrogen-free extract (NFE; i.e. carbohydrate) was determined by difference: NFE = 100 − (% protein + % lipid + % fiber + %moisture + % ash), energy (automated oxygen bomb calorimeter), amino acids (high-performance liquid chromatography, HPLC analysis, via AOAC methods 994.12, 985.28, 988.15, and 994.12) and fatty acid (fatty acids methyl esters, FAME analysis, via AOAC method 963.22.). Table 3 reports the fatty acid content of the two lipid sources used, menhaden fish oil and Sc whole dried cells; and Table 4reports the fatty acid content of the five experimental diets.

Statistical analysis

We conducted one-way analysis of variance (ANOVA) of growth performance and feed utilization parameters, whole body proximate composition, fillet fatty acid deposition and, when significant differences were found, compared the treatment means using Tukey's test of multiple comparisons with 95% level of significance. Data were expressed as the mean ± SE of three replicates. We conducted a repeated measure analysis within the general linear model (GLM) framework for 42 and 84 days of fillet fatty acids (%TFA) data to determine whether there were differences among dietary treatments, sampling time, or main effect interactions (diet x time). We carried out all statistical analyses using the IBM Statistical Package for the Social Sciences (SPSS) program for Windows (v. 21.0, Armonk, NY, USA). We computed the Pearson correlation coefficient between Sc inclusion level in a diet and tissue deposition of DHA, weight gain and FCR data.

RESULTS

We conducted a 84-day growth experiment with dried whole-cells of Sc and found (Table 5): significantly higher weight gain (g) and protein efficiency ratio (PER), and lower (improved) feed conversion ratio (FCR) and feed intake when fish oil was fully replaced by Sc (Sc100 diet) compared to control diet containing fish oil (Sc0 diet). Tilapia appeared healthy at the end of the experiment, and showed no difference in SGR and survival rate among all diets. Weight gain was in the range of 23.8–27.3 g. Feed intake ranged from 23.6 to 27.0 g/fish. Feed conversion ratios (FCR) were within the range 0.9–1.1 and protein efficiency ratios (PER) were within the range 2.4–3.1 among all dietary treatments. We found a strong proportional linear relationship between the Sc content of the diet and weight gain ($y = 0.0193x + 23.895$; $r = 0.970$; $P < 0.01$). The FCR decreased (improved) as the dietary Sc content increased, showing an inverse relationship ($y = -0.013x + 1.101$; $r = 0.970$; $P<0.05$).

Table 5: Initial weight, final weight, weight gain, percentage weight gain, feed conversion ratio (FCR), specific growth rate (SGR), protein efficiency ratio (PER), feed intake, and survival rate of tilapia fed experimental diets. https://doi.org/10.1371/journal.pone.0156684.t005

	Diet[1]					ANOVA	
	Sc0	Sc25	Sc50	Sc75	Sc100	*F* value	*P* value
Initial weight (g)	1.4 ± 0.1	1.42 ± 0.1	1.7 ± 0.2	1.5 ± 0.1	1.6 ± 0.1	1.5	0.27
Final weight (g)	25.3 ± 0.3bc	26.4 ± 0.4bc	27.2 ± 0.7ab	27.4 ± 0.3ab	28.8 ± 0.2a	9.9	<0.01
Weight gain (g)[2]	23.8 ± 0.4bc	24.9 ± 0.3bc	25.5 ± 0.7ab	25.8 ± 0.2ab	27.3 ± 0.2a	9.3	<0.01
FCR[3]	1.1 ± 0.0bc	1.0 ± 0.0bc	1.0 ± 0.1ab	0.9 ± 0.0ab	0.9 ± 0.0a	10.2	<0.01
SGR[4]	3.4 ± 0.1	3.5 ± 0.1	3.3 ± 0.0	3.4 ± 0.1	3.5 ± 0.1	0.8	0.52
PER[5]	2.4 ± 0.1bc	2.6 ± 0.1bc	2.7 ± 0.2ab	2.8 ± 0.1ab	3.1 ± 0.1a	9.6	<0.01
Feed intake (g/fish)	27.0 ± 0.3a	25.9 ± 0.4ab	25.1 ± 0.6bc	24.9 ± 0.2bc	23.6 ± 0.2c	10.4	<0.01
%Survival rate[6]	92.7 ± 0.37	95.2 ± 1.5	95.8 ± 0.8	92.3 ± 0.2	96.3 ± 2.6	0.9	0.57

Values are means of ±SE of three replicate groups (n = 3)

[1]Mean values not sharing a superscript letter in the same row differ significantly (P < 0.05).

[2]Weight gain (g) = final wet weight–initial wet weight

[3]FCR, feed conversion ratio = feed intake/ weight gain

[4]Specific growth rate SGR (%/day) = 100 x (ln final wet weight (g)–ln initial wet weight (g)) / Time (days);

[5]PER, protein efficiency ratio = weight gain (g)/protein fed (g)

[6]%Survival = (Final number of fish / Initial number of fish) x 100.

doi:10.1371/journal.pone.0156684.t005

The whole body proximate composition of Nile tilapia fillets did not differ among dietary treatments (data not shown). This included moisture (ranged from 70.5 to 71.0%), crude protein (ranged from 16.1 to 16.6.0%), ash (ranged from 4.5 to 4.8%) and total lipid (ranged from 6.2 to 6.7%).

The fillet fatty acids composition (% of total fatty acids) was significantly influenced either by the dietary treatment or the length of the experiment or both factors (Table 6). With the exception of 18:0, all SFA fractions showed significant time effects and were higher at the middle (42 days) than at the end of the experiment (84 days). The composition of four SFA fractions (15:0, 18:0, 20:0, and 22:0) did not differ across dietary treatments. One SFA, palmitic acid (16:0), had the highest final concentration in the fillet irrespective of dietary treatment, as well as significantly higher amounts deposited in the flesh of tilapia fed the Sc100 diet compared to fish fed the Sc0 diet (P<0.01). It also showed a significant diet and time interaction (P = 0.03). Concentrations of 14:0 and total saturated fatty acid (SFA) were significantly higher in the Sc100-fed fish than in Sc0-fed fish at 42 days, as well as at 84 days (P<0.01). These results were probably due to the higher supply of these two components in the Sc100 diet compared to the Sc0 diet (Table 4).

Table 6: Fatty acid (% of total fatty acids) content of fillets from Nile tilapia fed experimental diets for 84 days and 42 days sampling events; average ± SE for 15 fish per diet (5 fish/replicate with 3 replicates/diet)§. https://doi.org/10.1371/journal.pone.0156684.t006

	Control (Sc0)		Sc25		Sc50		Sc75		Sc100		P value‡		
Fatty acid (% TFA)	84 days	42 days	84 days	42 days	84 days	42 days	84 days	42 days	84 days	42 days	Diet	Time	Diet x Time interaction
14:00	5.6 ± 0.1^{b}	7.6 ± 0.7^{b}	6.1 ± 0.9^{b}	8.8 ± 0.2^{ab}	6.1 ± 0.3^{ab}	9.6 ± 0.4^{a}	6.9 ± 0.6^{b}	10.5 ± 0.2^{a}	7.5 ± 0.3^{b}	9.6 ± 0.1^{a}	0.01	<0.01	0.12
15:00	0.6 ± 0.0	0.7 ± 0.0	0.5 ± 0.0	0.8 ± 0.0	0.5 ± 0.0	0.8 ± 0.0	0.5 ± 0.0	0.6 ± 0.0	0.4 ± 0.0	0.6 ± 0.0	0.06	<0.01	0.47
16:00	19.9 ± 0.1^{c}	28.6 ± 1.1^{a}	24 ± 2.0^{bc}	32.2 ± 0.9^{b}	24.0 ± 0.6^{bc}	38.8 ± 0.4^{a}	25.6 ± 2.4^{bc}	41.2 ± 0.5^{a}	30.0 ± 1.6^{b}	42.3 ± 0.6^{a}	<0.01	<0.01	0.03
17:00	0.7 ± 0.0^{a}	0.8 ± 0.0^{a}	0.6 ± 0.0^{b}	0.8 ± 0.0^{a}	0.5 ± 0.0^{ab}	0.7 ± 0.0^{a}	0.4 ± 0.0^{c}	0.5 ± 0.0^{bc}	0.3 ± 0.0^{c}	0.4 ± 0.0^{c}	<0.01	<0.01	0.21
18:00	6.9 ± 0.2	7.9 ± 0.2	6.1 ± 0.5	8.0 ± 0.5	5.7 ± 0.2	8.3 ± 0.3	5.5 ± 0.4	8.2 ± 0.2	6.1 ± 0.3	7.9 ± 0.3	0.99	<0.01	0.65
20:00	0.3 ± 0.0	0.3 ± 0.0	0.3 ± 0.0	0.4 ± 0.0	0.3 ± 0.0	0.3 ± 0.0	0.2 ± 0.0	0.3 ± 0.0	0.3 ± 0.0	0.4 ± 0.0	0.41	0.08	0.2
22:00	0.1 ± 0.0	0.1 ± 0.0	0.1 ± 0.0	0.1 ± 0.0	0.1 ± 0.0	0.1 ± 0.0	ND	0.1 ± 0.0	0.1 ± 0.0	ND			
24:00:00	0.1 ± 0.0	ND	0.1 ± 0.0	0.1 ± 0.0	0.1 ± 0.0	ND	0.1 ± 0.0	0.1 ± 0.0	0.1 ± 0.0	0.1 ± 0.0	0.99	<0.01	0.65
Total SFA	34.2 ± 0.2^{c}	46.0 ± 2.3^{b}	37.8 ± 3.6^{c}	51.2 ± 1.8^{b}	37.3 ± 0.6^{c}	58.6 ± 0.6^{a}	39.2 ± 3.5^{c}	61.5 ± 0.5^{a}	44.8 ± 2.3^{b}	61.3 ± 0.9^{a}	<0.01	<0.01	0.07
16:1n9	0.4 ± 0.2^{a}	0.4 ± 0.0	0.3 ± 0.0^{ab}	0.4 ± 0.0	0.3 ± 0.0^{ab}	0.3 ± 0.0	0.3 ± 0.0^{ab}	0.3 ± 0.0	0.2 ± 0.0^{b}	0.2 ± 0.1	0.13	0.72	0.96
16:1n7	8.7 ± 0.1^{b}	11.1 ± 0.3^{a}	6.7 ± 1.1^{c}	9.5 ± 0.1^{a}	5.8 ± 0.2^{c}	7.6 ± 0.3^{b}	5.3 ± 0.6^{c}	5.8 ± 0.3^{c}	2.9 ± 0.0^{d}	2.9 ± 0.2^{d}	<0.01	<0.01	0.09
18:1n9	9.8 ± 0.2^{b}	14.4 ± 0.0^{a}	9.9 ± 0.6^{b}	12.8 ± 0.4^{a}	8.3 ± 0.5^{b}	12.1 ± 0.6^{a}	9.8 ± 1.0^{b}	12.2 ± 0.6^{a}	8.3 ± 0.4^{b}	9.1 ± 0.6^{b}	0.01	<0.01	0.05
18:1n7	4.1 ± 0.1^{b}	5.1 ± 0.0^{a}	3.5 ± 0.3^{bc}	4.5 ± 0.1^{a}	3.1 ± 0.1^{bc}	4.0 ± 0.1^{b}	2.8 ± 0.2^{c}	3.3 ± 0.1^{bc}	2.1 ± 0.0^{c}	2.5 ± 0.0^{c}	<0.01	<0.01	0.15
20:1n9	0.6 ± 0.0^{b}	0.8 ± 0.0^{a}	ND	0.8 ± 0.0^{a}	0.5 ± 0.0^{b}	0.6 ± 0.0^{b}	0.5 ± 0.0^{b}	0.7 ± 0.0^{a}	0.4 ± 0.0^{b}	0.4 ± 0.0^{b}	<0.01	<0.01	0.08
20:1n7	0.2 ± 0.0^{a}	0.1 ± 0.0^{b}	0.2 ± 0.0^{a}	0.1 ± 0.0^{b}	0.1 ± 0.0^{b}	ND	0.1 ± 0.0^{b}	0.1 ± 0.0^{b}	ND	ND	<0.01	<0.01	<0.01
24:1n9	0.3 ± 0.0^{a}	0.1 ± 0.0^{b}	0.2 ± 0.0^{ab}	0.1 ± 0.0^{b}	0.1 ± 0.0^{b}	ND	0.1 ± 0.0^{b}	0.1 ± 0.0^{b}	ND	ND	0.03	<0.01	<0.01
Total MUFA	24.1 ± 0.3^{b}	32.0 ± 0.7^{a}	21.2 ± 2.2^{b}	28.2 ± 0.4^{a}	18.2 ± 0.7^{bc}	24.6 ± 0.9^{b}	18.9 ± 1.9^{bc}	22.5 ± 1.1^{b}	14.0 ± 0.6^{c}	15.1 ± 0.9^{c}	<0.01	<0.01	0.13
18:2n6	8.6 ± 0.0^{a}	5.0 ± 0.8^{c}	6.9 ± 0.7^{b}	4.2 ± 0.4^{bc}	7.2 ± 0.2^{a}	2.8 ± 0.2^{c}	6.6 ± 0.2^{b}	2.3 ± 0.1^{c}	5.8 ± 0.3^{b}	4.0 ± 0.6^{c}	<0.01	<0.01	0.18
18:3n6	0.3 ± 0.0^{a}	ND	0.2 ± 0.0^{ab}	0.1 ± 0.0^{b}	0.2 ± 0.0^{ab}	ND	0.2 ± 0.0^{ab}	ND	0.1 ± 0.0^{b}	ND	0.02	<0.01	0.03
20:2n6	0.4 ± 0.0^{a}	0.2 ± 0.0^{ab}	0.3 ± 0.0^{a}	0.2 ± 0.0^{ab}	0.3 ± 0.0^{a}	ND	0.3 ± 0.0^{a}	0.1 ± 0.0^{b}	0.2 ± 0.0^{ab}	0.1 ± 0.0^{b}	<0.01	<0.01	<0.01
20:3n6	0.4 ± 0.0^{a}	0.2 ± 0.0^{b}	0.4 ± 0.0^{a}	0.2 ± 0.0^{b}	0.3 ± 0.0^{ab}	ND	0.4 ± 0.0^{a}	0.1 ± 0.0^{b}	0.3 ± 0.0^{ab}	ND	<0.01	<0.01	0.13
20:4n6 ARA	1.7 ± 0.0	0.9 ± 0.0	1.6 ± 0.1	0.9 ± 0.0	1.7 ± 0.1	0.8 ± 0.1	1.5 ± 0.1	0.8 ± 0.0	1.6 ± 0.0	1.1 ± 0.0	0.6	<0.01	0.07
22:4n6	0.5 ± 0.0^{a}	0.2 ± 0.0^{b}	0.4 ± 0.0^{a}	ND	0.4 ± 0.0^{a}	ND	0.4 ± 0.0^{a}	ND	0.3 ± 0.0^{ab}	ND	0.04	<0.01	0.22
22:5n6	0.8 ± 0.0^{d}	0.4 ± 0.0^{d}	4.2 ± 1.0^{b}	1.5 ± 0.1^{c}	5.6 ± 0.2^{ab}	2.0 ± 0.2^{c}	6.2 ± 1.0^{a}	2.3 ± 0.2^{c}	7.8 ± 0.5^{a}	3.9 ± 0.1^{ab}	<0.01	<0.01	<0.01
Total n6 PUFA	12.6 ± 0.1	6.9 ± 1.0	13.9 ± 1.8	7.1 ± 0.5	15.6 ± 0.5	5.6 ± 0.4	15.5 ± 1.4	5.6 ± 0.5	16.1 ± 0.9	9.1 ± 0.6	0.43	<0.01	<0.01
18:3n3 ALA	1.0 ± 0.0^{a}	0.4 ± 0.0^{b}	0.6 ± 0.1^{ab}	0.3 ± 0.0^{b}	0.7 ± 0.0^{ab}	0.1 ± 0.0^{c}	0.5 ± 0.0^{b}	0.1 ± 0.0^{c}	0.3 ± 0.0^{b}	0.2 ± 0.1^{c}	<0.01	<0.01	<0.01
18:4n3	0.7 ± 0.0^{a}	0.4 ± 0.0^{b}	0.5 ± 0.0^{a}	0.2 ± 0.0^{bc}	0.5 ± 0.0^{a}	0.1 ± 0.0^{c}	0.3 ± 0.0^{bc}	0.1 ± 0.0^{c}	0.1 ± 0.0^{c}	ND	0.03	<0.01	0.1
20:3n3	0.2 ± 0.0^{ab}	ND	0.1 ± 0.0^{b}	ND	0.2 ± 0.0^{ab}	ND	0.1 ± 0.0^{b}	ND	0.1 ± 0.0^{b}	ND	0.02	<0.01	0.02
20:4n3	0.9 ± 0.0^{a}	0.3 ± 0.0^{c}	0.6 ± 0.0^{b}	0.2 ± 0.0^{c}	0.6 ± 0.0^{b}	ND	0.5 ± 0.0^{a}	0.1 ± 0.0^{c}	0.3 ± 0.0^{c}	ND	<0.01	<0.01	0.03
20:5n3 EPA	3.9 ± 0.1^{a}	1.8 ± 0.3^{c}	2.2 ± 0.3^{b}	1.3 ± 0.1^{c}	1.9 ± 0.1^{b}	0.7 ± 0.1^{d}	1.3 ± 0.2^{c}	0.5 ± 0.0^{d}	0.6 ± 0.0^{d}	0.4 ± 0.1^{d}	<0.01	<0.01	<0.01
22:5n3 DPA	6.8 ± 0.1^{a}	2.7 ± 0.5^{c}	4.6 ± 0.7^{b}	1.9 ± 0.2^{c}	4.2 ± 0.1^{b}	1.0 ± 0.1^{cd}	3.0 ± 0.5^{bc}	0.7 ± 0.0^{d}	1.8 ± 0.1^{cd}	0.6 ± 0.0^{d}	<0.01	<0.01	<0.01
22:6n3 DHA	13.0 ± 0.3^{b}	7.7 ± 1.0^{c}	16.7 ± 2.7^{ab}	8.4 ± 0.6^{c}	18.9 ± 0.5^{a}	8.1 ± 0.8^{b}	19.0 ± 2.1^{a}	8.0 ± 0.7^{c}	20.5 ± 1.4^{a}	12.4 ± 0.7^{b}	<0.01	<0.01	<0.01
Total n3 PUFA	26.5 ± 0.3	13.3 ± 0.0	26.9 ± 3.6	12.3 ± 0.0	27.0 ± 0.7	10.0 ± 0.0	24.7 ± 3.7	10.4 ± 0.0	26.1 ± 1.7	13.6 ± 0.0	0.6	<0.01	0.15
Total PUFA	39.1 ± 0.4	20.2 ± 3.0	39.2 ± 5.6	19.5 ± 1.7	42.6 ± 1.2	15.6 ± 1.4	40.2 ± 5.2	16.0 ± 1.1	39.8 ± 2.7	22.7 ± 0.8	0.46	<0.01	0.14
Total n6 LCPUFA	3.8 ± 0.1^{c}	1.9 ± 0.1^{c}	6.8 ± 1.2^{b}	2.9 ± 0.1^{c}	8.3 ± 0.4^{ab}	2.8 ± 0.3^{c}	8.7 ± 1.3^{ab}	3.3 ± 0.2^{c}	10.2 ± 0.6^{a}	5.1 ± 0.1^{b}	<0.01	0	<0.01
Total n3 LCPUFA	24.8 ± 0.3	12.5 ± 1.8	24.2 ± 3.5	11.8 ± 1.0	25.8 ± 0.6	9.8 ± 0.9	23.9 ± 3.7	9.3 ± 0.8	23.3 ± 1.7	13.4 ± 0.8	0.67	0	0.06
n3:n6 PUFA ratio¶	2.1 ± 0.0^{a}	1.9 ± 0.0^{a}	1.8 ± 0.0^{b}	1.7 ± 0.0^{b}	1.7 ± 0.0^{b}	1.7 ± 0.0^{b}	1.5 ± 0.1^{bc}	1.8 ± 0.0^{b}	1.4 ± 0.0^{c}	1.4 ± 0.1^{c}	<0.01	0.07	<0.01
DHA:EPA ratio	3.3 ± 0.2^{c}	4.2 ± 0.3^{c}	7.6 ± 0.9^{bc}	6.5 ± 0.4^{bc}	9.9 ± 0.6^{b}	11.6 ± 2.1^{b}	14.6 ± 2.4^{b}	16 ± 0.7^{b}	34.2 ± 1.7^{a}	31 ± 4.9^{a}	<0.01	0.43	0.84

SFA, saturated fatty acids (sum of all fatty acids without double bonds); MUFA, monounsaturated fatty acids (sum of all fatty acids with a single bond); PUFA, polyunsaturated fatty acids (sum of all fatty acids with ≥2 double bonds); n6 PUFA, omega 6 polyunsaturated fatty acids (18:2, 18:3, 20:2, 20:3. 20:4, 22:4, 22:5); n6 LCPUFA, omega 6 long chain polyunsaturated fatty acids (20:2, 20:3, 20:4, 22:4, 22:5), n3 PUFA, omega 3 polyunsaturated fatty acids (18:3, 18:4, 20:3, 20:4. 20:5, 22:5, 22:6); n3 LCPUFA, omega 3 long chain polyunsaturated fatty acids (20:3, 20:4, 20:5, 22:5, 22:6); EPA, eicosapentaenoic acid; DHA, docosahexaenoic acid, and n3:n6 ratio calculated for total n3 PUFA: total n6 PUFA. Mean values across the row not sharing a common superscript were significantly different as determined by Tukey's HSD test, P<0.05. ND, not detectable (<0.1% of total fatty acids).

§ in many cases ± 0.0 (S.E) values are rounding error.

¶We first computed the ratio for each replicate and then computed an average and SEM for the diet treatment.

‡ Significance probability associated with F-statistics.

doi:10.1371/journal.pone.0156684.t006

With respect to MUFAs in the fillet, total MUFA content and all fractions, except for 16:ln9, had significant time effects and generally showed a higher content at the middle (42 days) of the experiment compared to the end of the experiment (84 days). The concentrations of 16:1n7, 18:1n7, 20:1n9, 20:1n7, and total MUFAs were significantly affected by dietary treatments and time. Fish fed the Sc0 diet displayed the highest amount of MUFA, which was directly related to the MUFA content in the experimental diets. Irrespective of the diet, oleic acid (18:1n9) was the most abundant monounsaturated fatty acid (MUFA) in the fillet. Most of the individual polyunsaturated fatty acids (PUFAs) varied greatly among five dietary treatments and time. Regarding n6 fatty acids, all n6 fractions including total n6 PUFA content had significant time effects and were higher at the end of the experiment (84 days) than at the middle (42 days). The total n6 PUFA content was not affected by the diet ($P = 0.43$) but showed a significant time effect and interaction between diet and time ($P<0.01$). From 42 days to 84 days, total n6 PUFA content increased by 12.6% in fish fed the Sc0 diet, and by 16.1% in fish fed the SC100 diet. The Sc0 fed fish contained the highest amounts of 18:2n6, 18:3n6, and 20:3n6. In contrast, fish fed the Sc0 diet had significantly decreased amounts of 22:5n6 compared to fish fed the Sc25, Sc50, Sc75, and Sc100 diet. At the end of the experiment, concentrations of 20:4n6 in the fillet did not differ among five dietary treatments ($P = 0.6$).

With respect to the n3 fatty acids, most of the n3 PUFA including 22:6n3 DHA and 20:5n3 EPA were significantly influenced by the dietary treatment or time or both. The n3 PUFA content was generally higher at the end of the experiment (84 days) than at the middle (42 days). For example, from 42 days to 84 days, the 22:6n3 DHA content increased by 13.0% in fish fed the Sc0 diet, and by 20.5% in fish fed the SC100 diet. Tilapia fed the Sc100 diet had the highest contents of 22:6n3 DHA in the fillet lipids at the end of the experiment, and reflected the higher 22:6n3 DHA supplied by this diet. Furthermore, increasing the levels of Sc (Sc50, Sc75 and Sc100), which corresponded to reduced levels of fish oil, resulted in significant increases in the fillet 22:6n3 DHA compared to the control diet (Sc0) at the end of the feeding experiment. Tilapia fed the Sc0 diet had significantly increased amounts of 18:3n3 compared to the Sc75 and the Sc100 diet. They also exhibited significantly increased amounts of 20:5n3 EPA and 22:5n3 DPA compared to the four Sc inclusion diets due to a higher concentration of 18:3n3 in the Sc0 diet. However, the amounts of total n3 PUFA, total PUFA, and total n3 LC PUFA were not significantly different ($P>0.01$) in any of the diets. The n3:n6 and DHA: EPA ratio were significantly influenced by the dietary treatment. The n3:n6 PUFA ratio in the fillet was the highest in fish fed the Sc0 diet. Tilapia fed the Sc100 diet had significantly increased DHA level, hence increased DHA:EPA ratio compared to the Sc0, Sc50, and Sc75 diet.

Amounts of 22:6n3 DHA deposited in the fish fillet (mg/100g fillet) significantly increased in fish fed the Sc100 diet compared to the Sc0 diet (Table 7). With increasing Sc inclusion levels in the diet, the deposition of DHA in the fillet increased from 143.5 mg/100g for the inclusion level of 0 g Sc /kg diet (Sc0 diet) to 261.8 mg/100g for the inclusion level of 161 g Sc /kg diet (Sc100 diet). The relationship between Sc inclusion level in the diet and deposition of DHA (mg/100g) in the fillets of tilapia was positively correlated ($y = 7.1423x + 155.8$; $r = 0.9459$; $P<0.01$; Fig 1).

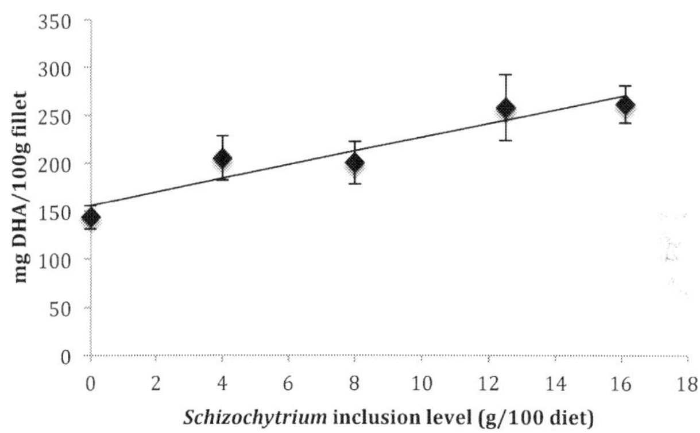

Figure 1: Linear correlation between the Sc inclusion levels in diet and DHA (mg/100g fillet) deposition in fish fillet. Each point represents the average value (±SE) of three tanks per diet. The amount of DHA is described by the function of $y = 7.1423x + 155.8$. The correlation coefficient (r) was 0.9459 ($p<0.01$). https://doi.org/10.1371/journal.pone.0156684.g001

Table 7: Amounts of lipid and major n3 and n6 PUFA in the fillet (wet weight basis) of Nile tilapia fed experimental diets for 84 days; average ± SE for 15 fish per diet (5 fish/replicate with 3 replicates/diet). https://doi.org/10.1371/journal.pone.0156684.t007

	Fillet					F value	P value
Composition	Control (FO)	SC25	SC50	SC75	SC100		
Lipid (g/100g)	2.2 ± 0.20	2.2 ± 0.74	2.3 ± 0.26	2.3 ± 0.18	2.2 ± 0.06	0.03	0.99
18:2n6 LA (mg/100g)	96.7 ± 13.9	70.7 ± 14.1	77.5 ± 11.2	76.2 ± 11.2	55.1 ± 7.2	1.6	0.24
20:4n6 ARA (mg/100g)	3.8 ± 0.6	3.3 ± 0.4	3.1 ± 0.2	2.9 ± 0.3	2.7 ±0.1	1.2	0.36
18:3n3 ALA (mg/100g)	13.6 ± 2.4	6.4 ± 1.4	6.9 ± 1.2	6.4 ± 0.9	3.2 ± 0.4	6.8	0.006
20:5n3 EPA (mg/100g)	52.0 ± 7.9^{a}	22.1 ± 5.0^{b}	20.5 ± 3.4^{b}	16.1 ± 2.9^{b}	5.9 ± 0.5^{c}	13.5	<0.01
22:6n3 DHA (mg/100g)	143.5 ± 12.2^{c}	205.3 ± 23.0^{b}	200.4 ± 22.3^{b}	258.0 ± 34.3^{a}	261.8 ± 19.3^{a}	4.2	0.02

LA, linoleic Acid; ARA, arachidonic acid; ALA, alpha linolenic acid; EPA, eicosapentaenoic acid; DHA, docosahexaenoic acid. Mean values across the row not sharing a common superscript were significantly different as determined by Tukey's HSD test, P<0.05.

doi:10.1371/journal.pone.0156684.t007

DISCUSSION

This study is the first report of improved feed utilization indices, weight gain, and beneficial fatty acid profiles in Nile tilapia when fish oil is fully replaced in tilapia diets by 16% of dried whole-cells of a marine microalga species, *Schizochytrium* sp. (Sc). Additionally, SGR and survival rates were high and showed no significant difference among all diets at the experiment's end. These results indicate that *Schizochytrium* sp. provides a high-quality substitute for fish oil and a supplement of long-chain polyunsaturated fatty acids (PUFA), especially 22:6n3 DHA that improves n3LC PUFA deposition in Nile tilapia fillets. These results also point to the possibility of formulating ecologically and socially sustainable aquafeeds, with greatly reduced or no fish oil (from marine fisheries) and without having to switch to vegetable oils (from industrially farmed crops). Commercial realization of this potential will require advances in strategies to reduce non-renewable inputs (e.g., inorganic fertilizers and fossil fuels) and monetary costs of large-scale production of marine microalgae.

Regarding improved growth, fish fed the Sc100 diet exhibited significantly higher final weight and weight gain compared to those fed the Sc0 diet. The specific growth rate SGR (%/day), however, did not differ among all diets, which is likely due to slight differences (not significant) in the initial body weight among treatments, and subsequently the final body weight and thereby the calculated weight gain was higher in fish a higher initial body weight, and the SGR remained almost unaltered by the diets. During this 84-day experiment, mean SGR ranged from 3.3 to 3.5, fitting with literature values for the exponential growth phase [24, 25].

Our results for improved weight gain may seem surprising in light of previous reports that a high dietary level of n3 LC PUFA depressed growth [26] or did not increase growth performance of tilapia [27]. We have identified two possible reasons for this difference between our results and previous research. Firstly, elevated levels of Sc exhibited improved FCR, consistent with our prior observation that the nutrient of Sc is highly digestible to tilapia [19]. Thus, the better feed conversion by Nile tilapia fed the Sc100 dietary nutrient was reflected in their improved weight gain; and this is similar to a recent report that dietary inclusion of Sc stimulated muscle or tissue development of Atlantic salmon [28].

Our observed effect of Sc on the growth of tilapia is also consistent with findings in shrimp and barramundi, which demonstrated an algal derived DHA stimulated growth performance [29,30]. Our followup studies are examining effects of Sc-inclusion diets in tilapia fed up to market size.

Secondly, dried whole cell Sc may contain high levels of a micronutrient [31, 32], such as the caroteniod astanxanthin and bioactive compounds, which we did not measure but could have contributed to the growth benefit of whole cell Sc.

The relatively high values of FCR and low value of PER in fish fed the control diet, compared to fish fed Sc inclusion diets, could be due to lower nutrient digestibility of the control diet containing fish oil compared to Sc-inclusion diets. In our previous digestibility study with Nile tilapia [19], we confirmed that the Sc diet, with high DHA amounts (302g kg^{-1} of total fatty acids), did not depress the digestibility of protein, amino acids, and unsaturated fatty acids in tilapia compared to a reference diet containing only fish oil. We also found high digestibility of lipid and all unsaturated fatty acid fractions in the Sc diet compared to the reference diet containing fishmeal and fish oil, and suggested this may be because tilapia more easily digest the polar fraction of phospholipids prevalent in marine microalgae than they digest neutral class lipids (triglycerides) prevalent in fish oil (reviewed in [19]).

Depositions of total n3 PUFA and total n3 LC PUFA did not differ among all dietary treatments. Thus, even the diet replacing 100% of the fish oil with Sc (Sc100) did not reduce these totals compared to the control diet (Sc0). Total n3 LC PUFAs in tilapia fillets increased from 13% to 20.5% of total fatty acids when we fed the Sc100 diet that replaced all the fish oil with dried Sc (at 16% of the Sc100 diet) compared to the Sc0 diet containing only fish oil (at 9% of the Sc0 diet), although this difference was not significant.

Length of feeding time significantly influenced the deposition of most of the n3 PUFA, including 22:6n3 DHA and 20:5n3 EPA. The fillet contents for all the n3 and n6 PUFA content was higher at the end of the experiment (84 days) than at the middle (42 days), whereas fillet contents of SFA and MUFA fractions was lower at 84 days. These results fit with the understanding that the fatty acid composition of the fish is largely determined by the digestible fatty acid intake of the animal. In fatty fish, the overall change in the fatty acid profile is determined by triglycerides while in lean fish, like tilapia, the muscle lipid is dominated by phospholipids (75.5%) [33, 34, 35]. Our prior results for digestibility of whole cells of Sc in tilapia showed that 22:6n3 DHA phospholipid is highly digestible (93%) [19]. The high digestibility of 22:6n3 DHA is reflected in this study, given that tilapia fed Sc100 deposited the highest amount of 22:6n3 DHA in muscle at the end of the experiment (84 days).

Our results regarding fillet deposition of two important n3 LC PUFAs, DHA and EPA, can be explained by prior research and the relative abundance of these two fatty acids in Sc. Regardless of the treatment, 22:6n3 DHA was the major n3 LC PUFA found in the fillet lipids, supporting the conclusions of [21] that DHA, rather than EPA, was preferentially incorporated into the fillet of tilapia. This result along with the significantly higher weight gain of tilapia fed the Sc100 diet, compared to the Sc0 diet, leads us to propose the following hypothesis: the Sc100 diet, containing elevated levels of dietary 22:6n3 DHA, reduces energy expenditure that would have been spent on *de novo* DHA biosynthesis and consequently supports the growth performance of tilapia. Thus, future research is needed to elucidate the mechanism by which dietary 22:6n3 DHA affects energy expenditure, fatty acid metabolism, and growth in tilapia.

The Sc0 diet's level of 20:5n3 was 11.1% of total fatty acids but tilapia fed the Sc0 diet deposited a lower level of 20:5n3 in the fillet (3.9% of total fatty acid), suggesting that 20:5n3 EPA will likely be extensively oxidized in Nile tilapia. This is in agreement with the notion that fish, including tilapia, selectively use 20:5n3 EPA as substrate for β-oxidation to produce sufficient energy [21].

Using dried whole-cells of a marine microalga as a rich source of LC n3 PUFA, instead of an extracted oil, offers several benefits regarding the human nutritional value of tilapia fillets. One benefit is reduced lipid oxidation in fillets, given that marine-derived fish oil is highly susceptible to oxidation. Aquafeeds enriched with fish oil increase the susceptibility of fish fillets to lipid oxidation, yielding undesirable lipid oxidation products that adversely affect fillet quality [36,37]. The benefit of using DHA-rich whole cell *Scizochytrium* sp. biomass in fish feeds is the natural encapsulation provided by the cell wall, which should protect valuable fatty acids from exposure to oxidative agents. Moreover, this alga is a potential source of natural antioxidants such as astaxanthin and carotenoid pigments [31,38], which could be readily deposited in fish tissue and enhance oxidative stability.

Inclusion of dried whole-cell Sc in aquafeeds offers a second human nutritional benefit of the fish product. Recent mammalian studies have revealed higher anti-inflammatory and immuno-enhancer properties of *Schizochytrium* sp., and attributed these findings to the docosapentaenoic acid (DPAn6) and DHA present in high levels in this microalga [38]. Such higher biological activity could also occur in tilapia fed diets with dried whole-cell Sc, and could translate to a human nutritional benefit of eating the fillet.

CONCLUSION

This is the first study to use Sc as a complete fish oil replacement for Nile tilapia. Total replacement of the fish oil used in commercial diets with dried whole-cell Sc resulted in significantly higher weight gain (g) and protein efficiency ratio (PER), and lower (improved) feed conversion ratio (FCR) and feed intake when compared to the control feed containing fish oil (Sc0); and no significant change in specific growth rate and survival rate among all diets. Our results are interesting for tilapia farming by showing how microalgae inclusion can help optimize the composition of tilapia starter and juvenile feed. Diet formulation with dried whole-cell Sc is also an innovative strategy to raise the n3 LC PUFAs level and thus rebalance the n3:n6 and DHA: EPA ratios in farmed tilapia, which will benefit the health of human consumers. In our study, complete substitution of fish oil with Sc led to n3:n6 and DHA:EPA ratios of 1.4:1 and 34.2:1, respectively in tilapia fillets, suggesting the relative ease by which inclusion of whole cell Sc in tilapia diets can manipulate and tailor the n3 LC PUFA composition of tilapia fillets. However, in this study we used juvenile fish over a narrow growth range. Feed manufacturers can explore this approach to develop aquafeeds for aquaculturists aiming to cater to the consumer willing to pay a premium for health enhanced foods. Niche markets may therefore be developed for these nutritionally enhanced tilapia fillets.

In order for such nutritionally enhanced tilapia to succeed in the market, researchers have to find the ways to cut the high production cost of microalgae [39, 40]. Towards this end, we are now focusing our research on reducing production costs by using organic waste streams, such as on-site fish culture effluent, as a partial replacement for expensive, external inputs of inorganic compounds normally used to grow microalgae. Furthermore, ongoing research to lower costs of producing and processing microalgae for biofuel and human nutraceuticals may lead to higher market availability and lower price of this microalga in the near future [41,42,43]. Our findings also open up a pathway to reduce or eliminate ingredients extracted from marine forage fish and from industrially farmed crops in aquafeeds. Our lab is now investigating combinations of different marine microalgae to reduce or eliminate fish oil, as well as fishmeal, in aquafeeds, towards the goal of improving life-cycle ecological and social sustainability of aquafeeds.

ACKNOWLEDGMENTS

We thank the faculty members and staff of the Department of Biology, Life Sciences Center for use of their facilities and for technical assistance. We also thank to Jianjun Hua, Statistical consultant, Dartmouth College, for helping us on the statistical analyses.

REFERENCES

1. Tacon AGJ, Hasan MR, Metian M (2011) Demand and Supply of Feed Ingredients for Farmed Fish and Crustaceans: Trends and Prospects (Food and Agriculture Organization, Rome).
2. Hall SJ, Delaporte A, Phillips MJ, Beveridge M, O'Keefe M (2011) Blue frontiers: managing the environmental costs of aquaculture. The WorldFish Center, Penang, Malaysia.
3. FAO (Food and Agriculture Organization of the United Nations) (2012) The state of world fisheries and aquaculture (SOFIA). FAO, Rome, Italy.
4. NRC (National Research Council) (2011) Nutrient Requirements of Fish and Shrimp. National Academies Press, Washington, D.C., USA.
5. Tacon AGJ, Metian M (2008) Global overview on the use of fish meal and fish oil in industrially compounded aquafeeds: Trends and future prospects. Aquaculture 285(1–4):146–58.
6. FAO (Food and Agriculture Organization of the United Nations) (2010) The State of World Fisheries and Aquaculture.

7. Weaver KL, Ivester P, Chilton JA, Wilson MA, Pandey P, Chilton FH (2008) The content of favorable and unfavorable polyunsaturated fatty acids found in commonly eaten fish. Journal of the American Dietetic Association 108: 1178–1185. pmid:18589026

8. Ng WK (2005) Lipid nutrition of farmed tilapia. Glob. Aquacult. Advocate, October, 60–61.

9. Turchini GM, Torstensen BE Ng WK (2009) Fish oil replacement in finfish nutrition. Reviews in Aquaculture 1: 10–57.

10. Menoyo D, Izquierdo MS, Robaina L, Gines R, Lopez-Bote CJ, Bautista JM (2004) Adaptation of lipid metabolism, tissue composition and flesh quality in gilthead seabream (*Sparus aurata*) to the replacement of dietary fish oil by linseed and soyabean oils. British Journal of Nutrition 92: 41–52. pmid:15230986

11. Mourente G, Good JE, Thompson KD, Bell JG (2007) Effects of partial substitution of dietary fish oil with blends of vegetable oils, on blood leucocyte fatty acid compositions, immune function and histology in European sea bass (*Dicentrarchus labrax* L.). British Journal of Nutrition 98: 770–779. pmid:17466094

12. Sarker PK, Bureau DP, Drew M, Hua K, Forster I, et al. (2013) Sustainability issues related to feeding salmonids: a Canadian perspective. Reviews in Aquaculture 5: 1–21.

13. Shiau S, Lin S, Yu S, Lin A, Kwok C (1990) Defatted and full-fat soybean meal as partial replacements for fish meal in tilapia (*Oreochromis niloticus* x *O. aureus*) diets at low protein level. Aquaculture 86: 401–407.

14. Thompson KR, Velasquez A, Patterson JT, Metts LS, Webster CD, et al. (2012) Evaluation of plant and animal protein sources as partial or total replacement of fish meal in diets for Nile tilapia fry and juvenile stages. North American Journal of Aquaculture 74: 365–375.

15. Hemaiswarya S, Raja R, Ravi-Kumar R, Ganesan V, Anbazhagan C (2011) Microalgae: a sustainable feed source for aquaculture. World Journal of Microbiology and Biotechnology 27: 1737–1746.

16. Ryckebosch E, Bruneel C, Muylaert K, Foubert I (2012) Microalgae as an alternative source of omega-3 long chain polyunsaturated fatty acids. Lipid Technology 24: 128–130.

17. Peet M, Stokes C (2005) Omega-3 fatty acids in the treatment of psychiatric disorders. Drugs 65(8): 1051–9. pmid:15907142
18. Brasky TM, Till C, White E, Neuhouser ML, Song X, et al. (2011) Serum Phospholipid Fatty Acids and Prostate Cancer Risk: Results From the Prostate Cancer Prevention Trial. American Journal of Epidemiology ().
19. Sarker PK, Gamble MM, Kelson S, Kapuscinski AR (2016) Nile tilapia (*Oreochromis niloticus*) show high digestibility of lipid and fatty acids from marine *Schizochytrium sp*. and of protein and essential amino acids from freshwater *Spirulina sp*. feed ingredients. Aquaculture Nutrition 22: 109–119.
20. Hussein EE, Dabrowski K, El-Saidy DMSD, Lee BJ (2012) Enhancing the growth of Nile tilapia larvae/juveniles by replacing plant (gluten) protein with algae protein. Aquaculture Research 44:937–949.
21. Karapanagiotidis IT, Bell MV, Little DC, Yakupitiyage A (2007) Replacement of dietary fish oils by alpha-linolenic acid-rich oils lowers omega-3 content in tilapia flesh. Lipids 42, 547–559. pmid:17473942
22. Trushenski JT, Boesenberg J, Kohler CC (2009) Influence of grow-out feed fatty acid composition on finishing success in Nile tilapia. North American Journal of Aquaculture 71:242–251.
23. Association of Official Analytical Chemists, AOAC (1995) Agriculture chemicals; contaminants, drugs, Vol. I. *In* Official Methods of Analysis of AOAC International, 16th edn. AOAC International, Arlington, VA.
24. Vechklang K, Boonanuntanasarn S, Ponchunchoovong S, Pirarat N, Wanapu C (2011) The potential for rice wine residual as an alternative protein source in a practical diet for Nile tilapia (*Oreochromis niloticus*) at the juvenile stage. Aquaculture Nutrition 17: 685–694.
25. Yones AM, El-Saidy DMSD, Abdel-Hakim NF (2013) Effects of fish oil substitution with vegetable oils in diets of juvenile Nile tilapia, *Oreochromis niloticus* (*L.*) on growth performance, nutrients utilization and muscle fatty acids contents. Merit Research Journal of Food Science and Technology 1 (1): 9–18.
26. Huang CH, Huang MC, Hou PC (1998) Effect of dietary lipids

on fatty acid composition and lipid peroxidation in sarcoplasmic reticulum of hybrid tilapia,*Oreochromis niloticus* x *O. aureus*. Comparative Biochemistry and Physiology 120: 331–336.

27. Takeuchi T, Satoh S, Watanabe T (1983) Requirement of Tilapia nilotica for essential fatty acids. Bulletin of the Japanese Society of Scientific Fisheries 49: 1127–1134.
28. Kousoulaki K, Mørkøre T, Nengas I, Berge RK, Sweetman J (2016) Microalgae and organic minerals enhance lipid retention efficiency and fillet quality in Atlantic salmon (*Salmo salar* L.). Aquaculture 451: 47–57
29. Glencross BD, Smith DM (2001) Optimising the essential fatty acids, eicosapentaenoic and docosahexaenoic acid in the diet of the prawn, Penaeus monodon. Aquaculture Nutrition 7, 101–112.
30. Glencross B.D. & Rutherford N. (2011) A determination of the quantitative requirements for docosahexaenoic acid for barramundi (Lates calcarifer). Aquaculture Nutrition 17, e536–e548.
31. Yokoyama R, Honda D (2007) Taxonomic rearrangement of the genus Schizochytriumsensulato based on morphology, chemotaxonomic characteristics, and18SrRNA gene phylogeny (Thraustochytriaceae, Labyrinthulomycetes) emendation for Schizochytrium and erection of Aurantiochytrum and Oblongichytrium gen.nov. Mycosciences 48: 199–211.
32. Raghukumar S (2008) Thraustochytrid Marine Protists: Production of PUFAs and Other Emerging Technologies. Marine Biotechnology 10(6): 631–640. Pmid: 18712565
33. Sargent J, Bell G, McEvoy L, Tocher D, Estevez A (1999) Recent developments in the essential fatty acid nutrition of fish. Aquaculture 177: 191–199
34. Bureau DP, Hua K, Harris AM (2008) The effect of dietary lipid and long-chain n-3 PUFA levels on growth, energy utilization, carcass quality and immune function of rainbow trout (*Oncorhynchus mykiss*). Journal of the World Aquaculture Society 39: 1–21.
35. Suloma A, Ogata HY, Garibay ES, Chavez DR el-Haroun ER (2008) Fatty acid composition of Nile Tilapia *Oreochromis niloticus* muscles: a comparative study with commercially important tropical freshwater fish in Philippines. 8th International Symposium

on Tilapia in Agriculture, Egypt Ministry of Agriculture, Cairo, Egypt, pp. 921–932.

36. Menoyo D, López-Bote CJ, Obach A, Bautista JM (2005) Effect of dietary fish oil substitution with linseed oil on the performance, tissue fatty acid profile, metabolism, and oxidative stability of Atlantic salmon. Journal of Animal Sciences 83: 2853–286.
37. Ng WK, Bahurmiz OM (2009) The impact of dietary oil source and frozen storage on the physical, chemical and sensorial quality of fillets from market-size red hybrid tilapia, *Oreochromis* sp. Food Chemistry 113: 1041–1048.
38. Nauroth JM, Liu YC, Van Elswyk M, Bell R, Hall EB, Chung G, et al. (2010) Docosahexaenoic acid (DHA) and docosapentaenoic acid (DPAn-6) algal oils reduce inflammatory mediators in human peripheral mononuclear cells in vitro and paw edema in vivo. Lipids 45 (5): 375–84. Pmid: 20364438
39. Walker DA (2009) Biofuels, facts, fantasy, and feasibility. Journal of Applied Phycology 21: 509–517.
40. Barrows FT, Silverstein J (2010) from fish meal-dependent to fish meal-free: feeds research is producing the alternative diets of the future for trout. p 15–16 In: Rust MB, Barrows FT, Hardy RW, Lazur A, Naughten K, Silverstein J, editors. The Future of Aquafeeds, NOAA/USDA Alternative Feeds Initiative, Maryland.
41. Chauton MS, Reitan KI, Norsker NH, Tveterås R, Kleivdal HT (2015) A techno-economic analysis of industrial production of marine microalgae as a source of EPA and DHA-rich raw material for aquafeed: Research challenges and possibilities. Aquaculture 436: 95–103.
42. Gross M, & Jarboe D, Wen Z (2015) Biofilm-based algal cultivation systems. Applied Microbiology and Biotechnology 99: 5781–5789. Pmid: 26078112
43. White RL, Ryan RA (2015) Long-Term Cultivation of Algae in Open-Raceway Ponds: Lessons from the Field. Industrial Biotechnology 11 (4): 213–220.

CITATION

CHAPTER 2

Jan-Willem F. Zijffers, Klaske J. Schippers, Ke Zheng, Marcel Janssen, Johannes Tramper and René H. Wijffels, Maximum Photosynthetic Yield of Green Microalgae in Photobioreactors, DOI 10.1007/s10126-010-9258-2

CHAPTER 3

Junfeng Wang, Jinli Liu and Tianzhong Liu, The difference in effective light penetration may explain the superiority in photosynthetic efficiency of attached cultivation over the conventional open pond for microalgae, DOI 10.1186/s13068-015-0240-0

CHAPTER 4

Wasana Arkronrat, Prapapron Deemark, and Vutthichai Oniam, Growth performance and proximate composition of mixed cultures of marine microalgae (Nannochloropsis sp. & Tetraselmis sp.) with monocultures, http://rdo.psu.ac.th/sjstweb/journal/38-1/38-1-1.pdf

CHAPTER 5

Hamed Safafar, Michael Z. Hass, Per Møller, Susan L. Holdt and Charlotte Jacobsen, High-EPA Biomass from Nannochloropsis salina Cultivated in a Flat-Panel Photo-Bioreactor on a Process Water-Enriched Growth Medium, doi:10.3390/md14080144

CHAPTER 6

Hamed Safafar, Michael Z. Hass, Per Møller, Susan L. Holdt and Charlotte Jacobsen, Astaxanthin-Producing Green Microalga Haematococcus pluvialis: From Single Cell to High Value Commercial Products, doi:10.3390/md14080144

CHAPTER 7

Victor Marco Emmanuel N. Ferriols, Riza O. Aguilar, Efficiency of various flocculants in harvesting the green microalgae Tetraselmis tetrahele (Chlorodendrophyceae: Chlorodendraceae), http://www.bioflux.com.ro/docs/2012.265-273.pdf

CHAPTER 8

Theoneste Ndikubwimana, Xianhai Zeng, Theophile Murwanashyaka, Emmanuel Manirafasha, Ning He, Wenyao Shao and Yinghua Lu, Harvesting of freshwater microalgae with microbial bioflocculant: a pilot-scale study, DOI 10.1186/s13068-016-0458-5

CHAPTER 9

Kevin J. Flynn, Philip Kenny, Aditee Mitra, Minimising losses to predation during microalgae cultivation, DOI 10.1007/s10811-017-1112-8

CHAPTER 10

Lum, K.K, Kim, J., Lei, X.G. (2013). potential of microalgae as a sustainable biofuel feedstock and animal feed. Journal of animal science and biotechnology, 4: 53. DOI: 10.1186/2049-1891-4-53.

CHAPTER 11

Edith Martinez-Guerra and Veera Gnaneswar Gude, Energy aspects of microalgal biodiesel production, Department of Civil and Environmental Engineering, Mississippi State University, Mississippi State, MS 39762, USA, DOI: 10.3934/energy.2016.2.347

CHAPTER 12

Marsullo M, Mian A, Ensinas AV, Manente G, Lazzaretto A and Marechal F (2015) Dynamic modeling of the microalgae cultivation phase for energy production in open raceway ponds and flat panel photobioreactors. Front. Energy Res. 3:41. doi: 10.3389/fenrg.2015.00041

CHAPTER 13

Niels Hempel, Ingolf Petrick and Frank Behrendt, Biomass productivity and productivity of fatty acids and amino acids of microalgae strains as key characteristics of suitability for biodiesel production, DOI 10.1007/s10811-012-9795-3

CHAPTER 14

Sukarni, Sudjito, Nurkholis Hamidi, Uun Yanuhar, and I. N. G. Wardana, Potential and properties of marine microalgae Nannochloropsis oculata as biomass fuel feedstock, DOI 10.1007/s40095-014-0138-9

CHAPTER 15

Xiaobin Wen, Kui Du, Zhongjie Wang, Xinan Peng, Liming Luo, Huanping Tao, Yan Xu, Dan Zhang,Yahong Geng and Yeguang Li, Effective cultivation of microalgae for biofuel production: a pilot-scale evaluation of a novel oleaginous microalga Graesiella sp. WBG-1, DOI 10.1186/s13068-016-0541-y

CHAPTER 16

Junping Lv, Jia Feng, Qi Liu and Shulian Xie, Microalgal Cultivation in Secondary Effluent: Recent Developments and Future Work, doi:10.3390/ijms18010079

CHAPTER 17

Removal of nitrate and phosphate from aqueous solutions by microalgae: An experimental study, M.H. Sayadi, N. Ahmadpour, M. Fallahi Capoorchali, M.R. Rezaei, DOI: 10.22034/gjesm.2016.02.04.005

CHAPTER 18

Kiron V, Sørensen M, Huntley M, Vasanth GK, Gong Y, Dahle D and Palihawadana AM (2016) Defatted Biomass of the Microalga, Desmodesmus sp., Can Replace Fishmeal in the Feeds for Atlantic salmon. Front. Mar. Sci. 3:67. doi: 10.3389/fmars.2016.00067

CHAPTER 19

Sarker PK, Kapuscinski AR, Lanois AJ, Livesey ED, Bernhard KP, Coley ML (2016) Towards Sustainable Aquafeeds: Complete Substitution of Fish Oil with Marine Microalga Schizochytrium sp. Improves Growth and Fatty Acid Deposition in Juvenile Nile Tilapia (Oreochromis niloticus). PLoS ONE 11(6): e0156684. https://doi.org/10.1371/journal.pone.0156684

INDEX

Symbols

A

B

C

D

O

P

R

S

T

U

Z